Bc

ENGINEERING
HEAT
TRANSFER

Second Edition

ENGINEERING HEAT TRANSFER

Second Edition

Dr. William S. Janna, Ph.D.

Associate Dean for Graduate Studies and Research
Herff College of Engineering
The University of Memphis
Memphis, Tennessee

CRC Press

Boca Raton   London   New York   Washington, D.C.

Library of Congress Cataloging-in-Publication Data

Janna, William S.
    Engineering heat transfer / William S. Janna.—2nd ed.
        p. cm.
    Includes bibliographical references and index.
    ISBN 0-8493-2126-3 (alk. paper)
    1. Heat—Transmission. 2. Heat exchangers. I. Title.
TJ260.J36 2000
621.402′2—dc21   99-048560
                                                                                    CIP

This book contains information obtained from authentic and highly regarded sources. Reprinted material is quoted with permission, and sources are indicated. A wide variety of references are listed. Reasonable efforts have been made to publish reliable data and information, but the author and the publisher cannot assume responsibility for the validity of all materials or for the consequences of their use.

Visit the CRC Press Web site at www.crcpress.com

© 2000 by CRC Press LLC

No claim to original U.S. Government works
International Standard Book Number 0-8493-2126-3
Library of Congress Card Number 99-048560
Printed in the United States of America   3 4 5 6 7 8 9 0
Printed on acid-free paper

Preface

In writing *Engineering Heat Transfer,* I have attempted to provide the reader with a foundation in the study of heat-transfer principles while also emphasizing some of the topic's practical applications. The mathematics presented in the text should not present undue learning difficulties to the student who has completed first courses in thermodynamics, fluid mechanics, and differential equations. The book is organized into three sections that cover conduction, convection, and radiation heat transfer.

Following an introductory chapter that presents fundamental concepts, Chapter 2 presents the general conduction equation and lays the groundwork for the material that follows in Chapters 3–6. Chapter 3 covers one-dimensional steady-state conduction, while Chapter 4, on extended surfaces, serves to present applications of heat-transfer principles to fin design. The last two chapters in this section concern steady-state conduction problems in which temperature varies with more than one independent space variable and heat transfer problems in which temperature varies with time.

The second section begins with an introductory chapter on convection and is followed by chapters that present convection heat transfer in closed conduits, flow past immersed bodies, and natural convection. Chapter 11, on heat exchangers, is another applications-oriented chapter, while Chapter 12 covers heat-transfer effects associated with condensation and vaporization.

In the final section (Chapters 13 and 14), the fundamental principles of radiation are introduced and then given practical application.

PREFACE TO THE SECOND EDITION

The philosophy taken in producing the second edition of *Engineering Heat Transfer* is the same as it was when the text was originally written: create a user-friendly text that provides many practical examples without overwhelming the student. Most people think in concrete terms, using mental pictures to envision what is being discussed. Thus, drawings, sketches, graphs, etc. are important to convey information. This is especially critical in an area like heat transfer, which is highly abstract. We cannot "see" the transfer of heat, but we can model it effectively using graphs and drawings.

As a user-friendly text that does not overwhelm the student, *Engineering Heat Transfer* contains many examples and a large number of confidence-builder problems. The examples amplify the theory and show how derived equations are used to model physical problems. Confidence-building problems are relatively simple and follow (somewhat) the examples. When a few of these types of problems are solved, the student feels comfortable with the information and can then apply it to situations that are not common, and that increase in complexity. Problem solving is a skill that must be developed slowly, methodically, and thoroughly. The emphasis in this text is on problems that are practical in nature. The applications make the text interesting.

The text has not been written at a highly advanced (mathematically) level. In years past, heat transfer was taught to seniors in the undergraduate mechanical engineering curriculum. The emphasis was on the mathematics, as if every student was on the way to graduate school. The emphasis in this text is on problem-solving skills using real-world examples. Many of the students graduating today look for employment, with graduate school considered as a part-time venture. Moreover, heat transfer in many curricula is being taught at the junior level. So a modern heat transfer text should be written at a lower level than those written years ago. It should pay particular attention to practical, real-world problems, and de-emphasize high-level mathematics in favor of effectively and accurately modeling what is physically occuring in the problem of interest.

Many heat transfer texts are very long, which is a trend set years ago, and propagated in subsequent writings. In fact, the trend in writing heat transfer texts has escalated to the point where they are approaching encyclopedic proportions. I believe this is not appropriate today. A good heat-transfer text can be written that is relatively short. It can be made less costly, and much of it can be used in its entirely in the classroom in a one-semester course.

One of the more recent directions in engineering education is in the implementation of "design throughout the curriculum." In addition, the design experience is to culminate in a capstone course, such as something like Design of Fluid Thermal Systems, or Mechanical Design. Such courses of necessity are taught at the senior level—either first or second semester. Moreover, these capstone design courses require knowledge of certain prerequisites, like fluid mechanics and heat transfer. Thus, in many curricula, heat transfer is being taught at the second semester junior or first semester senior level. Rather than being a course in itself, so to speak, heat transfer is now a backbone course for a senior design experience that is supposed to "put it all together." The niche that heat transfer now occupies is one that fits into the overall design experience as a backbone course.

The objective in writing the second edition of *Engineering Heat Transfer* now is to produce a revised text for a new market. It is time to move forward and keep pace with what is occurring in engineering education.

TEXT MODIFICATIONS AND DESCRIPTION, SECOND EDITION

Chapter 1 of the second edition of *Engineering Heat Transfer* remains essentially unchanged from the first edition. It provides an introduction to Heat Transfer and to the tables at the end of the text. Chapters 2, 3, and 4 have been combined into a single chapter (Chapter 2) on One-Dimensional Conduction. Heat flows through walls, cylinders, and fins are all considered.

Chapter 3 in the second edition is on Steady-State Conduction in Multiple Dimensions. The analytical method of solution applied to a simple problem is presented to illustrate the complexity of such problems. Next, alternative solution methods, including graphical and numerical methods, are presented.

Chapter 4 is on Unsteady-State Heat Conduction, and it contains the lumped capacitance method as well as Heisler charts for more complex problems.

Chapter 5 provides an Introduction to Convection. A discussion of basic fluid mechanics is given, which serves as an overview of the convection chapters that follow. Equations of fluid mechanics and the thermal energy equation are presented. Boundary layer flows are described, but boundary layer equations are not derived in great detail.

Chapter 6 is on Convection Heat Transfer in Closed Conduits. Circular and noncircular ducts are both discussed. Empirical correlations are also provided. Chapter 7 is on Convection Heat Transfer in Flow Past Immersed Bodies. Descriptions and correlations are provided for various flows: flow over a flat plate; flow past two- and three-dimensional bodies; and flow past a bank of tubes.

Chapter 8 is on Natural Convection. Problems considered are vertical surfaces, inclined surfaces, horizontal surfaces, cylinders, and arrays of fins. Chapter 9 is on Heat Exchangers, including double-pipe, shell and tube, and cross flow.

Chapter 11 gives an introduction to Radiation Heat Transfer. Topics include the electromagnetic radiation spectrum, emission and absorption, intensity, radiation laws, and characteristics of real surfaces. Chapter 12 is about Radiation Heat Transfer between Surfaces. The view factor and methods for evaluating it are presented. Heat transfer within enclosures of black and gray bodies is modeled.

Problems at the end of each chapter have been reorganized and classified according to topic. In many cases, the instructor may wish not to cover certain sections in the text, and such reorganization makes it simple to select the desired problems appropriately.

The writing style hopefully conveys my enthusiasm for the material, and the student should get as excited about solving heat transfer problems as I am.

TO THE INSTRUCTOR

Some deviations from traditional heat-transfer texts are evident in this book. First, because of space limitations, I have neither provided extensive references (which soon become dated anyway) at the end of each chapter nor discussed mass transfer. I feel that the engineer should look to the text for a presentation of the fundamentals of the topic and to the publications of professional and technical societies for a presentation of the most current research. (A bibliography of selected references is provided in the back matter.) Accordingly, I have tried here to provide a sound basis for understanding the fundamentals of heat transfer while emphasizing solving problems that have practical applications.

Rather than placing numerical methods in heat conduction in a single chapter, 1 have interspersed them in Chapters 2–4 so that the equations to be solved are handy and fresh in the reader's mind.

In the convection chapters, only one or two correlations for a specific geometry are presented in the text, while others are given in the exercises. The chapter summary contains the list of correlations that apply to various systems and where they can be found. This arrangement prevents unnecessarily cluttering the essentials of convection heat transfer.

Each chapter concludes with a section of problems that become progressively more difficult and are systematically designed to improve the reader's ability to understand and apply the principles of heat transfer. Wherever possible, problems with practical applications are formulated so that the student can become familiar with the process of using heat-transfer equations to model real-life situations.

Despite the best efforts of all involved, errors in a text are inevitable. I invite the reporting of errors to the publisher so that mistakes and misconceptions do not become taught as truth, and I also invite comments and suggestions on how the text might be improved.

Acknowledgments

I am greatly indebted to the many reviewers who read portions of the manuscript in its formative stages and made a number of helpful suggestions for its improvement: to Professors Jan F. Kreider, consulting engineer; F. L. Keating, U. S. Naval Academy; F. S. Gunnerson, University of Central Florida; W. C;. Steele, Mississippi State University; Roger A. Crane, University of South Florida; Satish Ramadhyani, Purdue University; Edwin I. Griggs, Tennessee Technical University; and Ozer A. Arnas, Louisiana State University. I am also indebted to Ms. Cindy Carelli, Engineering Editor for CRC, for her support in producing a second edition, and for providing much encouragement at critical times; to Susan Zeitz for excellent skills in producing the text; and to Felicia Shapiro for her help in expediting various matters.

I am indebted also to Melissa A. Cobb and Elizabeth B. London for their help in performing a number of secretarial tasks, and their attention to detail.

I am especially indebted to my lovely wife, Marla, whose unfailing moral support was always there when it was needed.

William S. Janna

Dedication

———

To Him who is our source of love and knowledge,
To Marla who transforms His love into strength,
and to the reader for whom knowledge becomes wisdom.

Contents

1 Fundamental Concepts

CONTENTS

1.1 INTRODUCTION

Heat transfer is the term applied to a study in which the details or mechanisms of the transfer of energy in the form of heat are of primary concern. Examples of heat transfer are many. Familiar domestic examples include broiling a turkey, toasting bread, and heating water. Industrial examples include curing rubber, heat treating steel forgings, and dissipating waste heat from a power plant. The analysis of such problems is the topic of study in this text.

1.2 MECHANISMS OF HEAT TRANSFER

Heat transfer is energy in transit, which occurs as a result of a temperature gradient or difference. This temperature difference is thought of as a driving force that causes heat to flow. Heat transfer occurs by three basic mechanisms or *modes:* conduction, convection, and radiation.

Conduction is the transmission of heat through a substance without perceptible motion of the substance itself. Heat can be conducted through gases, liquids, and solids. In the case of fluids in general, conduction is the primary mode of heat transfer when the fluid has zero bulk velocity. In opaque solids, conduction is the only mode by which heat can be transferred. The kinetic energy of the molecules of a gas is associated with the property we call *temperature*. In a high-temperature region, gas molecules have higher velocities than those in a low-temperature region. The random motion of the molecules results in collisions and an exchange of momentum and of energy. When this random motion exists and a temperature gradient is present in the gas, molecules in the high-temperature region transfer some of their energy, through collisions, to molecules in the low-temperature region. We identify this transport of energy as heat transfer via the diffusive or conductive mode.

Conduction of heat in liquids is the same as for gases—random collisions of high-energy molecules with low-energy molecules causing a transfer of heat. The situation with liquids is more complex, however, because the molecules are more closely spaced. Therefore, molecular force fields can have an effect on the energy exchange between molecules; that is, molecular force fields can influence the random motion of the molecules.

Conduction of heat in solids is thought to be due to motion of free electrons, lattice waves, magnetic excitations, and electromagnetic radiation. The motion of free electrons occurs only in substances that are considered to be good electrical conductors. The theory is that heat can be transported by electrons (known as the *electron gas*), which are free to move through the lattice structure of the conductor, in the same way that electricity is conducted. This is usually the case for metals.

The molecular energy of vibration in a substance is transmitted between adjacent molecules or atoms from a region of high to low temperature. This phenomenon occurs from lattice waves that can be considered as energy being transmitted by a gas composed of an integral number of quanta, known as *phonons*. Phonon motion is thought of as diffusing through the lattice in the same way as the electron gas does. The lattice-wave mechanism is usually not a significant factor in conduction of heat through metals. It is significant for nonmetals.

Magnetic dipoles of adjacent atoms in some cases provide effects between magnetic moments that may aid the conduction of heat in the solid. Electromagnetic radiation in translucent materials may have an effect on the conduction of heat. This is the case when the material has little capacity for absorbing energy.

Our concern here is not so much in describing the molecular or microscopic activity associated with conduction but in being able to describe mathematically the macroscopic effect of heat transfer via the conduction mode. An example of conduction heat transfer is the cooling or freezing of the ground during winter.

Convection is the term applied to heat transfer due to bulk movement of a fluid. Fluid mechanics therefore plays an important part in the analysis of convection problems. Consider a gas furnace for example. Combustion of natural gas yields products that heat a device known as a heat exchanger. A fan blows air over the opposite side of the exchanger. The air movement is caused by an external driving force—the fan. The air absorbs heat (gains energy) from the exchanger by convection. Moreover, to denote that an external device provides fluid movement that enhances heat transfer, the mode of energy transport is called *forced convection*.

Consider next a domestic gas-fired water heater. Gas is burned in the bottom of the heater. Combustion gases exhaust through tubes in the tank. The tubes are surrounded by water, which gains energy or absorbs heat from them. There will exist a movement of the water within the tank. As water near the tubes receives energy, the local water density decreases. The warmed water tends to rise to the top of the tank while cooler water is drawn to the tank bottom. Heat is transferred to the water by convection. Moreover, to denote that bulk fluid movement is caused by density differences resulting from the process of energy transfer, the mode of heat transfer in this case is called *natural* or *free convection.*

Radiation is the transfer of energy by electromagnetic radiation having a defined range of wavelengths. One example of radiant heat transfer is that of energy transport between the sun and the earth. Note that all substances emit radiant heat but that the *net* flow of heat is from the high- to low-temperature region. So the cooler substance will absorb more radiant energy than it emits.

Heat is usually transferred by a combination of conduction, convection, and radiation. As an example consider the sidewalk sketched in Figure 1.1. The sidewalk receives radiant energy from the sun. Some of this energy is absorbed, increasing the temperature of the sidewalk. Some of the radiant energy is reflected and, due to the opaque nature of the material, none of the radiant energy impacting the sidewalk is transmitted. As the sidewalk becomes warmer, it radiates more energy. Also, heat is transferred by natural and/or forced convection (depending on the wind) to the air in contact with it. Finally, heat is transferred by conduction through the sidewalk to the ground below. The general heat-transfer problem involving all three modes can be set up and described mathematically. Solving the equations analytically, however, is not always possible. In a number of cases,

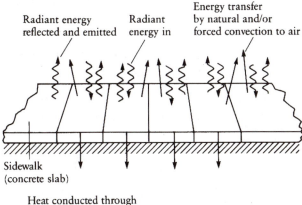

Radiant energy reflected and emitted

Radiant energy in

Energy transfer by natural and/or forced convection to air

Sidewalk (concrete slab)

Heat conducted through the sidewalk to the ground below

FIGURE 1.1   A multimode heat-transfer example.

one mode of heat transfer is dominant. It can then be identified and modeled satisfactorily to obtain a solution to what could be an otherwise insoluble problem. Consequently, conduction, convection, and radiation are presented here in separate discussions. In problems where combined models can be treated, an appropriate model is presented.

It is important to develop a method for solving heat-transfer problems. The flow of heat is not a visual experience in most cases. Consequently, it is advantageous to approach a problem by first *expressing what is physically happening* and then *stating the assumptions to be made.* These two steps will illustrate how the assumptions simplify the problem and how well the model fits the situation. Moreover, these steps show where sources of error will originate. One additional aid in setting up the problem is *sketching a temperature profile,* if appropriate. The gradients are then identified, and the direction of heat flow is discernible.

We emphasize that some thought should go into understanding the physics of the problem to identify what is really happening and then to make necessary assumptions so that a mathematical model of the phenomenon can be composed. This technique of model formulation will provide you with a consistent, methodical approach that leads to the development of an intuition or "feel" for heat-transfer problems. On the other hand, you are cautioned not to overcomplicate the problem but to keep in mind that the objective is to model a physical situation suitably and to obtain an appropriate and reasonable solution.

1.3   DIMENSIONS AND UNITS

A physical variable used to provide a specification of a particular system is a *dimension.* A dimension can be either fundamental or derived. In our study, we will select fundamental dimensions that will be used in combination to make up derived dimensions. The dimensions that are considered fundamental depend on the unit system selected. For example, the fundamental dimensions in the English Engineering System are as follows:

English Engineering System	
Dimension (symbol)	**Unit (abbreviation)**
length (L)	foot (ft)
time (T)	second (s)
mass (M)	pound-mass (lbm)
force (F)	pound-force (lbf)
temperature (t)	degree Rankine (°R) or degree Fahrenheit (°F) (°R is the absolute unit)

All physical quantities can be expressed in terms of the fundamental units. To relate the units for force and mass, Newton's Second Law of Motion is used. Newton's Second Law states that force is proportional to the time rate of change of momentum,

$$F \propto \frac{d(mV)}{dt}$$

where F is force, m is mass, V is velocity, and t is time. Introducing a proportionality constant, we get

$$F = K\frac{d(mV)}{dt}$$

For constant mass, the above equation becomes

$$F = Kma \tag{1.1}$$

where a is acceleration. By definition, a force of 1 lbf acting on an object having a mass of 1 lbm will accelerate the object at a rate of 32.174 ft/s². Substituting these values into Equation 1.1 yields

$$1 \text{ lbf} = K(1 \text{ lbm}) (32.174 \text{ ft/s}^2)$$

Solving for the reciprocal of the proportionality constant gives

$$\frac{1}{K} = g_c = 32.174 \frac{\text{lbm} \cdot \text{ft}}{\text{lbf} \cdot \text{s}^2} \tag{1.2}$$

Newton's Law thus becomes

$$F = \frac{ma}{g_c} \tag{1.3}$$

It is important to note that g_c is a proportionality constant or conversion factor that arises in the Engineering System of units.

The fundamental dimensions in SI (Système International) units are as follows:

SI Units	
Dimension (symbol)	**Unit (abbreviation)**
length (L)	meter (m)
time (T)	second (s)
mass (M)	kilogram (kg)
temperature (t)	degree Kelvin (K) or degree Celsius (°C) (K is the absolute unit)

Here again, all physical quantities can be expressed in terms of these fundamental units. Force in this unit system is not considered a fundamental dimension but, instead, is related to mass by the definition

$$1 \text{ N} = \frac{1 \text{ kg} \cdot \text{m}}{\text{s}^2}$$

where N is the abbreviation for newton, the force unit in SI. In the Engineering System, where force and mass are both fundamental units, a conversion factor must appear in the appropriate equations to make them dimensionally consistent. In SI units, the conversion factor g_c is not necessary because the units automatically incorporate the effect. Both of these unit systems will be used in this text. Although SI units have been adopted as a standard by many, if not all, professional and technical engineering societies, it will take considerable time for a complete conversion to occur. Therefore, it is premature at this time to exclude either unit system from this text. Equations will contain g_c in appropriate places; you are advised to ignore g_c in the equations when using SI units. Appendix Table A.1 provides a list of prefixes that have been adopted for use with the SI units for convenience Other unit systems that have been popular at one time or another are summarized in Table 1.1.

TABLE 1.1
Conventional System of Units

	English Engineering	SI	British Gravitational	British Absolute	CGS Absolute
Mass (fundamental)	lbm	kg	—	lbm	gram
Force (fundamental)	lbf	—	lbf	—	—
Mass (derived)	—	—	slug	—	—
Force (derived)	—	N	—	poundal	dyne
Length	ft	m	ft	ft	cm
Time	s	s	s	s	s
Temperature	°R (°F)	K (°C)	°R (°F)	°R (°F)	°R (°F)
g_c	$32.174\dfrac{\text{lbm}\cdot\text{ft}}{\text{lbf}\cdot\text{s}^2}$	—	—	—	—

*Degrees Kelvin is (by convention) properly written without the (°) symbol.

In a number of cases, it will be necessary to convert from one set of units to another. To make this task easier, a conversion-factor table is provided as Appendix Table A.2. The quantities are listed with conversion factors that transform a given unit to the SI equivalent.

Work and energy have the same dimension. The dimension for work arises from its definition—force times the distance through which the force acts (F·L). The ft·lbf is the unit for work in the Engineering System, and the joule is the unit for work in SI [1 J = 1 N·m = 1 kg·m²/s²].

The unit for heat in the Engineering System arises from a definition also. The British Thermal Unit (BTU) may be defined as the amount of heat necessary to raise the temperature of 1 lbm of water at 68°F by 1°F. In SI the unit for heat is the joule (J).

The First Law of Thermodynamics and the concept of *energy* relate work to heat. The term *energy* encompasses both heat and work. Consequently, the traditional engineering units used for heat and work can be considered as being just different energy units that are related by appropriate conversion factors. The BTU can be converted to other units by 1 BTU = 778.16 ft·lbf = 1 055 J = 1.055 kJ. (Note that there are certain conventions to be followed when using SI units. One convention is to leave a space where one might normally place a comma.) Power, work per unit time, has the dimensions of F·L/T (watt = 1 W = I J/s = 1 N·m/s = 1 kg·m²/s³; 1 HP = 550 ft·lbf/s).

1.4 FOURIER'S LAW OF HEAT CONDUCTION

Consider the experimental setup shown in Figure 1.2. A slab of material is in contact with a heat source, a hot plate, on its lower surface, which keeps that surface at an elevated temperature T_1. The upper surface of the slab is in contact with a cooling system that absorbs heat and maintains the upper surface at some lower temperature T_2. The area of contact between the plate and the devices is A. Heat is conducted through the slab in only one direction, perpendicular to the area of contact, from the hot plate to the cooling water. We now impose a coordinate system on the slab with x directed as shown. With the system at steady state, if temperature could be measured at each

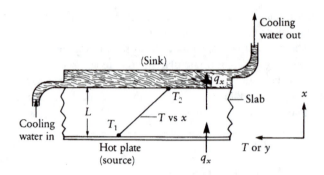

FIGURE 1.2   Experiment for observing Fourier's Law of Heat Conduction.

point within the slab and plotted as on the figure, the graph of T vs. x would result. The slope of this curve is dT/dx. If this procedure is repeated for different materials, the data would lead us to conclude that the flow of heat per unit area is proportional to the temperature difference per unit length. In differential form,

$$\frac{q_x}{A} \propto - \frac{dT}{dx} \tag{1.4}$$

where q_x is the heat flow in the x direction, A is the area normal to the heat flow direction, and dT/dx is the temperature gradient or slope of the temperature curve. Note that as x increases, temperature decreases so that the right-hand side of the above equation is actually a positive quantity. After substitution and insertion of a proportionality constant in Equation 1.4, we get

$$\frac{q_x}{A} = -k\frac{dT}{dx} \tag{1.5}$$

where k is the *thermal conductivity* of the material. The minus sign ensures that the equation properly predicts that heat flows in the direction of decreasing temperature. That is, if temperature decreases with increasing x, then (dT/dx) is actually a negative quantity, and the minus sign of Equation 1.5 ensures that q_x will be positive. Conversely, if temperature increases with increasing x, (dT/dx) is positive, and q_x is negative. In either case, heat flows in the direction of decreasing temperature. Equation 1.5 is known as Fourier's Law of Heat Conduction. It is an experimentally observed law and serves as a definition for a new property of substances called the *thermal conductivity*. The dimension of heat flow q_x is energy per unit time, F·L/T [W or BTU/hr; although BTU/s is appropriate, BTU/hr is customarily used instead]. The dimension of thermal conductivity k is energy per time per length per temperature, F·L/(L·T·t); [W/(m·K) or BTU/hr·ft·°R]. Note that the unit for time in the Engineering System is the second (s), but thermal-conductivity units are customarily expressed in terms of hours (hr). The quantity q_x/A is called the heat flux and is at times written as q''_x.

Equation 1.5 can be simplified and integrated to obtain an equation for the temperature:

$$\frac{q_x}{A}dx = -kdT$$

$$\int_0^L \frac{q_x}{A}dx = \int_{T_1}^{T_2} -kdT \tag{1.6}$$

where the temperature is T_1 at $x = 0$, and T_2 at $x = L$. For constant heat flux and constant thermal conductivity, we obtain after integration

$$\frac{q_x}{A}L = -k(T_2 - T_1)$$

$$\frac{q_x}{A} = k\frac{(T_2 - T_1)}{L}$$

(1.7)

Example 1.1

A fire wall separates two rooms. In one room, a fire has started. The fire wall is made of stainless steel that is 1/2 in thick. Heat in the room is sufficient to raise the outside surface temperature of one side of the wall to 350°F. If the thermal conductivity of the metal wall is 9.4 BTU/hr·ft·°R, and the heat flux is 6.3 BTU/s·ft², determine the surface temperature of the other side of the fire wall.

Solution

Before making any calculations, let us determine what is physically occurring. The fire is at a very high temperature, so radiation heat transfer from the chemical reaction to the wall is significant. As the air near the wall is warmed, convection to the wall may be important. On the other hand, depending on the wall and the air temperatures, the air may be receiving heat from the wall. Figure 1.3 illustrates the system, shown with the expected temperature profile.

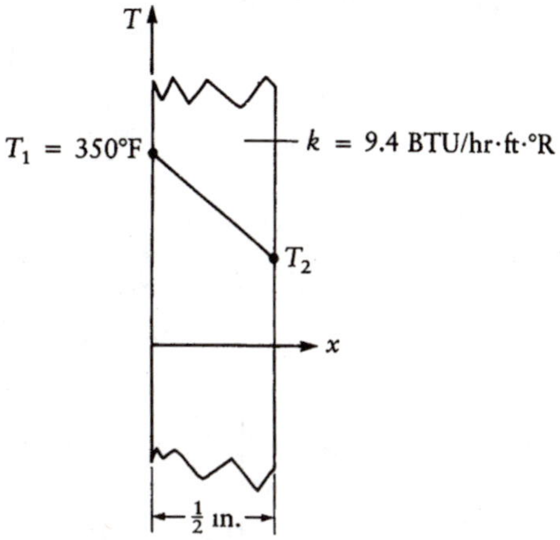

Assumptions

1. Net radiation and convection heat transfer has elevated the wall surface temperature to a constant and uniform 350°F.
2. The properties of the stainless steel are constant.
3. One-dimensional conduction through the fire wall is occurring.

FIGURE 1.3 The stainless steel fire wall of Example 1.1

Equation 1.7 therefore applies:

$$q''_x = -k\frac{dT}{dx} = k\frac{T_1 - T_2}{L}$$

where T_1 is 350°F and T_2 is the temperature of interest. All parameters are given in the problem statement. After substitution, we obtain

$$\left(6.3\frac{BTU}{s\cdot ft^2}\right)\left(3600\frac{s}{hr}\right) = \left(9.4\frac{BTU}{hr\cdot ft\cdot°R}\right)\frac{(350\ °F - T_2)}{(0.5/12\ ft)}$$

Solving, we find

$$T_2 = 249.5\ °F$$

1.5   THERMAL CONDUCTIVITY

As mentioned earlier, Equation 1.5 serves as a defining expression for thermal conductivity:

$$\frac{q_x}{A} = -k\frac{dT}{dx}$$

Thermal conductivity is a property of a material. The numerical value of thermal conductivity is an indication of how fast heat is conducted through a material and is a macroscopic representation of all the molecular effects that contribute to the conduction of heat through a material. Based on Equation 1.5, measurements can be made to determine thermal conductivity for various substances.

Kinetic theory can be applied to determine analytically the thermal conductivity of gases. This model predicts thermal conductivity fairly accurately only for monatomic gases at moderate pressures. In general, however, thermal conductivity of a material must usually be measured. For most substances, thermal conductivity varies with temperature. Figure 1.4 shows the variation with temperature of thermal conductivity for various metals. Figure 1.5 shows data for some liquids, and Figure 1.6 is for gases. Figure 1.7 is a line plot of thermal-conductivity ranges for classes of

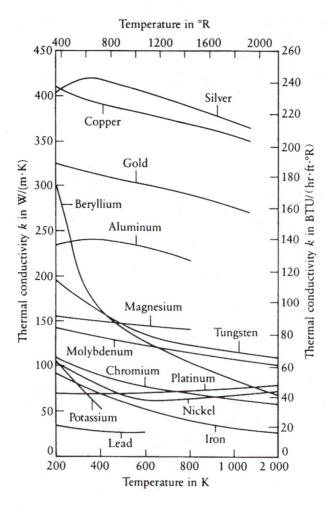

FIGURE 1.4   Variation of thermal conductivity with temperature for various metals. Data from several sources, see references at end of text.

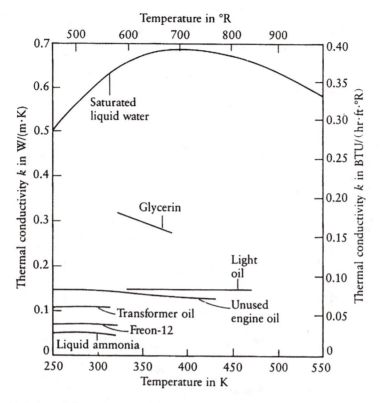

FIGURE 1.5 Variation of thermal conductivity with temperature for various liquids. Data from several sources, see references at end of text.

substances, showing a comparison of values of thermal conductivities for various materials. Mean values of thermal conductivity for a variety of substances are provided in the Appendix tables. More extensive data on thermal conductivity and its variation with temperature for many substances can be found in several reference texts.

Example 1.2

Figure 1.8 is a schematic of a device used to measure thermal conductivity. It consists of a hot plate as a heat source. In contact with the hot plate is a 2.5-cm-diameter rod made of stainless steel of known thermal conductivity that has thermocouples attached for obtaining temperature readings. Resting on the stainless steel is a 2.5-cm-diameter aluminum rod of unknown thermal conductivity that also contains thermocouples. On top of the aluminum is a water-cooled chamber that acts as a heat sink. The rods and water chamber are surrounded completely by thick insulation. The stainless steel rod has a thermal conductivity of 14.4 W/(m·K), and the following steady-state temperature data were obtained for one setting of the hot plate thermostat:

Thermocouple location	Temperature, °C
1	440
2	400
3	360
7	270
8	261.4
9	252.7

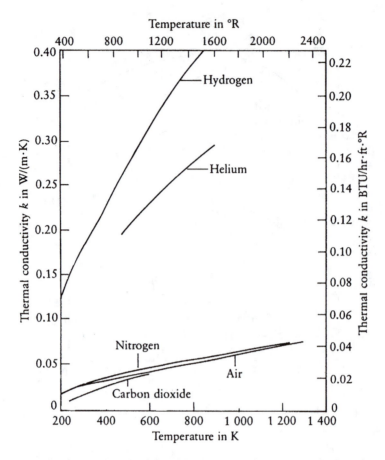

FIGURE 1.6 Variation of thermal conductivity with temperature for various gases. Data from several sources, see references at end of text.

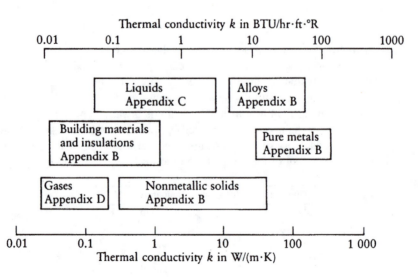

FIGURE 1.7 Thermal conductivity ranges for various classes of substances at moderate pressures and temperatures. Appendix tables that contain pertinent data are also shown.

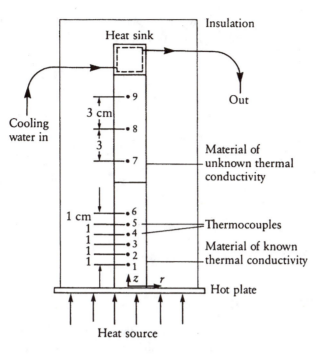

FIGURE 1.8 Schematic of a device used for measuring thermal conductivity of metals.

Determine the thermal conductivity of the aluminum.

Solution

The insulation is heated, as is the stainless steel. Just as a gradient in the axial direction exists in the metals, one will exist within the insulation. The difference is that the metal transfers heat far better than the insulation does. The existing axial gradients also produce temperature differences in the radial direction so that heat is transferred through the insulation to the surrounding air.

Assumptions

1. The insulation sufficiently retards losses such that one–dimensional heat flow exists in the axial direction through the metals.
2. The stainless steel has constant properties.
3. The aluminum has constant properties.

The heat moving upward is a constant, as is the area through which heat flows. Therefore,

$$\frac{q_z}{A} = -k\frac{dT}{dz}\bigg|_{\text{staines steel}} = -k\frac{dT}{dz}\bigg|_{\text{aluminum}}$$

Under steady-state conditions, the slope of the temperature profile *(dT/dz)* is a constant for each material. We can therefore write $dT/dz = \Delta T/\Delta z$, and the above equation becomes

$$k_{ss}\frac{\Delta T}{\Delta z}\bigg|_{ss} = k_{al}\frac{\Delta T}{\Delta z}\bigg|_{al}$$

where ΔT is the temperature difference between two thermocouples and Δz is the distance between those two thermocouples. Solving for the unknown thermal conductivity in the above equation, we get

$$k_{al} = k_{ss}\frac{\Delta T_{ss}\Delta z_{al}}{\Delta T_{al}\Delta z_{ss}}$$

Substituting using an average gradient of 8.65 K/3 cm brings

$$k_{al} = [14.4 \text{ W/(m·K)}]\frac{40 \text{ K}}{8.65 \text{ K}}\frac{3 \text{ cm}}{1 \text{ cm}}$$

Solving,

$$k_{al} = 200 \text{ W/(m·K)} \qquad \square$$

1.6   CONVECTION HEAT TRANSFER

Convection is the mode of heat transfer associated with fluid motion. If the fluid motion is due to an external motive source such as a fan or pump, the term *forced convection* applies. On the other hand, if fluid motion is due predominantly to the presence of a thermally induced density gradient, then the term *natural convection* is appropriate. Consider as an example an automobile that has been exposed to direct sunlight for a period of time. The metal body will receive heat by radiation from the sun. Besides reradiating heat in all directions, the body loses heat to the surrounding air by natural convection. When the car is moving, air surrounding the body absorbs heat by a forced-convection process. We know from experience that, in this case, the forced convection heat transfer is more effective than the natural convection process in transferring energy away from the car body. In our study of convection heat transfer, we wish to model such processes in a way that is simple and accurate. Let us then begin with a simple case.

Consider a plate immersed in a uniform flow as shown in Figure 1.9. The plate is heated and has a uniform wall temperature of T_w. The uniform flow velocity is U_∞, and the fluid temperature far from the plate (the *free-stream temperature*) is T_∞. At any location, we sketch the velocity distribution on the axes labeled V vs. y. The velocity at the wall is zero due to the nonslip condition;

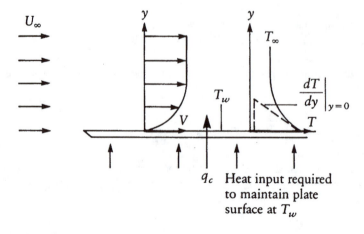

FIGURE 1.9   Uniform flow past a heated plate.

that is, fluid adheres to the wall due to friction or viscous effects. The velocity increases with increasing y from zero at the wall to nearly the free-stream value at some vertical distance away, known as the *hydrodynamic boundary layer.* jWe also sketch a temperature distribution, T vs. y, which is seen to decrease from T_w at the wall to T_∞ at some distance from the wall (assuming T_∞ < T_w). This distance is called the *thermal-boundary-layer thickness.*

Heat is transferred from the wall to the fluid. Within the fluid the mechanism of heat transfer at the wall is conduction because the fluid velocity at the wall is zero. However, the *rate* of heat transfer depends on the slope of the T vs. y curve at the wall—dT/dy at $y = 0$. A steeper slope is indicative of a greater temperature difference and is highly dependent on the flow velocity. The flow velocity will influence the distance from the wall that we must travel before we sense that the temperature is T_∞.

The heat transferred by convection is found to be proportional to the temperature difference. In the case of Figure 1.9, we have

$$\frac{q_c}{A} \propto T_w - T_\infty$$

where q_c is the heat transferred by convection.

Introducing a proportionality constant, we get

$$q_c = \bar{h}_c A (T_w - T_\infty) \tag{1.8}$$

in which $\bar{h}_c$ is called the *average convection heat transfer coefficient* or the *film conductance.* This coefficient accounts for the overall effects embodied in the process of convection heat transfer. The overbar notation indicates that the film conductance defined in Equation 1.8 is an average that is customarily assumed constant over the length of the plate. The dimension of $\bar{h}_c$ is energy per time per area per temperature, $F \cdot L/(L^2 \cdot T \cdot t)$; [$W/(m^2 \cdot K)$ or $BTU/(hr \cdot ft^2 \cdot {}^\circ R)$]. Equation 1.8 is known as Newton's Law of Cooling; it is the defining expression for the film conductance $\bar{h}_c$. For some systems, $\bar{h}_c$ can be determined analytically, but, for most geometries, it must be measured.

Example 1.3

The sloping roof of a house receives energy by radiation from the sun. The roof surface reaches a steady uniform temperature of 140°F, and the ambient air temperature is 85°F. For roof dimensions of 30 × 18 ft, and a convection heat-transfer coefficient of 3 $BTU/(hr \cdot ft^2 \cdot {}^\circ R)$, determine the heat transferred by convection to the air.

Solution

The roof receives energy from the sun. Heat is transferred through the roof by conduction; the roof reradiates heat in all directions and transfers heat to the air by natural convection (in the absence of wind). Figure 1.10 illustrates the system.

Assumptions

1. The net heat transferred to the roof results in a surface temperature of 140°F, which is constant and uniform over the entire roof area.
2. The average convection coefficient is applicable to the entire roof area.

For the case of convection heat transfer, Equation 1.8 applies:

$$q_c = \bar{h}_c A (T_w - T_\infty)$$

With all parameters given, we substitute directly:

$$q_c = (3\ \text{BTU/hr·ft}^2 \cdot {}^\circ\text{R})(30 \times 18)\ \text{ft}^2(140 - 85^\circ\ \text{F})$$

Solving,

$$q_c = 89{,}100\ \text{BTU/hr}$$

Note that the temperature unit on the conductance is °R, while $T_w - T_\infty$ is expressed in °F. If each temperature is first converted to °R, the *difference* will still have the same numerical value as the °F calculation. Consequently, the temperature unit in the denominator of the film conductance (also of thermal conductivity, specific heat, etc.) should be thought of as referring to a temperature difference. Remember that a 1°R *difference* equals a 1°F *difference*. Likewise, a 1 K difference = 1°C difference.   ❑

1.7   THE CONVECTION HEAT-TRANSFER COEFFICIENT

As mentioned in Section 1.6, Newton's Law of Cooling is the defining equation for the convection heat-transfer coefficient or the film conductance. Recall also that the fluid velocity influences the temperature gradient at a solid–fluid interface. Therefore, the fluid velocity influences $\bar{h}_c$. We can also infer that, because viscosity and density have an effect on the velocity profile (e.g., laminar vs. turbulent), $\bar{h}_c$ depends on the fluid viscosity and density. With conduction heat transfer occurring at the wall where the fluid velocity is zero, we conclude that $\bar{h}_c$ depends on the thermal conductivity of the fluid. Table 1.2 gives approximate values of the convection heat-transfer coefficient for various systems. Also shown in that table are values of $\bar{h}_c$ for boiling and condensation. Both of these phase-change phenomena are classified as convective problems. It should be remembered that $\bar{h}_c$ for all geometries may not be known, in which case a table such as 1.2 would be used as a guide to obtain an estimate.

Example 1.4

Figure 1.11 is a schematic of an apparatus from which to take data for determining the film conductance. Water flows through a stainless steel pipe having an inside diameter of 2.43 cm. The

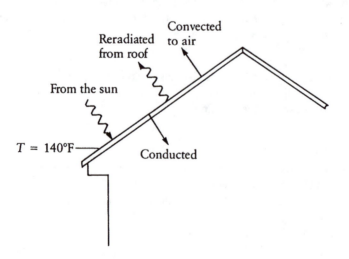

FIGURE 1.10   The roof of Example 1.3.

TABLE 1.2
Typical Values of the Convective Heat-Transfer Coefficient $\bar{h}_c$ for Various Fluids

Fluids and condition	$\bar{h}_c$ W/(m²·K)	$\bar{h}_c$ BTU/(hr·ft²·°R)
Air in natural convection	5–25	1–5
Superheated steam or air in forced convection	30–300	5–50
Oil in forced convection	60–1,800	10–300
Water in forced convection	300–6,000	50–2,000
Water during boiling	3,000–60,000	500–10,000
Steam during condensation	6,000–120,000	1,000–20,000

pipe is located within another pipe. A second fluid condenses on the exterior surface of the inner pipe. The process of condensation occurs at a constant temperature. With thermocouples appropriately placed, it is determined that the inside of the inner pipe wall temperature is a constant at 75°C. The incoming water temperature is 20°C. The bulk water temperature measured at section 2 is 21.4°C, and the flow meter reads 500 cm³/s. Determine the average film conductance between the inlet and section 2 (L = 20 cm). Take the specific heat of water to be approximately constant and equal to 4.2 kJ/(kg·K). Assume water density to be constant at 1 000 kg/m³.

Solution

As liquid flows through the heated tube, the liquid near the tube wall warms first. The heat is transferred to the liquid in the central portion of the tube by convective effects. Thus, at any section, temperature might vary with radial location. So we rely on bulk (T_b) or mixing cup temperature where it is assumed that the liquid at the section of interest is collected in a cup, mixed, and a uniform temperature is measured.

Assumptions

1. Bulk temperature gives an adequate representation of temperature at a section.
2. Newton's Law of Cooling applies.

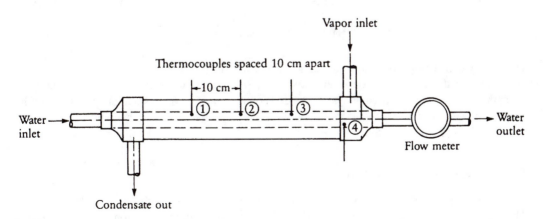

FIGURE 1.11 Schematic of a device used to determine the film conductance.

From the definition of specific heat, we can determine the amount of heat added to the water over the 20-cm length to section 2 as

$$q = \dot{m}c_p(T_{b2} - T_{bin}) \tag{i}$$

where $\dot{m}$ is the mass flow rate of water, c_p is the specific heat, T_{b2} is the bulk water temperature at section 2, and T_{bin} is the inlet water temperature. From Newton's Law of Cooling we have

$$q = \bar{h}_c A(T_w - T_{bin}) \tag{ii}$$

where $\bar{h}_c$ is the film conductance, A is the area of contact, and T_w is the wall temperature. Note that the temperature difference here is the maximum for the system. Moreover, the equation as written defines $\bar{h}_c$ as an average over the heated section. The area is given by

$$A = \pi D L$$

where D is the inside pipe diameter, and L is the heating length from inlet to section 2. Substituting for area and equating both expressions for q yields

$$\dot{m}c_p(T_{b2} - T_{bin}) = \bar{h}_c \pi D L(T_w - T_{bin})$$

Solving for $\bar{h}_c$, we obtain

$$\bar{h}_c = \frac{\dot{m}c_p(T_{b2} - T_{bin})}{\pi D L(T_w - T_{bin})}$$

The mass flow rate is equal to the product of density (ρ) and volume flow rate (Q):

$$\dot{m} = \rho Q = (1000 \text{ kg/m}^3)(500 \text{ cm}^3/\text{s})(1 \text{ m}^3/10^6 \text{ cm}^3) = 0.5 \text{ kg/s}$$

All other parameters were provided in the problem statement. Substituting,

$$\bar{h}_c = \frac{(0.5 \text{ kg/s})[4.2 \times 10^3 \text{ J/(kg·K)}](21.4 - 20) \text{ °C}}{\pi(0.024 \ 3 \text{ m})(0.20 \text{ m})(75 - 20)° \text{ C}}$$

$$\bar{h}_c = 3 \ 501 \text{ W/(m}^2\text{·K)}$$

The above convection coefficient is defined according to Equation ii of this example. Other definitions for the convection coefficient exist. (See Problem 20 for an example.) The selection of the appropriate $\bar{h}_c$ for a particular system depends on a number of factors that are discussed in the chapters on convection. It is therefore important to properly define the convection coefficient used in a specific problem.   ❏

1.8   RADIATION HEAT TRANSFER

Conduction and convection are heat-transfer mechanisms involving a material medium through which energy travels. However, energy can also be transferred through a region in which a perfect vacuum exists. This mechanism is commonly called *electromagnetic radiation*.

The many types of electromagnetic radiation include X-rays, gamma rays, the visible spectrum, radio waves, and microwaves. All thermal radiation is propagated at the speed of light in a vacuum. In this discussion, our concern is solely with the thermal component. Net heat transfer by radiation is a result of a temperature difference, but radiation propagates with or without a second body at some temperature. It is useful to define a surface or substance that emits radiation ideally. An ideal radiator is known conventionally as a black body, and it will emit energy at a rate that is proportional to the fourth power of its absolute temperature:

$$\frac{q_r}{A} \propto T^4$$

With a proportionality constant, the above equation becomes

$$q_r = \sigma A T^4 \qquad (1.9)$$

where σ is called the Stefan-Boltzmann constant and has the value of

$$\sigma = 5.67 \times 10^{-8} \ W/(m^2 \cdot K^4)$$

$$\sigma = 0.174 \times 10^{-8} \ BTU/(hr \cdot ft^2 \cdot {}^\circ R^4)$$

Temperature in Equation 1.9 must be expressed in absolute units. In the case of most real materials, surfaces do not emit electromagnetic radiation ideally; they are not ideal radiators as black bodies are. To account for the "gray" nature of real surfaces, we introduce a dimensionless factor, ε, called the *emissivity*. Emissivity values for various surfaces are provided in Appendix Table E.l. Modifying Equation 1.9 to account for gray-body behavior yields

$$q_r = \sigma A \varepsilon T^4 \qquad \varepsilon \le 1 \qquad (1.10)$$

where $\varepsilon = 1$ for an ideal radiator.

When a real body exchanges heat by radiation only with a black body, the net heat exchanged is proportional to the difference in T^4. For such a system, the net heat exchange is

$$q_r = \sigma A_1 \varepsilon_1 (T_1^4 - T_2^4) \qquad (1.11)$$

Another factor that must be considered in radiation heat transfer is that electromagnetic radiation travels in straight lines. As an illustration of this feature, consider a room with four sides, a floor, and a ceiling. One wall contains a small window. Someone looking through the window can see the ceiling, the floor, and each wall except the one containing the window. If the window is a radiative heat source, it would transfer heat *directly* to all surfaces in the room except to the wall containing the window. Furthermore, the size and orientation of the walls, ceiling, and floor as viewed from the window have an effect on how much heat each of these surfaces receives. We account for this phenomenon by introducing a term called the *view factor* (or configuration factor or shape factor), F, which is a function of the geometry of the configuration to be analyzed. If a real body exchanges radiant energy only with a black body, the net heat exchanged is given by

$$q_r = \sigma A_1 F_{1-2} \varepsilon_1 (T_1^4 - T_2^4) \qquad F_{1-2} \le 1 \qquad (1.12)$$

where the 1–2 subscript on F denotes how surface 2 is viewed by surface 1. More specifically, the shape factor is defined as the fraction of total radiant energy that leaves surface 1 and arrives directly on surface 2. Like emissivity, the shape factor is dimensionless and less than or equal to 1.

Example 1.5

An asphalt driveway is 14 ft wide and 30 ft long. During daylight hours the driveway receives enough energy to raise its temperature to 120°F. Determine the instantaneous heat loss rate into space by radiation from the upper surface of the driveway during nighttime, assuming the surface to be at 120°F and that the surface has an emissivity of 0.9. Neglect convection losses.

Solution

The driveway receives radiation from the sun during the daylight hours. As the asphalt temperature increases, heat is transferred by conduction through the asphalt to the ground below, by convection to the air, and by radiation to the surroundings.

Assumptions

1. The driveway transfers heat away only by radiation from its surface.
2. At night, space behaves as a black body, which has an emissivity of 1.
3. The view factor is 1, because the night sky is visible from any point on the driveway.
4. Driveway surface temperature is uniform at 120°F.
5. Temperature in space is 0°R (an oversimplified assumption for lack of more specific information).

Equation 1.12 applies:

$$q_r = \sigma A F_{1-2} \varepsilon_1 (T_1^4 - T_2^4)$$

The temperature of the driveway surface is

$$T_1 = 120 + 460 = 580° \text{ R}$$

The temperature in space is 0°R. Substituting, we obtain

$$q_r = (0.1714 \times 10^{-8} \text{ BTU/hr·ft}^2\text{·°R}^4)(14 \times 30) \text{ ft}^2(0.9)(580^4 - 0)°\text{R}^4$$

Solving,

$$q_r = 73,320 \text{ BTU/hr}$$

This value is an instantaneous radiative heat-transfer rate subject to the assumptions made. ❏

Equation 1.7 is the integrated form of Fourier's Law of Conduction in one dimension. That equation contains a temperature difference. Equation 1.8 is Newton's Law of Cooling; it too contains a temperature difference. In trying to simplify the expression for radiation heat transfer, Equation 1.12 is often written in terms of a temperature difference and a radiation thermal conductance. We therefore write

$$q_r = \bar{h}_r A(T_1 - T_2) \tag{1.13}$$

where $\bar{h}_r$ is a radiation thermal conductance, which has the same dimensions as the film conductance in convection: F·L/(L²·T·t); [W/(m²·K) or BTU/hr·ft²·°R].

Example 1.6

Determine the radiation thermal conductance for the data in Example 1.5. A summary of the parameters used there is

$$q_r = 73,320 \text{ BTU/hr} \qquad F_{1-2} = 1.0$$
$$T_1 = 580°\text{R} \qquad\qquad \varepsilon_1 = 0.9$$
$$T_2 = 0°\text{R}$$
$$A = 420 \text{ ft}^2$$

Solution

The heat transferred by radiation will vary from one instant in time to another for the driveway problem, as will the driveway's surface temperature. Thus, the radiation thermal conductance is not a constant.

Assumptions

1. Radiation is the only mode of heat transfer to consider.
2. Equation 1.13 applies.

Solving Equation 1.13 for $\bar{h}_r$ gives

$$\bar{h}_r = \frac{q_r}{A(T_1 - T_2)}$$

Substituting, we find

$$\bar{h}_r = \frac{73,320 \text{ BTU/hr}}{420 \text{ ft}^2 (580° \text{ R})}$$

$$\bar{h}_r = 0.30 \text{ BTU/(hr·ft}^2 \cdot °\text{R)}$$

This value applies only for the instantaneous conditions specified.   ❏

The radiation thermal conductance $\bar{h}_r$ is a simplification that is used in problems involving both convection and radiation. Using it allows us to more easily separate the convection from the radiation heat transfer modes in the problem of interest.

1.9 EMISSIVITY AND OTHER RADIATIVE PROPERTIES

Emissivity is a property that describes how radiant energy interacts with the surface of a material. Other properties that are important in radiation are the reflectivity, absorptivity, and transmissivity. As implied by these names,

- The *reflectivity* is the fraction of incident radiant energy reflected by the surface.
- The *absorptivity* is the fraction of incident radiant energy absorbed by the surface.
- The *transmissivity* is the fraction of incident radiant energy transmitted through a layer of the material.

Radiant energy is characterized by electromagnetic waves that travel at the speed of light. According to wave theory, radiation can be thought of as many waves, all oscillating at different wavelengths and frequencies. The radiative properties mentioned above (including emissivity) are, in general, functions of wavelength. Properties that describe surface behavior as a function of wavelength are called *monochromatic* or *spectral* properties. In addition, radiative properties can be a function of direction, specifically in reference to the direction at which radiation is incident upon the surface. Such properties are referred to as *directional* properties. In an analysis of radiation heat transfer, accounting for the exact behavior of a surface can be complex enough to make a solution elusive. Moreover, the properties may not all be known. A simplified approach must therefore be formulated. This often involves the use of radiative properties that are averages over all wavelengths and all directions. These properties are called *total* and *hemispherical* properties. The use of total properties is accurate enough in a majority of cases for engineering work. Measurement of radiation properties is beyond the scope of what we wish to cover. The interested reader can refer to any good text on engineering measurements.*

The surface of a substance highly influences its radiation characteristics and therefore the amount of radiative heat the surface can absorb, transmit, reflect, and emit. Consider the emissivity of aluminum as an example. At room temperature a polished aluminum plate has a normal emissivity (as determined from measurements normal to the surface) of 0.04. An aluminum plate with a rough surface has a normal emissivity of 0.07 at room temperature. Another significant case exists for gold. Gold having a polished surface has a normal emissivity of 0.025. An unpolished gold surface has an emissivity of 0.47. It is important to realize that such differences exist. It is also important to know that the surface treatment of a material is a very significant factor in how the material itself behaves in radiative heat-transfer studies. Sanding, polishing, electroplating, anodizing, and painting are processes that can have a controlling influence on radiative heat losses or gains under a variety of conditions.

1.10 THE THERMAL CIRCUIT

A simplified approach that can be adopted for the analysis of one-dimensional heat-transfer problems involves a concept called the *thermal circuit*. Suppose that we wrote the integrated form of Fourier's Law of Conduction (Equation 1.7) as follows:

$$\frac{q_x}{A} = \frac{k(T_1 - T_2)}{L}$$

$$q_x = \frac{(T_1 - T_2)}{L/kA} \tag{1.14}$$

Recall that this was written for a plane wall under steady-state conditions; T_1 and T_2 are temperatures of the surfaces, L is the distance between the surfaces, k is the thermal conductivity, and A is the area normal to the flow of heat that is conducted, q_x. With regard to the concept of a thermal circuit, the temperature difference can be identified as the potential or driving function that causes heat to flow against a resistance:

$$\text{Heat flow} = \frac{\text{Driving function}}{\text{Thermal resistance}} = \frac{\Delta T}{R_i}$$

The thermal resistance for the case of Equation 1.14 is

---

* Beckwith, T. G., N. L. Buck, and R. D. Marangoni, *Mechanical Measurements*, 3rd ed., Addison–Wesley Publishing Co., Reading, Mass, 1982, Chapter 16.

  J.P. Holman, *Experimental Methods for Engineers,* 4th ed., McGraw-Hill Book Co., New York, 1984, Chapter 12.

$$R_k = L/kA \qquad (1.15)$$

which is the defining equation for one-dimensional conduction in Cartesian coordinates.

A similar formulation for the convection equation can be deduced. Newton's Law of Cooling is

$$q_c = \bar{h}_c A (T_w - T_\infty)$$

or

$$q_c = \frac{(T_w - T_\infty)}{1/\bar{h}_c A}$$

The thermal resistance is therefore

$$R_c = 1/\bar{h}_c A \qquad (1.16)$$

which is the defining equation for the convective thermal resistance.

For radiation problems, using a thermal circuit with Equation 1.12 becomes cumbersome, because it is difficult to separate the temperature difference from the resistance terms. However, if we use Equation 1.13, we have

$$q_r = \bar{h}_r A (T_1 - T_2) = \frac{T_1 - T_2}{1/\bar{h}_r A}$$

The resistance then is

$$R_r = 1/\bar{h}_r A$$

The dimensions of thermal resistance are t·T/(F·L), [K/W or °R·hr/BTU].

Example 1.7

Figure 1.12 is half the top view of a chimney wall in cross section. On the left are moving exhaust gases, which are products of combustion. At this location the exhaust gases are at a temperature of 1 000 K. The chimney itself is made of common brick. The outside air temperature is 10°C. (a) Sketch the thermal circuit for the chimney at location A–A, assuming one-dimensional heat flow

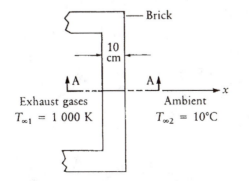

FIGURE 1.12 Chimney of Example 1.7.

there. (b) Identify all resistances and their values. (c) Estimate the heat transfer per unit area through the wall from exhaust gases to ambient air. (d) Determine the inside and outside wall temperatures.

Solution

We first impose a coordinate system on the chimney wall using A-A as one axis and temperature as another. Figure 1.13 shows these axes and the temperature variation with x. The exhaust gases are at 1 000 K at this location in the chimney. As they continue rising, they lose energy to the walls. In the vicinity of the wall shown in Figure 1.13, temperature decreases to some value T_1 on the inside surface. Temperature decreases linearly with x in the brick to some value T_2 on the outside surface. The temperature then decreases to the ambient value $T_{\infty 2} = 10°C + 273 = 283$ K. This occurs in the region about A–A. The corners have an effect on the temperature gradient also. At this stage in the problem, T_1 and T_2 are unknown.

Assumptions

1. One-dimensional conduction exists in the wall at section A–A.
2. The effect of the corners is negligible.
3. Properties of the material are constant, and average values can be used.
4. The values selected for the film conductances are constant and apply at section A–A.

(a) The thermal circuit is shown in Figure 1.13.
(b) Because area is unspecified, we can perform the calculations on a per unit area basis by selecting a 1 m² area for A for convenience. The resistance on the inside surface is

$$R_{c1} = \frac{1}{\bar{h}_{c1} A}$$

Table 1.2 is all the information we have available on $\bar{h}_c$ at this point. Data there indicate that $\bar{h}_c$ varies from 5 to 25 W/(m²·K) for free convection with air as the fluid. We will select 15 W/(m²·K), the average value. Therefore,

$$R_{c1} = \frac{1}{(15)1} = 0.067 \text{ K/W}$$

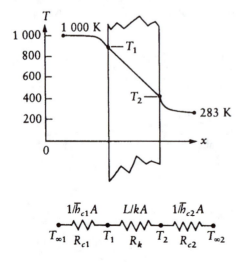

FIGURE 1.13   Temperature distribution and thermal resistance network for Example 1.7

Appendix Table B.3 lists thermal properties of building materials and insulations. Common brick has a thermal conductivity that varies from 0.38 to 0.52 W/(m·K). We use the average value of 0.45 W/(m·K). With $L = 10$ cm we find the thermal resistance to conduction as

$$R_k = \frac{L}{kA} = \frac{0.1}{0.45(1)}$$

$$R_k = 0.222 \ \text{K/W}$$

For the outside wall we again use a convection coefficient of 15 W/(m²·K). So,

$$R_{c2} = 0.067 \ \text{K/W}$$

(c) The heat transferred can be found by using the concept of thermal resistance applied from the inside gas temperature to the ambient air temperature:

$$q = \frac{T_{\infty 1} - T_{\infty 2}}{\Sigma R} = \frac{T_{\infty 1} - T_{\infty 2}}{R_{c1} + R_k + R_{c2}}$$

With all parameters known, we can substitute directly:

$$q = \frac{1000 - 283}{0.067 + 0.222 + 0.067}$$

Solving,

$$q = 2\,013 \ \text{W} = 2.013 \ \text{kW}$$

(d) The inside-wall temperature can be found by applying the resistance equation from $T_{\infty 1}$ to T_1:

$$q = \frac{T_{\infty 1} - T_1}{R_{c1}}$$

Rearranging and substituting gives

$$T_1 = T_{\infty 1} - R_{c1}q = 1\,000 - 0.067(2\,013)$$

$$T_1 = 865 \ \text{K}$$

Similarly,

$$q = \frac{T_2 - T_{\infty 2}}{R_{c2}}$$

$$T_2 = T_{\infty 2} + R_{c2}q = 283 + 0.067(2\,013)$$

$$T_2 = 418 \ \text{K} \qquad \qquad \square$$

1.11 COMBINED HEAT-TRANSFER MECHANISMS

It is interesting to model problems in which all three modes of heat transfer are taken into account. It was stated earlier that, in many problems, it is useful to determine which mode is predominant and to operate under that condition. In this section, we will consider a simple problem where all three modes are involved.

Example 1.8

A concrete sidewalk ($k = 0.604$ BTU/(hr·ft·°R)) is receiving heat from the ground. During nighttime the sidewalk loses heat by radiation to space and by convection to the surroundings. Ground temperature is such that the temperature at the bottom of the cement sidewalk is 80°F. Ambient air is at 20°F. For the system as sketched in Figure 1.14, determine the upper-surface temperature of the sidewalk if steady-state conditions exist.

Solution

An energy balance performed for the sidewalk would include

$$\text{Heat flow in} = \text{Energy stored} + \text{Heat flow out}$$

With no energy stored, we have

$$q_z = q_c + q_r$$

where q_z is the heat conducted into the sidewalk, and q_c and q_r combined represent the heat flowing out by convection and radiation, respectively.

Assumptions

1. One-dimensional conduction exists through the material.
2. The convective heat-transfer coefficient is a constant and applies to the entire surface.
3. The shape factor between the surface and space is 1.0 because all energy radiated is intercepted in space.
4. The space temperature is 0°R.

Evaluating each term in the heat-balance equation brings

$$q_z = \frac{kA(T_1 - T_w)}{L}$$

$$q_c = \bar{h}_c A(T_w - T_\infty)$$

$$q_r = \sigma A F_{w-r} \varepsilon_w (T_w^4 - T_r^4)$$

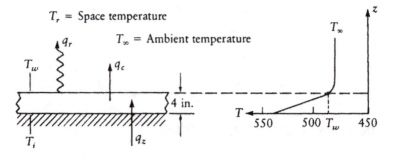

FIGURE 1.14 The sidewalk of Example 1.8.

Substituting, we obtain

$$\frac{kA}{L}(T_1 - T_w) = \bar{h}_c A(T_w - T_\infty) + \sigma A F_{w-r} \varepsilon_w (T_w^4 - T_r^4)$$

Dividing area out and rearranging gives

$$\frac{k}{L}T_w + \bar{h}_c T_w + \sigma F_{w-r} \varepsilon_w T_w^4 = \frac{k}{L}T_1 + \bar{h}_c T_\infty + \sigma F_{w-r} \varepsilon_w T_r^4$$

We now evaluate each variable:

From Table 1.2, the average value for natural convection $\bar{h}_c = 3.0$ BTU/(hr·ft²·°R).

We use Appendix Table E.2 for concrete, which, of the information available, is closest in surface composition to a sidewalk, $\varepsilon_w = 0.93$.

The shape factor we take as $F_{w-r} = 1$, assuming that all energy leaving by radiation is intercepted in space.

The Stefan-Boltzmann constant is 0.1714×10^8 BTU/(hr·ft²·°R).

Physical parameters: $L = 4/12 = 0.333$ ft, $T_1 = 80 + 460 = 540°R$, $T_\infty = 20 + 460 = 480°R$, $T_r = 0°R$.

Substituting gives

$$\frac{0.604}{0.333}T_w + 3.0T_w + \frac{0.1714}{10^8}(1)(0.93)T_w^4 = \frac{0.604}{0.333}(540) + 3.0(480) + 0$$

Simplifying, we get

$$4.814 T_w + \frac{0.1594}{10^8}T_w^4 = 2419$$

Solving by trial and error yields the following, where LHS is the left-hand side of the equation,

T_w	LHS
470	2340
480	2395
490	2450
485	2422
484.5	2420 close enough

The surface temperature then is

$$T_w = 484.5°R = 24.5°F \qquad \square$$

1.12 SUMMARY

In this chapter, we have examined conduction, convection, and radiation modes of heat transfer. We reviewed various systems of dimensions and made heat-transfer calculations for each mode. We wrote equations that defined thermal conductivity, film conductance, and emissivity. Although most heat-transfer problems involve all three modes, we remarked that, in many problems, the dominant mode could be identified, and an appropriate formulation could be used to obtain a solution. A summary of the equations used in conduction, convection, and radiation is provided in Table 1.3.

TABLE 1.3
Equation Summary

Mode or parameter	Equation	Property	Dimension and units
Conduction	$q_k = \dfrac{kA}{L}(T_1 - T_2)$	k, thermal conductivity	F·L/(L·T·t) W/(m·K) BTU/(hr·ft·°R)
Convection	$q_c = \bar{h}_c A(T_w - T_\infty)$	$\bar{h}_c$, film coefficient	F·L/(L²·T·t) W/(m²·K) BTU/(hr·ft²°R)
Radiation*	$q_r = \sigma A F_{1-2} \varepsilon_1 (T_1^4 - T_2^4)$	ε, emissivity	
Stefan-Boltzmann constant	$\sigma = 5.67 \times 10^{-8} \text{ W/(m}^2 \cdot \text{K}^4)$ $\sigma = 0.1714 \times 10^{-8} \text{ BTU/(hr} \cdot \text{ft}^2 \cdot °\text{R}^4)$		

*The radiation equation in the table applies only to the simplified cases of this chapter. A general radiation relationship is quite complex. The above question serves to introduce the property emissivity.

1.13 PROBLEMS

1.13.1 INTRODUCTION TO CONDUCTION

1. Equation 1.7 is the integrated form of Fourier's Law of Conduction in one dimension for constant thermal conductivity and steady-state operation. If a more accurate model is desired, the variation of thermal conductivity with temperature may be accounted for by using

$$k = k_0[1 + C_1(T - T_0)]$$

where k is thermal conductivity, k_0 is thermal conductivity at some reference temperature T_0, C_1 is a constant, and T is temperature. Determine k_0 and C_1 for copper if $T_0 = 650$ K. Use the graphs in this chapter to obtain the necessary data.

2. Repeat Problem 1 for glycerin if $T_0 = 170°$F.

3. Repeat Problem 1 for hydrogen if the reference temperature is 350 K.

4. Equation 1.7 was obtained by integration of Equation 1.6, assuming a constant thermal conductivity and steady-state operation. If thermal conductivity in Equation 1.6 is not a constant but instead is written as

$$k = k_0[1 + C_1(T - T_0)]$$

where k is thermal conductivity, k_0 is thermal conductivity at some reference temperature T_0, C_1 is a constant, and T is temperature, how is Equation 1.7 affected? Derive the new expression.

5. Figure P1.1 is a sketch of a guarded-hot-plate apparatus, which is used primarily for measuring thermal conductivity of an insulating material. The heat source is the main heater. To retard heat flow in any lateral direction (from the main heater), guard heaters are placed next to the main heater. Thus heat is made to flow in one direction from the main heater, through two samples of the material to be tested, to the cooled chambers. Temperature is measured with thermocouples at surfaces 1, 2, and 3. By measuring voltage and current into the main heater, the heat flow through both samples can be calculated.

   Suppose the thermal conductivity of glass fiber is to be measured. Two samples 3 cm thick are cut and placed in a guarded-hot-plate device. The voltage across the heater is 120 V, and at this value the heater draws 2 A. The surface temperature at 1 is determined to be 130°C. The temperatures at 2 and 3 are both 25°C. Determine the thermal conductivity. Take the area of contact to be 1 m².

6. Figure P1.2 shows a cross section of two materials. Also shown is a coordinate system with a graph of the steady-state temperature vs. distance profile. Based on T vs. x data, which material (A or B) has the higher thermal conductivity? Prove it.

7. The floor of a brick fireplace is to be appropriately insulated. It is proposed to use a 1/2–in thick sheet of asbestos. The upper-surface temperature of the asbestos can reach 1000°F while the underside temperature is only 400°F. How much heat per unit area is conducted through the asbestos?

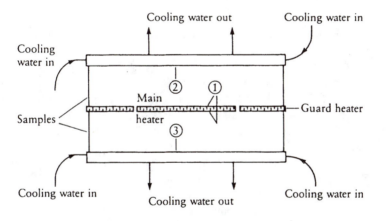

FIGURE P1.1

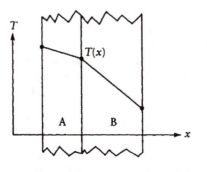

FIGURE P1.2

8. The guarded-hot-plate apparatus described in Problem 1.5 and Figure P1.1 is used to measure thermal conductivity of cork. By mistake, rather than obtaining two samples of the same cork, one sample each of a cork board and of expanded cork are used. A 1-in thick piece of expanded cork is placed in the top position, and a 1-in thick piece of cork board is placed in the lower position. The input to the main heater is 120 V with a current of 0.05 A. The temperature of surface 1 is 90°F, and the temperature at surface 2 is 70°F. Using values of thermal conductivity in Appendix Table B.3, determine the expected temperature reading at surface 3. Take the effective area to be 3 ft².

9. A guarded-hot-plate heater (described in Problem 1.5 and sketched in Figure P1.1) is used to measure thermal conductivity. A 2-cm thick sheet of Plexiglas® is placed in the top position, and a 4-cm thick sheet of Plexiglas is placed in the lower position. The power input is 120 V × 0.813 A, and the temperature at surface 1 is 35°C. What is the expected temperature at surface 3 if $T_2 = 30°C$? Take the area to be 1 m².

10. Thermal conductivity of insulating materials can be measured with a guarded-hot-plate heater (see Problem 1.5 and Figure P1.1). Suppose a 1/4-in thick piece of linoleum is placed in the top position and kapok in the lower position. The main heater operates at 120 V and draws 2 A. The temperature at surface 1 is 100°F. Determine the thickness of kapok required for the temperature at surface 3 to equal that at surface 2. Take the temperature at surface 2 to be 50°F and the area to be 1.3 ft².

11. The temperature distribution in a vertical stainless steel (type 304) wall is given by

$$T = 200 - 1500x^2$$

where x is measured from the left face and is expressed in m, and T in °C. The wall is 10 cm thick.
(a) Graph the temperature distribution.
(b) Is heat being applied to one or both sides of the metal, or is it being allowed to cool?

12. The apparatus of Figure 1.8 is used to measure thermal conductivity of platinum, located in place of the aluminum. The temperature data of the stainless steel show an average $\Delta T/\Delta x$ of 40°C/cm. Based on the data of Appendix Table B.1, what is the expected $\Delta T/\Delta x$ for the platinum?

13. A sweater made of wool is worn to "keep warm" during times of cold weather. Suppose a 1/4–in thick wool sweater is worn over a shirt. The temperature at the inside surface of the wool is 94°F while the outside-surface temperature of the wool is 90°F. Determine the heat transfer per unit area through the wool.

14. Figure P1.3 is a schematic of two copper rods set up in a one-dimensional heat-flow situation. On the left a copper rod of constant diameter is heated by a hot plate at the bottom and cooled with water at the top. The rod has 10 thermocouples attached, spaced 1 in apart. The rod has a 2-in diameter. The second rod is tapered. At its bottom, the diameter is 1 in, while at its top the diameter is 2 in. The rod is insulated, heated, and instrumented in the same way that the straight rod is. The heat input to both rods is the same, which results in a copper surface temperature at the hot plate of 150°F. At the water chamber of the straight copper rod, the temperature is recorded as 100°F. Construct a graph of temperature vs. distance for both rods. The quantity $\Delta T/\Delta x$ in the straight rod is constant. What quantity is constant in the tapered rod?

1.13.2 Introduction to Convection

15. It is desired to test the effectiveness of a viscous liquid lubricant that is to be used in sealed bearings. One of the tests to be conducted is to heat a ball bearing to 200°F and submerge the bearing in the lubricant, which is to be kept at 100°F. Determine the

convective heat-transfer rate when the bearing is first submerged. Take the convective heat-transfer coefficient between bearing and lubricant to be 150 BTU/hr·ft²·°R. The bearing diameter is 0.4375 in.

16. The side of a building convects heat away to the air in contact with it. The building's outside-wall temperature is 25°C, and the surrounding air is at 15°C. The wall is 3 m tall by 18 m wide. Estimate the heat convected away.

17. A bathtub is filled to a depth of 12 in with water whose temperature is 130°F. The side walls, front, and back of the tub have a surface temperature of 72°F, and they receive heat from the water. The tub sides can be approximated as vertical walls being 4 ft long. The front and back are 30 in wide. Estimate the heat convected to the tub walls. Sketch the temperature profile in the vicinity of a wall.

18. The temperature profile in a brick masonry wall is steady and linear, as shown in Figure P1.4. At surface 1, the temperature is 30°C. At surface 2, the temperature is 20°C. On

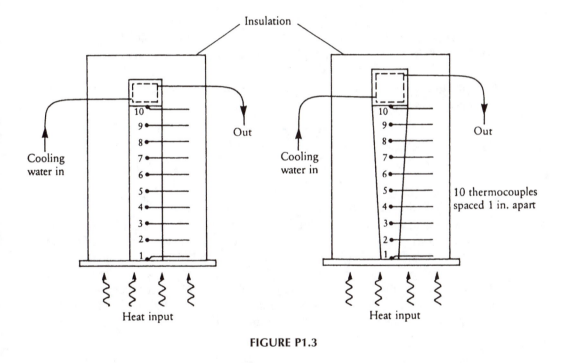

FIGURE P1.3

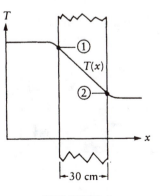

FIGURE P1.4

the left of the wall is water at a bulk temperature of 38°C. Determine the convective heat-transfer coefficient between the water and the wall. Take the area to be 1 m².

19. The temperature profile in a certain material is steady and linear, as shown in Figure P1.4. At surface 1, the temperature is 30°C. At surface 2, the temperature is 20°C. On the right side of the wall is air flowing at a free-stream velocity of 2 m/s. The area is 0.2×0.2 m, and the convective heat-transfer coefficient is 12 W/(m²·K). The free-stream air temperature is 10°C. Determine the thermal conductivity of the wall material.

20. The data presented in Example 1.4 in conjunction with the apparatus of Figure 1.11 allowed for finding a convective heat-transfer coefficient. The film coefficient calculated is the one defined in Newton's Law of Cooling. Suppose that we use a different definition for a convective heat-transfer coefficient. Specifically,

$$q = \bar{h}_a A \frac{(T_{win} - T_{bin}) + (T_{w2} - T_{b2})}{2}$$

where T_{win} is the wall temperature at the inlet, T_{w2} is the wall temperature at location 2, T_{bin} is the bulk fluid temperature at the inlet, and T_{b2} is the bulk fluid temperature at location 2. With the data of Example 1.4, determine the film coefficient $\bar{h}_a$. Is there any advantage of using $\bar{h}_a$ over $\bar{h}_c$?

21. Referring to Example 1.4 and Figure 1.11, determine $\bar{h}_c$ for each thermocouple location, given the following data:

$T_{b1} = 20.8°C$	$T_{b2} = 22.0°C$
$T_{b3} = 21.4°C$	$T_{b4} = 22.5°C$

Graph the bulk fluid temperature and the film coefficient vs. distance.

22. An experiment to determine the natural convection heat-transfer coefficient uses a device that is shown schematically in Figure P1.5. The device is a cylinder consisting of three sections. The center section is a test element. It is flanked by two guard heaters to eliminate heat losses via the ends from the test section. All three sections are heated electrically and instrumented appropriately, the objective being to operate the device so that heat is transferred by convection from the test section to the surrounding air uniformly. Neglecting radiation losses, the heat transfer is given by

$$q_c = \bar{h}_c A (T_w - T_\infty)$$

where the heat loss q_c is determined by measuring the voltage across and amperage through the test heater, $\bar{h}_c$ is the film conductance to be measured, A is the surface area

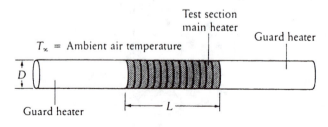

FIGURE P1.5

of the test section, T_w is the wall temperature of the test section, and T_a is the ambient air temperature. For the following data, determine $\bar{h}_c$:

Voltage = 120 V	Amperage = 0.29 A
$L = 4$ in	$D = 2$ in
$T_w = 120°F$	$T_\infty = 72°F$

23. A hydraulic elevator operates by a pump that moves hydraulic oil into or out of a telescoping cylinder that supports the elevator car. The oil is stored in a tank near the pump. When the elevator is not used much, such as during evenings, and when the ambient temperature is low, the oil loses more heat than is desired. The oil density decreases with temperature, and in the morning the elevator floor is not level with the hall floor. The elevator drops about 8–10 cm, which is unacceptable, and is due to the oil being cold. So several electric immersion heaters are used to maintain the oil temperature at 25°C. The tank is shown in Figure P1.6. The tank walls are thin enough so that the temperature within the wall is uniform and also at 25°C. The ambient air temperature is 15°C. Estimate the heat lost by convection through the four vertical tank walls. Use an average value for the film coefficient.

1.13.3 INTRODUCTION TO RADIATION

24. A black body having a surface area of 1 m² radiates heat to the surroundings, which are at 0 K. Determine the rate of heat transfer by radiation from the black body if its temperature is

(a) 0 K	(b) 200 K	(c) 400 K
(d) 600 K	(e) 800 K	(f) 1 000 K

Construct a graph of q_r vs. T.

25. A tungsten filament at 5900°R radiates heat in all directions. Assuming the heat receiver is a black body, how much heat per unit area is radiated from the filament? The receiver temperature is 0°R.

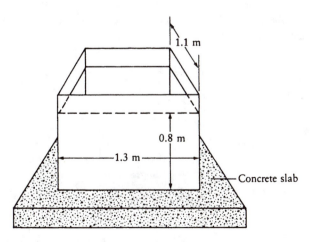

FIGURE P1.6

26. There are various methods for evaluating shape factor. Consider the sketch of Figure P1.7. The enclosure consists of three planes that are infinite in length (into the page). In this problem we will determine the shape factor for various combinations of surfaces. Using one particular method of evaluating shape factor, it is determined that

$$F_{1-2} = \frac{1}{2L_1}(L_2 + L_1 - L_3)$$

which is written assuming unit depth into the page.

Recall that shape factor is defined as the fraction of total energy leaving one surface that impinges directly upon another surface—F_{1-2} is the fraction of total energy leaving surface 1 that arrives directly on surface 2. There exist various algebraic relations regarding shape factors. One such relationship deals with *reciprocity*. For example, we can write

$$A_i F_{i-j} = A_j F_{j-i}$$

(a) Given the definition of shape factor, what is the of F_{1-1}, F_{1-2} and F_{1-3}?
(b) Can surface 1 "see itself"? What is the numerical value of F_{1-1}?
(c) Evaluate F_{1-2} and use the results of part (a) to evaluate F_{1-3}.
(d) Evaluate F_{2-1} using reciprocity.
(e) Evaluate F_{3-1} using reciprocity.
(f) Determine F_{2-3} and F_{3-2}.
Leave answers in fractional form.

27. The porcelain surface of a kitchen stove receives heat during the time that something is cooking inside. With the kitchen lights off, the surroundings behave as though they were a black body. If the stove-top area is considered plane with dimensions of 2.5 × 2.5 ft, determine the net rate of heat transfer between the stove top and the surroundings. The porcelain surface temperature is 130°F, and the surroundings are at 68°F. Neglect convection losses.

28. Use the data of Problem 27 to determine the radiation film conductance $\bar{h}_r$.

29. Two infinite black plates at 500°C and 300°C exchange heat by radiation. Determine the net heat flux from the hot plate to the cold plate.

30. A 2-cm thick plate of steel (1C) receives a radiant heat flux of 1 000 W/m². The steel is in an environment where convection can be neglected. The surface temperature on the opposite side of the plate is 25°C. Determine the temperature of the surface on which the radiant energy is incident.

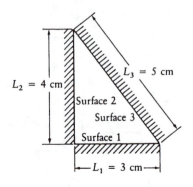

FIGURE P1.7

31. A satellite in orbit uses solar cells to convert sunlight to power. A preliminary analysis determines that the plate will radiate 14 000 W of heat into space when not facing the sun. If the plate is a 6 × 6 ft square, determine the surface temperature of the plate. Assume $\varepsilon = 1$.

1.13.4 COMBINED MECHANISM PROBLEMS

32. Figure P1.8 is a cross-sectional view of a wall as found in a domestic dwelling. The wall itself is silica brick. Outside (to the left) of the wall the ambient temperature is 37°C. The brick surface temperature is 30°C, and the film coefficient on the outside is 5 W/(m²·K). The temperature profile in the brick is steady and decreases linearly with distance, with 30°C being the maximum.
    (a) Determine the temperature of the air in the air space if $\bar{h}_c$ on that side is 3.5 W/(m²·K).
    (b) Draw the thermal circuit from the ambient air to the air space.
    (c) Evaluate all resistances.
    (d) With what resistance does the largest temperature drop correspond? Neglect the effect of the wall stud.
33. A radiant heat flux of 30 BTU/hr·ft² is absorbed by a surface, which in turn transfers heat away only by convection to surrounding air. For an ambient air temperature of 85°F and a film coefficient of 0.9 BTU/hr·ft²·°R, determine the equilibrium temperature of the surface.
34. Rework Example 1.8, neglecting the radiation heat loss.
35. Rework Example 1.8, neglecting the convection heat loss.
36. A new design of an electric water heater for a home is being analyzed. The heater is to consist of a length of tubing wrapped with an electrical conducting tape, which in turn is well insulated. When electric current passes through the tape, it heats the tube, which heats the water inside. The system is set up such that heating occurs only when water flows through the tube. The inside diameter of the tube is 20 mm, and the tube length is 3 m. The convection coefficient between the tube and water is 1 000 W/(m²·K), which is an average over the 3-m length, based on an effective temperature difference of 340 K (assumed constant).
    (a) Determine the heat transferred to the water, and the downstream temperature of the water if the average water velocity is 1.2 m/s and its density is 1 000 kg/m³.
    (b) If the voltage applied to the heater is 120 V, will a 15-A fuse be sufficient in the electrical line? Take the initial water temperature to be 20°C, and the specific heat of water to be 4.2 kJ/(kg·K).

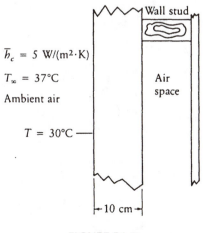

FIGURE P1.8

37. One way of using stove-burner heat to heat a kitchen effectively is to utilize a hot plate. Consider a lighted gas burner. Heat is transferred from the flame to the surroundings by what mode(s) of heat transfer? Suppose a hot plate consisting of cast iron with a 6-in diameter is placed over the burner. The plate absorbs a significant fraction of the energy from the flame and, in turn, the plate radiates heat in all directions. It is observed that use of the hot plate heats the surrounding room (kitchen) very well compared to the burner alone.

Consider a hot plate over a burner. The flame temperature is 1800°F. The convective heat-transfer coefficient between the rising exhaust gases and the plate is 35 BTU/hr·ft²·°R. The plate transfers heat to the surroundings primarily by radiation. Determine the equilibrium temperature of the plate, assuming the surroundings to be at 65°F and to behave as a black body. The plate surface has $\varepsilon = 0.9$.

2 Steady-State Conduction in One Dimension

CONTENTS

2.1 INTRODUCTION

In this chapter, we examine various problems in steady-state conduction. The discussion, however, is restricted to those geometries in which the temperature is a function of only one space variable. The simplest example of such a system is the plane wall. A pipe wall is the corresponding example in polar cylindrical coordinates, and a hollow sphere is the example in spherical coordinates.

As we will see in the analytical method of solution of conduction problems, the first objective will be to determine the temperature distribution. It is convenient to begin with the appropriate form of the general conduction equation and solve it subject to appropriate boundary conditions to obtain the temperature distribution, if possible. Once the variation of temperature within the substance is known, the final objective of finding the heat-transfer rate can be accomplished.

In a number of problems, it seems as if the amount of information provided is overwhelming. A method of solution is most important to organize the data into a form that aids in streamlining the calculations. As indicated in the examples of this chapter (and reiterated from Chapter 1), the method recommended includes the following steps:

1. Select the system for analysis.
2. Sketch the expected temperature profile based on a physical interpretation of what is happening.
3. List the assumptions to be made in the analysis.
4. Sketch the thermal circuit if appropriate.
5. Write the descriptive equations
6. Obtain the solution to the problem.

Such a method, when developed, will provide the reader with a consistent way to solve problems efficiently.

Our objective is to derive an equation for conduction of heat in a material. We will do this by performing a heat balance on an elemental volume through which heat is being transferred by conduction. We then combine the heat-balance equation with Fourier's Law of Heat Conduction and obtain a relation for temperature within the material.

Before proceeding, however, it is useful to discuss some pertinent definitions. The temperature distribution existing within a material can at most depend on three space variables and on time. If indeed the temperature is a function of time, the problem is called *unsteady* and the temperature distributions is referred to as being *transient*. If the temperature is not a function of time, the problem is referred to as *steady* or the temperature distribution as being in a (or at) *steady state*. If temperature depends only on a single space coordinate, the problem, or the temperature distribution, is referred to as one dimensional. When temperature depends on two or three space variables, the problem is referred to as a two- or three-dimensional problem, respectively. In a one-dimensional, unsteady problem, temperature is a function of one space variable and of time.

In our analysis, we will assume that thermal conductivity is constant. Even though thermal conductivity for most materials does vary with temperature, the dependence in a majority of cases is not a strong one. A material in which thermal conductivity does not vary with direction is called *isotropic*. On the other hand, a fibrous material such as wood will have a thermal conductivity in the grain direction that is different from that measured across the grain. Thus, wood is *anisotropic*.

The notation for heat flow can be a source of confusion. In this text we will use q to denote a heat-flow rate, F·L/T [BTU/hr or W]. Furthermore, the term q'' will represent a heat-flow rate per unit area normal to the heat-flow direction, or a *heat flux*, F·L/(T·L^2) [BTU/hr·ft^2 or W/m^2]. We will also encounter a heat-flow term written as q''', which denotes a heat rate per unit volume, F·L/(T·L^3) [BTU/(hr·ft^3) or W/m^3].

2.2 THE ONE-DIMENSIONAL CONDUCTION EQUATION

In Chapter 1, Fourier's Law of Heat Conduction was written in one dimension as

$$q_x = -kA\frac{dT}{dx}$$

For the general case where conduction in more than one dimension is to be modeled, temperature can vary in more than one direction. So we would rewrite Fourier's Equation in partial derivative notation as

$$q''_x = -k\frac{\partial T}{\partial x} \tag{2.1a}$$

where the double-prime notation denotes a heat transfer per unit area, and the area is normal to the direction of heat flow. In the y and z directions, we have

$$q_y'' = -k\frac{\partial T}{\partial y} \tag{2.1b}$$

$$q_z'' = -k\frac{\partial T}{\partial z} \tag{2.1c}$$

Let us consider a one-dimensional coordinate system, as shown in Figure 2.1. The system consists of a plane wall on which we impose the T vs. x axes. A slice of the material dx thick is selected for study. The fact that energy can be generated within the material is taken into account

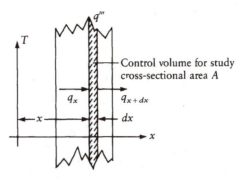

FIGURE 2.1 A one-dimensional system in rectangular coordinates.

by a term denoted as q'''. Typically, internal heat generation within a solid can be from chemical reactions, as the result of an electric current passing through the material (a wire for example), or by a nuclear reaction. Some of the energy passing through the control volume may be stored, thus increasing the internal energy of the material, which is sensed physically as an increase in the temperature of the material. This is the case for an unsteady problem.

For any geometry, we can write

$$
\begin{bmatrix}
\text{Rate of energy} \\
\text{conducted into} \\
\text{control volume}
\end{bmatrix}
+
\begin{bmatrix}
\text{Rate of energy} \\
\text{generated inside} \\
\text{control volume}
\end{bmatrix}
=
\begin{bmatrix}
\text{Rate of energy} \\
\text{conducted out of} \\
\text{control volume}
\end{bmatrix}
+
\begin{bmatrix}
\text{Rate of energy} \\
\text{stored inside} \\
\text{control volume}
\end{bmatrix}
$$

For the system of Figure 2.1, this equation becomes

$$ q_x + q''' A \, dx = q_{x+dx} + \rho A \, dx \frac{\partial u}{\partial t} \tag{2.2} $$

where A is the cross-sectional area, $A \, dx$ is the volume of the element, q''' is the internal heat generated per unit volume, ρ is density, $\rho A \, dx$ is mass, and $\partial u / \partial t$ represents the rate of change in internal energy per unit mass of the control volume. The rate of energy conducted into the control volume is q_x. The rate of energy conducted out of the control volume becomes

$$ q_{x+dx} = q_x + dq_x $$

where we denote a change as dq_x. Alternatively, we could write

$$ q_{x+dx} = q_x + \frac{dq_x}{dx} dx $$

to denote more generally that the change is a function of the x coordinate. If it is anticipated that q_x will be a function of more than one variable, the following partial derivation notation is appropriate:

$$ q_{x+dx} = q_x + \frac{\partial q_x}{\partial x} dx \tag{2.3} $$

Another way of obtaining this expression is to expand q_{x+dx} in a Taylor series and write only the first two terms.

Of the various expressions written above for q_{x+dx}, Equation 2.3 is the form to be used here. Combining with Equation 2.2 gives

$$q_x + q'''A\ dx\ =\ q_x + \frac{\partial q_x}{\partial x}dx + \rho A\ dx\frac{\partial u}{\partial t}$$

so that

$$-\frac{\partial q_x}{\partial x}dx + q'''A\ dx\ =\ \rho A\ dx\frac{\partial u}{\partial t} \tag{2.4}$$

This is the general energy equation for unsteady heat conduction in one dimension with internal heat generation. It is desirable to substitute something more easily measured for the various terms in this equation. Fourier's Equation (2.1a) is used for the heat conducted:

$$q_x\ =\ -kA\frac{\partial T}{\partial x} \tag{2.1a}$$

where k is thermal conductivity, T is temperature, and because we anticipate that temperature will be a function of more than one variable, the partial derivative notation is used. Recall from thermodynamics the following definition:

$$h\ =\ u + pv$$

where h is specific enthalpy, p is pressure, and v is specific volume. Rearranging and implicitly differentiating yields

$$du\ =\ dh - p\ dv - v\ dp$$

Conduction is the principal mode of energy transfer in solids; the preceding equation reduces to

$$du\ =\ dh$$

Substituting from the definition of specific heat, we obtain

$$c_v dT\ =\ c_p dT$$

or

$$c_v\ =\ c_p\ =\ c$$

where c_v is the specific heat at constant volume, and c_p is the specific heat at constant pressure. They are equal in a solid. So the internal energy change can be written as

$$\frac{\partial u}{\partial t}\ =\ c\frac{\partial T}{\partial t} \tag{2.5}$$

Combining Equations 2.1a and 2.5 with 2.4 yields

$$-\frac{\partial}{\partial x}\left(-kA\ \frac{\partial T}{\partial x}\right)dx + q'''A\ dx = (\rho A\ dx)c\frac{\partial T}{\partial t}$$

Dividing by $A\ dx$ and simplifying, we obtain

$$\frac{\partial}{\partial x}\left(k\frac{\partial T}{\partial x}\right) + q''' = \rho c\frac{\partial T}{\partial t} \tag{2.6}$$

For constant thermal conductivity the above becomes

$$\frac{\partial^2 T}{\partial x^2} + \frac{q'''}{k} = \frac{\rho c}{k}\frac{\partial T}{\partial t} = \frac{1}{\alpha}\frac{\partial T}{\partial t} \tag{2.7}$$

which is the general energy equation for unsteady heat conduction in one dimension with internal heat generation, written in terms of temperature. The term $\alpha = k/\rho c$ is introduced as the *thermal diffusivity* of the material, with dimensions of L^2/T (m^2/s or ft^2/s). Values of thermal diffusivity are provided in the property tables of the Appendix.

The conduction equation in three dimensions for Cartesian, polar cylindrical, and spherical coordinates in provided in the Summary section of this chapter. Also provided is a vector form of the conduction equation as well as Fourier's Law.

2.3   PLANE GEOMETRY SYSTEMS

In this section we will consider problems of heat flow through various examples of plane walls. We will consider not only plane walls of a single substance (as was done in Chapter 1) but composite plane walls involving series and/or parallel arrangements of materials. Figure 2.2 is a sketch of a plane wall with a T vs. x set of axes imposed. The temperature axis is coincident with the left face of the wall. The surface temperatures are T_1 and T_2. Also shown below the sketch is the thermal resistance circuit for the system. Our concern here will be with conduction heat transfer in the wall. For constant thermal conductivity Equation 2.7 applies:

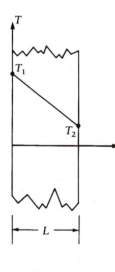

$T_1 - \mathord{\wedge\mkern-6mu\wedge\mkern-6mu\wedge} - T_2$
$\qquad L/kA$

FIGURE 2.2   Steady-state temperature distribution within a plane wall.

$$\frac{\partial^2 T}{\partial x^2} + \frac{q'''}{k} = \frac{1}{\alpha}\frac{\partial T}{\partial t}$$

Temperature varies only with x; there is no internal heat generation. The above equation reduces to

$$\frac{\partial^2 T}{\partial x^2} = 0$$

Because temperature varies only with x, the above partial differential equation can be rewritten as an ordinary differential equation:

$$\frac{d^2 T}{dx^2} = 0$$

Methods for solving ordinary differential equations now apply. Integrating yields

$$\frac{dT}{dx} = C_1$$

$$T = C_1 x + C_2 \tag{2.8}$$

where C_1 and C_2 are constants of integration. To evaluate them, we apply the boundary conditions:

1. at $x = 0$, $T = T_1$
2. at $x = L$, $T = T_2$

Substituting these conditions into Equation 2.8 gives

1. $T_1 = 0 + C_2$

or

$$C_2 = T_1$$

2. $T_2 = C_1 L + C_2 = C_1 L + T_1$

or

$$C_1 = \frac{T_2 - T_1}{L}$$

Equation 2.8 now becomes

$$T = \frac{T_2 - T_1}{L}x + T_1 \tag{2.9}$$

which simplifies to

$$\frac{T - T_1}{T_2 - T_1} = \frac{x}{L} \qquad (2.10)$$

The heat flow can now be determined with Fourier's Law:

$$q_x = -kA \frac{\partial T}{\partial x} = -kA \frac{dT}{dx}$$

where the exact differential is written because one-dimensional heat flow exists. From Equation 2.9

$$\frac{dT}{dx} = \frac{T_2 - T_1}{L}$$

Thus, the heat flow is

$$q_x = -kA \frac{T_2 - T_1}{L} = \frac{T_1 - T_2}{(L/kA)}$$

2.3.1 MATERIALS IN SERIES

Figure 2.3 is a sketch of a composite wall that consists of three materials placed next to each other. With regard to heat flow through the wall, the materials are said to be placed in series. The thermal circuit is shown below the sketch. We apply Equation 2.10 to each material, subject to appropriate boundary conditions. The results become

$$\frac{T - T_0}{T_1 - T_0} = \frac{x}{L_1} \qquad \text{for } 0 \le x \le L_1$$

$$\frac{T - T_1}{T_2 - T_1} = \frac{x - L_1}{L_2 - L_1} \qquad \text{for } L_1 \le x \le L_2$$

$$\frac{T - T_2}{T_3 - T_2} = \frac{x - L_2}{L_3 - L_2} \qquad \text{for } L_2 \le x \le L_3$$

The heat flow can be determined, again, by applying Fourier's Law to each material:

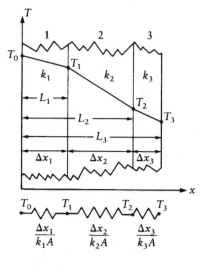

FIGURE 2.3 Composite wall with materials in series.

$$q_x = -kA \; \frac{\partial T}{\partial x} = -kA \; \frac{dT}{dx}$$

$$q_x = -k_1 A \frac{T_1 - T_0}{L_1} = \frac{T_0 - T_1}{L_1 / k_1 A} = \frac{T_0 - T_1}{\Delta x_1 / k_1 A} \qquad (2.11a)$$

$$q_x = \frac{T_1 - T_2}{\Delta x_2 / k_2 A} \qquad (2.11b)$$

$$q_x = \frac{T_2 - T_3}{\Delta x_3 / k_3 A} \qquad (2.11c)$$

The heat flows written in Equations 2.11 are all equal. Moreover, those equations can be combined to obtain still another equation for the heat flow through the wall:

$$q_x = \frac{T_0 - T_3}{\dfrac{\Delta x_1}{k_1 A} + \dfrac{\Delta x_2}{k_2 A} + \dfrac{\Delta x_3}{k_3 A}} = \frac{\text{Overall temperature difference}}{\text{Sum of the thermal resistances}} \qquad (2.11d)$$

Example 2.1

A cold-storage warehouse is to be constructed of a number of individual rooms that are maintained at various temperatures. The walls of one room are constructed of 4-in brick masonry on the outside and 1/2-in-thick plywood on the inside. Sandwiched between the brick and plywood is glass fiber insulation that is 3-1/2 in thick. The inside-wall temperature is to be maintained at −10°F, and the outside-wall temperature is 70°F. As a first step in sizing the required cooling unit, determine the heat flow through the composite wall on a per-square-foot basis.

Solution

Figure 2.4a is a sketch of the wall with the accompanying expected temperature profile and thermal circuit. The temperature within the room will probably not be uniform, so the −10°F temperature will not be the inside-wall temperature at all locations. The heat flow near the corners is a two-dimensional problem, which will also have an effect on temperature.

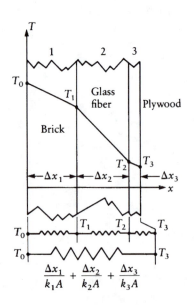

FIGURE 2.4a   The composite wall of Example 2.1.

Assumptions

1. The inside- and outside-wall temperatures are uniform and constant.
2. One-dimensional steady heat conduction exists through the wall.
3. Material properties are constant.

The heat flow is given by Equation 2.11d as

$$q_x = \frac{T_0 - T_3}{\dfrac{\Delta x_1}{k_1 A} + \dfrac{\Delta x_2}{k_2 A} + \dfrac{\Delta x_3}{k_3 A}}$$

From Appendix Table B.3, we have

$$k_1 = 0.38 \text{ BTU/(hr·ft·°R)} \qquad \text{(brick masonry)}$$

$$k_2 = 0.02 \text{ BTU/(hr·ft·°R)} \qquad \text{(glass fiber)}$$

$$k_3 = 0.063 \text{ BTU/(hr·ft·°R)} \qquad \text{(plywood)}$$

The wall temperatures are uniform and constant. They are given as $T_0 = 70°F$, $T_3 = -10°F$. The material thicknesses are $\Delta x_1 = 4/12 = 0.333$ ft, $\Delta x_2 = 3.5/12 = 0.292$ ft, $\Delta x_3 = 0.5/12 = 0.042$ ft. Each resistance is

$$R_{k1} = \frac{\Delta x_1}{k_1 A} = \frac{0.333}{0.38(1)} = 0.877 \text{ hr·°R/BTU}$$

$$R_{k2} = \frac{0.292}{0.02(1)} = 14.6 \text{ hr·°R/BTU}$$

$$R_{k2} = \frac{0.042}{0.063(1)} = 0.667 \text{ hr·°R/BTU}$$

Thus, the glass fiber represents the greatest resistance to heat flow and so, compared to the other materials, we expect the greatest temperature difference to exist across the glass fiber. The heat flow becomes

$$q_x = \frac{70 - (-10)}{0.877 + 14.6 + 0.667}$$

$$q_x = 4.96 \text{ BTU/hr}$$

for every square foot of wall area ❏

2.3.2 Materials in Parallel

Figure 2.4b is a sketch of a composite wall that consists of several materials placed next to each other. Below the sketch is a one–dimensional model of the heat flow through the wall. In the true situation, there may be heat flow normal to the x direction. Consider, for example, the two materials that occupy Δx_2. If one is wood and the other insulation, then all along the interface between the two there will be a heat flow from one to the other because a temperature difference will exist. This possibility is not shown on the thermal circuit. The one-dimensional model, although not strictly accurate, is often employed in engineering practice for the calculation of overall heat flow through the wall of Figure 2.4b or for a similar configuration.

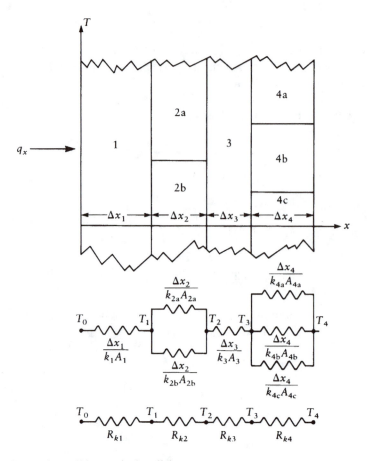

FIGURE 2.4b Composite wall in a series/parallel arrangement.

To obtain an expression for the overall heat flow through the wall of Figure 2.4b, we must first find an equivalent single resistance for the parallel configurations. For the materials in section 2, we see that the heat entering from section 1 follows two paths—one through 2a, the other through 2b. We therefore write

$$q_x = q_{2a} + q_{2b}$$

Evaluating each of these, we obtain

$$q_{2a} = \frac{T_1 - T_2}{\Delta x_2 / k_{2a} A_{2a}}$$

$$q_{2b} = \frac{T_1 - T_2}{\Delta x_2 / k_{2b} A_{2b}}$$

The heat flow through Δx_2 becomes

$$q_x = (T_1 - T_2)\left(\frac{1}{\Delta x_2 / k_{2a} A_{2a}} + \frac{1}{\Delta x_2 / k_{2b} A_{2b}}\right)$$

$$q_x = (T_1 - T_2)\left(\frac{k_{2a}A_{2a} + k_{2b}A_{2b}}{\Delta x_2}\right) = \frac{T_1 - T_2}{Rk_2}$$

We conclude that

$$R_{k2} = \frac{\Delta x_2}{k_{2a}A_{2a} + k_{2b}A_{2b}} \tag{2.12}$$

Similarly, for the materials occupying section 4,

$$R_{k4} = \frac{\Delta x_4}{k_{4a}A_{4a} + k_{4b}A_{4b} + k_{4c}A_{4c}} \tag{2.13}$$

The heat transfer through the wall based on the overall temperature difference is

$$q_x = \frac{T_0 - T_4}{\dfrac{\Delta x_1}{k_1 A_1} + \dfrac{\Delta x_2}{k_{2a}A_{2a} + k_{2b}A_{2b}} + \dfrac{\Delta x_3}{k_3 A_3} + \dfrac{\Delta x_4}{k_{4a}A_{4a} + k_{4b}A_{4b} + k_{4c}A_{4c}}} \tag{2.14}$$

Example 2.2

Figure 2.5 is a cross-sectional view of a wall found in a typical home dwelling. The outer wall is common brick 10 cm thick. The wall studs are made of 2 × 4 pine (2 × 4 in is a nominal size; actual size is 1-1/2 × 3-1/2 in = 3.8 × 8.9 cm). The inside wall is plaster board 1.3 cm thick. When the outside-wall temperature is 0°C and the inside-wall temperature is 25°C, determine the heat

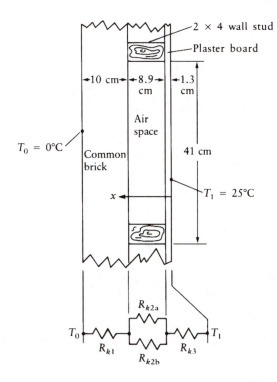

FIGURE 2.5 Typical wall construction in conventional home.

flow through the wall. Neglect the convective effects occurring in the air space. Take the wall stud spacing to be 41 cm point-to-point and the wall height to be 3 m (perpendicular to the page). Determine the heat flow through the wall.

Solution

The area we select for study is 3 m tall and 41 cm wide and includes only one wall stud. The thermal circuit for the system is shown also in Figure 2.5.

Assumptions

1. Convective effects occurring in the air space are neglected.
2. One-dimensional heat flow exists.
3. Material properties are constant.

Modifying Equation 2.14 appropriately yields the following equation for the heat flow:

$$q_x = \frac{T_1 - T_0}{R_{k1} + R_{k2} + R_{k3}} \tag{i}$$

where

$$R_{k1} = \frac{\Delta x_1}{k_1 A_1} \qquad \text{(brick)}$$

$$R_{k1} = \frac{\Delta x_2}{k_{2a} A_{2a} + k_{2b} A_{2b}} \qquad \text{(wall stud, 2a; air, 2b)}$$

$$R_{k1} = \frac{\Delta x_3}{k_3 A_3} \qquad \text{(plaster board)}$$

From Appendix Table B.3, we obtain

$k_1 = 0.45$ W/(m·K)   (average value for common brick)
$k_{2a} = 0.15$ W/(m·K)   (pine)
$k_3 = 0.814$ W/(m·K)   (plaster board)

From Appendix Table D.1, for air, we find

$k_{2a} = 0.025\ 1$ W/(m·K)   (air at 20°C, which is a reasonably representative temperature to select)

The areas needed for evaluating q_x are

$A_1 = 41$ cm × 3 m = 1.23 m²   (brick)
$A_{2a} = 3.8$ cm × 3 m = 0.114 m²   (wall stud)
$A_{2b} = (41 - 3.8$ cm) × 3 m = 1.116 m²   (air)
$A_3 = 41$ cm × 3 m = 1.23 m²   (plaster)

All other parameters are given in the problem statement. The resistances are calculated as

$$R_{k1} = \frac{0.1}{0.45(1.23)} = 0.181 \text{ K/W}$$

$$R_{k2} = \frac{0.089}{0.15(0.114) + 0.025\ 1(1.116)} = 1.97\ \text{K/W}$$

$$R_{k3} = \frac{0.013}{0.814(1.23)} = 0.013\ \text{K/W}$$

Substituting into Equation (i) yields

$$q_x = \frac{25 - 0}{0.181 + 1.97 + 0.013}$$

$$q_x = 11.5\ \text{W}$$

These resistance values indicate that the greatest resistance to heat flow is the air space. In the real situation, the air moves due to convective effects, which in turn cause a different resistance to heat flow to exist in the air space. Notice, however, that the thermal conductivity of air is far lower than that of the solids discussed in this example. Consequently, a material that contains a high number of entrapped air or gas pockets will tend to have a low thermal conductivity and would therefore be a good insulator. Styrofoam® is an example of such a material. Loose sawdust would also exhibit this property. ❏

2.3.3 PLANE WALL WITH HEAT GENERATION

In this section, we will investigate the effect of uniform internal heat generation on the one-dimensional temperature distribution within a plane wall. Figure 2.6 shows a plane wall in which there is internal heat generation per unit volume, q'''. The T vs. x axes are shown, but in this case $x = 0$ is at the center of the wall. The heat generated is uniform, and we expect a temperature profile that is symmetric about the center, assuming the wall temperatures at $x = \pm L$ are equal. At this point, the temperature distribution is not known, although an expected profile is sketched in the figure. Assuming constant thermal conductivity, Equation 2.7 applies:

$$\frac{\partial^2 T}{\partial x^2} + \frac{q'''}{k} = \frac{1}{\alpha}\frac{\partial T}{\partial t} \tag{2.7}$$

Temperature varies only with x. In the steady state, the above equation reduces to

$$\frac{\partial^2 T}{\partial x^2} + \frac{q'''}{k} = 0$$

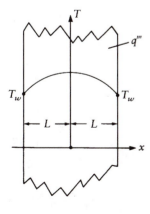

FIGURE 2.6 Plane wall with internal heat generation.

or

$$\frac{d^2T}{dx^2} = -\frac{q'''}{k}$$

For constant and uniform internal heat generation, integrating yields

$$\frac{dT}{dx} = -\frac{q'''}{k}x + C_1$$

$$T = -\frac{q'''x^2}{2k} + C_1 x + C_2$$

(2.15)

The simplest boundary conditions for this case are

   1. at $x = L, T = T_w$
   2. at $x = L, T = T_w$

Applying the boundary conditions to Equation 2.15 gives

$$C_2 = T_w + \frac{q'''L^2}{2k}$$

and

$$C_1 = 0$$

Substituting for C_1 and C_2 gives, after simplification:

$$T - T_w = \frac{q'''L^2}{2k}\left(1 - \frac{x^2}{L^2}\right)$$

(2.16)

This is a parabolic distribution that is symmetric about $x = 0$. Thus at $x = 0$, $dT/dx = 0$. The variable x/L varies from 0 to 1. Maximum temperature exists at $x = 0$ and is determined to be

$$T_{max} = \frac{q'''L^2}{2k} + T_w$$

$$T_{max} - T_w = \frac{q'''L^2}{2k}$$

Combining with Equation 2.16 yields yet another form of the temperature distribution:

$$\frac{T - T_w}{T_{max} - T_w} = 1 - \frac{x^2}{L^2}$$

(2.17)

The distribution sketched in Figure 2.6 is an adequate representation.

2.3.4   The Overall Heat-Transfer Coefficient

Heat transfer by convection from a boundary can be expressed by Newton's Law of Cooling, as seen in Chapter 1:

$$q_c = \bar{h}_c A(T_w - T_\infty) = \frac{(T_w - T_\infty)}{1/\bar{h}_c A}$$

where the convective resistance is

$$R_c = \frac{1}{\bar{h}_c A} \tag{2.18}$$

Figure 2.7 is a sketch of a plane wall. The temperature of the left face is T_1, and it receives heat by convection. The right face is at T_2 and transfers heat away by convection. The left-face and right-face convective coefficients are $\bar{h}_{c1}$ and $\bar{h}_{c2}$, respectively. The thermal circuit for the system also appears in the figure. From basic principles, the heat transfer through the wall based on the fluid-to-fluid or overall temperature difference is

$$q = \frac{T_{\infty 1} - T_{\infty 2}}{\dfrac{1}{\bar{h}_{c1} A} + \dfrac{L}{kA} + \dfrac{1}{\bar{h}_{c2} A}} = \frac{T_{\infty 1} - T_{\infty 2}}{\displaystyle\sum_i R_i} \tag{2.19}$$

In some cases, it is convenient to express the heat transfer in terms of a single value that accounts for both conduction and convection resistances. We could express the heat transfer as

$$q = UA(T_{\infty 1} - T_{\infty 2}) \tag{2.20}$$

where U is defined as an overall heat-transfer coefficient based on an area A. Comparing Equations 2.19 and 2.20 leads to the conclusion that for the plane wall,

$$U = \frac{1}{\dfrac{1}{\bar{h}_{c1}} + \dfrac{L}{k} + \dfrac{1}{\bar{h}_{c2}}} = \frac{1}{A(R_{c1} + R_k + R_{c3})} \tag{2.21}$$

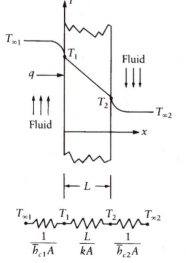

FIGURE 2.7 Steady-state temperature distribution within a plane wall.

In general, we can write for plane geometries

$$UA = 1/\sum_i R_i$$

The dimensions of the overall heat-transfer coefficient U are $F{\cdot}L/(L^2{\cdot}t{\cdot}T)$ [$W/(m^2{\cdot}K)$ or $BTU/hr{\cdot}ft^2{\cdot}{}^\circ R$].

Example 2.3

An ice chest contains a mixture of ice and water at 32°F, a temperature which is constant during the melting of the ice. The walls of the chest are made of three materials sandwiched together as shown in Figure 2.8. The outside shell is made of stamped sheet metal: 0.040-in-thick, low-carbon (1C) steel. The insulating material in the wall is 3/4-in-thick styrofoam ($k = 0.02$ BTU/hr·ft·°R). The inside material is a fiberglass liner ($k = 0.09$ BTU/hr·ft·°R), 1/4 in thick.

   Outside the ice chest is air at 90°F, and the convection coefficient between the air and the vertical steel wall is 0.79 BTU/(hr·ft²·°R). Inside, the convection coefficient between the ice water and the fiberglass is 150 BTU/(hr·ft²·°R). Determine the heat-transfer rate through the wall of the ice chest based on a per-square-foot area. Determine also the overall heat-transfer coefficient U for the wall.

Solution

The thermal circuit and an expected temperature profile are shown in Figure 2.8.

Assumptions

   1. One-dimensional steady heat flow exists.
   2. Material properties are constant.
   3. Convection coefficients are constant and apply to the appropriate wall.

The thermal conductivity for 1C steel is 24.8 BTU/(hr·ft·°R) from Appendix Table B.2. We can first determine U and then calculate the heat transferred. Equation 2.21 applies:

$$UA = 1/\sum_i R_i = 1/(R_{c1} + R_{k1} + R_{k2} + R_{k3} + R_{c2})$$

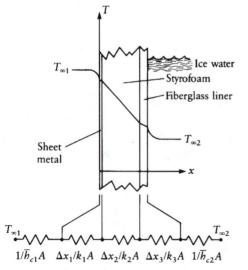

FIGURE 2.8   The wall of an ice chest.

where

$$R_{c1} = \frac{1}{\bar{h}_{c1}A} = \frac{1}{0.79(1)}$$

$$= 1.266 \, °F \cdot hr/BTU \qquad \text{(air to sheet metal)}$$

$$R_{k1} = \frac{\Delta x_1}{k_1 A} = \frac{0.04/12}{24.8(1)}$$

$$= 0.0001 \, °F \cdot hr/BTU \qquad \text{(sheet metal)}$$

$$R_{k2} = \frac{\Delta x_2}{k_2 A} = \frac{0.75/12}{0.02(1)}$$

$$= 3.125 \, °F \cdot hr/BTU \qquad \text{(styrofoam)}$$

$$R_{k3} = \frac{\Delta x_3}{k_3 A} = \frac{0.25/12}{0.09(1)}$$

$$= 0.231 \, °F \cdot hr/BTU \qquad \text{(fiberglass)}$$

$$R_{c2} = \frac{1}{\bar{h}_{c2}A} = \frac{1}{150(1)}$$

$$= 0.0067 \, °F \cdot hr/BTU \qquad \text{(ice water to fiberglass)}$$

As can be seen, the sheet metal offers a negligible resistance to the heat flow. The overall coefficient thus becomes

$$U = 1/\sum_i R_i A = 1/(1.266 + 0.0001 + 3.125 + 0.231 + 0.0067)(1)$$

$$U = 0.216 \, BTU/hr \cdot ft^2 \cdot °F$$

The heat transferred can be calculated with Equation 2.20:

$$q = UA(T_{\infty 1} - T_{\infty 2})$$

Substituting and solving yields

$$q = 0.216(1)(90 - 32)$$

$$q = 12.5 \, BTU/hr \qquad \qquad \Box$$

2.4 POLAR CYLINDRICAL GEOMETRY SYSTEMS

In this section we will consider problems of heat flow through various configurations of cylindrical geometries. We will also examine cylinders made up of composite materials. We will begin with an analytical approach in which the appropriate differential equation is solved subject to the applicable boundary conditions.

Figure 2.9 is a sketch of a volume element in polar cylindrical coordinates. The element is of dimensions $r\,d\theta$ by dr by dz. Internal heat generated per unit volume is q'''. Again, we write

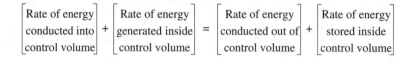

Assuming heat flow only in the radial direction, we write

$$q_r + q'''(rd\theta)dr\ dz = q_r + \frac{\partial q_r}{\partial r}dr + \rho(r\ d\theta)dr\ dz\frac{\partial u}{\partial t}$$

Simplifying yields

$$-\frac{\partial q_r}{\partial r}dr + q'''r\ dr\ d\theta dz = \rho r\ dr\ d\theta\ dz\frac{\partial u}{\partial t} \tag{2.22}$$

Fourier's Law of Conduction in polar coordinates is

$$q_r = -kA_r\frac{\partial T}{\partial r} = -k(rd\theta\ dz)\frac{\partial T}{\partial r} \tag{2.23a}$$

$$q_\theta = -\frac{k}{r}A_\theta\frac{\partial T}{\partial \theta} = -\frac{k}{r}(dr\ dz)\frac{\partial T}{\partial \theta} \tag{2.23b}$$

$$q_z = -kA_z\frac{\partial T}{\partial z} = -k(r\ d\theta dr)\frac{\partial T}{\partial z} \tag{2.23c}$$

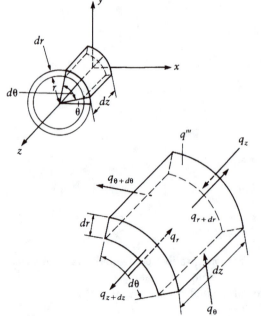

FIGURE 2.9 Volume element in cylindrical coordinates.

Internal energy is again written as $\partial u = c\partial T$. Combining with Equation 2.22 and simplifying gives

$$\frac{1}{r}\frac{\partial}{\partial r}\left(r\frac{\partial T}{\partial r}\right) + \frac{q'''}{k} = \frac{1}{\alpha}\frac{\partial T}{\partial t} \tag{2.24}$$

Equation 2.24 is the general one-dimensional conduction equation in cylindrical coordinates. (The three-dimensional equation in polar cylindrical coordinates is provided in the Summary section at the end of this chapter.)

Figure 2.10 is a sketch of a hollow cylinder of outside radius R_2 and inside radius R_1. The inside- and outside-wall temperatures are T_1 and T_2, respectively. Also shown in the figure is the thermal circuit. Our concern here is in deriving an equation for the heat flow through the wall. Assuming constant thermal conductivity and that the system is at steady state without internal heat generation, Equation 2.24 becomes

$$\frac{1}{r}\frac{d}{dr}\left(r\frac{dT}{dr}\right) = 0$$

$$\frac{d}{dr}\left(r\frac{dT}{dr}\right) = 0$$

Integrating once gives

$$r\left(\frac{dT}{dr}\right) = C_1$$

Dividing by r and integrating again yields

$$\frac{dT}{dr} = \frac{C_1}{r} \tag{2.25a}$$

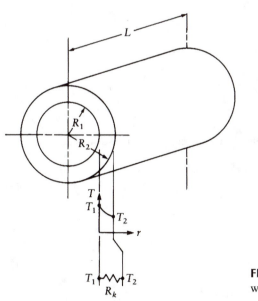

FIGURE 2.10 Heat conduction through a cylindrical wall.

$$T = C_1 \ln(r) + C_2 \tag{2.25b}$$

The boundary conditions are

    1.   at $r = R_1$, $T = T_1$ (2.26a)

    2.   at $r = R_2$, $T = T_2$ (2.26b)

Applying the boundary conditions to Equation 2.25, we obtain

    1.
$$T_1 = C_1 \ln(R_1) + C_2 \tag{2.27a}$$
$$C_2 = T_1 - C_1 \ln(R_1)$$

    2.
$$T_2 = C_1 \ln(R_2) + C_2 \tag{2.27b}$$
$$C_2 = T_2 - C_1 \ln(R_2)$$

Equating both expressions for C_2 allows for determining C_1:

$$T_1 - C_1 \ln(R_1) = T_2 - C_1 \ln(R_2)$$
$$T_1 - T_2 = C_1[\ln(R_1) - \ln(R_2)]$$

or

$$C_1 = \frac{T_1 - T_2}{\ln(R_1/R_2)} \tag{2.28a}$$

Combining with Equation 2.27a gives

$$C_2 = T_1 - \frac{T_1 - T_2}{\ln(R_1/R_2)} \ln(R_1) \tag{2.28b}$$

Substituting Equations 2.28 into 2.25 yields, for the temperature distribution,

$$\frac{T - T_1}{T_2 - T_1} = \frac{\ln(R_1/r)}{\ln(R_1/R_2)} = \frac{\ln(r/R_1)}{\ln(R_2/R_1)} \tag{2.29}$$

The heat flow can be calculated with the r component of Fourier's Law of Heat Conduction:

$$q_r'' = -k\frac{\partial T}{\partial r} \tag{2.23a}$$

where the area normal to heat flow is

$$A = 2\pi r L$$

Substituting for area and for the temperature gradient from Equation 2.25a,

$$q_r = -k(2\pi rL)\left[\frac{1}{r}\frac{T_1 - T_2}{\ln(R_1/R_2)}\right]$$

Simplifying and introducing the concept of thermal resistance,

$$q_r = \frac{2\pi kL}{\ln(R_2/R_1)}(T_1 - T_2) = \frac{T_1 - T_2}{R_k} \qquad (2.30)$$

We can now evaluate the thermal resistance to conduction in polar cylindrical coordinates, obtaining

$$R_k = \frac{\ln(R_2/R_1)}{2\pi kL} \qquad (2.31)$$

2.4.1 PIPE AND TUBE SPECIFICATIONS

Common examples of hollow cylinders are pipes and tubes, which are used extensively in industry to convey fluids from location to location. A number of the problems that we will model involve pipes and tubes, so it is worthwhile to digress from our main course to become familiar with industrial practice regarding pipes and copper water tubing. Other types of tubes have commercial applications, and specification of the other types is reserved for later portions of this text, where appropriate.

Pipes are usually specified according to a nominal size and a schedule. For example, a specification might read "1 nominal, schedule 40." The "1 nominal" refers to a nominal size in inches, in this case 1 inch. The "inch" is dropped here, because the nominal size can refer to either an English Engineering unit or to an SI unit. In the smaller pipe sizes (12 nominal or less), the nominal size is only remotely related to the actual size.

The pipe schedule refers to the wall thickness of the pipe. Usually, the higher the schedule number, the thicker the wall. All 1 nominal pipe, regardless of schedule, has the same outside diameter. The schedule will determine the inside dimension.

Wrought steel and wrought iron pipe specifications are provided in Appendix F.1. Although the table stipulates wrought steel and wrought iron, other materials such as aluminum, PVC (polyvinyl chloride), and stainless steel pipe usually have the same or similar specifications.

The discussion above refers to pipe. We will now consider copper water tubing or, more plainly, copper tubing. Note that copper can be a pipe material, and we would expect to find the same specifications as for other pipes. So it is necessary to draw the distinction between copper *pipe* and copper *tubing*. Tubing is like pipe in that they are both hollow cylinders but, in general, tubing has thinner walls. Copper tubing used for conveying water has different specifications from pipe. Copper tubing is manufactured in standard sizes and types, for example, 1-1/2 type K. The 1-1/2 is a nominal size, which is *indicative* of the outside diameter. The type is a specification of the wall thickness. Type K is for underground service and general plumbing. Type L is for interior plumbing, while type M is for use only with soldered fittings. Specifications for seamless copper tubing are provided in Appendix Table F.2.

Example 2.4

A 12 nominal, schedule 80 stainless steel (type 304) pipe carries water condensate from a condenser to a pump. The inside-wall temperature is 40°C, and the outside-wall temperature is 38°C. Determine the heat transfer through the pipe wall per unit length of pipe.

Solution

The given wall temperatures will probably exist for only a short distance in the pipe. As fluid flows downstream, its temperature will decrease, and the outside-wall temperature will increase.

Assumptions

1. One-dimensional, steady-state, radial heat flow exists.
2. Material properties are constant.

Equation 2.30 applies:

$$q_r = \frac{2\pi kL}{\ln(R_2/R_1)}(T_1 - T_2)$$

For 304 stainless steel, from Table B.2,

$$k = 14.4 \ \text{W/(m·K)}$$

From Table F.1, for 12 nominal, schedule 80 pipe,

$$D_2 = 32.39 \ \text{cm}$$

$$D_1 = 29.53 \ \text{cm}$$

Substituting into our equation for heat flow, we obtain

$$\frac{q_r}{L} = \frac{2\pi(14.4)}{\ln(32.39/29.53)}(40 - 38)$$

Solving,

$$\frac{q_r}{L} = 1\ 957.4 \ \text{W/m} = 1.96 \ \text{kW/m} \qquad \square$$

2.4.2 Materials in Series

Figure 2.11 is a sketch of a composite cylindrical wall that consists of three materials placed in series, with regard to the direction of heat flow. The thermal circuit is shown below the sketch. As was done for the plane wall, we apply Equation 2.29 to the system, specifically to each material, subject to appropriate boundary conditions. The results are:

$$\frac{T - T_1}{T_2 - T_1} = \frac{\ln(r/R_1)}{\ln(R_2/R_1)} \quad R_1 \le r \le R_2 \tag{2.32a}$$

$$\frac{T - T_2}{T_3 - T_2} = \frac{\ln(r/R_2)}{\ln(R_3/R_2)} \quad R_2 \le r \le R_3 \tag{2.32b}$$

and

$$\frac{T - T_3}{T_4 - T_3} = \frac{\ln(r/R_3)}{\ln(R_4/R_3)} \quad R_3 \le r \le R_4 \tag{2.32c}$$

Equation 2.31 can be applied to the thermal circuit to evaluate each resistance. The results become:

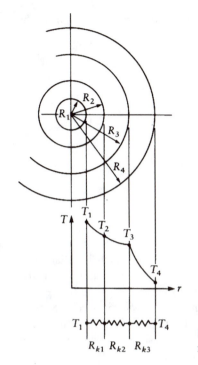

FIGURE 2.11   Composite cylindrical wall and the corresponding thermal circuit.

$$R_{k1} = \frac{\ln(R_2/R_1)}{2\pi k_1 L} \tag{2.33a}$$

$$R_{k2} = \frac{\ln(R_3/R_2)}{2\pi k_2 L} \tag{2.33b}$$

and

$$R_{k3} = \frac{\ln(R_4/R_3)}{2\pi k_3 L} \tag{2.33c}$$

The heat flow is determined from Fourier's Law applied to each material:

$$q = -kA\frac{\partial T}{\partial r} = -kA\frac{dT}{dr}$$

where $A = 2\pi r L$. The results are

$$q_r = \frac{2\pi k_1 L}{\ln(R_2/R_1)}(T_1 - T_2) \tag{2.34a}$$

$$q_r = \frac{2\pi k_2 L}{\ln(R_3/R_2)}(T_2 - T_3) \tag{2.34b}$$

$$q_r = \frac{2\pi k_3 L}{\ln(R_4/R_3)}(T_3 - T_4) \tag{2.34c}$$

The heat transfer based on the overall temperature difference can be obtained directly from the thermal circuit by using

$$q_r = \frac{T_1 - T_4}{\sum_i R_i}$$

or by algebraically manipulating Equations 2.34. The result is

$$q_r = \frac{2\pi L(T_1 - T_4)}{\dfrac{\ln(R_2/R_1)}{k_1} + \dfrac{\ln(R_3/R_2)}{k_2} + \dfrac{\ln(R_4/R_3)}{k_3}} \qquad (2.34d)$$

Example 2.5

Refrigeration tubing is used in a domestic air conditioning system. In one such installation, 1 standard type M copper tubing is soldered in place from the compressor/condenser unit outside the building to the expansion valve/evaporator unit in the attic. The refrigerant keeps the inside wall of the tubing at 40°F. To ensure that little heat is gained through the tube wall, it is covered with a sponge type of insulation that is 1/2 in thick and has a thermal conductivity of 0.02 BTU/(hr·ft·°R). The outside surface temperature of the insulation is 70°F. Determine the heat gain per unit length of tubing.

Solution

Copper refrigeration tubing has specifications on size that are similar to those for copper water tubes. The difference is that refrigeration tubing is ductile enough that it can be bent by hand, while the water tubing is quite rigid. Figure 2.12 is a sketch of the cross section of the insulated tube.

Assumptions

1. One-dimensional steady heat flow in the radial direction exists.
2. Material properties are constant.

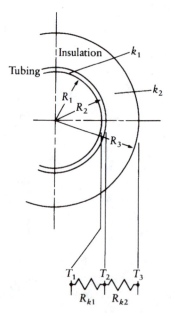

FIGURE 2.12   Insulated copper tubing.

From Table B.1 for copper, $k = 231$ BTU/(hr·ft·°R). From Table F.2, for 1 standard type M copper tubing,

$$D_2 = 1.125 \text{ in.}$$

$$D_1 = 0.08792 \text{ ft}$$

We therefore calculate

$$R_2 = 1.125/(12)(2) = 0.0469 \text{ ft}$$

$$R_1 = 0.044 \text{ ft}$$

The wall thickness of the insulation is 0.5 in. Thus,

$$R_3 = R_2 + 0.5 = 0.0469 + 0.5/12 = 0.0886 \text{ ft}$$

Each resistance is calculated with Equation 2.33:

$$R_{k1} = \frac{\ln(R_2/R_1)}{2\pi k_1 L} = \frac{\ln(0.0469/0.044)}{2\pi(231)L}$$

or

$$LR_{k1} = 4.4 \times 10^{-5}$$

Similarly,

$$LR_{k2} = \frac{\ln(0.0886/0.0469)}{2\pi(0.02)} = 5.06$$

As expected, the insulation offers the greater resistance to heat flow. The heat transferred by conduction then is

$$\frac{q_r}{L} = \frac{T_1 - T_3}{L\sum_i R_i} = \frac{40 - 70}{5.06}$$

$$\frac{q_r}{L} = -5.93 \text{ BTU/hr·ft}$$

where the negative sign indicates that the direction of heat flow is opposite to that assumed in the equation. Heat flow is thus from outside to inside. ❏

2.4.3 CYLINDER WITH HEAT GENERATION

In this section, we will consider the effect of uniform steady internal heat generation on the one-dimensional temperature distribution existing in a cylinder. One example of this type of system is a cylindrical nuclear fuel rod. Another example is electricity passing through a wire such as the tungsten wire in an electric light bulb. Electricity passing through the wire heats it, and the wire transfers the heat by radiation to the surroundings. A significant portion of the heat transferred is radiated over wavelengths to which our eyes are sensitive.

Figure 2.13 is a sketch of a solid cylinder that uniformly generates heat. Also shown are the r-θ-z coordinate axes and an expected, although as yet undetermined, temperature distribution. For a sufficiently long cylinder, axial symmetry, and steady conditions, temperature will depend only on the radial coordinate. Equation 2.24 applies:

$$\frac{1}{r}\frac{d}{dr}\left(r\frac{dT}{dr}\right) + \frac{q'''}{k} = 0 \tag{2.24}$$

Rearranging and integrating once with respect to r yields

$$\frac{d}{dr}\left(r\frac{dT}{dr}\right) = -\frac{rq'''}{k}$$

$$r\frac{dT}{dr} = -\frac{r^2 q'''}{2k} + C_1 \tag{2.35}$$

Dividing by r and integrating again gives

$$\frac{dT}{dr} = -\frac{rq'''}{2k} + \frac{C_1}{r}$$

$$T = -\frac{r^2 q'''}{4k} + C_1 \ln(r) + C_2 \tag{2.36}$$

We can write several boundary conditions, depending on how heat is transferred away. Here we will assume that only the surface temperature of the wire is known and that the slope of the T vs. r curve is a maximum at the centerline. Thus,

1.   at $r = R$, $T = T_w$

2.   at $r = 0$, $\dfrac{dT}{dr} = 0$

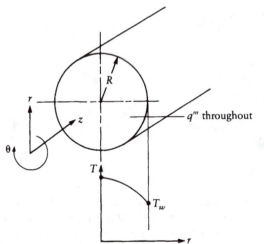

FIGURE 2.13 Solid cylinder with uniform internal heat generation.

Applying these boundary conditions to Equation 2.36, we get

$$C_1 = 0$$

and

$$C_2 = T_w + \frac{R^2 q'''}{4k}$$

The temperature now becomes

$$T - T_w = \frac{R^2 q'''}{4k}\left(1 - \frac{r^2}{R^2}\right) \tag{2.37a}$$

The maximum temperature in the system exists at the centerline, where $r = 0$:

$$T_{max} - T_w = \frac{R^2 q'''}{4k}$$

Combining this result with Equation 2.37a gives a dimensionless equation for the temperature distribution:

$$\frac{T - T_w}{T_{max} - T_w} = 1 - \frac{r^2}{R^2} \tag{2.37b}$$

A similar analysis can be formulated for a hollow cylinder. A prime example of this system is a pipe carrying fluid. Electricity passing through the pipe heats the wall, which in turn heats the fluid. For a hollow cylinder of outside radius R_2 and inside radius R_1 with uniform heat generation and constant thermal conductivity, we can follow the same lines of reasoning as for the solid cylinder. The conduction equation reduces to Equation 2.24, which has a general solution expressed in Equation 2.36:

$$T = -\frac{r^2 q'''}{4k} + C_1 \ln(r) + C_2 \tag{2.38}$$

For the hollow cylinder the boundary conditions are

1.   at $r = R_1$, $T = T_1$

2.   at $r = R_2$, $T = T_2$

Applying these conditions to Equation 2.38 yields the following solution for temperature:

$$\frac{T_1 - T}{T_1 - T_2} = \frac{R^2 q'''}{4k(T_2 - T_1)}\left(2 - \frac{r^2}{R_1^2} - \frac{R_2^2}{R_1^2}\right) + \frac{\ln(R_1/r)}{\ln(R_1/R_2)} \tag{2.39}$$

2.4.4 THE OVERALL HEAT-TRANSFER COEFFICIENT

In many problems having a polar cylindrical geometry, heat transfer by convection is quite common. Examples include steam flowing through an insulated pipe, cold Freon flowing through a copper tube, and heated crude oil flowing up from underground at an oil well. It is therefore convenient to include convective effects in the one-dimensional conduction problem for cylinders.

Figure 2.14 is a sketch of a pipe or tube containing a fluid at temperature $T_{\infty 1}$. Heat is transferred to the pipe by convection, through the pipe wall by conduction, then to the fluid outside, which is at temperature $T_{\infty 2}$. Also shown in the figure is the thermal circuit. From previous discussions, we can write equations for each resistance as

$$R_{c1} = \frac{1}{\bar{h}_{c1} A_1}$$

$$R_k = \frac{\ln(R_2/R_1)}{2\pi k L}$$

$$R_{c2} = \frac{1}{\bar{h}_{c2} A_2}$$

In these equations, the convection coefficients, assumed constant, on the inside and outside surfaces are $\bar{h}_{c1}$ and $\bar{h}_{c2}$, respectively. The areas for heat transfer on the inside and outside surfaces are

$$A_1 = 2\pi R_1 L$$

and

$$A_2 = 2\pi R_2 L$$

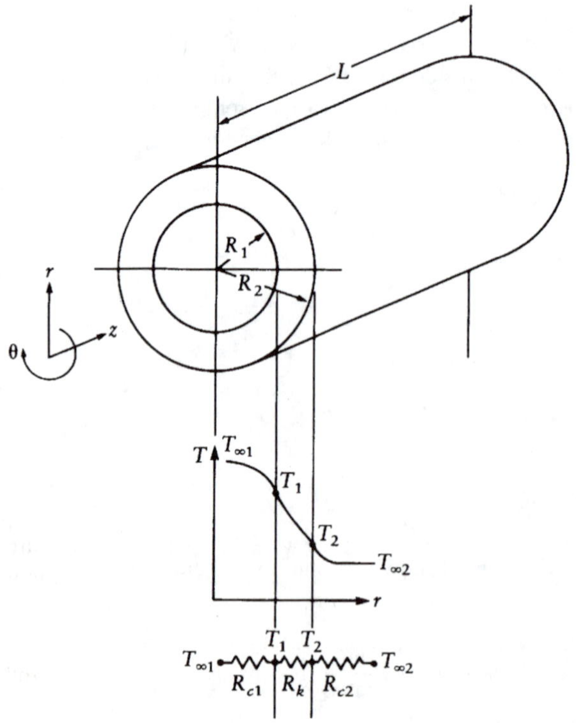

FIGURE 2.14   Heat flow through a cylinder with convection.

The heat flow from fluid to fluid, based on the overall temperature difference, is

$$q_r = \frac{T_{\infty 1} - T_{\infty 2}}{\dfrac{1}{\bar{h}_{c1} 2\pi R_1 L} + \dfrac{\ln(R_2/R_1)}{2\pi k L} + \dfrac{1}{\bar{h}_{c2} 2\pi R_2 L}}$$

or

$$q_r = \frac{2\pi L(T_{\infty 1} - T_{\infty 2})}{\dfrac{1}{\bar{h}_{c1} R_1} + \dfrac{\ln(R_2/R_1)}{k} + \dfrac{1}{\bar{h}_{c2} R_2}} \tag{2.40}$$

In some problems, it is desirable to express this equation in terms of an overall heat-transfer coefficient that accounts for the combined effects of convection at both surfaces and conduction:

$$q_r = UA(T_{\infty 1} - T_{\infty 2}) \tag{2.41}$$

where U is an overall heat-transfer coefficient and A is some area on which to base U. The question arises as to which area to select. It is customary practice to select either the outside surface area or the inside surface area of the cylinder. Based on A_2, U_2 can be determined by equating 2.40 and 2.41:

$$U_2(2\pi R_2 L)(T_{\infty 1} - T_{\infty 2}) = \frac{2\pi L(T_{\infty 1} - T_{\infty 2})}{\dfrac{1}{\bar{h}_{c1} R_1} + \dfrac{\ln(R_2/R_1)}{k} + \dfrac{1}{\bar{h}_{c2} R_2}}$$

Solving,

$$U_2 = \frac{1}{\dfrac{R_2}{\bar{h}_{c1} R_1} + \dfrac{R_2 \ln(R_2/R_1)}{k} + \dfrac{1}{\bar{h}_{c2}}} \tag{2.42a}$$

Based on the inside surface area, we have

$$U_1 = \frac{1}{\dfrac{1}{\bar{h}_{c1}} + \dfrac{R_1 \ln(R_2/R_1)}{k} + \dfrac{R_1}{R_2 \bar{h}_{c2}}} \tag{2.42b}$$

As seen in Equations 2.42, the overall coefficient is independent of length.

Example 2.6

A 6 nominal steel (1C) pipe in the drying part of a car wash carries warmed air at a temperature of 140°F. The pipe is insulated with a 1-in-thick layer of glass wool. The convection coefficient between the air and the pipe wall is 12 BTU/(hr·ft²·°R). Outside the insulation the air temperature is 68°F, and the convection coefficient between the glass wool and the ambient air is 1.5 BTU/(hr·ft²·°R). Determine the overall heat transfer coefficient for the system, based on the inside diameter of the pipe.

Solution

A sketch of the system is given in Figure 2.15 along with the thermal circuit and the expected temperature profile.

Assumptions

1. One-dimensional, steady, radial heat flow exists.
2. Material properties are constant.
3. Convection coefficients are constant.

Equation 2.42b, modified slightly to account for the insulation, can be applied to the problem:

$$U_1 = \frac{1}{\dfrac{1}{\bar{h}_{c1}} + \dfrac{R_1 \ln(R_2/R_1)}{k_{12}} + \dfrac{R_1 \ln(R_3/R_2)}{k_{23}} + \dfrac{R_1}{\bar{h}_{c2} R_3}}$$

Note that the ratio of radii appearing with $1/\bar{h}_{c2}$ is R_1/R_3 because A_1 and A_3 are the areas through which convection takes place. From Table B.2, for 1C steel, $k = 24.8$ BTU/(hr·ft·°R). From Table B.3, for glass wool insulation, $k = 0.023$ BTU/(hr·ft·°R). From Table F.1, for 6 nominal, schedule 40 pipe (no schedule was specified, so the standard is assumed), $D_2 = 6.625$ in $= 0.552$ ft, $D_1 = 0.5054$ ft. Because the insulation is 1 in thick,

$$D_3 = D_2 + 2 \text{ in} = 8.625 \text{ in} = 0.71875 \text{ ft}$$

Substituting into the equation for U_1 gives

$$U_1 = \frac{1}{\dfrac{1}{12} + \dfrac{0.5054 \ \ln(0.552/0.5054)}{24.8} + \dfrac{0.5054 \ \ln(0.71875/0.552)}{0.023} + \dfrac{1(0.5054)}{1.5(0.71875)}}$$

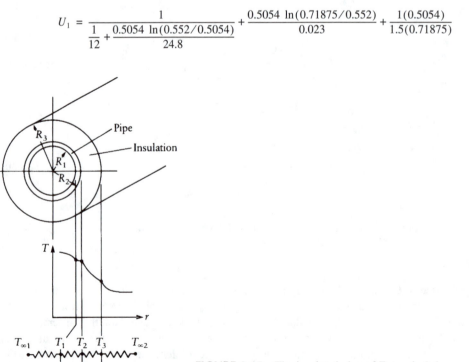

FIGURE 2.15 The insulated pipe of Example 2.6.

Evaluating each term in the denominator, we get

$$U_1 = \frac{1}{0.0833 + 0.0018 + 5.80 + 0.469}$$

The resistance of the insulation, 5.80, is the greatest of the individual components, so we expect the greatest temperature drop to exist across the insulation. Solving,

$$U_1 = 0.157 \; \text{BTU/(hr·ft}^2\text{·°F)} \qquad \square$$

2.4.5 CRITICAL THICKNESS OF INSULATION

Several example problems in this discussion have dealt with insulated pipe. In industry, insulating a pipe is a common practice because it is an inexpensive method of retarding heat losses. In some cylindrical geometry cases, however, adding insulation causes an *increase* in the heat loss. In this section we will investigate this effect.

Consider a 1 nominal pipe (OD = 3.340 cm) covered with kapok insulation [$k = 0.035$ W/(m·K)]. Assume that the outside-pipe-wall temperature is 200°C and that the insulation covering the pipe transfers heat to ambient air at 20°C, with $\bar{h}_c = 1.7$ W/(m²·K). We have not stated an insulation thickness but instead will allow the thickness to vary from 0 (no insulation) to 2.5 cm. A sketch and thermal circuit for the system are shown in Figure 2.16.

The effect we are examining is how the insulation thickness affects the heat-transfer rate (neglecting radiation losses). For no insulation the heat loss per unit length is

$$q = \frac{T_1 - T_\infty}{\dfrac{1}{\bar{h}_c A_1}}$$

With $A_1 = 2\pi R_1 L = \pi D_1 L$, we obtain after rearranging and solving,

$$\frac{q}{L} = \bar{h}_c (\pi D_1)(T_1 - T_\infty) = 1.7\pi(0.033\,40)(200 - 20)$$

$$\frac{q}{L} = 32.1 \; \text{W/m} \quad \text{(no insulation)}$$

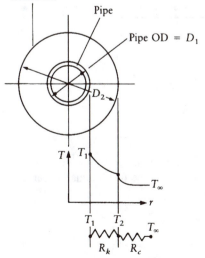

FIGURE 2.16 An insulated pipe.

If insulation is added, the heat transfer can be calculated with

$$q = \frac{T_1 - T_\infty}{\dfrac{\ln(R_2/R_1)}{2\pi k L} + \dfrac{1}{\bar{h}_c A_2}}$$

With $A_2 = 2\pi R_2 L$, the above is simplified to

$$\frac{q}{L} = \frac{2\pi(T_1 - T_\infty)}{\dfrac{\ln(R_2/R_1)}{k} + \dfrac{1}{\bar{h}_c R_2}} \tag{2.43a}$$

Substituting, we get

$$\frac{q}{L} = \frac{2\pi(200 - 20)}{\dfrac{\ln(D_2/3.340)}{0.035} + \dfrac{2}{1.7 D_2}} \tag{2.43b}$$

We have left the outer diameter of the insulation as a variable. The insulation thickness is found with

$$\text{Thickness} = R_2 - R_1 = \frac{D_2 - D_1}{2}$$

We now select values of the thickness and calculate D_2 and then q/L. A summary of the results is provided in Table 2.1 and graphed in Figure 2.17. As shown in the figure, the addition of insulation that is 0.5 cm thick actually increases the heat transferred above the value for no insulation. If the insulation thickness is 1 cm, the heat transferred is about the same as that for no insulation. The heat transferred decreases steadily as the thickness of insulation increases beyond 1 cm. The maximum heat transferred occurs at approximately 0.5 cm.

TABLE 2.1
Summary of Calculations Made with Equation 2.43

Thickness, cm	D_2, cm	q/L in W/m Equation 2.43
0	3.340	32.1
0.5	4.340	32.7
1.0	5.340	31.9
1.5	6.340	30.7
2.0	7.340	29.4
2.5	8.340	28.1

The exact value of insulation thickness that corresponds to maximum heat transferred can be found by differentiating Equation 2.43a with respect to R_2 and setting the result equal to zero:

$$\frac{q}{L} = \frac{2\pi(T_1 - T_\infty)}{\dfrac{\ln(R_2/R_1)}{k} + \dfrac{1}{\bar{h}_c R_2}}$$

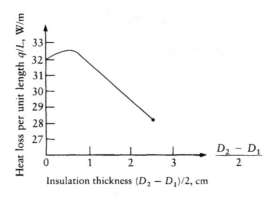

FIGURE 2.17   Heat loss from an insulated pipe as a function of insulation thickness.

$$\frac{d(q/L)}{dR_2} = 0 = -\frac{2\pi(T_1 - T_\infty)(1/kR_2 - 1/\bar{h}_c R_2^2)}{\left(\dfrac{\ln(R_2/R_1)}{k} + \dfrac{1}{\bar{h}_c R_2}\right)^2}$$

Solving for the critical value of R_2 gives

$$R_{cr2} = \frac{k}{\bar{h}_c} * \tag{2.44}$$

For the example just discussed,

$$D_2 = 2R_2 = \frac{2k}{\bar{h}_c} = \frac{2(0.035)}{1.7}$$

$$D_{cr2} = 0.0412 \text{ m} = 4.12 \text{ cm}$$

The maximum heat transferred is determined by substituting this value into Equation 2.43b:

$$\left.\frac{q}{L}\right|_{max} = 32.7 \text{ W/m}$$

The critical insulation thickness then is

$$\frac{D_2 - D_1}{2} = 0.39 \text{ cm}$$

If the outer radius of insulation is less than $k/\bar{h}_c$, then adding insulation increases the heat transferred. If the outlet radius is greater than $k/\bar{h}_c$, adding insulation decreases the heat transferred. The two effects that influence the heat-transfer rate are the insulation thickness and the corresponding surface area. As insulation is added, the conduction resistance is increased, but so is the surface area. The increased surface area causes more heat to be transferred by convection. When the insulation thickness is greater than $k/\bar{h}_c$, the conduction-resistance effect dominates. Note that, in

---

* That this R_{cr2} corresponds to a maximum can be proven by taking the second derivative, substituting, and evaluating the sign.

the above calculations, the convection coefficient was assumed constant for all insulation outer diameters selected. This assumption does not necessarily apply. Also, the effect of increasing the heat transferred by the addition of insulation occurs usually for small diameters and low values of the convection coefficient.

For steam flowing in a pipe, it is in general not desirable to lose heat through the pipe wall. So the critical thickness should be calculated and the pipe insulated appropriately. For electricity flowing through an insulated wire, insulation might be added in order to *help* dissipate heat that is generated internally. In this case the insulation is used to keep the wire from overheating as well as from short-circuiting.

2.5   SPHERICAL GEOMETRY SYSTEMS

In this section, we will consider the problems of heat flow through a spherical wall. Figure 2.18 is a sketch of a hollow sphere of inside-wall radius R_1 and outside-wall radius R_2. The inside- and outside-wall temperatures are T_1 and T_2, respectively. Also shown in the figure is the thermal circuit. Assuming constant thermal conductivity, it can be shown that the differential equation that applies is

$$\frac{1}{r^2}\frac{\partial}{\partial r}\left(r^2\frac{\partial T}{\partial r}\right)+\frac{q'''}{k} = \frac{1}{\alpha}\frac{\partial T}{\partial t} \tag{2.45}$$

For steady-state conditions with no internal heat generation, temperature varies only with r. The above reduces to

$$\frac{1}{r}\frac{d}{dr}\left(r^2\frac{dT}{dr}\right) = 0$$

$$\frac{d}{dr}\left(r^2\frac{dT}{dr}\right) = 0$$

Integrating once gives

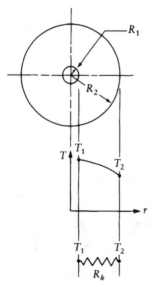

FIGURE 2.18   A hollow sphere.

$$r^2 \frac{dT}{dr} = C_1$$

$$\frac{dT}{dr} = \frac{C_1}{r^2}$$

Integrating again yields

$$T = -\frac{C_1}{r} + C_2 \tag{2.46}$$

The simplest boundary conditions are

1. $\qquad\qquad$ at $r = R_1, T = T_1$

2. $\qquad\qquad$ at $r = R_2, T = T_2$

Applying these to Equation 2.46 brings

$$C_1 = \frac{T_1 - T_2}{\dfrac{1}{R_2} - \dfrac{1}{R_1}}$$

$$C_2 = T_1 - \frac{1}{R_1}\left(\frac{T_1 - T_2}{\dfrac{1}{R_1} - \dfrac{1}{R_2}}\right)$$

Substituting into Equation 2.46 gives for the temperature distribution

$$\frac{T - T_1}{T_2 - T_1} = \frac{R_2}{r}\left(\frac{r - R_1}{R_2 - R_1}\right) \tag{2.47}$$

The heat transferred is found by using the r component of Fourier's Law in spherical coordinates,

$$q_r'' = -k\frac{\partial T}{\partial r}$$

The area normal to the direction of heat transfer is

$$A = 4\pi r^2$$

Substituting gives

$$q_r = 4\pi k \frac{R_1 R_2}{R_2 - R_1}(T_1 - T_2) = \frac{T_1 - T_2}{R_k} \tag{2.48}$$

The resistance to heat transfer in spherical coordinates then is evaluated as

$$R_k = \frac{R_2 - R_1}{4\pi k R_1 R_2} \tag{2.49}$$

It is possible to continue the discussion of heat transfer in spherical geometry problems and closely parallel the developments of the previous sections for plane and cylindrical systems. However, results of such problems are reserved for the exercises.

2.6   THERMAL CONTACT RESISTANCE

We have considered several problems in which two materials were placed in series, and the heat transferred was calculated. One factor that was not mentioned is the *thermal contact resistance* discussed here.

Consider two solid bars in series as shown in Figure 2.19. Heat flows in the axial direction only. The left face of material A is at temperature T_1. The right face of material B is at T_3. The temperature distribution is as shown in the figure. Note that there is a temperature drop at the interface between the bars. The right face of material A is at T_{2A}, while the left face of material B is at T_{2B}. In general,

$$T_{2A} \neq T_{2B}$$

The existence of a temperature difference is due to imperfections in the touching surfaces. Figure 2.20 is a close-up view of two surfaces in contact. Each surface has high and low spots. Where the two surfaces actually touch (contact spots), heat is transferred by conduction. In the void spaces, heat is transferred by convection and radiation through the trapped gas.

Surface roughness can be measured in several ways. The most commonly used way is the tracer method, in which a stylus is moved back and forth over the surface. This method yields quantitative results, which are expressed in terms of what is known as an *rms roughness*. The rms roughness is the root-mean-square of the actual deviations of the surface from a reference plane.

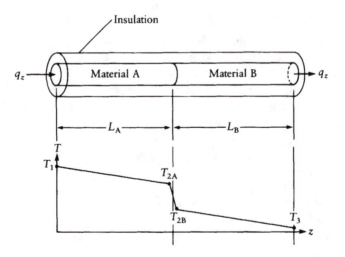

FIGURE 2.19   Two solid bars in contact.

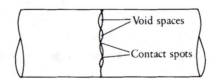

FIGURE 2.20   An exaggerated view of the interface between two bars.

For example, a mirror surface will have an rms roughness of 0.1 μm (or 4 μin), while a rough finish has a value of 25.4 μm (= 1000 /μin). Polishing removes the high spots and would therefore reduce the radiation resistance across the void spaces when two surfaces are in contact.

Another method of reducing contact resistance is to apply a high pressure to the materials and place them under a vacuum. Table 2.2 gives thermal resistance values for various metallic surfaces

TABLE 2.2

Thermal Resistance Values for Various Metal-to-Metal Interfaces under a Vacuum

Metal	Thermal Resistance	
	m²·K/W	hr·ft²·°R/BTU
Contact pressure = 100 kPa = 14.7 psia		
Stainless steel–stainless steel	0.000 6–0.002 5	0.003 4–0.015
Copper–copper	0.000 1–0.001 0	0.000 6–0.006
Magnesium–magnesium	0.000 15–0.000 35	0.000 85–0.000 3
Aluminum–aluminum	0.000 15–0.000 5	0.000 85–0.000 3
Contact pressure = 10 000 kPa = 1 470 psia		
Stainless steel–stainless steel	0.000 07–0.000 4	0.000 4–0.002 3
Copper–copper	0.000 01–0.000 05	0.000 06–0.000 3
Magnesium–magnesium	0.000 02–0.000 04	0.000 1–0.002 3
Aluminum–aluminum	0.000 02–0.000 04	0.000 1–0.000 04

placed under a vacuum. As indicated, the contact resistance decreases with increasing external pressure exerted on the metals to push them together.

A third way of reducing contact resistance is to coat the surfaces with a liquid before putting them into contact with each other. The void spaces ordinarily filled with air would be filled instead with a liquid that has a higher thermal conductivity than air. Table 2.3 gives values of the contact resistance for an aluminum-aluminum interface using various interfacial fluids.

TABLE 2.3

Thermal Resistance of a Medium to Semi-rough (10 μm = 400 μin rms roughness) Aluminum–Aluminum Interface for a Variety of Fluids. Contact Pressure = 100 kPa = 14.7 psia.

Interfacial fluid	Thermal resistance	
	m²·K/W	hr·ft²·°R/BTU
Air	0.000 275	0.0016
Helium	0.000 105	0.0006
Hydrogen	0.000 072 0	0.0004
Silicone oil	0.000 052 5	0.0003
Glycerine	0.000 026 5	0.00015

The thermal contact resistance can be determined from measurements made on a device similar to that shown in Figure 2.19. The heat transferred in the axial direction is calculated with

$$\frac{q_z}{A} = \frac{T_1 - T_{2a}}{L_a/k_a} = \frac{T_{2a} - T_{2b}}{R_{tc}} = \frac{T_{2b} - T_3}{L_b/k_b}$$

where R_{tc} is the thermal contact resistance. Because the interface temperatures are very difficult to measure accurately, the above equations are usually combined into a form that is more suitable for use:

$$\frac{q_z}{A} = \frac{T_1 - T_3}{\dfrac{L_a}{k_a} + R_{tc} + \dfrac{L_b}{k_b}} \qquad (2.50)$$

Generally, the thermal contact resistance must be measured.

Example 2.7

Figure 2.21 is a schematic of a device that can be used to measure thermal contact resistance. It consists of three rods 3.175 cm in diameter. The stainless steel is heated by electrical resistance heaters. Heat flows axially through the aluminum and magnesium to the cooling water chamber. Ten thermocouples are located as shown. Distances between adjacent thermocouples are given in the following table.

Δz	Distance, cm
1–2	3.5
2–interface	3.18
interface–3	2.54
3–4	4.45
4–5	4.45
5–6	4.45
6–interface	3.81
interface–7	2.54
8–9	4.45
9–10	4.45

Temperature data obtained from the device are as follows:

$$
\begin{array}{ll}
T_1 = 543 \text{ K} & T_6 = 366 \text{ K} \\
T_2 = 460 \text{ K} & T_7 = 349 \text{ K} \\
T_3 = 391 \text{ K} & T_8 = 337 \text{ K} \\
T_4 = 382 \text{ K} & T_9 = 325 \text{ K} \\
T_5 = 374 \text{ K} & T_{10} = 319 \text{ K}
\end{array}
$$

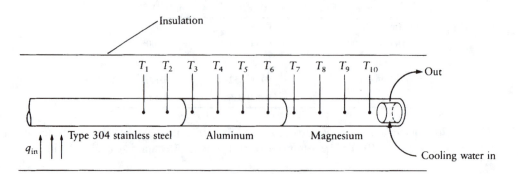

FIGURE 2.21 One-dimensional heat flow in a heat-conduction apparatus.

Determine the thermal contact resistance between the aluminum and the magnesium, assuming each has constant thermal conductivity. (Based on actual data obtained in the laboratory.)

Solution

Stainless steel is used at the heat source because its thermal conductivity is essentially constant over a very wide temperature range. So we can calculate the heat flow with T_1, T_2, and the thermal conductivity of the stainless steel. Heat will be lost or transferred through the insulation, which will affect the axial heat-flow rate.

Assumptions

1. One-dimensional steady heat flow exists in the axial direction.
2. Material properties are constant.
3. The insulation eliminates the radial heat losses.

From Appendix Table B.2, for type 304 stainless steel, $k = 14.4$ W/(m·K). The heat flow is calculated with Fourier's law:

$$\frac{q_z}{A} = -k\frac{dT}{dz} = -k\frac{\Delta T}{\Delta z}$$

$$= +k\frac{T_1 - T_2}{\Delta z_{12}} = 14.4\frac{(543 - 460)}{0.035}$$

Solving,

$$\frac{q_z}{A} = 34.15 \text{ kW/m}^2$$

The slope dT/dz within the aluminum is a constant. We can therefore extrapolate beyond T_6 to find the temperature of the aluminum at the interface (T_{ial}). Thus,

$$\frac{dT}{dz} = \text{a constant}$$

Arbitrarily selecting any two temperatures within the aluminum brings

$$\frac{T_5 - T_6}{\Delta z_{56}} = \frac{T_5 - T_{ial}}{\Delta z_{5i}}$$

Rearranging and solving for T_{ial} yields

$$T_{ial} = T_5 - \frac{\Delta z_{5i}}{\Delta z_{56}}(T_5 - T_6)$$

$$T_{ial} = 374 - \frac{(4.45 + 3.81)}{4.45}(374 - 366)$$

or

$$T_{ial} = 359.2 \text{ K}$$

Similarly, for the magnesium,

$$\frac{dT}{dz} = \text{a constant}$$

$$\frac{T_{img} - T_8}{\Delta z_{i8}} = \frac{T_7 - T_8}{\Delta z_{78}}$$

Rearranging and solving,

$$T_{img} = \frac{\Delta z_{i8}}{\Delta z_{78}}(T_7 - T_8) + T_8$$

$$T_{img} = \frac{2.54 + 4.45}{4.45}(349 - 337) + 337$$

$$T_{img} = 355.8 \text{ K}$$

Fourier's Law applied at the interface is

$$\frac{q_z}{A} = \frac{T_{ial} - T_{img}}{R_{tc}}$$

or

$$R_{tc} = \frac{359.2 - 355.8}{34\ 150}$$

Solving,

$$R_{tc} = 9.96 \times 10^{-5} \text{ K·m}^2/\text{W} \qquad \qquad \square$$

2.7   HEAT TRANSFER FROM EXTENDED SURFACES

In this section, we will examine heat transfer from what is known as an *extended surface,* such as a fin. Extended surfaces are usually added to a structure to increase the rate of heat removal. The extended surface in effect provides a greater area through which energy can flow.

A familiar example of an extended surface is a spoon placed in a cup of hot coffee. The spoon handle extends beyond the free surface of the liquid. Heat is lost by the coffee through the cup sides, the bottom, and the free surface. Heat is also conducted along the spoon handle, causing the handle to become warmer than the surrounding air. The heat conducted to the handle is then transferred to the surrounding air by convection.

In industry, fins are used in numerous applications, such as in electrical equipment, to help dissipate unwanted or potentially harmful heat. Fins designed for cooling electronic equipment are shown in Figure 2.22. Another application of fins is in single- and double-pipe heat exchangers, as found in boilers and in radiators, examples of which are in Figure 2.23. Perhaps a more familiar application of fins is found in air-cooled engines or compressors, where circumferential fins are

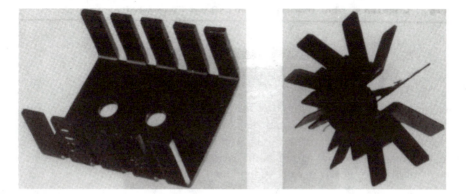

FIGURE 2.22   Fins used in cooling of electronic equipment.

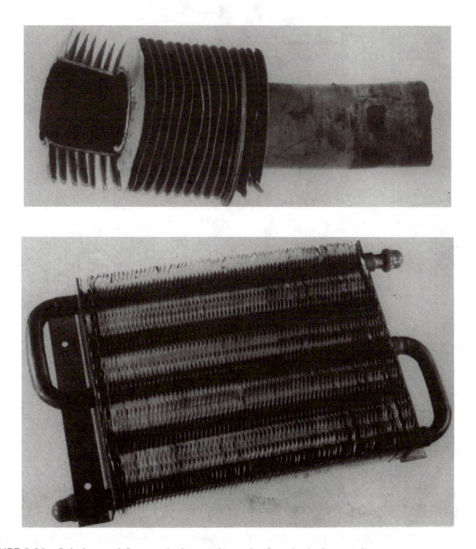

FIGURE 2.23   Spiral wound fins attached to a tube, and a finned tube heat exchanger.

integrally cast as part of a cylinder wall (see Figure 2.24). Figure 2.25 contains sketches of several types of fin configurations.

Convection is important to the process of heat transfer from an extended surface. A plane wall at an elevated temperature will lose heat to a cooler fluid in contact with the wall. The heat lost by the wall can be increased or enhanced by increasing the convection film coefficient $\bar{h}_c$ or by increasing the surface area A_s. Increasing the film coefficient can involve use of a fan or pump, which may introduce additional initial and operating costs to an installation. Increasing area by means of an extended surface or a fin may be a less costly way to increase the heat-transfer rate.

For any installation in which a fin is used, it should provide maximum cooling efficiency with a minimum amount of material and of cost to manufacture. Other considerations may include weight and space requirements, ambient cooling fluid, and proper strength.

FIGURE 2.24   Cylinder wall of a compressor.

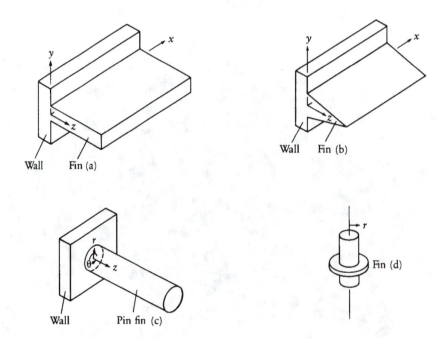

FIGURE 2.25   Some typical fin types: (a) straight rectangular fin, (b) straight triangular fin, (c) pin fin or spine, and (d) circumferential fin of rectangular cross section.

2.7.1 THE GENERAL DIFFERENTIAL EQUATION FOR EXTENDED SURFACES

The purpose of adding an extended surface is to help dissipate heat. As was the case for conduction problems previously discussed, if the temperature distribution is known, then the heat-transfer rate can be determined. So again, the method is to formulate a differential equation for the system, solve for the temperature distribution, then differentiate appropriately as per Fourier's Law to obtain the heat-transfer rate. In this section, we will develop a general differential equation for extended surfaces. In following sections, we will examine solutions for specific cases.

Figure 2.26 is of an extended surface of arbitrary cross section that varies with length. Temperature within the volume will vary with r and with z. The variation with r, however, is small, and we will assume that it can be neglected. So in our model we expect temperature to be constant across the cross section at any axial location. This is a reasonable assumption because in practice fins are quite thin, and a one-dimensional model is satisfactory for a description of the behavior. Also, the system is at steady state.

We select a slice dz wide of the volume for analysis. Heat is conducted into the control volume at a rate of q_z; q_{z+dz} is conducted out while dq_c is convected away through the surface area dA_s. Assuming steady-state conditions, constant thermal conductivity, and no internal heat generation, and neglecting radiation, we can write

$$
\begin{bmatrix} \text{Rate of energy} \\ \text{conducted into} \\ \text{control volume} \end{bmatrix} = \begin{bmatrix} \text{Rate of energy} \\ \text{conducted out of} \\ \text{control volume} \end{bmatrix} + \begin{bmatrix} \text{Rate of energy} \\ \text{convected away} \\ \text{from control volume} \end{bmatrix}
$$

For the system of Figure 2.26, this equation becomes

$$
q_z = q_{z+dz} + dq_c
$$

or

$$
q_z = \left(q_z + \frac{dq_z}{dz} dz \right) + dq_c
$$

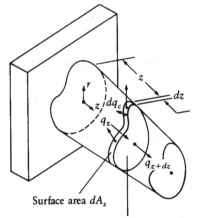

Surface area dA_s

Cross-sectional area A

FIGURE 2.26 An extended surface of arbitrary shape and cross section.

Simplifying and rearranging yields

$$-\frac{dq_z}{dz}dz = dq_c \tag{2.51}$$

From Fourier's Law, we have

$$q_z = -kA\frac{dT}{dz}$$

where k is thermal conductivity of the fin material, A is cross-sectional area that may vary with z, and dT/dz is the temperature gradient. Differentiating with respect to z gives

$$\frac{dq_z}{dz} = -k\frac{d}{dz}\left(A\frac{dT}{dz}\right) \tag{2.52}$$

Heat transfer by convection is evaluated with Newton's Law:

$$dq_c = \bar{h}_c dA_s(T - T_\infty)$$

where $\bar{h}_c$ is the convective coefficient, which is assumed constant over the entire surface of the fin, dA_s is the surface area, and $(T - T_\infty)$ is the temperature difference between the fin and the ambient. To account for the energy convected away, we use Newton's Law of Cooling. Accordingly, the convective coefficient $\bar{h}_c$ is assumed constant. This is almost a necessity to obtain a closed-form solution. The radiation effect or loss is neglected but could be included by introduction of a radiation coefficient $\bar{h}_r$, analogous to the convection coefficient $\bar{h}_c$. Of course, $\bar{h}_c$ is temperature dependent and, traditionally, radiation is neglected in the analysis. It should be remembered, however, that in some cases the energy transferred away by radiation is comparable in magnitude to that transferred by convection.

Note that the original assumption of temperature being invariant with r means that, at any z, the fin surface temperature equals the temperature within. Substituting Newton's Equation and Equation 2.52 into Equation 2.51, we obtain

$$k\frac{d}{dz}\left(A\frac{dT}{dz}\right)dz = \bar{h}_c dA_s(T - T_\infty)$$

Dividing by $k\,dz$ and carrying out the differentiation on the left-hand side yields

$$\frac{dA}{dz}\frac{dT}{dz} + A\frac{d^2T}{dz^2} = \frac{\bar{h}_c dA_s}{k\,dz}(T - T_\infty)$$

Rearranging and simplifying gives

$$\frac{d^2T}{dz^2} + \frac{1}{A}\frac{dA}{dz}\frac{dT}{dz} - \frac{\bar{h}_c}{k}\frac{1}{A}\frac{dA_s}{dz}(T - T_\infty) = 0 \tag{2.53}$$

This is the general form of the energy balance in terms of temperature for an extended surface. To solve it for temperature, it is necessary first to select a specific cross section and evaluate the areas.

In a number of cases, it is helpful to simplify the differential equation by introducing new variables. This procedure is not always necessary, but it makes the mathematics easier. Consider Equation 2.53. The dependent variable in the first and second terms is temperature T. In the third term the dependent variable is $T - T_\infty$. We know that temperature difference is significant, because it is the driving force for heat transfer. Moreover, measurements of temperature made with thermocouples, for example, are usually in terms of a temperature difference between that which is desired and some reference temperature (of an ice-water mixture, for example). So we wish to rewrite Equation 2.53 to make it mathematically easier to work with, while preserving the concept that temperature difference is significant. Based on experience with differential equations, we introduce a new variable θ, which is defined as

$$\theta = T - T_\infty \tag{2.54a}$$

The derivative of temperature with respect to z is found as

$$T = \theta + T_\infty$$

$$\frac{dT}{dz} = \frac{d\theta}{dz} \tag{2.54b}$$

Also,

$$\frac{d^2T}{dz^2} = \frac{d^2\theta}{dz^2} \tag{2.54c}$$

Substituting Equations 2.54 into 2.53 yields

$$\frac{d^2\theta}{dz^2} + \frac{1}{A}\frac{dA}{dz}\frac{d\theta}{dz} - \frac{\bar{h}_c}{k}\frac{1}{A}\frac{dA_s}{dz}\theta = 0 \tag{2.55}$$

Now a temperature difference appears in each term without affecting the descriptive quality of the equation.

2.7.2 ANALYSIS OF A PIN FIN

In this section, we will select a specific fin and analyze it. The fin we choose is known as a *pin fin* or a *spine*. It is a rod attached to a wall as shown in Figure 2.27. The pin fin has a length L, diameter D with cross-sectional area A, perimeter P, convection coefficient $\bar{h}_c$, and thermal conductivity k. The ambient temperature is T_∞. We will assume that little or no heat is transferred from the end of the fin, because the tip area is considered to be small compared to the remaining surface area of the fin (that is, a long fin). This is equivalent to assuming that the fin tip is insulated, which is a boundary condition. At the root of the fin (where it attaches to the wall), the wall temperature is T_w. With constant cross-sectional area,

$$\frac{dA}{dz} = 0 \tag{2.56}$$

The fin perimeter P can be written in terms of the surface area for the element as

$$P\,dz = dA_s$$

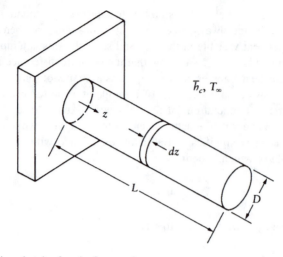

FIGURE 2.27 Definition sketch of a pin fin or spine.

or

$$\frac{dA_s}{dz} = P \tag{2.57}$$

Substituting Equations 2.56 and 2.57 into 2.55 brings

$$\frac{d^2\theta}{dz^2} - \frac{\bar{h}_c P}{kA}\theta = 0 \tag{2.58}$$

It is customary to define a new parameter m as

$$m^2 = \frac{\bar{h}_c P}{kA} \, * \tag{2.59}$$

Equation 4.9 now becomes

$$\frac{d^2\theta}{dz^2} - m^2\theta = 0 \tag{2.60}$$

The general solution to Equation 2.60 can be written in two ways:

$$\theta = C_1 \cosh(mz) + C_2 \sinh(mz) \tag{2.61a}$$

or

$$\theta = C_3 e^{mz} + C_4 e^{-mz} \tag{2.61b}$$

---

* m^2 *is* selected over m because experience shows that the square root of $\bar{h}_c P/kA$ will arise in the solution. It is easier to write m than $\sqrt{m}$ or $\sqrt{\bar{h}_c P/kA}$.

where C_i variables are constants of integration yet to be evaluated. Mathematically, there is no difference in these equations. Physically, however, we prefer Equation 2.61a. Equation 2.61b, written in exponential terms, is more suitable for geometries that extend to infinity or are very large (such as a semi-infinite wall, or a volume like that of Earth). Our problem is that of a pin fin having a finite length L.

The constants C_1 and C_2 can be evaluated with the boundary conditions. These are

1.   at $z = 0$, $T = T_w$

2.   at $z = L$, $\dfrac{dT}{dz} = 0$

Boundary condition 1 (B.C.1) specifies wall temperature, while B.C.2 says that no heat is transferred across the tip area. This is a reasonable assumption, because the tip area is small. In terms of $\theta = T - T_\infty$, the boundary conditions become

1.   at $z = 0$, $\theta_w = T_w - T_\infty$

2.   at $z = L$, $\dfrac{d\theta}{dz} = 0$

where the notation θ_w has been introduced and defined. Applying B.C.1 to Equation 2.61a gives

$$\theta = C_1 \cosh(mz) + C_2 \sinh(mz)$$

$$\theta_w = C_1(1) + C_2(0)$$

or

$$C_1 = \theta_w$$

Before applying B.C.2, we must first determine $d\theta/dz$:

$$\theta = \theta_w \cosh(mz) + C_2 \sinh(mz)$$

$$\frac{d\theta}{dz} = \theta_w m\ \sinh(mz) + C_2 m\ \cosh(mz)$$

Applying B.C.2, we get

$$0 = \theta_w m\ \sinh(mL) + C_2 m\ \cosh(mL)$$

$$C_2 = -\frac{\theta_w \sinh(mL)}{\cosh(mL)}$$

The solution then is

$$\frac{\theta}{\theta_w} = \frac{\cosh(mL)\cosh(mz) - \sinh(mL)\sinh(mz)}{\cosh(mL)} \tag{2.62a}$$

An alternative form makes use of a trigonometric identity applied to the above equation to yield

$$\frac{\theta}{\theta_w} = \frac{\cosh[mL(1 - z/L)]}{\cosh\,(mL)} \tag{2.62b}$$

In terms of the physical variables of the problem,

$$\frac{T - T_\infty}{T_w - T_\infty} = \frac{\cosh[(L\sqrt{\bar{h}_c P/kA})(1 - z/L)]}{\cosh(L\sqrt{\bar{h}_c P/kA})} \tag{2.62c}$$

The heat transferred away by the fin is equal to the heat conducted into it at the wall. Using Fourier's Law, we have

$$q_z = -kA\frac{dT}{dz}\Big|_{z = 0} = -kA\frac{d\theta}{dz}\Big|_{z = 0}$$

Equation 2.62a differentiated and substituted into the above equation gives

$$q_z = -\frac{kA\theta_w}{\cosh(mL)}$$
$$\times [m\,\cosh(mL)\sinh(mz) - m\,\sinh(mL)\,\cosh(mz)]\big|_{z = 0}$$
$$= -\frac{kA\theta_w}{\cosh(mL)}[m(0) - m\,\sinh(mL)]$$

Simplifying,

$$\frac{q_z}{kAm\theta_w} = \tanh(mL) \tag{2.63a}$$

In terms of the physical variables,

$$\frac{q_z}{\sqrt{kA\bar{h}_c P}(T_w - T_\infty)} = \tanh(L\sqrt{\bar{h}_c P/kA}) \tag{2.63b}$$

It is instructive to graph Equations 2.62 for the temperature distribution and 2.63 for the heat transferred. These graphs are provided in Figures 2.28 and 2.29, respectively. Figure 2.29 shows that the heat flow in a pin fin cannot be substantially increased past $mL = 3$. Practically, a fin length of over $L = 3/m$ will not improve the heat-transfer rate. The $mL = 4$ and $mL = 5$ lines in Figure 2.28 thus are of overdesigned fins that can be shortened without significantly decreasing the amount of heat that can be transferred.

In many cases, the shape of the fin or the boundary conditions lead to a temperature distribution that is quite complex. It therefore becomes tedious to determine the heat-transfer rate and to find an optimum length for a pin fin. So we introduce two parameters that will help characterize fins. These are the fin efficiency η_e and the fin effectiveness η_f.

The fin efficiency is defined as

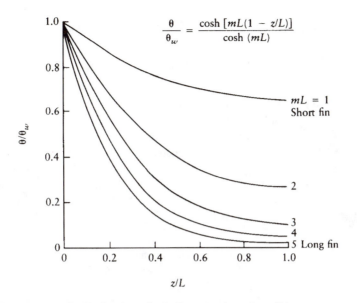

FIGURE 2.28   Temperature distribution in a pin fin for a number of conditions.

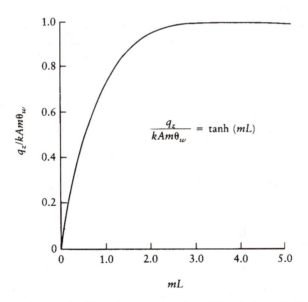

FIGURE 2.29   Dimensionless graph of heat flow as a function of length for a pin fin.

$$\eta_e = \frac{\text{Actual heat transferred from wall with fin attached}}{\text{Heat that would be transferred if entire fin}\atop\text{were at wall temperature}} \qquad (2.64)$$

Evaluating this expression for the pin fin considered above, we obtain

$$\eta_e = \frac{\sqrt{kA\bar{h}_c P}(T_w - T_\infty)\tanh(mL)}{\bar{h}_c(PL)(T_w - T_\infty)}$$

or

$$\eta_e = \frac{\tanh(mL)}{mL} \tag{2.65}$$

This equation is plotted in Figure 2.30. As can be seen, when $mL = 0$, the pin fin is most efficient. If the efficiency curve is available on a certain fin, then mL can be calculated, η_e can be read from the graph, and q_z can be determined with Equation 2.65.

The addition of a fin to a surface is made to increase the heat transfer from that surface. When a fin is added, however, the fin itself is another resistance to heat transfer from the wall—a conduction resistance. So even though the surface area of the original wall is increased, in some cases adding a fin will actually decrease the heat transfer from the surface. Fin effectiveness gives an indication of this phenomenon, and is defined as

$$\eta_f = \frac{\text{Heat flux from wall after adding fin}}{\text{Heat flux from wall before adding fin}} \tag{2.66}$$

The heat flux from the wall after adding the fin is found either by equation or by an efficiency curve. The heat flux from the wall before adding the fin is $\bar{h}_c A(T_w - T_\infty)$. For the pin fin analyzed in this section, the effectiveness becomes

$$\eta_f = \frac{q_z}{\bar{h}_c A(T_w - T_\infty)}$$

Substituting from Equation 4.14b, we obtain

$$\eta_f = \sqrt{\frac{kP}{\bar{h}_c A}} \, \tanh(mL) \tag{2.67}$$

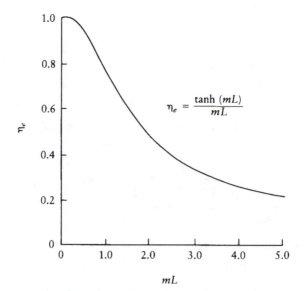

FIGURE 2.30 Efficiency of a pin fin with an insulated end.

This equation is plotted in Figure 2.31. All curves on this graph have the same general shape-multiples of tanh *(mL)*. The trend with each curve shows again that the fin performance cannot be substantially improved over that of *mL* = 3. Under normal circumstances, the effectiveness should be as high as possible and practical. The fin effectiveness is related to the efficiency by

$$\eta_f = \eta_e \frac{\text{Surface area of fin}}{\text{Cross-sectional area of fin}} \tag{2.68}$$

Example 2.8

A barbecue grill consists of a base with a hinged cover. The cover is opened by lifting a metal rod bolted onto the cover (see Figure 2.32). During times of use, the cover reaches a uniform elevated temperature of 200°F. The rod then becom-*es too hot to hold when it is desired to raise the cover. Analyze the rod (a) by determining the temperature profile and (b) by calculating how much heat the rod transfers away. Determine also (c) the efficiency and (d) the effectiveness. (e) Suggest a method for making the rod more suitable for its intended use as a lifting handle. The rod has a 5/16-in diameter, is 4-1/2 in long, and is made of chrome-plated low-carbon 1C) steel. The ambient temperature is 70°F. Assume $\bar{h}_c = 1$ BTU/(hr·ft²·°R) .

Solution

The rod will be considered as a fin attached to a wall of temperature 200°F.

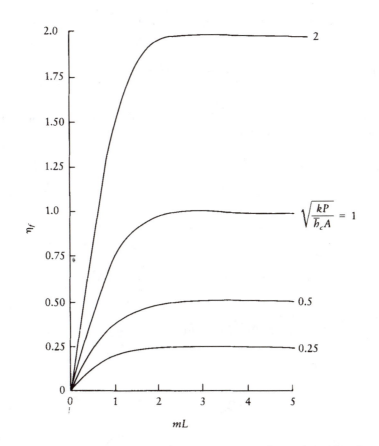

FIGURE 2.31   Effectiveness of a pin fin. Note that the effectiveness is not always less than 1.0

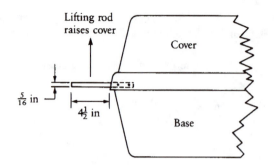

FIGURE 2.32 Barbecue grill detail showing cover lifting handle.

Assumptions

1. We can neglect the effect of the chrome plating because it is thin and put on to prevent corrosion rather than to affect the heat-transfer rate.
2. The rod gains heat by conduction from the cover and loses heat only by convection to the surroundings. Radiation losses are neglected.
3. Material properties are constant.
4. One-dimensional conduction exists along the rod.
5. The system is at steady state.

(a) The temperature profile can be calculated with Equation 2.62:

$$\frac{T - T_\infty}{T_w - T_\infty} = \frac{\cosh\ [m(L-z)]}{\cosh\ (mL)}$$

where $m = \sqrt{\bar{h}_c P / kA}$. We now evaluate each term. *(Note: If $\bar{h}_c$ was not provided, we would use an estimated value from Table 1.2.)* With $D = 5/16$ in, we find

$$P = \pi D = \pi\left(\frac{5}{16}\right)\left(\frac{1}{12}\right) = 0.0818 \text{ ft}$$

$$A = \frac{\pi D^2}{4} = \frac{\pi}{4}\left(\frac{5}{16(12)}\right)^2 = 0.000533 \text{ ft}^2$$

From Appendix Table B.2, for low-carbon (1C) steel, we know that $k = 24.8$ BTU/(hr·ft·°R). Proceeding,

$$m = \sqrt{1(0.0818)/[24.8(0.000533)]} = 2.488/\text{ft}$$

The rod length is given as

$$L = 4\frac{1}{2} \text{ in} = 0.375 \text{ ft}$$

Substituting into the temperature equation, we get

$$\frac{T-70}{200-70} = \frac{\cosh[2.488(0.375-z)]}{\cosh[2.488(0.375)]}$$

$$T = 70 + 88.57 \cosh[2.488(0.375-z)]$$

A graph of this equation is provided in Figure 2.33.

(b) The heat transferred is calculated with Equation 2.63:

$$q_z = kAm(T_w - T_\infty) \tanh (mL)$$

$$= 24.8(0.000533)(2.488)(200-70) \tanh (2.488(0.375))$$

$$q_z = 3.13 \text{ BTU/hr}$$

The heat transferred could have been determined by using the graph of $q_z/kAm\theta_w$ vs. mL from Figure 2.29.

(c) The fin efficiency likewise can be found either by using Figure 2.30 or by equation. From Figure 2.30, with $mL = 2.488(0.375) = 0.933$, we get

$$\eta_e = 0.78$$

Thus, the actual heat transferred is 0.78 times what it would be if the fin were entirely at 200°F.

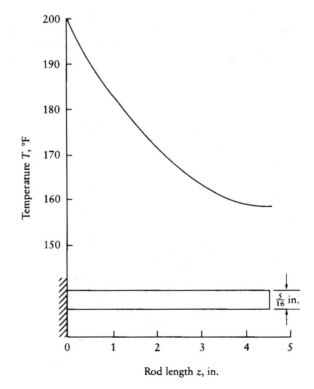

FIGURE 2.33   Temperature distribution within the rod of Example 2.8.

(d) The effectiveness can be determined with Equation 2.67 as

$$\eta_f = \sqrt{\frac{kP}{\bar{h}_c A}} \tanh(mL) = \sqrt{\frac{24.8(0.0818)}{1(0.000533)}} \tanh(0.933)$$

Solving,

$$\eta_f = 45.2$$

Thus, the heat flux through a 5/16-in-diameter spot on the wall (before adding the fin) has been increased by a factor of 45.2 by adding the fin.

(e) The rod is too hot to touch. An insulating material that will not melt at 200°F could be added as a covering over the handle.   ❏

The pin fin considered in this section was analyzed, assuming that the tip was insulated. The effect that this boundary condition has on the temperature distribution is seen in Figure 2.33. The temperature profile is tangent to a horizontal line at the tip. We will refer to the insulated-tip formulation for the pin fin as Case #1, because there are at least two other cases that can be derived for the pin-fin geometry:

Case #2—The pin fin or spine is very long, and the temperature far from the wall (or at the tip) equals that of the surrounding fluid.
Case #3—The fin is of finite length and does transfer heat from its end by convection.

The applicable differential equation for these cases is still Equation 2.55; the general solution is still Equation 2.61. The difference is in the boundary conditions used to evaluate the constants of integration. Rather than discussing each of these problems in detail, only the solutions are presented here, with the derivations left as exercises.

Case #2

B.C.1.   at $z = 0$, $T = T_w$

B.C.2.   at $z = \infty$, $T = T_\infty$

$$\frac{T - T_\infty}{T_w - T_\infty} = e^{-mz} \tag{2.69a}$$

$$m^2 = \bar{h}_c P / kA$$

$$q_z = \sqrt{\bar{h}_c PkA}(T_w - T_\infty) \tag{2.69b}$$

$$\eta_e = \frac{1}{mL} \tag{2.69c}$$

$$\eta_f = \sqrt{\frac{kP}{\bar{h}_c A}} \tag{2.69d}$$

Case #3

B.C.1.   at $z = 0$, $T = T_w$

B.C.2.   at $z = L$, $-k\dfrac{dT}{dz} = \bar{h}_c(T - T_\infty)$

$$\frac{T - T_\infty}{T_w - T_\infty} = \frac{\cosh[mL(1 - z/L)] + (\bar{h}_c/mk)\sinh[mL(1 - z/L)]}{\cosh(mL) + (\bar{h}_c/mk)\sinh(mL)} \tag{2.70a}$$

$$m^2 = \bar{h}_c P/kA$$

$$q_z = \sqrt{\bar{h}_c PkA}(T_w - T_\infty)\frac{\sinh(mL) + (\bar{h}_c/mk)\cosh(mL)}{\cosh(mL) + (\bar{h}_c/mk)\sinh(mL)} \tag{2.70b}$$

$$\eta_e = \frac{1}{mL}\frac{\sinh(mL) + (\bar{h}_c/mk)\cosh(mL)}{\cosh(mL) + (\bar{h}_c/mk)\sinh(mL)} \tag{2.70c}$$

$$\eta_f = \sqrt{\frac{kP}{\bar{h}_c A}}\frac{\sinh(mL) + (\bar{h}_c/mk)\cosh(mL)}{\cosh(mL) + (\bar{h}_c/mk)\sinh(mL)} \tag{2.70d}$$

To expedite evaluating the above equations for specific problems, a table of hyperbolic functions and identities is provided in Appendix Table A.4.

2.7.3   ANALYSIS OF A STRAIGHT FIN OF RECTANGULAR PROFILE

A straight fin of rectangular profile is sketched in Figure 2.34. Such fins can be found on conventional single-cylinder lawn mower engines. As seen in the figure, the fin is of length L and width W. The fin thickness is 2δ. For our analysis, we assume that the fin is attached to a wall of uniform

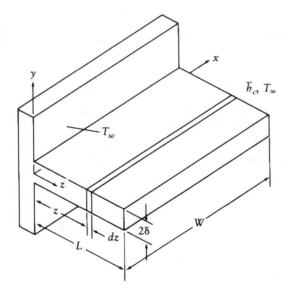

FIGURE 2.34   Definition sketch of a straight rectangular fin.

temperature T_w, the ambient fluid temperature is T_∞, the convection coefficient between the fin and the ambient is $\bar{h}_c$, and the fin material has a thermal conductivity of k. Our interest is in the performance of such a fin. The slice of material that is $dz \times 2\delta \times W$ is selected for analysis.

If we were to analyze this fin, subject to suitable assumptions, we would obtain a solution for temperature as a function of distance. The differential equation that applies is Equation 2.55. The boundary conditions are the same as those for the pin fin, and so the solution we get will be the same as that for Case #1 or Case #3 of the previous section. Case #2 is not usually used to model straight fins, because such fins are seldom made infinitely long. The difference between the results for the rectangular fin and the pin fin is in the area and perimeter terms.

The temperature profile for Case #3 appears rather cumbersome to use, and so it is desirable to have a way to simplify the evaluation of temperature in Equation 2.70a. Such a method has been developed for the rectangular fin, which involves using the results of Case #1, Equation 2.62c. We define a "corrected fin length" L_c as

$$L_c = L + \delta = \text{actual fin length} + \frac{1}{2} \times \text{fin thickness}$$

Substituting the corrected fin length into the equations of Case #1, we obtain a result that models Case #3 with very small error. If $(\bar{h}_c/mk) \le 1/2$, the maximum error between the two solutions is less than 8%. This method of modifying length in the Case #1 solution to model the Case #3 results is known as the *corrected tip* or *corrected length* solution.

Example 2.9

The head of a single-cylinder lawn mower engine has a number of fins. Consider one such fin, with dimensions of $L = 1\text{-}1/8$in, $W = 2\text{-}1/4$ in, and $2\delta = 1/16$ in. The fin is made of aluminum and while the engine is running at steady state, the root temperature reaches 1000°F. The ambient temperature locally (around the fins) is 90°F. The convection coefficient is estimated as 0.8 BTU/(hr·ft²·°R). (a) Determine the heat transferred by using Case #3, Equation 2.70b, assuming heat is transferred at the tip. (b) Determine the heat transferred by using Case #1, Equation 2.63b, assuming the tip is insulated. (c) Determine the heat transferred by using Equation 2.63b, which is the heat transferred in part (b) corrected for length, where $L_c = L + \delta$.

Solution

The proximity of other fins will affect the convection heat transfer from, and the ambient temperature around, the fin of interest.

Assumptions

1. The fin of interest can be treated as though it were isolated.
2. Heat is transferred along the fin by conduction.
3. One-dimensional conduction exists.
4. Heat is transferred away from the fin only by convection; no radiation.
5. Material properties are constant.
6. The system is at steady state.

From Appendix Table B.1, for aluminum, $k = 136$ BTU/(hr·ft·°R). The information given in the problem is

$$L = 1\text{-}1/8 \text{ in} \quad = 0.09375 \text{ ft}$$
$$W = 2\text{-}1/4 \text{ in} \quad = 0.1875 \text{ ft}$$
$$\delta = 1/32 \text{ in} \quad = 0.002604 \text{ ft}$$

$\bar{h}_c = 0.8$ BTU/(hr·ft^2·°R)
$T_w = 1000$°F
$T_\infty = 90$°F

With $A = 2\delta W$ and $P = 2W$, then $P/A = 2W/2\delta W = 1/\delta$ and

$$m = \sqrt{\bar{h}_c P/kA} = \sqrt{0.8/136(0.002604)}$$

$$m = 1.503$$

(a) Equation 2.70b is

$$q_z = \sqrt{\bar{h}_c PkA}(T_w - T_\infty)\frac{\sinh(mL) + (\bar{h}_c/mk)\cosh(mL)}{\cosh(mL) + (\bar{h}_c/mk)\sinh(mL)}$$

$$= \sqrt{0.8(2)^2(0.1875)^2(136)(0.002604)}(1000 - 90)$$

$$\times \frac{\sinh[(1.503)(0.09375)] + [0.8/1.503(136)]\cosh[(1.503)(0.09375)]}{\cosh[(1.503)(0.09375)] + [0.8/1.503(136)]\sinh[(1.503)(0.09375)]}$$

Solving,

$$q_x = 26.12 \text{ BTU/hr} \qquad \text{(includes heat loss from tip)}$$

(b) Equation 2.63b is

$$q_z = \sqrt{kA\bar{h}_c P}(T_w - T_\infty)\tanh(mL)$$

$$= \sqrt{136(2)^2(0.002604)(0.1875)^2}(1000 - 90)$$

$$\times \tanh[(1.503)(0.09375)]$$

Solving,

$$q_z = 28.43 \text{ BTU/hr} \qquad \text{(insulated tip)}$$

This result may seem insignificantly different from that in part (a), but it should be remembered that only one fin is being considered.

(c) Equation 2.63b with length $= L + \delta = 0.09635 = L_c$

$$q_z = kAm(T_w - T_\infty)\tanh[mL(1 + \delta/L_c)]$$

$$= 136(2)(0.002604)(0.1875)(1.503)(1000 - 90)$$

$$\times \tanh[(1.503)(0.09635)(1 + 0.002604)(0.09635)]$$

Solving,

$$q_z = 26.12 \text{ BTU/hr} \qquad \text{(corrected length solution)}$$

We can draw at least two conclusions from these results. First, the equation in part (a) is very cumbersome. Second, the corrected tip solution provides an answer that has excellent agreement with the exact solution of part (a) and is easier to compute.   ❏

2.7.4   STRAIGHT FINS OF TRIANGULAR AND OF PARABOLIC PROFILE

Figure 2.35a is a sketch of a straight triangular fin. The fin has straight sloping upper and lower boundaries. The fin thickness at the wall ($z = 0$) is 2δ. The fin length × width dimensions are $L \times W$. Figure 2.35b shows a straight inverse parabolic fin. Again, the width of the fin at the wall is 2δ; length × width dimensions are $L \times W$.

The solution of the differential equation that describes these cases is beyond the scope of this text. The mathematics involved yields series solutions that are more conveniently summarized in graphical rather than in equation form. Figure 2.36 is a graph of efficiency vs. $L_c^{3/2}(\bar{h}_c/kA_p)^{1/2}$ for the straight fins discussed thus far, including the rectangular fin. So, instead of providing an equation for all cases, Figure 2.36 is used to determine efficiency, from which the heat transferred can be calculated. Note that beyond 1.5 on the horizontal axis of the graph, the inverse parabolic fin has the highest efficiency, because this fin has a higher proportion of its material at the wall where the temperature is a maximum. Although the inverse parabolic profile is efficient, the triangular and rectangular fins are usually used in practice instead, because both are easier to manufacture and to

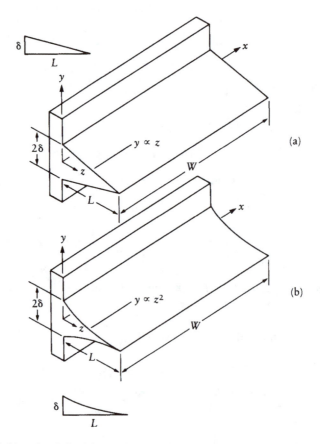

FIGURE 2.35   Definition sketch for (a) a straight triangular fin and (b) an inverse parabolic fin.

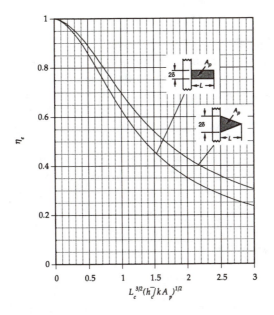

FIGURE 2.36 Efficiency of fins having equal length L and profile area A_p.

maintain. Accompanying Figure 2.36 is Table 2.4, where the corrected fin length L_c and the profile area A_p are listed for each case. Also appearing in the table is the optimum condition or dimension that corresponds to maximum heat transfer.

TABLE 2.4
Straight–Fin Parameters

Profile	Corrected length L_c	Profile area A_p	Optimum condition for maximum heat transfer	Surface area A_s for heat–transfer calculation
	$L_c = L + \delta$	$A_p = 2\delta\, L_c$	$\delta_{opt} = \dfrac{\bar{h}_c L_c^2}{2.014k}$ $mL_c\big|_{opt} = 1.419\,2$	$A_s = PL_c = 2WL_c$
	$L_c = L$	$A_p = \delta L_c$	$\delta_{opt} = 0.583\bar{h}_c L_c^2/k$	$A_s = 2W\sqrt{L^2 + \delta^2}$
	$L_c = L$	$A_p = 2\delta L_c/3$	$\delta_{opt} = 0.488\bar{h}_c L_c^2/k$	$A_s \cong 2.05\,W\sqrt{L^2 + \delta^2}$

Example 2.10

A commercial air-conditioning unit in a motel rejects heat ordinarily to the atmosphere. It is proposed that a heat exchanger be installed in which water from the swimming pool will absorb the heat rejected by the air conditioner. The heat exchanger itself will be stainless steel (type 304) because of its anticorrosive properties. In one section straight triangular fins are to be machined as part of the exchanger walls. The fins transfer heat to the pool water as the water passes them. It is determined that the convection coefficient between the fins and the water is 400 BTU/(hr·ft²·°R). The fins are confined to being 12 in long, and the design calls for six of them. Each fin has a base width (2δ) of 1/2 in. For this proposed design (a) determine the corresponding optimum fin length. (b) Find the heat transferred by one fin if the wall temperature is 190°F, and the ambient water temperature is 58°F.

Solution

Besides keeping pool water warmed somewhat, the device described will help make the air-conditioning unit run more efficiently.

Assumptions

1. Heat is transferred by conduction through the fin from the base wall.
2. One-dimensional steady conduction exists within the fin.
3. Heat is transferred from the fin by convection.
4. Material properties are constant.

(a) From Appendix Table B.2 for Type 304 stainless steel, $k = 8.32$ BTU/(hr·ft·°R). Use the equation in Table 2.4 to calculate the optimum dimensions of one fin:

$$\delta_{opt} = 0.583\bar{h}_c^2 L_c/k$$

or

$$L_c^2 = \frac{\delta_{opt}k}{0.583\bar{h}_c}$$

With the wall width given as

$$2\delta_{opt} = 0.5 \text{ in} = 0.04167 \text{ ft}$$

we find

$$L_c^2 = \frac{(0.04167/2)(8.32)}{0.583(400)}$$

So, for optimum heat transfer

$$L_c = 0.0273 \text{ ft} = 0.327 \text{ in}$$

(b) We can find the heat transferred with Figure 2.36. First, calculate

$$L_c^{3/2}(\bar{h}_c/kA_p)^{1/2} = (0.0273)^{3/2}\left[\frac{400}{(8.32)(0.04167/2)(0.0273)}\right]^{1/2} = 1.31$$

Entering Figure 2.36 at the above value, we read

$$\eta_e = 0.6$$

The actual heat transferred is

$$q_{ac} = \eta_e \bar{h}_c A_s (T_w - T_\infty) = \eta_e \bar{h}_c 2W(L^2 + \delta^2)^{1/2}(T_w - T_\infty)$$

Substituting,

$$q_{ac} = \eta_e \bar{h}_c A_s (T_w - T_\infty) = \eta_e \bar{h}_c 2W(L^2 + \delta^2)^{1/2}(T_w - T_\infty)$$

Solving,

$$q_{ac} = 2175 \text{ BTU/hr} \qquad \square$$

2.7.5 Circular Fin of Rectangular Profile

Figure 2.37 is a definition sketch of a circular fin having a rectangular profile. Fins of this type are found on the outside walls of single-cylinder and multicylinder air-cooled motorcycle engines. As seen in the figure, the fin thickness is 2δ, while the fin length is L. It is convenient to express the fin dimensions in terms of radii from the centerline of the cylinder. The radius R_2 is from the centerline to the fin tip, and R_1 is to the fin root. It is possible to formulate the differential equation and solve it, but again the solution for the temperature profile is in terms that are beyond the scope of this text. Instead, a graph of efficiency vs. $L_c^{3/2}(\bar{h}_c/kA_p)^{1/2}$ is provided for the circular fin (Figure 2.38). The concept of a corrected length is incorporated into the parameters of that graph. With the efficiency curve, the actual heat transferred for a particular fin can be calculated.

Example 2.11

A two-stage reciprocating compressor consists of two piston-cylinder arrangements of different size. The connecting rod from each piston is bolted to a common crankshaft onto which a flywheel is attached. An external electric motor turns the flywheel, which in turn causes the pistons to reciprocate and provide compressed air. Each cylinder has fins for cooling. The flywheel has ribs

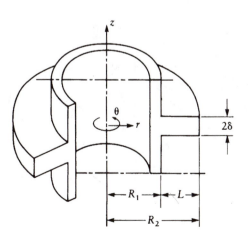

FIGURE 2.37 A circular fin of rectangular profile.

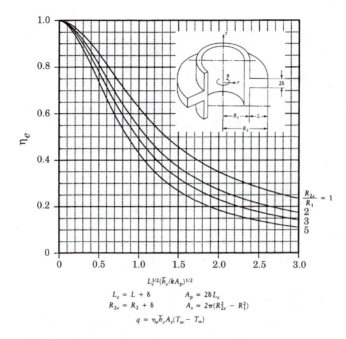

$$L_c^{3/2}(\bar{h}_c/kA_p)^{1/2}$$

$L_c = L + \delta$ $A_p = 2\delta L_c$

$R_{2c} = R_2 + \delta$ $A_s = 2\pi(R_{2c}^2 - R_1^2)$

$q = \eta_e \bar{h}_c A_s (T_w - T_\infty)$

FIGURE 2.38 A circular fin of rectangular profile.

that are inclined at an angle so that when the flywheel rotates, the ribs act as fan blades that force air past the finned cylinder walls. (See Figure 2.39.)

Figure 2.40 is a schematic of the cross section of the larger cylinder. There are nine fins all taken to be rectangular in cross section. Each fin is 3 mm thick and 2.5 cm long. The outside diameter of the cylinder, fin tip to tip, is 21.9 cm. The distance between adjacent fins is 1.27 cm. (In the real cylinder, the fins are cast with a slight taper, because this aids in the casting process. Also, the fins get shorter near the bottom of the cylinder; piston movement is nearer the top, and that is where the heat is generated.)

The cylinder is made of cast iron. During operation the root wall temperature reaches 260°C. Ambient air temperature is 27°C. With air moving past the fins at several meters per second, $\bar{h}_c = 15 \text{ W/(m}^2\cdot\text{K)}$. (a) Determine the amount of heat transferred from the cylinder to the air. (b) Repeat the calculations for a plain cylinder wall without fins. (c) Determine the fin effectiveness.

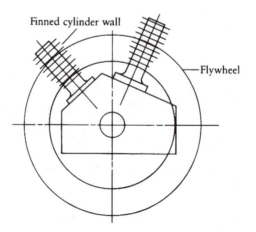

FIGURE 2.39 Two-cylinder air compressor.

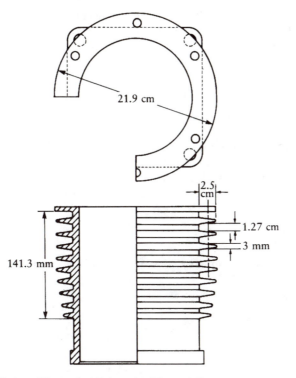

FIGURE 2.40   The cylinder of Example 4.7 (see also Figure 4.3).

Solution

The fins are not identical. Piston movement is near the top of the cylinder so, near the bottom, where there is less heat to dissipate, the fins are not as long as those near the top. Moreover, in the real cylinder, extra material is cast as part of the cylinder wall for bolt holes, which are needed to attach the cylinder head.

Assumptions

1. One-dimensional steady conduction heat transfer exists within the fin.
2. Material properties are constant.
3. All fins are identical, having the same dimensions, efficiency, and surface area.
4. Heat is transferred away by convection; radiation is neglected.
5. The presence of the material into which bolt holes are drilled and threaded is neglected.
6. The fin root temperature is constant over the cylinder height.

(a) The heat transferred by the cylinder will equal that transferred from 9 fins (q_f) plus that transferred by the wall sections between adjacent fins (q_w); or

$$q = q_f + q_w$$

The fin heat-transfer rate is

$$q_f = N\eta_e q_{max}$$

where N is the number of fins, η_e is the fin efficiency to be found from Figure 2.38, and $q_{max} = \bar{h}_c A_s (T_w - T_\infty)$. It is assumed that all fins are identical, having the same dimensions, efficiency, and surface area. It is also assumed that the fin root temperature is constant. The surface area of each fin is $A_s = 2\pi(R_{2c}^2 - R_1^2)$. The parameters of the problem are

$2\delta = 3$ mm

or

$\delta = 0.001\ 5$ m
$L_c = L + \delta = 0.025 + 0.001\ 5 = 0.026\ 5$
$R_{2c} = R_2 + \delta = 0.219/2 + 0.001\ 5 = 0.111$ m
$R_1 = R_2 - L = 0.219/2 - 0.025 = 0.084\ 5$ m
$T_w = 260°C$
$T_\infty = 27°C$
$\bar{h}_c = 15$ W/(m²·K)

From Appendix Table B.2 for cast iron, $k = 52$ W/(m·K). We now calculate the areas of importance:

Profile area and surface area of one fin

$$A_p = 2\delta L_c = 2(0.001\ 5)(0.026\ 5) = 7.95 \times 10^{-5}\ \text{m}^2$$

$$A_s = 2\pi(R_{2c}^2 - R_1^2) = 3.26 \times 10^{-2}\ \text{m}^2$$

To use Figure 2.38 for finding efficiency, we first determine

$$R_{2c}/R_1 = 1.3$$

and

$$L_c^{3/2}(\bar{h}_c/kA_p)^{1/2} = 0.26$$

With these parameters, Figure 2.38 shows

$$\eta_e = 0.93$$

So the heat transferred by the nine fins is

$$q_f = N\eta_e\bar{h}_cA_s(T_w - T_\infty) = 9(0.93)(15)(3.26 \times 10^{-2})(260 - 27)$$

$$q_f = 953.7\ \text{W}$$

The heat transferred by the exposed cylinder surface (between the fins) is

$$q_w = \bar{h}_cA_{sw}(T_w - T_\infty)$$

The exposed surface area we will take to consist of nine sections or rings of radius 0.084 5 m (= R_1) and height 12.7 mm. We find

$$A_{sw} = 2\pi R_1(S_p)(9) = 2\pi(0.084\ 5)(0.012\ 7)(9)$$
$$= 6.07 \times 10^{-2}\ m^2$$

where S_p is fin spacing. The heat transferred is

$$q_w = 15(6.07 \times 10^{-2})(260 - 27)$$
$$q_w = 212\ W$$

The total heat transferred from the cylinder is $q_f + q_w$ or

$$q = 1\ 166\ W$$

(b) In the case of no fins, the surface area is the total cylinder height $\times 2\pi R_1$.

The height is

$$H = 9(0.012\ 7 + 0.003) = 0.141\ 3\ m$$

The surface area without fins is

$$A_{so} = 2\pi(0.084\ 5)(0.141\ 3) = 7.502 \times 10^{-2}\ m^2$$

The heat transferred without fins is

$$q_o = \bar{h}_c A_{so}(T_w - T_\infty) = 15(7.502 \times 10^{-2})(260 - 27)$$
$$q_o = 262\ W$$

(c) Fin effectiveness is defined in Equation 2.66:

$$\eta f = \frac{\text{Heat flux with fins}}{\text{Heat flux without fins}}$$

For this case,

$$\eta_f = \frac{q}{q_o} = \frac{1\ 166}{262}$$
$$\eta_f = 4.45$$

The fins used here are highly effective.   ❏

2.8   SUMMARY

In this chapter, various one-dimensional problems in steady-state conduction were considered (see Table 2.5). The discussion was restricted to one-dimensional problems in which temperature varied only with one space variable, and in some cases the effect of internal energy release was considered.

TABLE 2.5
Conduction Equations

	Fourier's Law of Heat Conduction	The general heat-conduction equation including internal heat generation
Vector form	$\tilde{q}'' = -k\tilde{\nabla}T$	$\nabla^2 T + \dfrac{q'''}{k} = \dfrac{1}{\alpha}\dfrac{\partial T}{\partial t}$
Cartesian coordinates	$q_x'' = -k\dfrac{\partial T}{\partial x}$ $q_y'' = -k\dfrac{\partial T}{\partial y}$ $q_z'' = -k\dfrac{\partial T}{\partial z}$	$\dfrac{\partial^2 T}{\partial x^2} + \dfrac{\partial^2 T}{\partial y^2} + \dfrac{\partial^2 T}{\partial z^2} + \dfrac{q'''}{k} = \dfrac{1}{\alpha}\dfrac{\partial T}{\partial t}$
Polar cylindrical coordinates	$q_r'' = -k\dfrac{\partial T}{\partial r}$ $q_\theta'' = -\dfrac{k}{r}\dfrac{\partial T}{\partial \theta}$ $q_z'' = -k\dfrac{\partial T}{\partial z}$	$\dfrac{1}{r}\dfrac{\partial}{\partial r}\left(r\dfrac{\partial T}{\partial r}\right) + \dfrac{1}{r^2}\dfrac{\partial^2 T}{\partial \theta^2} + \dfrac{\partial^2 T}{\partial z^2} + \dfrac{q'''}{k} = \dfrac{1}{\alpha}\dfrac{\partial T}{\partial t}$
Spherical coordinates	$q_r'' = -k\dfrac{\partial T}{\partial r}$ $q_\theta'' = -\dfrac{k}{r}\dfrac{\partial T}{\partial \theta}$ $q_\phi'' = -\dfrac{k}{r\sin\theta}\dfrac{\partial T}{\partial \phi}$	$\dfrac{1}{r^2}\dfrac{\partial}{\partial r}\left(r^2\dfrac{\partial T}{\partial r}\right) + \dfrac{1}{r^2\sin\theta}\dfrac{\partial}{\partial \theta}\left(\sin\theta\dfrac{\partial T}{\partial \theta}\right)$ $+ \dfrac{1}{r^2\sin^2\theta}\dfrac{\partial^2 T}{\partial \phi^2} + \dfrac{q'''}{k} = \dfrac{1}{\alpha}\dfrac{\partial T}{\partial t}$

Plane, cylindrical, and spherical coordinate systems were examined. The concept of thermal contact resistance was also introduced.

We have also examined heat transfer from extended surfaces. The general one-dimensional conduction equation for a fin with a convective surface was derived. The general equation was applied to a number of fin problems, including pin fins or spines and straight rectangular fins. Straight fins of other cross sections were examined. Other fins, such as circumferential or annular fins, were discussed and solutions were presented in graphical form.

2.9 PROBLEMS

(*Note:* Contact resistance for composite materials is to be neglected unless otherwise stated.)

2.9.1 One–Dimensional Planar Conduction

1. A 2-in-thick aluminum plate is placed over a coal-fueled fire. The plate, which is 4 ft square, provides a hear-transfer area for a warming surface. The temperature at the lower surface of the aluminum is 600°F, while the upper surface temperature is 250°F.
   (a) Graph the temperature distribution within the plate.
   (b) Graph the dimensionless temperature distribution (Equation 2.10) within the plate as $(T - T_1)/(T_2 - T_1)$ vs. x/L.
   (c) Determine the heat flow through the plate.

2. Chrome steel (5 Cr) is to be cut into strips and rolled to form tubing. Just before it is rolled into a thin sheet, the chromium steel exists as a 4-cm-thick plate, which is heated in an oven. The upper surface of the plate is at 600 K, while the lower surface is at 800 K.
   (a) Graph the temperature distribution within the plate.
   (b) Graph the dimensionless temperature distribution (Equation 2.10) within the plate.
   (c) How does the dimensionless graph vary with material properties?
   (d) Determine the heat-flow rate through the plate per unit area.

3. Type 347 stainless steel is to be used in the wall of a nuclear reactor. The stainless steel is 3 in thick and is designed to transfer 10,000 BTU/(hr·ft²) of energy during times of emergency. If the maximum allowable temperature at one surface of the wall is 1000°F, determine the temperature at the other surface.

4. In regions where there is a low elevation, houses are usually built on cement slabs because basements cannot be dug. A typical cement slab for a conventional home is 10 cm thick. Consider such a slab to be covered with linoleum 1 cm thick. The lower surface of the slab is at 10°C. The upper surface of the linoleum is 23°C.
   (a) Determine the heat flow per unit area through the floor.
   (b) Calculate the interface temperature between the cement and the linoleum.

5. The second floor of a home dwelling consists of a 3/4-in-thick plywood, covered with 1/4-in-thick linoleum. Wool carpeting 1/2 in thick is installed over the linoleum. The temperature of the lower surface of the plywood is 75°F. The heat flow upward through the floor is 5 BTU/(hr·ft²). Determine the interface temperatures.

6. The wall of a laboratory furnace used in heat-treating metal consists of one layer of firebrick [(k = 0.85 BTU/(hr·ft·°R)] 3-1/2 in thick, with 1-in-thick glass fiber insulation surrounding it. The exterior jacket is low-carbon steel (1C) sheet metal 0.0625 in thick. When the furnace is in use, the interior wall temperature is 3000°F, and data from the electrical meters indicate a heat flow rate of 215 BTU/(hr·ft²). For an area normal to the hear flow of 48 in², determine the outside-wall surface temperature.

7. A wall found in a home is constructed of 2 × 4s (actual dimensions 1-1/2 × 3-1/2 in) made of pine and spaced 12 in apart (center to center), 12 ft tall. Nailed to the 2 × 4s on the outside is 3/4-in plywood, while 1/2-in fiber sheets are on the inside. Insulation made of glass wool fills the air space in the wall. For a heat-flow rate through the wall of 40 BTU/(hr·ft²), determine the overall temperature drop from wall surface to wall surface, inside to outside.

8. Figure P2.1 shows the cross section of a wall that consists of 4-in-thick masonry brick, 2 × 4 (1-1/2 × 3-1/2 actual size) wall studs spaced 16 in apart (center to center), and 1/2-in-thick plaster board. The external wall temperature of the brick is 98°F. The temperature of the plaster board inside is 60°F. Determine the heat transferred through the wall for the following two cases:
   (a) Air within the air space is motionless.
   (b) Air within the air space can move, and the convection heat-transfer coefficient between the air and brick, and the air and the plaster board is 3 BTU/(hr·ft²·°R).
   Take the wall height to be 10 ft in both cases.

9. A refrigerator wall consists of low-carbon steel (1C) 1 mm thick on the outside, 2-mm-thick polystyrene inside, with 2-cm-thick expanded cork between as an insulation, as shown in Figure P2.2. Outside the refrigerator wall, the air temperature is 25°C, and the heat-transfer coefficient is 3 W/ (m²·K). Inside the refrigerator, the air temperature is 10°C, and the film coefficient at the wall is 4.5 W/ (m²·K).
   (a) Determine the overall heat-transfer coefficient for the wall.
   (b) Determine the hear flow through the wall.

10. The front windshield of a car is 1/4 in thick. During cold weather, warmed air is directed at the windshield to "defrost" it. The warmed air is at 120°F, and as it travels past the

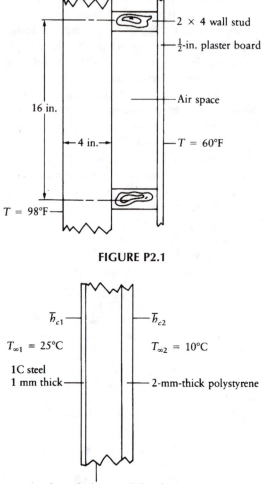

FIGURE P2.1

FIGURE P2.2

windshield, heat is transferred to the glass; the convective heat-transfer coefficient is 2 BTU/(hr·ft²·°R). Outside, the air temperature is 30°F, and the convective hear-transfer coefficient depends (in the velocity of the ear according to

$$\bar{h}_c = V - 0.008V^2$$

where $\bar{h}_c$ is in BTU/(hr·ft²·°R) and V is in ft/s. Determine and graph the outside glass temperature of the windshield over the velocity range of 5 to 55 mph.

2.9.2 ONE-DIMENSIONAL CONDUCTION IN POLAR COORDINATES

11. A steam pipe is made of 24 nominal, schedule 160 stainless steel (type 347). The inside-wall temperature of the pipe is 200°C. The outside-wall temperature is 190°C. Determine the heat loss through the wall per length of pipe.

12. Water goes from a hot water heater to a bathtub in 1 standard type M copper tubing. At one section of the tube, the water elevates the inside-wall temperature of the copper to

140°F. If the heat loss through the wall at this section is 10,000 BTU/(hr·ft of tube), determine the temperature of the outside tube surface.

13. Heater hose in the engine compartment of an automobile is 16 mm in diameter (OD). The wall thickness is 3 mm. The hose conveys heated water to the heater core in the passenger compartment. The inside-wall temperature of the hose is 90°C, and it is determined that 100 W/(m of tubing) is lost through the wall. Determine the outside-wall temperature of the hose. Take the hose properties to be the same as those for buna rubber. Graph the temperature distribution within the tube wall.

14. Crude oil is warmed to reduce its viscosity and make it easier (less costly) to pump. In one section of the pipeline, the pipe itself is made of low-carbon (1C), 4 nominal, schedule 40 wrought steel. The inside-wall temperature of the pipe is 50°C. The convection coefficient between the outside pipe surface and the surrounding air is 3 W/(m²·K). The surrounding air temperature is 25°C.

    (a) Determine the heat loss per unit length from the pipe.

    (b) A 3-cm-thick insulating layer of kapok is added. Determine the heat loss per unit length of the insulated pipe, assuming $\bar{h}_c$ is unchanged.

15. Determine the overall heat-transfer coefficient for Problem 2.14a and b, based on the inside surface area.

16. Determine the overall heat-transfer coefficient for Problem 2.14a and b, based on the outside surface area.

17. Sheet-metal ductwork is found in an attic and used in the air-handling unit to deliver cooled air to various rooms of a house. One such duct has a 12-in diameter (ID) and is made of rolled sheet metal (low-carbon, 1C), having a thickness of 0.040 in. The duct is insulated with 1-in-thick glass fiber. The air inside the duct is at 40°F, and the heat-transfer coefficient between the air and the duct surface is 7 BTU/(hr·ft²·°R). The attic air temperature is 110°F, and the heat-transfer coefficient between the air and the insulation surface is 2 BTU/(hr·ft²·°R). What is the heat transferred per unit length of duct?

18. The critical thickness of insulation discussed in this chapter involved heat loss by convection, neglecting radiation. Suppose instead that convection is neglected and radiation is not. The heat transfer by radiation is expressed as

$$q = \sigma A \varepsilon_1 F_{1-2}(T_1^4 - T_2^4)$$

or in terms of a radiation coefficient,

$$q = \bar{h}_r A(T_1 - T_2)$$

Express the critical radius in terms of the Stefan-Boltzmann constant σ, the surface emissivity ε_1, and the shape factor F_{1-2}.

19. A 1 nominal pipe is covered with glass-wool insulation. The outside-wall temperature of the pipe is 100°C, and the insulated pipe is exposed to air at 25°C. The convection heat-transfer coefficient between the insulation and the air is 1 W/(m²·K). Graph the heat transferred as a function of the insulation thickness, and determine the critical value of the insulation thickness.

20. A wire in the recharging system of an electric vehicle has a diameter of 1 cm. It will carry 250 amps at 115 volts. The wire is pure copper. The wire's insulation is a plastic that has the same properties as ebonite rubber. Heat will be transferred from the wire through the insulation to ambient air at 20°C. If the insulation is 1 mm thick, has adding it increased or decreased the heat loss to the ambient? Take the convection coefficient to be 4 W/(m²·K).

2.9.3 INTERNAL HEAT GENERATION

21. In the development concerning internal heat generation within a plane wall, the boundary conditions

$$1. \quad x = L, T = T_w$$

$$2. \quad x = -L, T = T_w$$

led to the following temperature distribution:

$$\frac{T - T_w}{T_{max} - T_w} = 1 - \frac{x^2}{L^2}$$

Show that if the boundary conditions

$$1. \quad x = 0, T = T_{max}$$

$$2. \quad x = \pm L, T = T_w$$

are used instead, then the resulting temperature distribution becomes

$$\frac{T - T_{max}}{T_w - T_{max}} = \frac{x^2}{L^2}$$

22. The plane wall in Figure P2.3 generates heat internally. On the left the wall is in contact with a cooling fluid that is at a temperature of $T_{\infty 1}$. the convective heat transfer coefficient on the left face is $\bar{h}_{c1}$. Similarly, on the right, there is a fluid at $T_{\infty 2}$, and the convective heat-transfer coefficient is $\bar{h}_{c2}$. Write the differential equation to be solved and the boundary conditions. (It is not necessary to solve it.) Does the maximum temperature necessarily exist at the midpoint of the material?

23. A cylindrical fuel element is used in a nuclear reactor is 5 cm in diameter and 40 cm long. It is supported by aluminum cladding that is a surrounding layer 1 cm thick. The outside surface of the aluminum is in contact with a liquid that keeps the surface temperature at 120°C. The aluminum cladding must not exceed a temperature of 200°C.

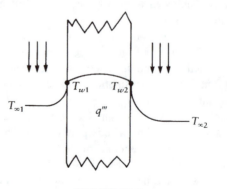

FIGURE P2.3

(a) Determine the maximum allowable value of the heat that can be generated in the core (that is, q''').

(b) Calculate the corresponding maximum temperature with the core if the core material has a thermal conductivity of 25 W/(m·K).

24. Construct a graph of the dimensionless temperature distribution in Equation 2.37b for a cylinder with internal heat generation.

25. An immersion heater is a device that can be submerged in a liquid to heat it. The heater is plugged into an electrical socket and in effect transforms electrical to thermal energy. One such heater is made of a 1/2-in metal rod 5 in long that draws 5 W. When the heater is submerged in 50°F oil, the heat-transfer coefficient between the rod and oil is 1.0 BTU/(hr·ft·²·R). Determine the maximum temperature within the rod and the exterior surface temperature. Take the thermal conductivity of the metal to be 7.0 BTU/(hr·ft·°R).

26. The development in Section 2.5 was for a hollow sphere. Derive the temperature distribution for a solid sphere with internal heat generation. Use the following boundary conditions:

$$\text{1. at } r = 0, \frac{dT}{dr} = 0$$

$$\text{2. at } r = R, T = T_w$$

2.9.4   ONE-DIMENSIONAL CONDUCTION IN SPHERICAL COORDINATES

27. An orange is placed in a refrigerator. At one instant in time, the air surrounding the orange is at 30°F, and the convection coefficient between the air and the orange surface is 0.2 BTU/(hr·ft²·°R). The exterior surface temperature of the orange is at 65°F. If the orange peel is 1/4 in thick, determine the temperature at the interface between the peel and the edible remainder of the fruit. Assume the orange outside diameter is 4 in, and the thermal conductivity of the orange is 0.25 BTU/(hr·ft·°R).

28. Derive an equation for heat transfer through a three-layered sphere.

29. Develop an equation for the critical thickness of insulation for a sphere.

2.9.5   CONTACT RESISTANCE

30. Rework Example 2.7 by using the following steps:

(a) Use the stainless steel data to determine the heat-flow rate.

(b) Using the results of part a, determine the thermal conductivity of the aluminum for the intervals between 3 and 4, 4 and 5, and 5 and 6.

Average them together to obtain a single value for the aluminum.

(c) Repeat part b to determine the thermal conductivity of the magnesium.

(d) Use Equation 2.50 to solve for R_{tc}. Select suitable values of L_A and L_B, keeping in mind that Equation 2.50 applies to Figure 2.19.

(e) Aside from errors in temperature measurement, determine the source of any discrepancy between the result obtained here and that in the example. (Without more details on the effectiveness of the insulation, it is difficult to tell which answer is more accurate.)

31. Figure P2.4 is a sketch of a device used to determine the thermal contact resistance. Temperature is measured with thermocouples placed in the metals. Distances between thermocouples are as follows:

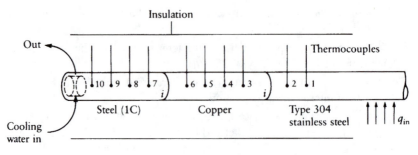

FIGURE P2.4

$$\Delta z_{12} = 1\frac{3}{8} \text{ in} \qquad \Delta z_{6i} = 1\frac{3}{4} \text{ in}$$

$$\Delta z_{2i} = 1\frac{1}{4} \text{ in} \qquad \Delta z_{i7} = 1 \text{ in}$$

$$\Delta z_{i3} = 1 \text{ in} \qquad \Delta z_{78} = 1\frac{3}{4} \text{ in}$$

$$\Delta z_{34} = 1\frac{3}{4} \text{ in} \qquad \Delta z_{89} = 1\frac{3}{4} \text{ in}$$

$$\Delta z_{45} = 1\frac{3}{4} \text{ in} \qquad \Delta z_{910} = 1\frac{3}{4} \text{ in}$$

$$\Delta z_{56} = 1\frac{3}{4} \text{ in}$$

For the following data determine the contact resistance between the copper and the steel. (Actual data obtained in the laboratory.)

$$
\begin{aligned}
T_1 &= 544.2°F & T_6 &= 232.0°F \\
T_2 &= 376.0°F & T_7 &= 201.6°F \\
T_3 &= 252.0°F & T_8 &= 168.6°F \\
T_4 &= 245.0°F & T_9 &= 137.0°F \\
T_5 &= 238.2°F & T_{10} &= 105.8°F
\end{aligned}
$$

2.9.6   PIN FINS

32. Figure 2.28 states that $mL = 1$ is a short fin and that $mL = 5$ is a long fin. Suppose an aluminum pin fin or rod 1 cm in diameter is in an environment where the convection coefficient $\bar{h}_c = 4.5 \text{ W/(m}^2\cdot\text{K)}$. Evaluate L for $mL = 1, 2, 3, 4,$ and 5.

33. Repeat Problem 32 for copper.

34. Repeat Problem 32 for glass (assume window pane).

35. Figure 2.29, for a pin fin, shows that the heat transferred cannot be substantially improved when mL is made larger than 3. For a pin fin with a 3/8-in diameter, determine L for $mL = 3$ for the following materials:

    (a) Aluminum

    (b) (1C) Steel

    (c) Stainless steel (Type 304).

    Use $\bar{h}_c = 0.8 \text{ BTU/(hr}\cdot\text{ft}^2\cdot°F)$.

36. A 1-1/2-quart aluminum saucepan has a handle that is riveted to its wall (see Figure P2.5). The handle itself is made of cast aluminum (with properties similar to those of Duralumin) and is to have attached a plastic grip that is comfortable to grasp. Before selecting a plastic, it is necessary to have information on the temperature of the aluminum handle. The aluminum handle can be considered as a rod 11 mm in diameter and 45 mm long. When being used over a stove burner, the ambient temperature is 44°C, and the temperature at the base of the handle reaches 110°C. For a convection coefficient $\bar{h}_c = 8 \text{ W/(m}^2 \cdot \text{K)}$, determine and graph the temperature profile. Determine also the heat transferred by the handle.

37. A number of bicycle frames are manufactured by using lugs. Lugs are sleeves into which tubes are placed and soldered or brazed in position. There is always discussion of how heating the tube affects the strength of the metal. Consider Figure P2.6, which shows a tube of 1 in outside diameter, heated at one end to 1700°F as in a brazing process. The tube wall thickness is 1/16 in, and the tube material is 1Mn steel. In an environment where the ambient temperature is 75°F and $\bar{h}_c$ is 2 BTU/(hr·ft^2·°R), determine and graph the temperature profile in the 2-ft-long tube. Also, determine the efficiency. (*Hint:* The pin-fin-solution equations can be used if the cross-sectional area is modified for an annular configuration. Other parameters can be considered unaffected.)

38. A chrome steel (1Cr) shaft rotates within bushings as shown in Figure P2.7. The shaft diameter is 12 mm. During times of high rotational speed, friction causes the shaft temperature at the bushings to reach 175°C. The shaft is in a warm engine room where the ambient temperature is 35°C, and the convection coefficient between the shaft and the air is 5 W/(m²·K). A pulley will be located on the shaft at the midpoint between the

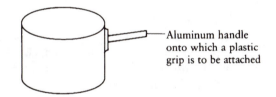

FIGURE P2.5

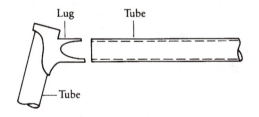

FIGURE P2.6

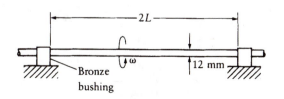

FIGURE P2.7

bushings, and for long belt life it has been determined that the shaft temperature (at the midpoint) cannot exceed 100°C. What minimum distance (2L) must exist between the bushings to satisfy this condition?

39. For a pin fin long enough such that the rod-end temperature equals that of the surrounding fluid, the solution for temperature is expressed in Equations 2.69. Verify that Equation 2.69a satisfies the differential equation and boundary conditions.

40. Derive Equation 2.69b.

41. Derive Equation 2.69c

42. Derive Equation 2.69d

43. The brazing operation is a metal-joining process that is done by heating a base metal to a high temperature and applying a brazing material to the heated joint. The heated base metal melts the brazing alloy, which fills the joint, and then solidifies when allowed to cool. The brazing metal is a copper alloy in the shape of a rod of 1/8 in diameter and 3 ft length. A schematic of the operation is provided in Figure P2.8. The end of the brazing rod reaches an average temperature of 1600°F. How far from the heated joint can the rod be grasped by hand—that is, where the temperature is 100°F? Take the rod to have the same properties as brass, the ambient temperature to be 80°F, and the convection coefficient to be 0.9 BTU/(hr·ft²·°R).

44. Verify that Equation 2.70a satisfies the appropriate differential equation and boundary conditions.

45. Derive Equation 2.70b.

46. Derive Equation 2.70c.

47. Derive Equation 2.70d.

2.9.7 STRAIGHT FINS

48. Figure P2.9 is a sketch of a device used to dissipate heat from a transistor. The device consists of a total of 20 straight rectangular fins. (Actually the thickness of each fin decreases with length; the cross section is trapezoidal. The taper is slight, however, and it is possible to approximate the cross section as a rectangle, with the thickness at $L/2$ being representative.) Each fin has a (mean) thickness of 0.75 mm, a length of 14 mm, and a width of 38 mm. The wall common to all fins reaches a temperature of 65°C—too hot to touch. Analyze one fin, assuming the ambient temperature to be 30°C and the convection coefficient to be 4 W/(m²·K). The entire device is (extruded) aluminum.

    (a) Determine the heat transferred by one fin.

    (b) Determine the fin efficiency.

    (c) Determine the fin effectiveness.

    (d) Determine whether or not the fin has the optimum thickness.

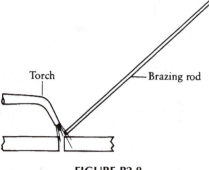

FIGURE P2.8

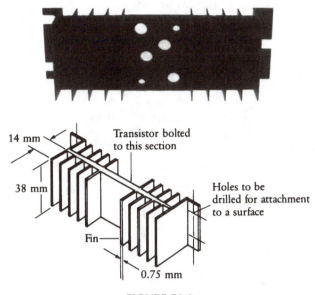

14 mm

38 mm

Transistor bolted
to this section

Holes to be
drilled for attachment
to a surface

Fin

0.75 mm

FIGURE P2.9

49. Repeat Problem 48 for the electronic cooling device made of aluminum and shown in Figure P2.10. Use $T_\infty = 75°F$, $T_w = 150°F$, and $\bar{h}_c = 15$ BTU/(hr·ft²·°R).

50. A straight rectangular fin is 4 in wide and 1 in long. It is made of wrought iron and is to be designed for an environment where $T_\infty = 60°F$ and $T_w = 600°F$; $\bar{h}_c$ can vary from 1 to 50 BTU/(hr·ft²·°R). The fin must transfer 30 BTU/hr. Determine its optimum thickness as a function of $\bar{h}_c$.

51. In a commercial air-handling system, fresh air is brought into a building through a duct as building air is exhausted. It is desired to recover some of the heat in the exhaust air and transfer the heat to the cooler incoming air. A heat exchanger is to be used for this purpose. The inlet air is to be passed over triangular fins within the exchanger. The exchanger is 1.2 m long, which is the fin width. Each fin is 4 cm wide at its base and is made of steel (1C). The inlet air velocity is such that $\bar{h}_c = 75$ W/(m²·K). The air temperature is 5°C, and the fin-root temperature is 25°C.

    (a) Determine corresponding optimum fin length.

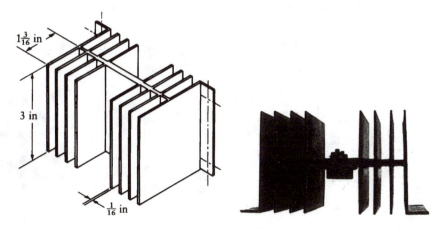

$1\frac{3}{16}$ in

3 in

$\frac{1}{16}$ in

FIGURE P2.10

(b) Determine the heat transferred by one fin.

(c) Find the fin efficiency.

(d) Calculate the fin effectiveness.

52. Construct a graph of δ_{opt}/L_c (vertical axis) vs. $\bar{h}_c L_c/k$ for straight rectangular, triangular, and inverse parabolic fins. Let $\bar{h}_c L_c/k$ vary from 0 to 10.

53. A straight fin is to be used in an application where $\bar{h}_c = 10$ BTU/(hr·ft²·°R). If the material to be used is aluminum and the base width of the fin is 1/4 in, determine the (corrected) length that will exist under optimum conditions for each of the following:

(a) A rectangular profile.

(b) A triangular profile.

(c) An inverse parabolic profile.

(d) Calculate also the profile areas (which give a relative indication of the volume of material required for each fin).

54. The straight-fin-efficiency graph of Figure 2.36 has

$$L_c^{3/2}(\bar{h}_c/kA_p)^{1/2}$$

as its horizontal axis label. Evaluate this term for optimum dimensions for each of the following types of fins:

(a) Rectangular profile

(b) Triangular profile

(c) Inverse parabolic profile

55. Figure P2.11 shows a transistor-cooling device having longitudinal fins. All 15 fins are identical; each is 1 mm thick, 6 mm wide, and 1.5 mm long. As indicated, the attached fins transfer heat away from the transistor to keep it cool enough so that its electrical properties do not change drastically. The fin-root temperature is 50°C, and the ambient air temperature is 28°C. The convection coefficient can vary depending on whether or not a cooling fan is used. Determine the variation of fin efficiency with the convection coefficient over the range of 1 to 20 W/(m²·K). Determine the amount of heat that can be transferred by the 15 fins for $\bar{h}_c = 10$ W/(m²·K).

56. Determine the effectiveness of the proposed fin in Example 2.10 by using the efficiency calculations there. What can be concluded about using the fins?

57. Figure P2.12 is a sketch of a device used to cool transistors. The device is made of stamped aluminum. It has 10 straight rectangular fins. Each fin is 3/16 in thick, 31/32 in

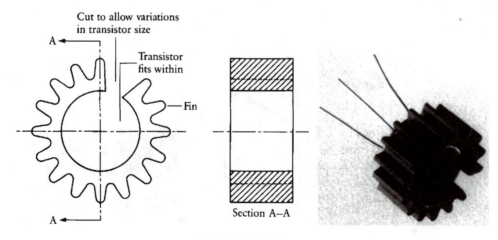

FIGURE P2.11

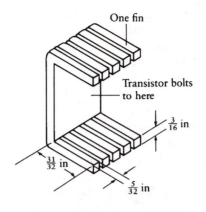

FIGURE P2.12

long, and 5/32 in wide. A transistor would be bolted to the flat surface that connects each bank of fins. The transistor generates enough heat to raise the fin-root temperature to 140°F. The ambient temperature is 80°F. The circuit board is located in a cabinet with other heat sources. The convection coefficient between all fins and the surrounding air is 0.85 BTU/(hr·ft²·°R). Determine the fin efficiency and effectiveness, and the heat transferred from all 10 fins.

58. The fin profile area A_p was introduced in the development that led to an expression for optimum fin thickness. Also, the efficiency curves for straight fins contain A_p in the label for the horizontal axis:

$$L_c^{3/2}(\bar{h}_c/kA_p)^{1/2}$$

Express this label in terms of $m = (\bar{h}_c/kA_p)^{1/2}$ for a rectangular fin.

2.9.8 CIRCULAR FINS

59. Figure P2.13 is a sketch of 1 nominal type K copper tube that has fins attached. The tube is designed to convey compressed air. The fins consist of a metal strip spirally

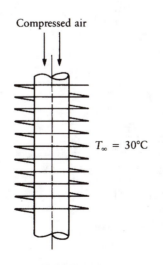

FIGURE P2.13

wound around the tubing. (See also Figure 2.23.) Each fin is made of sheet metal (1C steel), has a mean thickness of 0.063 5 cm, and a fin tip-to-tip diameter of 5.56 cm. The convection coefficient between the surrounding ambient air and the fins is 4 W/(m²·K). The ambient air is at 30°C. At one section, the hot air within the tube causes the tube wall at the fin root to reach 95°C.

(a) Determine the amount of heat transferred by one fin.

(b) Determine the fin effectiveness.

(c) Determine the fin efficiency.

60. Figure P2.14 shows a finned device used to transfer heat away from a transistor, which fits into the internal cavity. For the dimensions (in inches) shown in the figure, determine how much heat is transferred away by each individual fin. Find the total heat transferred by the device. Take the ambient temperature to be 95°F, the fin-root temperature to be 170°F, and the convection coefficient as 1.05 BTU/(hr·ft²·°R). The device is made of (cold-rolled) aluminum.

61. Figure 2.40 is a cross-sectional view of a cylinder used in an air compressor. It contains nine fins, 3 mm thick, separated from each other by 12.7 mm. Suppose that the distance between fins is reduced (or increased) so that the number of fins can be increased (or decreased) while the total height and the fin dimensions are maintained constant. Repeat the calculations in Example 2.11 for each of the following:

(a) 7 fins

(b) 11 fins

(c) 13 fins

(d) 15 fins

62. Figure P2.15 is a sketch of a device used to keep a transistor from overheating. A transistor fits into the cavity, and the 4 identical fins transfer heat away. The material is cold-rolled aluminum. When in use, the fin root reaches a temperature of 150°F. The ambient temperature is 100°F, and the convection coefficient is 15 BTU/(hr·ft²·°R).

(a) Determine the amount of heat transferred from the device to the air.

(b) Determine the fin effectiveness.

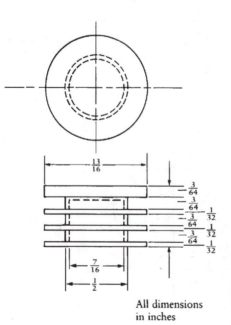

All dimensions
in inches

FIGURE P2.14

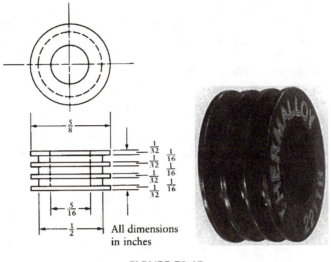

FIGURE P2.15

63. The second cylinder of the two-stage compressor described in Example 2.11 is shown schematically in Figure P2.16. Each of the nine fins is 0.32 cm thick and 2.5 cm long. The cylinder is made of cast iron. During operation the root-wall temperature reaches 260°C, and the ambient air temperature is 27°C. The convection coefficient is 15 W/(m²·K).

   (a) Determine the heat transferred by the cylinder to the air.

   (b) Calculate the fin effectiveness.

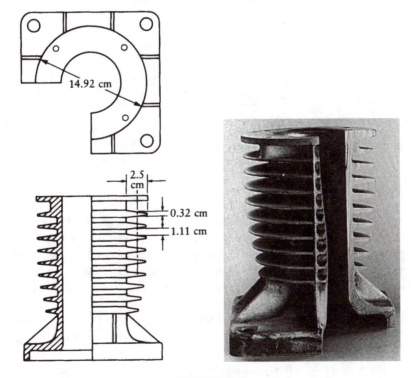

FIGURE P2.16

64. In Problem 63, the fin thickness is 0.32 cm, and the fin spacing is 1.11 cm. Suppose that the fin thickness is increased (or decreased) but that the number of fins is constant, as is the spacing plus fin thickness (1.11 + 0.32 = 1.43 cm = a constant). Repeat Problem 63 for fin thicknesses of 0.16, 0.32, 0.48, and 0.64 cm. Tabulate heat transferred vs. fin thickness and fin effectiveness vs. thickness.

65. Figure P2.17 is a sketch of the cross section of a Volkswagen engine cylinder. It consists of 20 circular fins of rectangular cross section that vary in length from cylinder top to midpoint. For the dimensions shown, determine the heat transferred by the cylinder to the air. Analyze each fin individually and add the results. The cylinder is made of hardened steel (use 1.5C carbon steel properties). the root temperature of each fin varies with height and is given in the following chart:

Fin#	Root temperature, °F
1 (top)	1800
2	1700
3	1650
4	1600
5	1550
6	1500
7	1400
8	1300
9	1200
10	1175
11	1150
12	1150
13	1150
14	1150
15	1150
16	1150
17	1100
18	1000
19	900
20 (bottom)	875

The ambient air temperature is 110°F, and the convection coefficient is 20 BTU/(hr·ft²·°R). All fins are 2 mm thick, and the distance between adjacent fins is 3 mm.

2.10 PROJECT PROBLEMS

1. The information on Fiberglas® insulation provided in Figure P2.18 is taken from an advertising circular. The drawing is of an insulated tube with important features of the Fiberglas insulation shown and explained. The chart lists available pipe-insulation thicknesses and the corresponding iron pipe sizes. The graph shows thermal conductivity vs. temperature for the insulation. Note that the units appearing in this excerpt from the circular (published August 1983) are English Engineering Units.

   (a) What is the significance of a "vapor barrier" as described in the drawing?

   (b) The 1/2-in insulation thickness is available for iron pipe sizes that range from 1/2 to 6 nominal. If this insulation is applied to a pipe that is exposed to air at some temperature, heat flows radially through the insulation to the air (or vice versa). For whichever insulation and corresponding nominal pipe sizes are assigned, determine

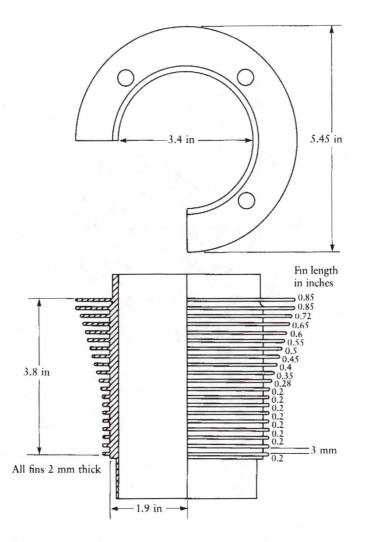

FIGURE P2.17

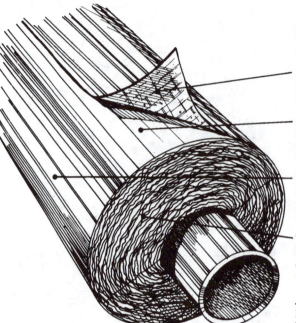

Open the insulation, pull away a release strip, and close the insulation over the pipe.

A quick rub of the hand presses an adhesive strip on the jacket lap ...

against a second strip of adhesive on the jacket, creating a permanent, positive vapor seal.

The tough, reinforced all-service jacket meets requirements of model building codes.

The heavy density ASJ/SSL-II Fiberglas insulation comes in sizes to fit piping up to 15" IPS.

Just peel and seal—no staples or mastic—no hassle with Fiberglas ASJ/SSL-II pipe insulation.

Insulation thickness, in.	Iron pipe size range
$\frac{1}{2}''$	$\frac{1}{2}''$ to 6"
1"	$\frac{1}{2}''$ to 15"
$1\frac{1}{2}''$	$\frac{1}{2}''$ to 14"
2"	$\frac{1}{2}''$ to 12"
$2\frac{1}{2}''$	$\frac{1}{2}''$ to 11"
3"	$\frac{1}{2}''$ to 10"
$3\frac{1}{2}''$	$\frac{1}{2}''$ to 9"
4"	$\frac{1}{2}''$ to 8"
$4\frac{1}{2}''$	$\frac{1}{2}''$ to $3\frac{1}{2}''$
5"	$\frac{1}{2}''$ to $2\frac{1}{2}''$

Thermal conductivity (ASTM C335-69):

Courtesy of Owens-Corning Fiberglas®.

FIGURE P2.18

the allowable limit or limits on the convection coefficient between the insulation and the ambient air.

(c) The thermal-conductivity graph refers to a test or standard labeled ASTM C335-69. Report on the method of measurement described in this standard.

(d) What is the significance of the unit used for thermal conductivity in the graph? We have been using BTU/(hr·ft·°R), but the graph shows BTU·in/(hr·ft^2·°F).

2. Consider the composite* shown in Figure P2.19a, in which heat is transferred through the materials from one fluid at $T_{\infty 1}$ to another at $T_{\infty 2}$. According to the information in this chapter and countless other sources, a suitable thermal circuit for determining the steady-state heat flow would be that shown in Figure P2.19b. Recent work performed in this area, however, indicates that the circuit of Figure P2.19b is inappropriate and that the

---

* Problem and schematics provided by Prof. E.P. Russo.

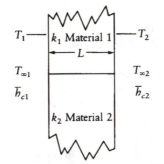

(a) Composite; materials in parallel

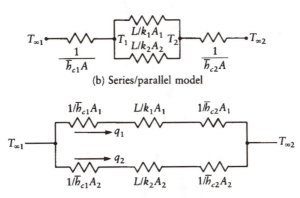

(b) Series/parallel model

(c) Parallel/series model

FIGURE P2.19

circuit of Figure P2.19c gives a better description. Both circuits are examined and compared in this problem.

(a) Use elementary circuit theory applied to parallel resistors and show that $1/\bar{h}_{c1}A_1$ and $1/\bar{h}_{c1}A_2$, when combined, result in an equivalent resistance of value $1/\bar{h}_{c1}A$, where $A_1 + A_2 = A$, the total area normal to the heat flow. Involved in this combination is the assumption that the entire left surface is isothermal at temperature T_1. Performing a similar analysis for $1/\bar{h}_{c2}A_1$ and $1/\bar{h}_{c2}A_2$, it is clear that the circuit in P2.19b is a special case of the one in P2.19a.

(b) If the entire left surface is isothermal at T_1, and the right surface is isothermal at T_2, then show that the circuit in b is correct only if $k_1 = k_2$ by performing the following steps:

  (1) Using the circuit in Figure P2.19c, write an equation for q_1/A_1 in terms of $\bar{h}_{c1}$, $T_{\infty1}$, and T_1.

  (2) Write an equation for q_1/A_1 in terms of k_1, T_1, and T_2.

  (3) Set the two equations from the above steps equal to one another and show that

$$k_1 = \bar{h}_{c1}L\frac{T_{\infty1} - T_1}{T_1 - T_2}$$

  (4) Perform steps (1) and (2) for q_2/A_2 and show that

$$k_2 = \bar{h}_{c1}L\frac{T_{\infty1} - T_1}{T_1 - T_2}$$

(5) Comparing the equations for k_1 and k_2, what can be concluded about the assumption that the material surfaces are isothermal?

(c) Write an equation for the total heat flow through the circuit in Figure P2.19b. Repeat for the circuit in Figure P2.19c.

(d) Use a circuit similar to the one shown in Figure P2.19c to solve the problem illustrated in Figure 2.5. Determine the outside surface temperatures for all heat-flow paths, and the heat-flow rate through each path. Calculate percentage errors between this method and the one in the text.

(e) Repeat part (d) for Problem 2.8.

(f) Formulate the parallel/series circuit for the system illustrated in Figure 2.4.

3 Steady-State Conduction in Multiple Dimensions

CONTENTS

3.1   INTRODUCTION

One-dimensional conduction was introduced in Chapter 2. For the one-dimensional problem, the general equation was derived. The variables on which temperature depended were identified, and appropriate terms were cancelled from the general equation. The boundary conditions were written, and the equation was then solved.

In this chapter, where we will consider cases in which temperature depends on more than one space variable, it is expected that the same technique will be followed. For a given geometry we will begin with the general equation, cancel terms appropriately, and obtain a solution. As we will see, this analytical method is conveniently applied only to very simple geometries. For complex shapes and systems, it will be necessary to rely on alternative methods. These include graphical, shape-factor, and numerical techniques.

3.2   GENERAL CONDUCTION EQUATION

Before proceeding with the analytical method, let us first derive an equation for conduction of heat in a material. We will do this by performing a heat balance on an elemental volume through which heat is being transferred by conduction. We then combine the heat-balance equation with Fourier's law of Heat Conduction and obtain a relation for temperature within the material. (The three-dimensional equations are provided in the summary section of Chapter 2.)

3.2.1   CARTESIAN COORDINATES

Consider the volume element shown in Figure 3.1. The element has dimensions of $dx \times dy \times dz$. Shown are heat flows into and out of the element, and these are by conduction. As was done in Chapter 2, we write

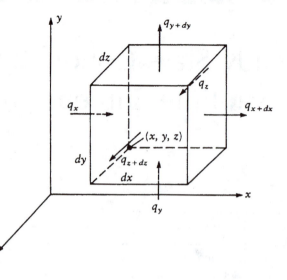

FIGURE 3.1 A control volume for deriving the three-dimensional conduction equation in Cartesian coordinates.

$$
\left(\begin{array}{c} \text{Rate of energy} \\ \text{conducted into} \\ \text{control volume} \end{array}\right) + \left(\begin{array}{c} \text{Rate of energy} \\ \text{generated inside} \\ \text{control volume} \end{array}\right) = \left(\begin{array}{c} \text{Rate of energy} \\ \text{conducted out of} \\ \text{control volume} \end{array}\right) + \left(\begin{array}{c} \text{Rate of energy} \\ \text{stored inside} \\ \text{control volume} \end{array}\right)
$$

The equation becomes

$$
q_x + q_y + q_z + q'''dx\,dy\,dz = \left(q_x + \frac{\partial q_x}{\partial x}dx\right) + \left(q_y + \frac{\partial q_y}{\partial y}dy\right) + \left(q_z + \frac{\partial q_z}{\partial z}dz\right) + \rho\,dx\,dy\,dz\frac{\partial u}{\partial t}
$$

Simplifying yields

$$
-\frac{\partial q_x}{\partial x}dx - \frac{\partial q_y}{\partial y}dy - \frac{\partial q_z}{\partial z}dz + q'''\,dx\,dy\,dz = \rho\,dx\,dy\,dz\frac{\partial u}{\partial t} \tag{3.1}
$$

Fourier's Law of Heat Conduction can be written in each direction as

$$
q_x = -kA\frac{\partial T}{\partial x} = -k(dy\,dz)\frac{\partial T}{\partial x} \tag{3.2a}
$$

$$
q_y = -k(dx\,dz)\frac{\partial T}{\partial y} \tag{3.2b}
$$

$$
q_z = -k(dx\,dy)\frac{\partial T}{\partial z} \tag{3.2c}
$$

Assuming constant thermal conductivity, substituting into Equation 3.1 and simplifying gives

$$
k\left(\frac{\partial^2 T}{\partial x^2} + \frac{\partial^2 T}{\partial y^2} + \frac{\partial^2 T}{\partial z^2}\right) + q''' = \rho c\frac{\partial T}{\partial t} \tag{3.3}
$$

Equation 3.3 is the general three-dimensional unsteady-conduction equation in Cartesian coordinates. The first three terms on the left-hand side of Equation 3.3 represent the net rate of heat conducted into the control volume per unit volume (net ratio into = in minus out). The q''' term is heat generated per unit volume inside the control volume. The right-hand side of Equation 3.3 is the rate of increase of internal energy of the control volume per unit volume. Equation 3.3 is usually written as

$$\frac{\partial^2 T}{\partial x^2} + \frac{\partial^2 T}{\partial y^2} + \frac{\partial^2 T}{\partial z^2} + \frac{q'''}{k} = \frac{1}{\alpha}\frac{\partial T}{\partial t} \tag{3.4}$$

In vector notation, the above equation is written as

$$\nabla^2 T + \frac{q'''}{k} = \frac{1}{\alpha}\frac{\partial T}{\partial t} \tag{3.5}$$

The operator ∇^2 is called the Laplacian; in Cartesian coordinates, this can be determined to be

$$\nabla^2 = \frac{\partial^2}{\partial x^2} + \frac{\partial^2}{\partial y^2} + \frac{\partial^2}{\partial z^2} \tag{3.6}$$

3.3 ANALYTICAL METHOD OF SOLUTION

In this section, we analytically solve an example of a two-dimensional conduction problem. Consider a rectangular plate of dimensions W by H as shown in Figure 3.2. Also shown are the x, y axes imposed. The plate could represent the cross section of a column within a structure. The left, lower, and right sides are maintained at some temperature T_1. The upper surface has some elevated temperature distribution impressed upon it. The temperature distribution could be a constant temperature, which then causes a problem at the upper corners where each corner is to exist at two temperatures. While mathematically this is not too difficult to accept, a more physically appealing temperature distribution is

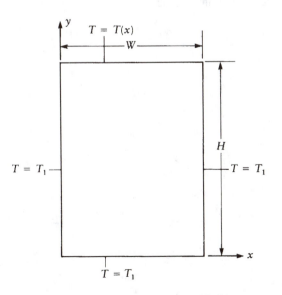

FIGURE 3.2 A rectangular plate with boundary conditions specified.

$$T = T_1 + T_2 \sin[\pi(x/W)]$$

This distribution is graphed in Figure 3.3. Temperature varies from T_1 at the corners to a maximum of $T_1 + T_2$ in the center of the upper surface of the plate.

Because a temperature difference exists across the plate, heat will flow from the upper surface to the other surfaces. Our objective will be to determine the temperature distribution within the plate, from which we will find the heat-flow rate.

For a rectangular geometry in which thermal conductivity is constant, Equation 3.4 applies:

$$\frac{\partial^2 T}{\partial x^2} + \frac{\partial^2 T}{\partial y^2} + \frac{\partial^2 T}{\partial z^2} + \frac{q'''}{k} = \frac{1}{\alpha}\frac{\partial T}{\partial t} \tag{3.4}$$

Temperature varies only with x and y, and for no internal heat generation, Equation 3.4 becomes

$$\frac{\partial^2 T}{\partial x^2} + \frac{\partial^2 T}{\partial y^2} = 0 \tag{3.7}$$

The boundary conditions are

1. $\qquad\qquad\qquad x = 0,\ T = T_1$ $\hfill$ (3.8a)

2. $\qquad\qquad\qquad x = W,\ T = T_1$ $\hfill$ (3.8b)

3. $\qquad\qquad\qquad y = 0,\ T = T_1$ $\hfill$ (3.8c)

4. $\qquad\qquad\qquad y = H,\ T = T_1 + T_2 \sin[\pi(x/W)]$ $\hfill$ (3.8d)

Equation 3.7 is a partial differential equation that can be solved by the separation-of-variables method. Before proceeding, a change of variables can be made that simplifies the algebra involved considerably. Otherwise, we will get to the point where evaluating constants of integration becomes quite tedious. Simplifying the boundary conditions will make the task easier. Inspection of the boundary conditions suggests the following substitution:

$$\theta = T - T_1$$

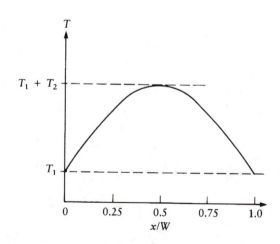

FIGURE 3.3 Graph of the temperature distribution imposed on the upper surface of the plate in Figure 3.2.

The differential equation (3.7) in terms of θ becomes

$$\frac{\partial^2 \theta}{\partial x^2} + \frac{\partial^2 \theta}{\partial y^2} = 0 \qquad (3.9)$$

The boundary conditions become

1. $\qquad\qquad\qquad x = 0,\ \theta = 0 \qquad\qquad\qquad (3.10a)$

2. $\qquad\qquad\qquad x = W,\ \theta = 0 \qquad\qquad\qquad (3.10b)$

3. $\qquad\qquad\qquad y = 0,\ \theta = 0 \qquad\qquad\qquad (3.10c)$

4. $\qquad\qquad\qquad y = H,\ \theta = T_2 \sin\left[\pi(x/W)\right] \qquad\qquad (3.10d)$

The separation-of-variables method involves assuming a solution of the following form:

$$\theta(x, y) = X(x)Y(y) \qquad (3.11)$$

The function X depends only on x, and Y depends only on y. Evaluating the derivatives of θ gives

$$\frac{\partial \theta}{\partial x} = Y\frac{dX}{dx}$$

$$\frac{\partial^2 \theta}{\partial x^2} = Y\frac{d^2 X}{dx^2}$$

$$\frac{\partial \theta}{\partial y} = X\frac{dY}{dy}$$

$$\frac{\partial^2 \theta}{\partial y^2} = X\frac{d^2 Y}{dy^2}$$

Substituting into Equation 3.9 yields

$$Y\frac{d^2 X}{dx^2} + X\frac{d^2 Y}{dy^2} = 0$$

or

$$-\frac{1}{X}\frac{d^2 X}{dx^2} = +\frac{1}{Y}\frac{d^2 Y}{dy^2} \qquad (3.12)$$

The left-hand side is a function only of x, and the right-hand side is a function only of y. If these functional dependencies exist and satisfy Equation 3.12, then each side must equal a constant, known as the separation constant. Equation 3.12 can now be rewritten as two ordinary differential equations:

$$\frac{d^2 X}{dx^2} + \lambda^2 X = 0 \qquad (3.13)$$

$$\frac{d^2Y}{dy^2} - \lambda^2 Y = 0 \tag{3.14}$$

where λ^2 is the separation constant* whose value is yet to be determined. The solution forms of Equations 3.13 and 3.14 depend on whether λ^2 is negative, zero, or positive. So we consider the three cases as follows:

Case 1

If $\lambda^2 < 0$, then the general solutions to 3.13 and 3.14 are†

$$X = C_1 \cosh(\lambda x) + C_2 \sinh(\lambda x)$$

$$Y = C_3 \cos(\lambda y) + C_4 \sin(\lambda y)$$

where the C_i variables are constants of integration. The temperature distribution is the product of X and Y. The 4th boundary condition (Equation 3.10d) cannot be satisfied by the resulting equation form. So we reject the possibility of a negative λ^2.

Case 2

If $\lambda^2 = 0$, then the general solutions to Equations 3.13 and 3.14 become

$$X = C_1 x + C_2$$

$$T = C_3 y + C_4$$

Recognizing again that the 4th boundary condition cannot be satisfied, we reject the possibility of $\lambda^2 = 0$.

Case 3

If $\lambda^2 > 0$, the general solutions to Equations 3.13 and 3.14 become

$$X = C_1 \cos(\lambda x) + C_2 \sin(\lambda x)$$

$$Y = C_3 \cosh(\lambda y) + C_4 \sinh(\lambda y)$$

This solution form can satisfy all the boundary conditions. The temperature distribution becomes

$$\theta = T - T_1 = [C_1 \cos(\lambda x) + C_2 \sin(\lambda x)] \times [C_3 \cosh(\lambda y) + C_4 \sinh(\lambda_y)] \tag{3.15}$$

After applying the first boundary condition ($x = 0$, $\theta = 0$), we get

B. C. 1 $\qquad\qquad\qquad 0 = C_1[C_3 \cosh(\lambda y) + C_4 \sinh(\lambda y)]$

We conclude that $C_1 = 0$. Applying the second boundary condition ($x = W$, $\theta = 0$) gives

---

* The square of the term is used, because experience has shown that the square root of the separation constant will appear in the solution. λ is easier to write than $\sqrt{\lambda}$.

† Alternatively, $X = C_1 \exp(\lambda x) + C_2 \exp(-\lambda x)$. However, the exponential form is more suitable for infinite geometries Because our region is finite, the hyperbolic functions are selected instead.

B.C. 2
$$0 = C_2 \sin(\lambda W)[C_3 \cosh(\lambda y) + C_4 \sinh(\lambda y)]$$

or
$$0 = \sin(\lambda W)[C_{23} \cosh(\lambda y) + C_{24} \sinh(\lambda y)]$$

where C_{23} and C_{24} are new constants formed from the constants of integration. We conclude from the above equation that

$$\sin(\lambda W) = 0 \tag{3.16}$$

This equation is satisfied if

$$\lambda W = 0$$
(This gives the trivial solution $\theta = 0$, $T = T_1$ everywhere.)

or

$$\lambda W = \pi$$

or

$$\lambda W = 2\pi, \text{ etc.}$$

In general, then, λW must equal an integral multiple of π. Therefore,

$$\lambda_n = \frac{n\pi}{W}$$

where n can be 1 or 2 or 3, etc. The temperature distribution becomes

$$\theta = T - T_1 = \sin(\lambda_n x)[C_{23} \cos(\lambda_n y) + C_{24} \sinh(\lambda_n y)] \tag{3.17}$$

Because λ can take on any number of values, a different solution will result for each consecutive integer n, and associated with each solution will be a different integration constant. The temperature distribution is therefore more appropriately written as a summation of all possible solutions for all positive integral values of n:

$$\theta = T - T_1 = \sum_{n=1}^{\infty} \sin(\lambda_n x) \times C_{n23} \cosh(\lambda_n y) + C_{n24} \sinh(\lambda_n y) \tag{3.18}$$

Applying the third boundary condition ($y = 0$, $\theta = 0$), Equation 3.18 becomes

$$0 = \sum_{n=1}^{\infty} \sin(\lambda_n x)(C_{n23} + 0)$$

from which we conclude that $C_{n23} = 0$. We are thus left with

$$\theta = T - T_1 = \sum_{n=1}^{\infty} C_{n24} \sin(\lambda_n x) \sinh(\lambda_n y) \tag{3.19}$$

Applying the fourth boundary condition, $y = H$, $\theta = T_2 \sin[\pi(x/W)]$, we get

$$T_2 \sin(\pi x/W) = \sum_{n=1}^{\infty} C_n \sin(\lambda_n x) \sinh(\lambda_n H) \qquad (3.20a)$$

where the 24 subscript on the coefficient has been dropped to simplify the notation. To facilitate the evaluation of the constant C_n, we now expand the right-hand side of the series to obtain

$$\begin{aligned} T_2 \sin(\pi x/W) = \; & C_1 \sin(\pi x/W)\sin(\pi H/W) \\ & + C_2 \sin(2\pi x/W)\sinh(2\pi H/W) \\ & + C_3 \sin(3\pi x/W)\sinh(3\pi H/W) + \ldots \end{aligned} \qquad (3.20b)$$

We conclude that all constants except C_1 are zero. Furthermore,

$$C_1 = \frac{T_2}{\sinh(\pi H/W)}$$

and with $n = 1$ only, $\lambda_1 = \pi/W$. Substituting into Equation 3.19 gives the solution as

$$\theta = T - T_1 = T_2 \frac{\sin(\pi x/W)\sinh(\pi y/W)}{\sinh(\pi H/W)} \qquad (3.21)$$

A graph of this equation will allow us to examine this two-dimensional temperature profile in detail. It is conventional practice to draw constant temperature lines in the region. Such lines are also known as *isothermal lines* or *isotherms*. By way of illustration, let $T_1 = 100°C$, $T_2 = 200°C$, and $H = 1.5\ W$ in Equation 3.21. The equation becomes

$$T = T_1 + T_2 \sin(\pi x/W)\frac{\sinh(\pi y/W)}{\sinh(\pi H/W)}$$

$$= 100 + 200\sin(\pi x/W)\frac{\sinh[1.5\pi(y/H)]}{\sinh(1.5\pi)}$$

or

$$T = 100 + 3.594\sin(\pi x/W)\sinh[1.5\pi(y/H)] \qquad (3.22)$$

A value of temperature T is selected, and an equation relating x/W to y/H results, which, when plotted, gives an isotherm. A graph of Equation 3.22 for $T = 100, 125, 150, 175,$ and $250°C$ is provided in Figure 3.4. As indicated, the isotherms are a family of lines, each of which satisfies the temperature profile, Equation 3.22. All isotherms are symmetric about $x/W = 0.5$.

After obtaining the temperature profile, it is now possible to differentiate to get the heat-flow rate. Fourier's Law is

$$q_x'' = -k\frac{\partial T}{\partial x}$$

and

$$q_y'' = -k\frac{\partial T}{\partial y}$$

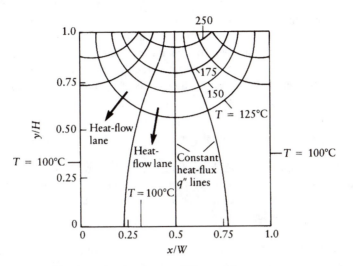

FIGURE 3.4 Isotherms and adiabats for the system of Figure 3.2.

The magnitude of the heat-flux vector then is

$$q'' = \sqrt{q_x''^2 + q_y''^2} = +k\sqrt{(\partial T/\partial x)^2 + (\partial T/\partial y)^2} \qquad (3.23)$$

Differentiating Equation 3.21 brings

$$T = T_1 + T_2 \sin(\pi x/W)\sinh(\pi y/W)/\sinh(\pi H/W)$$

$$\frac{\partial T}{\partial x} = \frac{T_2\pi}{W}\cos(\pi x/W)\frac{\sinh(\pi y/W)}{\sinh(\pi H/W)} \qquad (3.24)$$

and

$$\frac{\partial T}{\partial y} = \frac{T_2\pi}{W}\sin(\pi x/W)\frac{\cosh(\pi y/W)}{\sinh(\pi H/W)} \qquad (3.25)$$

Squaring these equations, substituting into Equation 3.23, and simplifying gives

$$q'' = \frac{kT_2\pi}{W\sinh(\pi H/W)}[\sinh^2(\pi y/W) + \sin^2(\pi x/W)]^{1/2} \qquad (3.26)$$

The heat flux thus depends on x and y. Given temperature T_2, thermal conductivity k, and plate dimensions H and W, constant values of q'' can be selected to reduce Equation 3.26 to a family of equations relating x/W to y/H. Each curve represents a line of constant heat flux called an *adiabatic line* or an *adiabat*. Adiabats are shown in Figure 3.4. Note that the adiabats are symmetric about $x/W = 0.5$, and that no heat flows across an adiabat. Another feature of note, although not proven here, is that the adiabatic and isothermal lines are everywhere perpendicular to each other.

The region between two adjacent constant-heat-flux lines is defined as a *heat-flow lane* (see Figure 3.4). The adiabats and isotherms plotted together (as in Figure 3.4) make up what is known as a *heat-flow net*. To calculate the total heat flow from the upper surface of the plate to the other surfaces, the heat flowing through each individual heat-flow lane is evaluated, then they are added together.

The above development was for a two-dimensional heat-flow problem in which a very simple geometry was analyzed. The derived temperature distribution, however, is quite complex in form and tedious to evaluate. The heat flux is also difficult to obtain. For different geometries, such as a buried pipeline or a round duct in a square chimney, the thought of using the analytical technique becomes unpopular. Fortunately, alternative methods exist. A sampling of several alternative methods now follows.

3.4 GRAPHICAL METHOD OF SOLUTION

The graphical method of solution to two-dimensional conduction problems is a versatile technique that can be applied to some very irregular geometries. The method involves the development of a *flow net* (also called a *flux plot*) by freehand sketching. The actual sketching is aided by paying strict attention to some key concepts:

1. Heat-flow lines and isotherms are perpendicular to each other.
2. A heat-flow line is parallel or coincident to an insulated surface or line of symmetry.
3. An isotherm will intersect an insulated surface or line of symmetry at a right angle.

Making sure that the above concepts are incorporated into the sketch is a matter of trial and error. Once the flow net or flux plot is completed, the heat transferred from one boundary to another can be approximated. A closed form equation for the temperature distribution is not found—only an estimate of the heat transferred. The accuracy of the estimate depends on how fine the mesh or net is and on how well it is drawn. It should be noted that before computers and calculators were developed, analogical methods such as the graphical technique were the only alternatives to the analytical method.

The graphical method will be illustrated by an example. Consider the chimney shown in cross section in Figure 3.5. The chimney cross section is rectangular, with the longer side having a wall thickness double that of the shorter side. At this location in the chimney the inside surface temperature is 200°F, and the outside surface temperature is 30°F. We wish to determine the heat transferred through the chimney walls. (One way to solve this problem is to assume one-dimensional heat flow through each of the four sides and neglect or estimate the effect of the corners.)

Before constructing the flow net for the entire cross section, we first identify lines of symmetry. As evident from Figure 3.5, the centerlines are indeed lines of symmetry. The advantage now is that we need work only with one quarter of the cross section, as shown in Figure 3.6. We next draw the one-quarter section to a scale that is convenient with which to work. The lines of symmetry

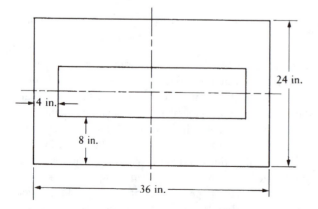

FIGURE 3.5 Cross section of a rectangular chimney.

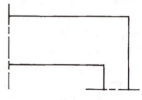

FIGURE 3.6 One quarter of the chimney of Figure 3.5.

will have no heat crossing them, so they behave as insulated boundaries. Figure 3.7a shows heat-flow lines that were drawn in freehand as a first trial. The lines intersect the inner surface at approximately equal intervals. Also, each of these lines intersects the surfaces at right angles. Figure 3.7b shows the completed flow net with constant temperature lines drawn in. Each isotherm intersects the constant-heat boundaries and the heat-flow lines at right angles. Note that Figure 3.7b is only one solution and that another analyst will probably draw up a totally different flow net.

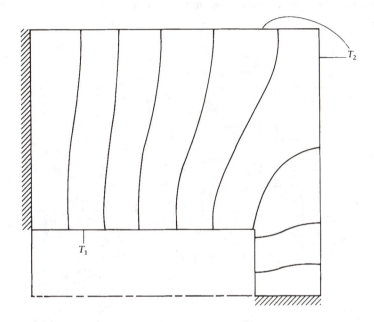

FIGURE 3.7a Scale drawing of the chimney quarter cross section and heat-flow lines.

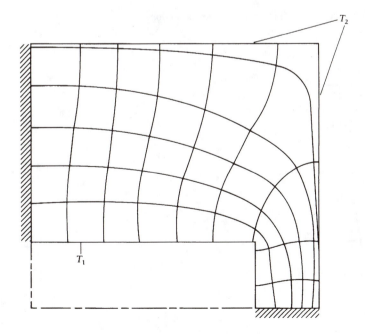

FIGURE 3.7b Flow net for the chimney quarter cross section.

Figure 3.8 is another solution. If the rules of orthogonality are observed during the freehand construction of both flow nets, then each solution will provide a suitable *estimate* of the heat flow through the walls. In developing the flow net for a given system or region, the orthogonality condition (among other requirements) must be satisfied. Producing an acceptable solution is largely a matter of trial and error, which, in turn, is a function of the experience and patience of the analyst.

For purposes of developing an equation to calculate the heat transferred, consider the third flow lane from the left in Figure 3.7b, which is redrawn in Figure 3.9. The regions labeled a through e are known as *curvilinear squares,* a term that implies that the sides of each four-sided figure are

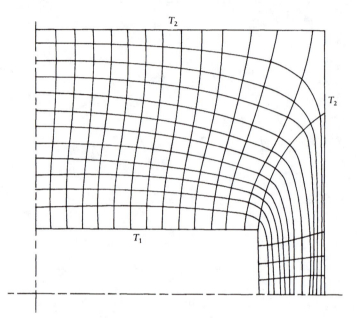

FIGURE 3.8 Flow net for the chimney quarter cross section.

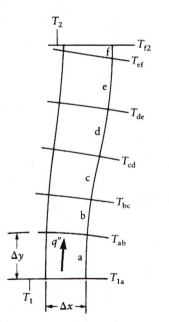

FIGURE 3.9 One heat-flow lane from Figure 3.7b.

approximately equal and curved. In most cases, the freehand sketch will not have a flow net composed entirely of perfect curvilinear squares. In Figure 3.8, the "squares" are quite rectangular in the corner. In a majority of cases, the number of heat-flow lanes or squares will not be an integer. Consequently, the relative size of a fraction of a heat-flow lane or of a square will have to be estimated. If the heat-flow lanes are made to be integral, then the number of squares per lane will probably be a fraction in the final picture. The converse is also true.

Returning now to Figure 3.9, the region labeled f is a fraction of a curvilinear square. It arises as a consequence of trying to make regions a through e square. We estimate that region f is 1/4 of a square. Thus, heat flowing in this flow lane from the boundary at T_1 to the boundary at T_2 passes through 5-1/4 curvilinear squares. The heat flow is found by assuming one-dimensional flow through the lane; that is,

$$q_3 = -kA\frac{\partial T}{\partial y}$$

where the 3 subscript denotes lane 3. The area perpendicular to the heat-flow direction is $\Delta x(L)$, where L is the depth into the page. Therefore,

$$q_3 = k\Delta x L\frac{(T_{ab} - T_{1a})}{\Delta y} = k\Delta x L\frac{(T_{bc} - T_{ab})}{\Delta y} = \dots \tag{3.27}$$

These heat-flow equations reflect the assumption that every square is Δx by Δy. Another expression for the total heat flow from the inside surface to the outside surface along lane 3 is

$$q_3 = k\Delta x L\frac{(T_1 - T_2)}{N\Delta y} = k\frac{(T_1 - T_2)L}{N} \tag{3.28}$$

where N is the total number of curvilinear squares in lane 3 (in this case 5-1/4) and $\Delta x \approx \Delta y$. An equation like Equation 3.28 can be written for each flow lane. So for M flow lanes, we obtain for the total heat flow in one quarter of the chimney

$$q = k\frac{ML}{N}(T_1 - T_2) = kS(T_1 - T_2) \tag{3.29}$$

The notation S is defined as the *conduction shape factor* for the geometry. Applied to Figure 3.7b, we find

$$S = \frac{ML}{N} = \frac{\text{Number of heat-flow lanes} \times \text{Depth into the page}}{\text{Number of curvilinear squares per lane}} \tag{3.30}$$

$$\frac{S}{L} = \frac{9}{5.25} = 1.71$$

For Figure 3.8 of the same chimney but of finer mesh,

$$\frac{S}{L} = \frac{18}{12} = 1.5$$

In general, the conduction shape factor is a function of only the geometry of the system.

The question will arise as to which of the estimates (Figure 3.7b or 3.8) is more accurate. It is generally assumed that the finer mesh gives a more accurate result. This is not necessarily the case. The best way to verify the accuracy is to set up the system and make measurements. It is important to remember that we are trying to model physical phenomena, and the most accurate model is the one that most closely approximates the real-life situation. This is the case with any mathematical model.

The two-dimensional conduction problem, when solved graphically, involves determining the conduction shape factor. The steps in the graphical solution are summarized as follows:

1. Identify lines of symmetry and reduce the working area appropriately.
2. Sketch uniformly spaced lines of constant heat flux (or of constant temperature, whichever is more convenient).
3. Complete the network by sketching in the isotherms (or adiabats).
4. Determine the shape factor.
5. Calculate the heat-flow rate, taking into account that only a portion of the total area may have been used (due to symmetry considerations).

Sketches should be made by keeping in mind that isotherms and adiabats are everywhere perpendicular. Also, isotherms intersect lines of symmetry and insulated boundaries at right angles. Finally, isotherms cannot cross each other and, likewise, adiabats cannot cross each other. In spite of the inherent shortcomings of the method, it still provides surprisingly good results.

Example 3.1

A concrete driveway is 4.2 m wide. Within the concrete slab are 1 standard type K copper tubes through which heated water is pumped to keep the concrete warm during winter months. Snow that falls or ice that forms on the concrete will melt and flow off the surface. The copper tubes are 6 cm apart (center to center), and the line of centers of the tubes is 4 cm below the slab surface, as shown in Figure 3.10. Water within the tubes heats each tube surface to 85°C. The concrete surface should be at least 0°C. The ground beneath the slab is also at 0°C. Determine the heat-flow rate from one tube at this location.

Solution

Heat flows from each pipe to the upper and lower surfaces. As fluid flows through the pipes, the fluid temperature may change with length (for example, into the page). We are concerned only with the section where the temperatures shown in Figure 3.10 apply.

Assumptions

1. Copper tube wall offers negligible resistance to heat flow, so the outside surface is at 85°C.
2. The system is at steady state.
3. Only two-dimensional conduction effects are considered. Heat transferred into the page is neglected.
4. Material properties are constant.

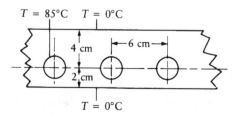

FIGURE 3.10   Copper tubes embedded within a concrete slab.

From Table F.2 for 1 standard type K, OD = 2.858 cm. We now follow the steps enumerated earlier and solve this problem graphically.

Lines of symmetry are identified, resulting in Figure 3.11, which shows the area of interest. Also shown is the resulting flow net. The total number of heat-flow lanes is $M = 8$. The number of squares per lane is $N = 6$. The conduction shape factor is

$$\frac{S}{L} = \frac{M}{N} = \frac{8}{6} = 1.333$$

From Appendix Table B.3, $k = 0.128$ W/(m·K) for concrete. The heat flow per unit length from one half of one tube is

$$\frac{q_{1/2}}{L} = \frac{kS}{L}(T_1 - T_2) = 0.128(1.3333)(85 - 0) = 14.5 \text{ W/m}$$

The total heat flow per tube is $q = 2q_{1/2}$ or

$$\frac{q}{L} = 29.0 \text{ W/m (one tube)} \qquad \square$$

Note that the heat transferred in the preceding problem is temperature dependent. The calculated value of the heat transferred applies only to the section where the temperatures are the same as those given in Figure 3.10. Elsewhere along the cross section where the pipe-wall temperature is different, heat transferred is different. The conduction shape factor, however, is a function of only the geometry. The shape factor determined for Figure 3.10 applies at any location where the cross section is the same.

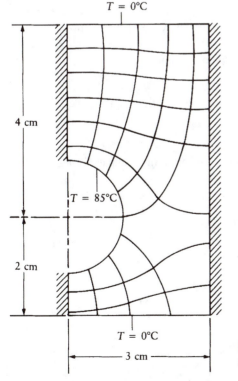

FIGURE 3.11 Graphical solution to the heated concrete slab of Example 3.1.

3.5   CONDUCTION SHAPE FACTOR

The conduction shape factor is a factor that can be found for a number of systems. Recall from Chapter 2, for example, where an equation for heat conducted through a cylinder wall was written as

$$q = \frac{2\pi kL(T_1 - T_2)}{\ln(R_2/R_1)} \tag{3.31}$$

where k is thermal conductivity, L is cylinder length, $T_1 - T_2$ is the overall temperature difference, and R_1 and R_2 are inside and outside cylinder radii, respectively. As evident from Section 3.4, heat flow in terms of shape factor is

$$q = kS(T_1 - T_2) \tag{3.32}$$

which is the defining equation for S. The shape factor has dimensions of length. Comparing Equations 3.31 and 3.32, we have for a cylindrical geometry

$$S = \frac{2\pi L}{\ln(R_2/R_1)} \tag{3.33}$$

In a similar way, the conduction shape factor has been worked out for a number of geometries. Table 3.1 provides a summary of shape factors for several systems.

Example 3.2

Crude oil is often heated before pumping it to reduce the liquid viscosity and pumping costs. Consider a 10 nominal, schedule 80 pipe buried underground [assume the ground to have a thermal conductivity of 0.072 BTU/(hr·ft·°R)], with its center line 18 in below the surface, conveying heated crude oil. The outside surface temperature of the pipe is 140°F. The surface temperature of the soil is 65°F. Determine the heat transferred from the buried pipe per unit length.

Solution

The outside pipe-surface temperature will probably change with length (into the page).

Assumptions

1. Only two-dimensional effects are considered; heat transferred into the page is neglected.
2. The system is at steady state.
3. Material properties are constant.

From Table 3.1 we select shape factor #8,

$$S = \frac{2\pi L}{\cosh^{-1}(d/R)}$$

The heat transferred then is

$$q = kS(T_1 - T_2) = \frac{2\pi kL}{\cosh^{-1}(d/R)}(T_1 - T_2)$$

where $T_1 = 140°F$ and $T_2 = 65°F$. Thermal conductivity is given as $k = 0.072$ BTU/(hr·ft·°R). From Table F.1 for 10 nominal, schedule 80 pipe

$$OD = 10.74 \text{ in}$$

or

$$R = (10.75/2)(1/12) = 0.448 \text{ ft}$$

Also,

$$d = 18/12 = 1.5 \text{ ft}$$

By substitution,

$$\frac{q}{L} = \frac{2\pi(0.072)(140 - 65)}{\cosh^{-1}(1.5/0.448)}$$

or

$$\frac{q}{L} = 18.1 \text{ BTU/(hr·ft)}$$

The use of the conduction shape factor makes the calculations simple.   ❏

The conduction shape factors given in Table 3.1 can be combined to obtain an overall conduction shape factor for a composite. The shape factors must be combined appropriately while paying attention to symmetrical considerations. Adding shape factors is justifiable because the heat-transfer equation (Fourier's Law) is a linear relationship. Thus if two individual shape-factor expressions satisfy the equation, so will their sum. The technique is illustrated in the following example.

Example 3.3

A heat-treating furnace has outside dimensions of $150 \times 150 \times 200$ mm. The walls are 6 mm thick and made of fireclay brick (assumed to have the same properties as silica brick). (a) For an inside wall temperature of 550°C and an outside-wall temperature of 30°C, determine the heat lost through the walls, using the shape-factor method. (b) Repeat the calculations but neglect the effects of the corners; that is, assume only one-dimensional effects through all the walls.

Solution

From Appendix Table B.3 for silica brick, $k = 1.07$ W/(m·K). We can break the furnace walls up into separate parts where each component has a known conduction shape factor. Figure 3.12 shows an isometric view of the box and some of its components.

Assumptions

1. The system is at steady state.
2. Material properties are constant.

We can evaluate the conduction shape factor for each component. A summary of the calculations is provided in the following table:

TABLE 3.1
Shape Factors for a Number of Conduction Heat-Transfer Systems. *(Data from several sources; see references at end of chapter.)*

System	Figure	Shape factor	Conditions
1. One-dimensional heat flow through a plane wall.		$S = \dfrac{WH}{L}$	1-D heat flow
2. Conduction through the edge of two adjoining walls.		$S = 0.54W$	$W > L/5$
3. Conduction through corner of three adjoining walls having a temperature difference of $T_1 - T_2$ across the walls.		$S = 0.15L$	$L \ll$ length and width of walls
4. Isothermal rectangular parallelepiped buried in a semi-infinite medium that has an isothermal surface		$S = 1.685L\left[\log\left(1 + \dfrac{d}{W}\right)\right]^{-0.59}$ $\times \left(\dfrac{d}{H}\right)^{-0.078}$	—

5.	Thin rectangular plate buried in a semi-infinite medium having an isothermal surface.		$S = \dfrac{\pi W}{\ln(4W/L)}$ ... $d = 0$ $S = \dfrac{2\pi W}{\ln(4W/L)}$ ... $d \gg W$
6.	Cylinder centered in a square, both of length L.		$S = \dfrac{2\pi L}{\ln(0.54W/R)}$ ... $L \gg W$
7.	Hollow cylinder of length L.		$S = \dfrac{2\pi L}{\ln(R_2/R_1)}$ ... $L \gg R$
8.	Isothermal cylinder buried in semi-infinite medium.		$S = \dfrac{2\pi L}{\cosh^{-1}(d/R)}$ ... $L \gg R$ $S = \dfrac{2\pi L}{\ln(2d/R)}$ ... $L \gg R$, $d > 3R$ $S = \dfrac{2\pi L}{\ln(L/R)\left[1 - \dfrac{\ln(L/2d)}{\ln(L/R)}\right]}$ ... $d \gg R$, $L \gg d$
9.	Isothermal cylinder with vertical axis placed in a semi-infinite medium.		$S = \dfrac{2\pi L}{\ln(2L/R)}$ ... $L \gg 2R$

(continued)

TABLE 3.1
(continued)

System	Figure	Shape factor	Conditions
10. Horizontal circular cylinder of length L placed midway between parallel planes of equal length and infinite width.		$S = \dfrac{2\pi L}{\ln(8z/D)}$	$z > D/2$
11. Conduction between two isothermal cylinders, both of length L, buried in an infinite medium.	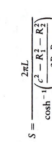	$S = \dfrac{2\pi L}{\cosh^{-1}\left(\dfrac{c^2 - R_1^2 - R_2^2}{2R_1 R_2}\right)}$	$L \gg R$ $L \gg c$
12. Eccentric isothermal cylinders of length L.	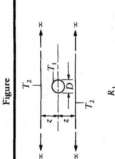	$S = \dfrac{2\pi L}{\cosh^{-1}\left(\dfrac{R_1^2 + R_2^2 - e^2}{2R_1 R_2}\right)}$	$L \gg R_2$
13. Isothermal disc on an isothermal semi-infinite medium of thermal conductivity k.	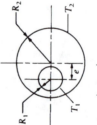	$S = 2D$	—
14. Thin horizontal isothermal disk buried in a semi-infinite medium having an isothermal surface.	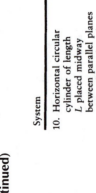	$S = 8R$	$d \gg 2R$

15. Two parallel isothermal disks buried in an infinite medium.

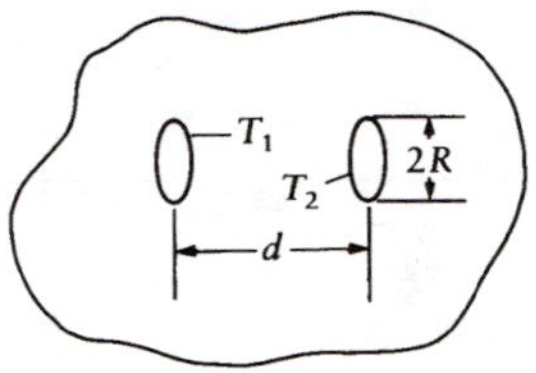

$$S = \frac{4\pi d}{2\left(\dfrac{\pi}{2} - \tan^{-1}(R/d)\right)}$$

$d > 5R$

16. A hollow sphere.

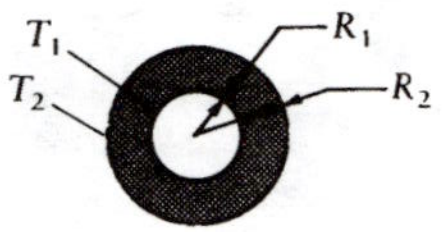

$$S = \frac{4\pi R_2 R_1}{R_2 - R_1}$$

—

17. Isothermal sphere buried in an infinite medium.

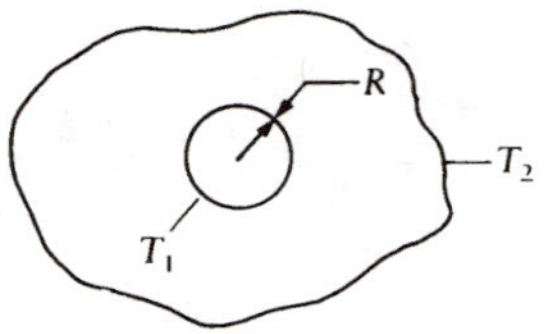

$$S = 4\pi R$$

—

18. Isothermal sphere buried in a semi-infinite medium having isothermal surface.

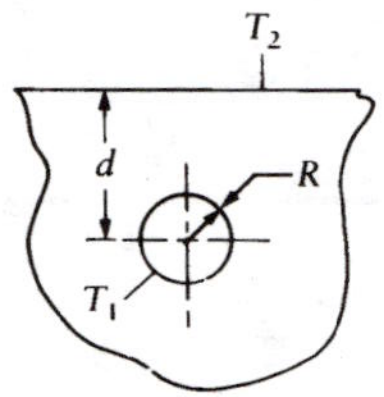

$$S = \frac{4\pi R}{1 - R/2d}$$

—

(continued)

TABLE 3.1
(continued)

System	Figure	Shape factor	Conditions
19. Hemisphere buried in a semi-infinite medium.	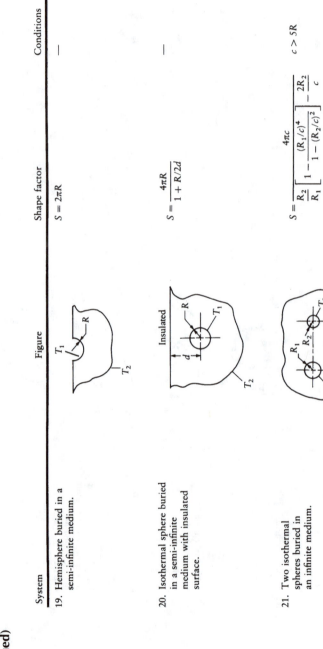	$S = 2\pi R$	—
20. Isothermal sphere buried in a semi-infinite medium with insulated surface.		$S = \dfrac{4\pi R}{1 + R/2d}$	—
21. Two isothermal spheres buried in an infinite medium.		$S = \dfrac{4\pi c}{\dfrac{R_2}{R_1}\left[1 - \dfrac{(R_1/c)^4}{1 - (R_2/c)^2}\right] - \dfrac{2R_2}{c}}$	$c > 5R$

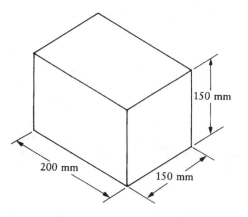

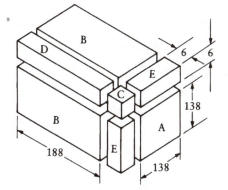

FIGURE 3.12   The furnace of Example 3.3 and some of its components.

Component	Shape factor from Table 3.1	How many of this component	Total shape factor
A	#1; $S_1 = \dfrac{0.138(0.138)}{0.006} = 3.174$	2	6.348
B	#1; $S_1 = \dfrac{0.138(0.188)}{0.006} = 4.324$	4	17.296
C	#3; $S_3 = 0.15(0.006) = 0.000\ 9$	8	0.007 2
D	#2; $S_2 = 0.54(0.188) = 0.101\ 52$	4	0.406 08
E	#2; $S_2 = 0.54(0.138) = 0.074\ 52$	8	0.596 16
		Total	24.65

(a) The heat transferred through the walls of the furnace is

$$q = kS(T_1 - T_2) = 1.07(24.65)(550 - 30) = 13\ 717\ \text{W} = 13.7\ \text{kW}$$

(b) Neglecting the effects of the edges and corners, the shape factor for all walls is found as

$$S = 6.348 + 17.296 = 23.64$$

The heat transferred is

$$q = kS(T_1 - T_2) = 1.07(23.64)(550 - 30) = 13\ 155\ \text{W} = 13.2\ \text{kW}$$

The error introduced by neglecting heat flow through the edges and corners is

$$\text{error} = \frac{13.717 - 13.155}{13.171} = 0.041; \quad \text{percent error} = 4.1\% \qquad \square$$

Example 3.4

It is proposed to cool a certain volume of air by piping it underground. The cooled air would then supplement the air-conditioning system of a dwelling and reduce costs. Figure 3.13 illustrates the proposed piping system. Determine the conduction shape factor for the underground portion of the configuration if the pipe is 4 nominal, schedule 40.

Solution

Heat is lost from the pipe to the surroundings. As air travels through the pipe and loses energy, the air temperature varies with pipe length.

Assumptions

1. Air temperature is a constant throughout the pipe and equal to T_1.
2. The system is at steady state.
3. Material properties are constant.

From Appendix Table F.1 for 4 nominal pipe

$$OD = 4.5 \text{ in} = 0.375 \text{ ft}$$

and so

$$R = 0.1875 \text{ ft}$$

We have data for the straight portions of the piping system but not the elbows. The elbows can be accounted for by including them with the length of the vertical sections. The length we use for pipes labeled A (Figure 3.13) then is 4.5 ft. So from Table 3.1, for each pipe we find

- pipe A (one pipe):

$$\#9; \ S_A = \frac{2\pi(4.5)}{\ln[2(4.5)/0.1875]} = 7.3$$

- pipe B:

$$\#8; \ S_B = \frac{2\pi(18)}{\cosh^{-1}(4.5/0.1875)} = 29.2$$

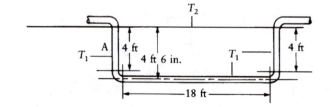

FIGURE 3.13 The underground piping system of Example 3.4

The total conduction shape factor for the system is

$$S = 2S_A + S_B = 2(7.3) + 29.2 = 43.8$$

For relatively simple geometries the conduction shape factor is thus a powerful tool to use as an alternative to the analytical method.   ❏

3.6   SOLUTION BY NUMERICAL METHODS (FINITE DIFFERENCES)

Numerical methods can be applied to a great number of problems. The technique involves rewriting the descriptive differential equation in terms of a set of simultaneous algebraic equations. The accuracy of the results is influenced by the number of algebraic equations written.

The simultaneous equations can be solved in a variety of ways. Some methods are well suited for solution with a calculator. Other methods are better suited for solution with a computer. The numerical method of solution provides the temperature at a discrete number of points in the region. The heat transferred can be estimated from the temperature distribution.

3.6.1   NORMALIZATION OF EQUATIONS

It is advantageous in many cases to simplify a differential equation before solving it. This is an especially useful technique when the number of parameters is large, such as in heat-transfer problems. The method of simplification discussed here is known as *normalization*. Normalizing a differential equation reduces the number of parameters required by combining them into physically significant ratios or products. The technique is illustrated by applying it to the differential equation for a pin fin with an insulated boundary (see Section 2.7).

The descriptive expression for a pin fin is

$$\frac{d^2T}{dz^2} - \frac{\bar{h}_c P}{kA}(T - T_\infty) = 0 \tag{3.34}$$

The dimensions of this equation are t/L^2 (°C/m² or °F/ft²). The boundary conditions are

1. At $z = 0$, $T = T_w$
2. At $z = L$, $\dfrac{dT}{dz} = 0$

The procedure for normalizing the variables is first to define two new dimensionless terms:

$$\theta = \frac{T - T_\infty}{T_0} \tag{3.35}$$

$$\xi = \frac{z}{z_0} \tag{3.36}$$

where T_0 and z_0 are yet to be determined and, hopefully, will represent significant physical parameters. The temperature difference selected is $T - T_\infty$ instead of, say, just T_∞ or T, because it is realized that a temperature difference is important in heat transfer. We now work toward expressing Equation 3.34 in terms of θ and ξ. Evaluating the first derivative gives

$$\frac{dT}{dz} = \frac{d}{dz}(T_0\theta + T_\infty) = T_0\frac{d\theta}{dz}$$

From the chain rule,

$$\frac{d\theta}{dz} = \frac{d\theta}{d\xi}\frac{d\xi}{dz} = \frac{d\theta}{d\xi}\frac{1}{z_0}$$

So by substitution.

$$\frac{dT}{dz} = \frac{T_0}{z_o}\frac{d\theta}{d\xi}$$

Evaluating the second derivative, we get

$$\frac{d^2T}{dz^2} = \frac{d}{dz}\left(\frac{dT}{dz}\right) = \frac{d}{d\xi}\left(\frac{dT}{dz}\right)\frac{d\xi}{dz} = \frac{d}{d\xi}\left(\frac{T_0}{z_0}\frac{d\theta}{d\xi}\right)\frac{1}{z_0}$$

or

$$\frac{d^2T}{dz^2} = \frac{T_0}{z_0^2}\frac{d^2\theta}{d\xi^2}$$

The differential equation (3.34) now becomes

$$\frac{T_0}{z_0^2}\frac{d^2\theta}{d\xi^2} - \frac{\bar{h}_cP}{kA}T_0\theta = 0$$

Substituting $m^2 = \bar{h}_cP/kA$ and simplifying, we have

$$\frac{d^2\theta}{d\xi^2} - m^2 z_0^2\theta = 0 \qquad\qquad (3.37)$$

This equation is dimensionless. We still must evaluate z_0 and T_0. We refer next to the boundary conditions, which reveal that z will vary from 0 to L. So physically, we could select $z_0 = L$. Boundary condition 1 shows that at $z = 0$, $T = T_w$; so we could let $T_0 = T_w - T_\infty$. The values for z_0 and T_0 are thus made on physical grounds revealed through the boundary conditions. We see that ξ varies from 0 to 1; 0 to ∞ are other suitable and commonly encountered limits. Also, θ varies from 1 to 0. With the above choices for z_0 and T_0, the differential equation (3.37) and the boundary conditions become as shown below:

Dimensional	Normalized	
$T = T(z)$	$\theta = \theta(\xi)$	
$\dfrac{d^2T}{dz^2} - \dfrac{\bar{h}_cP}{kA}(T - T_\infty) = 0$	$\dfrac{d^2\theta}{d\xi^2} - (mL)^2\theta = 0$	(3.38)
B.C.1. $z = 0$, $T = T_w$	$\xi = 0$, $\theta = 1$	
B.C.2. $z = L$, $\dfrac{dT}{dz} = 0$	$\xi = 1$, $\dfrac{d\theta}{d\xi} = 0$	(3.39)
where,	$\xi = \dfrac{z}{L}$; $\theta = \dfrac{T - T_\infty}{T_w - T_\infty}$	(3.40)

Equations 3.38 through 3.40 are the normalized form of the differential equation for a pin fin with an insulated end. Normalization should be thought of as a mathematical convenience. It is notarequiredtechniqueinthecourseofsolvingaproblem,althoughitcansimplifytheequationsconsiderably.

3.6.2 Numerical Method of Solution for One-Dimensional Problems

The one-dimensional steady-state equation written for fins is a boundary-value problem in which information is specified at the boundary of the geometry of interest. Given the conditions at the boundary, the problem is to formulate an equation to describe the behavior within the region. The technique of solution of the boundary-value problem that we will investigate is the method of finite differences. It involves approximating the differential equation for the system with algebraic equations. In many cases where the differential equation has no known explicit solution, numerical methods can be used to transform the equation into a form suitable for solution with a computer. We will solve the pin-fin problem numerically and compare the numerical solution to the exact solution.

One preliminary concept that is the key to the method is in approximating derivatives in terms of what are called *differences*. Consider the region shown in Figure 3.14, in which T vs. x coordinates have been imposed. The region is divided into a number of equal increments Δx wide. (The width of Δx at this point is not significant.) The interest is in selecting a value of x, such as x_n, and evaluating the corresponding derivative dT/dx at x_n. The definition of the derivative is

$$\frac{dT}{dx}\bigg|_{x=x_n} = \lim_{\Delta x \to 0} \frac{T(x_n + \Delta x) - T(x_n)}{\Delta x}$$

We can approximate this definition by not applying the limiting process. Then

$$\frac{dT}{dx}\bigg|_{x=x_n} = \frac{T(x_n + \Delta x) - T(x_n)}{\Delta x}$$

or, in a more compact notation form,

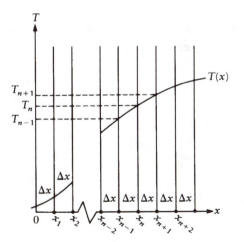

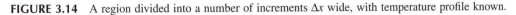

FIGURE 3.14 A region divided into a number of increments Δx wide, with temperature profile known.

$$\frac{dT}{dx}\bigg|_{x_n} = \frac{T_{n+1} - T_n}{\Delta x}$$

in which the subscripts on temperature refer to specific x locations. This equation is recognized as the slope at an intermediate point midway between x_n and x_{n+1}. This is illustrated in Figure 3.15. It would therefore be more appropriate to write the equation as

$$\frac{dT}{dx}\bigg|_{n+1/2} = \frac{T_{n+1} - T_n}{\Delta x} \tag{3.41}$$

Following similar lines of reasoning, the slope at $x_{n-1/2}$ can be written as

$$\frac{dT}{dx}\bigg|_{n-1/2} = \frac{T_n - T_{n-1}}{\Delta x} \tag{3.42}$$

This equation also appears in Figure 3.15. Adding Equations 3.41 and 3.42 gives the derivative at n as

$$\frac{dT}{dx}\bigg|_{n} = \frac{T_{n+1} - T_{n-1}}{\Delta x} \tag{3.43}$$

The second derivative is the rate of change of the first derivative. With the foregoing equations we can approximate the second derivative as

$$\frac{d^2T}{dx^2}\bigg|_{n} = \frac{(dT/dx)_{n+1/2} - (dT/dx)_{n-1/2}}{\Delta x}$$

Substituting Equations 3.41 and 3.42, we get

$$\frac{d^2T}{dx^2}\bigg|_{n} = \frac{T_{n+1} - 2T_n + T_{n-1}}{(\Delta x)^2} \tag{3.44}$$

We now have approximations for the first and second derivatives.

With the equations developed thus far, it is possible to solve numerically the pin-fin problem discussed in Section 2.7. The differential equation for the case where the free end is insulated is

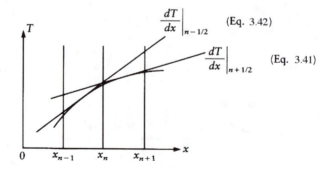

FIGURE 3.15 Illustration of the various forms of the derivative dT/dx.

$$\frac{d^2T}{dz^2} - \frac{\bar{h}_c P}{kA}(T - T_\infty) = 0$$

with

B.C.1. $z = 0, T = T_w$

B.C.2. $z = L, \dfrac{dT}{dz} = 0$

However, rather than working with these equations, we will begin with the normalized equations as derived earlier:

$$\frac{d^2\theta}{d\xi^2} - (mL)^2\theta = 0 \tag{3.38}$$

with

B.C.1 $\xi = 0, \theta = 1$

B.C.2. $\xi = 1, \dfrac{d\theta}{d\xi} = 0$ $\tag{3.39}$

The normalization parameters are

$$\xi = z/L$$

and

$$\theta = \frac{T - T_\infty}{T_w - T_\infty} \tag{3.40}$$

To formulate the problem by finite differences, we first divide the fin into an arbitrary number of sections $\Delta\xi$ long, as shown in Figure 3.16. It is important to remember that $\Delta\xi$ is dimensionless and equal to $\Delta(z/L)$, so that $\Delta\xi = 0.25$ in Figure 3.16. The midpoint of each section is called a *node*. The temperature at each node is what the solution will provide, which is shown schematically in Figure 3.16. Note the node locations, especially at the wall and the tip, which are the boundaries of our region.

Next, we substitute the approximation for second derivative into Equation 3.38, obtaining

$$\frac{\theta_{n+1} - 2\theta_n + \theta_{n-1}}{(\Delta\xi)^2} - (mL)^2\theta_n = 0$$

Rearranging,

$$\theta_{n-1} - [2 + (mL\Delta\xi)^2]\theta_n + \theta_{n+1} = 0$$

Letting $K = 2 + (mL\Delta\xi)^2$, we simplify the equation, obtaining

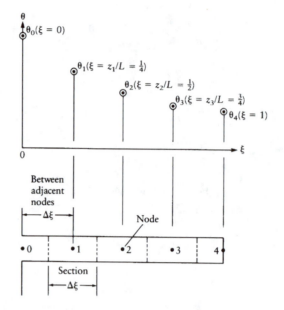

FIGURE 3.16 Pin fin divided into four sections and the indicated output.

$$\theta_{n-1} - K\theta_n + \theta_{n+1} = 0 \qquad (3.45)$$

The above equation relates θ at the interior nodes to each other and eventually to each boundary. Note that if $n \le 0$, we get subscripts on θ that are less than 0 that are not part of our region. So n must be equal to or greater than 1. For $n = 1$, the above equation becomes

$$\theta_0 - K\theta_1 + \theta_2 = 0$$

But θ_0 is θ evaluated at $\xi = 0$, and according to B.C.1, $\theta_0 = 1$. Thus,

$$1 - K\theta_1 + \theta_2 = 0$$

or

$$K\theta_1 - \theta_2 = 1 \quad (n = 1)$$

For $n = 2$ and 3, Equation 3.45 becomes

$$\theta_1 - K\theta_2 + \theta_3 = 0 \quad (n = 2)$$

$$\theta_2 - K\theta_3 + \theta_4 = 0 \quad (n = 3)$$

At the end we have an insulated tip, and a condition of this type can be easily analyzed by considering Figure 3.17. Shown is one pin fin attached to two walls. Both walls are at the same temperature, and the slope of the temperature curve at the midpoint is zero. No heat is transferred across the midpoint, which is equivalent mathematically to an insulated boundary. Due to symmetry, then, we conclude that

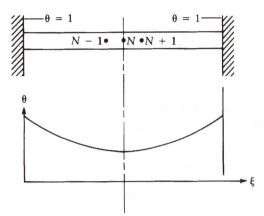

FIGURE 3.17 Pin fin attached to two walls, with corresponding temperature profile.

$$\theta_{N-1} = \theta_{N+1}$$

Equation 3.45 then becomes for the insulated end

$$\theta_{n-1} - K\theta_n + \theta_{n-1} = 0$$

or

$$2\theta_{n-1} - K\theta_n = 0$$

For $n = 4$, then,

$$2\theta_3 - K\theta_4 = 0 \quad (n = 4)$$

Summarizing, our nodal equations are

$$K\theta_1 - \quad \theta_2 \qquad\qquad\qquad = 1 \qquad\qquad (3.46a)$$

$$\theta_1 - \quad K\theta_2 + \quad \theta_3 \qquad\qquad = 0 \qquad\qquad (3.46b)$$

$$\theta_2 - \quad K\theta_3 + \quad \theta_4 = 0 \qquad\qquad (3.46c)$$

$$2\theta_3 - \quad K\theta_4 = 0 \qquad\qquad (3.46d)$$

The original differential equation has thus been transformed into four algebraic equations that must be solved simultaneously. Note that these algebraic equations already incorporate the boundary conditions.

Equations 3.46 can be solved in a number of ways. There exists a variety of computer software for solving simultaneous algebraic equations. The method that we select to illustrate is known as Gaussian elimination. The technique involves eliminating all terms below the main diagonal of the coefficient matrix. This technique now follows.

The first equation can be multiplied by $1/K$ and subtracted from the second equation to give

$$K\theta_1 \quad - \quad \theta_2 \quad = \quad 1$$
$$-(K - 1/K)\theta_2 \quad + \theta_3 \quad = -1/K$$
$$\theta_2 - \quad K\theta_3 + \quad \theta_4 = \quad 0$$
$$2\theta_3 - K\theta_4 = \quad 0$$

So θ_1 has been eliminated from the second equation. Next, multiplying the second equation by $1/(K - 1/K)$ and adding it to the third equation gives

$$K\theta_1 \quad - \quad \theta_2 \quad = \quad 1$$
$$-(K - 1/K)\theta_2 \quad + \theta_3 \quad = \quad -1/K$$
$$-\left(K - \frac{1}{K - 1/K}\right)\theta_3 \quad + \theta_4 = -\frac{1}{K(K - 1/K)}$$
$$2\theta_3 - \quad K\theta_4 = \quad 0$$

Simplifying gives

$$K\theta_1 \quad - \quad \theta_2 \quad = \quad 1$$
$$-\frac{K^2 - 1}{K}\theta_2 \quad + \theta_3 \quad = \quad -1/K$$
$$-\frac{K^3 - 2K}{K^2 - 1}\theta_3 \quad + \theta_4 = -\frac{1}{K^2 - 1}$$
$$2\theta_3 - \quad K\theta_4 = \quad 0$$

Multiplying the third equation by $2(K^2 - 1)/(K^3 - 2K)$ and adding it to the fourth equation gives, after simplification,

$$K\theta_1 \quad - \quad \theta_2 \quad = \quad 1$$
$$-\frac{K^2 - 1}{K}\theta_2 \quad + \theta_3 \quad = \quad -1/K$$
$$-\frac{K^3 - 2K}{K^2 - 1}\theta_3 \quad + \theta_4 = -\frac{1}{K^2 - 1}$$
$$+ (K^4 - 4K^2 + 2)\theta_4 = \quad 2$$

Thus, all terms below the main diagonal have been systematically eliminated. The fourth equation yields for θ_4

$$\theta_4 = \frac{2}{K^4 - 4K^2 + 2} \tag{3.47}$$

This can be substituted into the third equation, which is then solved for θ_3. The result is

$$\theta_3 = \frac{K}{K^4 - 4K^2 + 2} \tag{3.48}$$

Substituting into the second equation and solving for θ_2 brings

$$\theta_2 = \frac{K^2 - 1}{K^4 - 4K^2 + 2} \tag{3.49}$$

Substituting into the first equation for θ_1, finally, gives

$$\theta_1 = \frac{K^3 - 3K}{K^4 - 4K^2 + 2} \tag{3.50}$$

The heat transferred to the fin at the wall equals the heat convected away by each section. Thus,

$$q_z = \sum_{n \to 0}^{N} q_{cn} \tag{3.51}$$

where N is the total number of sections. Applying Equation 3.51 to Figure 3.16 for $N = 4$ gives

$$q_z = \bar{h}_c \left(\frac{P\Delta z}{2} \right)(T_0 - T_\infty)$$

$$+ \bar{h}_c(P\Delta z)(T_1 - T_\infty) + \bar{h}_c(P\Delta z)(T_2 - T_\infty)$$

$$+ \bar{h}_c(P\Delta z)(T_3 - T_\infty) + \bar{h}_c \left(\frac{P\Delta z}{2} \right)(T_4 - T_\infty)$$

The form of this equation presumes that the temperature of a node is representative of its entire section. Simplifying, and noting that $T_0 = T_w$, we get

$$q_z = \bar{h}_c P\Delta z \left[\frac{T_w - T_\infty}{2} + (T_1 - T_\infty) + (T_2 - T_\infty) + (T_3 - T_\infty) + \frac{T_4 - T_\infty}{2} \right]$$

Rearranging,

$$\frac{q_z}{\bar{h}_c P\Delta z(T_w - T_\infty)} = \left(\frac{1}{2} + \frac{T_1 - T_\infty}{T_w - T_\infty} + \frac{T_2 - T_\infty}{T_w - T_\infty} + \frac{T_3 - T_\infty}{T_w - T_\infty} + \frac{1}{2}\frac{T_4 - T_\infty}{T_w - T_\infty} \right)$$

Recognizing that $L = 4\Delta z$, the above equation in terms of θ becomes

$$\frac{q_z}{\bar{h}_c PL(T_w - T_\infty)} = \eta_e = \frac{1}{4}\left(\frac{1}{2} + \theta_1 + \theta_2 + \theta_3 + \frac{\theta_4}{2} \right) \tag{3.52}$$

where η_e is the fin efficiency defined in Chapter 2. Each θ has been solved previously in terms of K. Substituting Equations 3.47 through 3.50 into 3.52 gives

$$\eta_e = \frac{K}{8}\frac{(K^2 - 2)(K + 2)}{K^4 - 4K^2 + 2} \tag{3.53}$$

Example 3.5

A glass beaker contains some chemicals that undergo a heat-producing reaction. A glass rod is used to stir the mixture. When the rod is stationary, as shown in Figure 3.18, heat is conducted

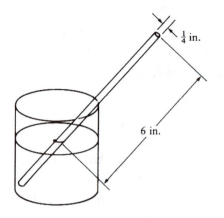

FIGURE 3.18 Heated glass rod.

axially along the rod and transferred by convection to the surrounding air. A rod length of 6 in extends above the liquid surface. The rod diameter is 1/4 in. At the liquid surface, the rod temperature is 200°F. (a) Using the pin-fin equations for the case where the exposed tip is assumed insulated, graph the temperature distribution existing within the rod. (b) Use the numerical formulation of this section to obtain the temperature distribution. (c) Compare the two models to determine how well the numerical results approximate the exact results. Use a convection coefficient of 1.1 BTU/(hr·ft²·°R) and an ambient temperature of 68°F.

Solution

As heat is produced by the mixture, air in the region about the portion of the rod within the beaker is warmer than the surrounding air. Thus, the convection coefficient $\bar{h}_c$ is not really constant along the rod. In addition, the rod is in contact with the beaker, which affects the air flow past the rod and conduction along the rod.

Assumptions

1. One-dimensional steady conduction exists along the rod.
2. The convection coefficient $\bar{h}_c$ is constant over the rod length.
3. Material properties are constant.
4. Heat is transferred from the rod to the air only by convection; radiation effects are neglected.

(a) From Table B.3 for glass (window pane for lack of a more specific type), $k = 0.47$ BTU/(hr·ft·°R). The applicable differential equation and boundary conditions are

$$\frac{d^2T}{dz^2} - \frac{\bar{h}_cP}{kA}(T - T_\infty) = 0$$

$$\text{B.C.1.} \quad z = 0, \ T = T_w$$

$$\text{B.C.2.} \quad z = L, \ \frac{dT}{dz} = 0$$

Introducing $\theta = (T - T_\infty)/(T_w - T_\infty)$, $\xi = z/L$, and $m^2 = \bar{h}_c P/kA$, these equations become, in normalized form,

$$\frac{d^2\theta}{d\xi^2} - (mL)^2\theta = 0$$

B.C.1. $\xi = 0$, $\theta = 1$

B.C.2. $\xi = 1$, $\dfrac{d\theta}{d\xi} = 0$

From Section 2.7, the solution is

$$\theta = \frac{\cosh\left[mL\left(1 - \dfrac{z}{L}\right)\right]}{\cosh(mL)}$$

The fin cross-sectional area and perimeter are

$$A = \pi D^2/4 = \frac{\pi}{4}\left(\frac{0.25}{12}\right)^2 = 3.409 \times 10^{-4}\ \text{ft}^2$$

$$P = \pi D = P\left(\frac{0.25}{12}\right) = 6.545 \times 10^{-2}\ \text{ft}$$

So the product mL is

$$mL = L\sqrt{\frac{\bar{h}_c P}{kA}} = \frac{6}{12}\left[\frac{1.1(6.545 \times 10^{-2}\ \text{ft})}{0.47(3.409 \times 10^{-4}\ \text{ft}^2)}\right]^{\frac{1}{2}}$$

$$mL = 10.60$$

The solution equation now becomes

$$\frac{T - T_\infty}{T_w - T_\infty} = \frac{\cosh\left[10.60\left(1 - \dfrac{z}{6}\right)\right]}{\cosh(10.60)}$$

$$T = 68 + (200 - 68)\frac{\cosh\left[10.60\left(1 - \dfrac{z}{6}\right)\right]}{\cosh(10.60)} \tag{i}$$

We now select values of z (which are in inches), substitute into the above equation, and determine T.

(a) The results are shown in the second column of Table 3.2.

  The numerical formulation begins with the differential equation applied to Figure 3.19, in which the rod is broken up into four sections 1.5 in (= $L/4 = 6/4$) long. The differential equation in finite-difference form is Equation 3.46:

$$\begin{aligned}
K\theta_1 - \theta_2 && &= 1 \\
\theta_1 - K\theta_2 &+ \theta_3 && = 0 \\
&\theta_2 - K\theta_3 &+ \theta_4 &= 0 \\
&&2\theta_3 - K\theta_4 &= 0
\end{aligned}$$

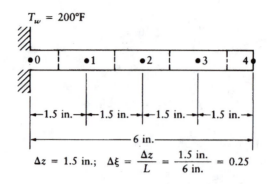

FIGURE 3.19   The glass rod divided into four sections, with nodal points.

TABLE 3.2
A Comparison of Exact to Numerical Results for the Data of Example 3.5.

z, in.	Exact (e) T (Eq. i) (°F)	Numerical (n) T (°F)	Percent error (e − n)/e
0	200.00	200.00	0
1.5	77.33	82.81	−7.09
3.0	68.66	69.66	−1.46
4.5	68.05	68.19	−0.21
6.0	68.00	68.04	−0.059

The boundary conditions are incorporated into the above equations. With $\Delta\xi = 1/4 = 0.25 \ (= \Delta z/L)$ the parameter K is

$$K = 2 + (mL\Delta\xi)^2 = 2 + [(10.60)(0.25)]^2 = 9.0225$$

We can solve this system of simultaneous algebraic equations in any convenient way. Here we use the explicit equations for each θ derived earlier as Equations 3.47 through 3.50:

$$\theta_4 = \frac{T_4 - T_\infty}{T_w - T_\infty} = \frac{2}{K^4 - 4K^2 + 2}$$

$$T_4 = 68 + (200 - 68)\frac{2}{6.303 \times 10^3}$$

Solving,

$$T_4 = 68.04\,°F$$

Likewise,

$$\theta_3 = \frac{T_3 - T_\infty}{T_w - T_\infty} = \frac{K}{6.303 \times 10^3}$$

from which we get

$$T_3 = 68.19\,°F$$

Also,

$$\theta_2 = \frac{T_2 - T_\infty}{T_w - T_\infty} = \frac{K^2 - 1}{6.303 \times 10^{-3}}$$

and

$$T_2 = 69.66°F$$

Finally,

$$\theta_1 = \frac{T_1 - T_\infty}{T_w - T_\infty} = \frac{K^3 - 3K}{6.303 \times 10^{-3}}$$

$$T_1 = 82.81°F$$

(b) These results are displayed in column 3 of Table 3.2. A graphical comparison is provided in Figure 3.20.   ❏

Numerical Method of Solution for Two-Dimensional Problems

Consider the two-dimensional region illustrated in Figure 3.21. If a temperature difference is imposed on the region, then heat will flow from the high-temperature surface to the low-temperature surface. For two-dimensional conduction with constant thermal conductivity, Equation 3.7 applies:

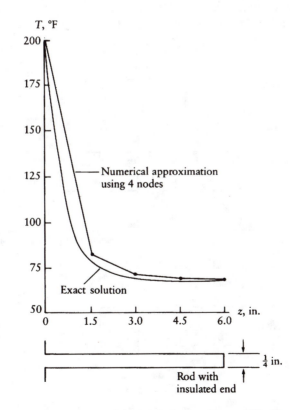

FIGURE 3.20   A comparison of the exact to the numerical temperature profiles for the pin fin of Example 3.5.

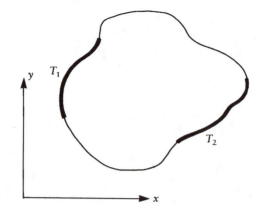

FIGURE 3.21   Two-dimensional region with a temperature difference imposed.

$$\frac{\partial^2 T}{\partial x^2} + \frac{\partial^2 T}{\partial y^2} = 0 \tag{3.7}$$

We will rewrite this equation, using a finite-difference scheme. The first step is to divide the region into a number of nodal points whose temperature will be determined numerically. This is shown in Figure 3.22.

The dimensions Δx and Δy are selected arbitrarily, depending on the desired fineness of the grid. Each node has associated with it values m and n. The m value indicates the relative location in the x direction, and the n value corresponds to the y location. The temperature at node m, n would thus be written as $T_{m,n}$. Using the same lines of reasoning as in the preceding section, we evaluate the derivatives of the temperature as

$$\left.\frac{\partial T}{\partial x}\right|_{m+\frac{1}{2},\, n} = \frac{T_{m+1,\, n} - T_{m,\, n}}{\Delta x}$$

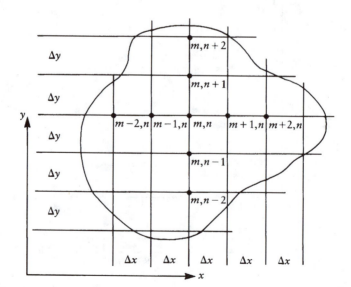

FIGURE 3.22   Two-dimensional region divided up into a grid.

$$\frac{\partial T}{\partial x}\bigg|_{m-\frac{1}{2},\,n} = \frac{T_{m,n} - T_{m-1,n}}{\Delta x}$$

$$\frac{\partial T}{\partial x}\bigg|_{m,\,n+\frac{1}{2}} = \frac{T_{m,n+1} - T_{m,n}}{\Delta y}$$

$$\frac{\partial T}{\partial x}\bigg|_{m,\,n-\frac{1}{2}} = \frac{T_{m,n} - T_{m,n-1}}{\Delta y}$$

The second derivatives are evaluated as

$$\frac{\partial^2 T}{\partial x^2}\bigg|_{m,n} = \frac{(\partial T/\partial x)\big|_{m+\frac{1}{2},n} - (\partial T/\partial x)\big|_{m-\frac{1}{2},n}}{\Delta x} = \frac{\dfrac{T_{m+1,n} - T_{m,n}}{\Delta x} - \dfrac{T_{m,n} - T_{m-1,n}}{\Delta x}}{\Delta x}$$

or

$$\frac{\partial^2 T}{\partial x^2}\bigg|_{m,n} = \frac{T_{m+1,n} - 2T_{m,n} + T_{m-1,n}}{(\Delta x)^2} \tag{3.54}$$

Similarly,

$$\frac{\partial^2 T}{\partial y^2}\bigg|_{m,n} = \frac{T_{m,n+1} - 2T_{m,n} + T_{m,n-1}}{(\Delta y)^2} \tag{3.55}$$

Substituting Equations 3.54 and 3.55 into 3.7 gives

$$\frac{\partial^2 T}{\partial x^2} + \frac{\partial^2 T}{\partial y^2} = 0 \tag{3.7}$$

$$\frac{T_{m+1,n} - 2T_{m,n} + T_{m-1,n}}{(\Delta x)^2} + \frac{T_{m,n+1} - 2T_{m,n} + T_{m,n-1}}{(\Delta y)^2} = 0 \tag{3.56}$$

This equation can be simplified greatly if we choose our grid dimensions such that $\Delta x = \Delta y$. The above equation becomes

$$T_{m+1,n} + T_{m-1,n} + T_{m,n+1} + T_{m,n-1} - 4T_{m,n} = 0$$

or, solving for $T_{m,n}$,

$$T_{m,n} = \frac{T_{m+1,n} + T_{m-1,n} + T_{m,n+1} + T_{m,n-1}}{4} \tag{3.57}$$

Thus, when $\Delta x = \Delta y$, the temperature at any interior node is the arithmetic average of the temperatures of the four surrounding nodes.

Equation 3.57 was derived by a mathematical approach in which the differential equation was finite differenced. Equation 3.57 can be derived, using a more physically oriented approach, by performing an energy balance on a node. Consider the node m, n and surrounding nodes illustrated in Figure 3.23. For conduction at steady state (no energy stored) without internal heat generation, node m, n exchanges energy with the four adjacent nodes according to

$$\text{Energy flow in} = \text{Energy flow out}$$

To evaluate the energy flow, we first recognize that the three adjacent squares in either the x or y direction make up a heat-flow lane. We assume that Fourier's Law applies to both lanes. So the heat transferred by conduction from node $m - 1, n$ to m, n is

$$q_{(m-1,n)-(m,n)} = kA\frac{\Delta T}{\Delta x} = k\Delta y(1)\frac{T_{m-1,n} - T_{m,n}}{\Delta x} \tag{3.58a}$$

where a unit depth has been assumed, and Δx is the distance between the nodes. Likewise, we can write

$$q_{(m+1,n)-(m,n)} = k\frac{\Delta y}{\Delta x}(T_{m+1,n} - T_{m,n}) \tag{3.58b}$$

$$q_{(m,n+1)-(m,n)} = k\frac{\Delta x}{\Delta y}(T_{m,n+1} - T_{m,n}) \tag{3.58c}$$

$$q_{(m,n-1)-(m,n)} = k\frac{\Delta x}{\Delta y}(T_{m,n-1} - T_{m,n}) \tag{3.58d}$$

Noting that $\Delta x = \Delta y$ and that the q_i values added together equal zero, Equations 3.58, when summed, yield

$$0 = (T_{m-1,n} - T_{m,n}) + (T_{m+1,n} - T_{m,n}) + (T_{m,n+1} - T_{m,n}) + (T_{m,n-1} - T_{m,n})$$

or, after simplification,

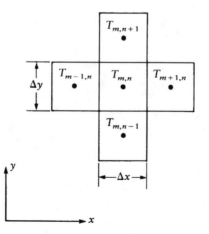

FIGURE 3.23 An interior node and four surrounding nodes.

$$0 = T_{m-1,n} + T_{m+1,n} + T_{m,n+1} + T_{m,n-1} - 4T_{m,n}$$

This equation is a restructured Equation 3.57.

Equation 3.57, the starting point for the numerical method, must be written for every node within the region. As an example, consider the square slab of Figure 3.24. The slab is shown divided, with $\Delta x = \Delta y = 1/3$ the length of one side. The temperature along each side is specified, and each node is labeled. Equation 3.57 written for each unknown nodal temperature—(1,1), (2,1), (1,2), (2,2)—is

$$T_{1,1} = \frac{0 + T_{1,2} + T_{2,1} + 0}{4} = \frac{T_{1,2} + T_{2,1}}{4}$$

$$T_{2,1} = \frac{T_{1,1} + T_{2,2} + 0 + 0}{4} = \frac{T_{1,1} + T_{2,2}}{4}$$

$$T_{1,2} = \frac{0 + 100 + T_{2,2} + T_{1,1}}{4} = \frac{100 + T_{2,2} + T_{1,1}}{4}$$

$$T_{2,2} = \frac{T_{1,2} + 100 + 0 + T_{2,1}}{4} = \frac{T_{1,2} + 100 + T_{2,1}}{4}$$

(Solution techniques will be illustrated later. The emphasis here is on developing the nodal equations.)

It is not difficult to write the nodal equations when the boundary temperatures are specified. It is necessary, however, to be able to handle convective, radiative, and insulated boundaries. Figure 3.25 is a sketch of a plane wall that is part of a divided region. We wish to develop a finite-difference equation for the temperature of node m, n. The boundary condition in this case is

$$-k\frac{\partial T}{\partial x} = \bar{h}_c(T_w - T_\infty)$$

But this equation arises from an energy balance performed at the wall:

Energy conducted to the wall = Energy convected away

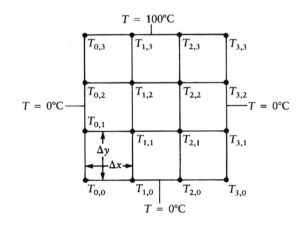

FIGURE 3.24 A square slab divided into nine squares.

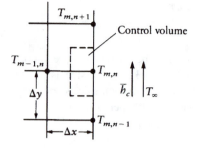

FIGURE 3.25   Plane wall with a convective boundary.

We now apply this energy balance to the control volume about node m, n (see Figure 3.25). The heat conducted to node m, n from node m, $n - 1$ is

$$q_{(m, n-1)-(m, n)} = kA\frac{\Delta T}{\Delta y} = k\frac{\Delta x}{2\Delta y}(1)(T_{m, n-1} - T_{m, n}) \tag{3.59a}$$

The heat conducted to m, n from the other surrounding nodes is

$$q_{(m-1, n)-(m, n)} = k\frac{\Delta y}{\Delta x}(1)(T_{m-1, n} - T_{m, n}) \tag{3.59b}$$

$$q_{(m, n+1)-(m, n)} = k\frac{\Delta x}{2\Delta y}(1)(T_{m, n+1} - T_{m, n}) \tag{3.59c}$$

The heat transferred by convection from node m, n to the ambient fluid is

$$q_c = \bar{h}_c A\Delta T = \bar{h}_c \Delta y(1)(T_{m, n} - T_\infty) \tag{3.60}$$

This equation reflects the assumption that the entire wall of the control volume exposed to the fluid has temperature $T_{m,n}$. Summing Equations 3.59 and setting the result equal to Equation 3.60 gives

$$\frac{k}{2}(T_{m, n-1} - T_{m, n}) + k(T_{m-1, n} - T_{m, n}) + \frac{k}{2}(T_{m, n+1} - T_{m, n}) = \bar{h}_c \Delta y(T_{m, n} - T_\infty)$$

in which $\Delta x = \Delta y$ has been used. Simplifying,

$$0 = (T_{m, n-1} + 2T_{m-(1, n)} + T_{m, n+1}) + \frac{2\bar{h}_c\Delta y}{k}T_\infty - 2T_{m, n}\left(2 + \frac{\bar{h}_c\Delta y}{k}\right) \tag{3.61}$$

We can now extend this result. To obtain an equation applicable to an insulated boundary, we set $\bar{h}_c$ equal to zero. So, for an insulated boundary or a surface of symmetry, Equation 3.61 becomes

$$0 = (T_{m, n-1} + 2T_{m-1, n} + T_{m, n+1}) - 4T_{m, n} \tag{3.62}$$

Following similar lines of reasoning, equations analogous to 3.61 can be derived for interior and exterior corners. The results are summarized in Table 3.3. In geometries where a curved boundary is encountered, the grid usually cannot be made to coincide with the surface. A finite-difference equation has been developed for such a configuration, and the equation is provided (without derivation here) also in Table 3.3.

TABLE 3.3
Summary of Finite-Difference Equations for Some Common Configurations

Configuration	Item	Equation
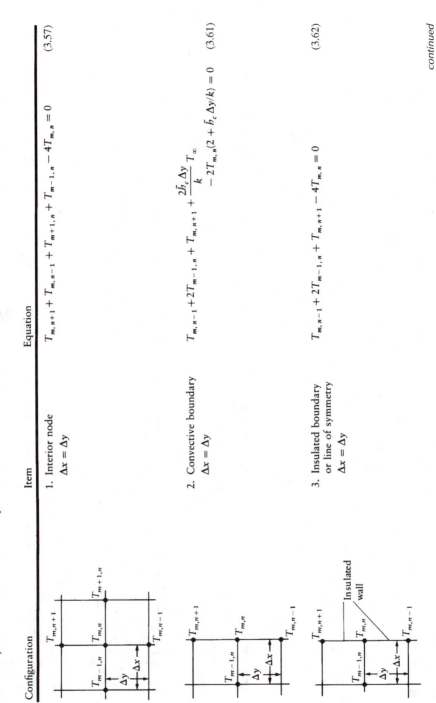	1. Interior node $\Delta x = \Delta y$	$T_{m,n+1} + T_{m,n-1} + T_{m+1,n} + T_{m-1,n} - 4T_{m,n} = 0$ (3.57)
	2. Convective boundary $\Delta x = \Delta y$	$T_{m,n-1} + 2T_{m-1,n} + T_{m,n+1} + \dfrac{2\bar{h}_c \, \Delta y}{k} T_{\infty}$ $- 2T_{m,n}(2 + \bar{h}_c \, \Delta y / k) = 0$ (3.61)
	3. Insulated boundary or line of symmetry $\Delta x = \Delta y$	$T_{m,n-1} + 2T_{m-1,n} + T_{m,n+1} - 4T_{m,n} = 0$ (3.62)

continued

TABLE 3.3
(continued)

4. Interior corner with convection
   $\Delta x = \Delta y$

$$2T_{m-1,n} + 2T_{m,n+1} + T_{m+1,n} + T_{m,n-1} + \frac{2\bar{h}_c \Delta x}{k} T_\infty - 2T_{m,n}(3 + \bar{h}_c \Delta x/k) = 0$$

5. Exterior corner with convection
   $\Delta x = \Delta y$

$$T_{m,n-1} + T_{m-1,n} + \frac{2\bar{h}_c \Delta x}{k} T_\infty - 2T_{m,n}(1 + \bar{h}_c \Delta x/k) = 0$$

6. Interior node m, n near a curved surface
   $\Delta x = \Delta y$

$$\frac{2T_2}{b(b+1)} + \frac{2}{c+1} T_{m+1,n} + \frac{2}{b+1} T_{m,n-1}$$
$$+ \frac{2T_1}{c(c+1)} - 2T_{m,n}(1/c + 1/b) = 0$$

7. Radiative boundary
   $\Delta x = \Delta y$

$$T_{m,n-1} + 2T_{m-1,n} + T_{m,n+1} + \frac{2\sigma \Delta y \varepsilon F}{k} T_r^4$$
$$- 4T_{m,n} - \frac{2\sigma \Delta y \varepsilon F}{k} T_{m,n}^4 = 0$$

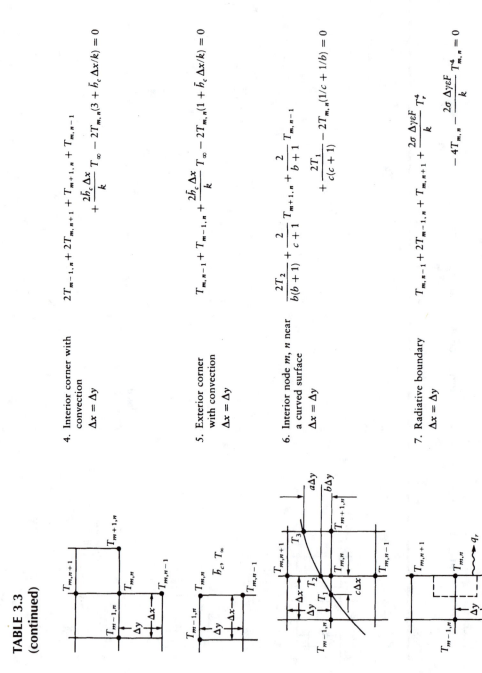

A radiation boundary can also be analyzed. Consider the node m, n of Case 7 in Table 3.3. The node receives heat by conduction from its surrounding nodes, as described by Equations 3.59. The node m, n transfers heat by radiation according to

$$q_r = \sigma A \varepsilon F(T_{m,n}^4 - T_r^4)$$

where T_r is the receiver temperature, σ is the Stefan-Boltzmann constant, ε is the surface emissivity, and F is the shape factor from node m, n to the receiver. Note that absolute temperatures must be used in this expression. With $A = \Delta y(1)$, the radiation heat-transfer equation becomes

$$q_r = \sigma \Delta y(1) \varepsilon F(T_{m,n}^4 - T_r^4)$$

Summing Equations 3.59 and setting the result equal to the above equation gives

$$\frac{k}{2}(T_{m,n-1} - T_{m,n}) + k(T_{m-1,n} - T_{m,n}) + \frac{k}{2}(T_{m,n+1} - T_{m,n}) = \sigma \Delta y \varepsilon F(T_{m,n}^4 - T_r^4)$$

in which $\Delta x = \Delta y$. Rearranging and simplifying gives the result shown in Table 3.3.

Example 3.6

Figure 3.26 shows the cross section of a fireplace chimney built along the wall of a house. The chimney and the house exterior wall are made of masonry brick. During use of the fireplace, the warm exhaust gases heat the interior surface of the chimney to 200°C. The temperature inside the house wall is 30°C. Outside the chimney the ambient temperature is 5°C. The convection coefficient

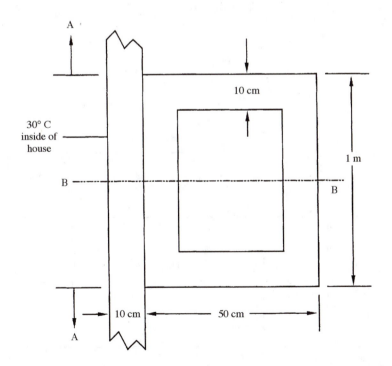

FIGURE 3.26 Cross section of a fireplace chimney.

between the chimney wall and the air is 4 W/(m²·K). Develop finite-difference equations for temperature within the cross section of the chimney and house wall. Consider the brick wall beyond lines labeled A to be insulated.

Solution

The exhaust gases lose heat by convection to the chimney walls. The exhaust-gas temperature therefore varies with height, and the analysis here will apply only to the region where the temperatures are those given in the problem statement.

Assumptions

1. Two-dimensional heat flow exists.
2. The brick wall beyond lines labeled A is insulated.
3. The system is at steady state.
4. Material properties are constant.

Due to symmetry about B–B, we need consider only half the cross section, as shown in Figure 3.27. Arbitrarily, we select $\Delta x = \Delta y = 10$ cm and divide up the region. Each node is labeled. The boundary temperatures and conditions are also shown in the figure. From Table B.3 for masonry brick, $k = 0.658$ W/(m·K).

We now write the finite-difference equations for each node, referring to Table 3.3:

node 1, 0 (line of symmetry, item 3)

$$30 + 2T_{1,1} + 200 - 4T_{1,0} = 0$$

node 1, 1 (interior node, item 1)

$$T_{1,0} + 30 + T_{1,2} + 200 - 4T_{1,1} = 0$$

node 1, 2 (interior node, item 1)

$$T_{1,1} + 30 + T_{1,3} + 200 - 4T_{1,2} = 0$$

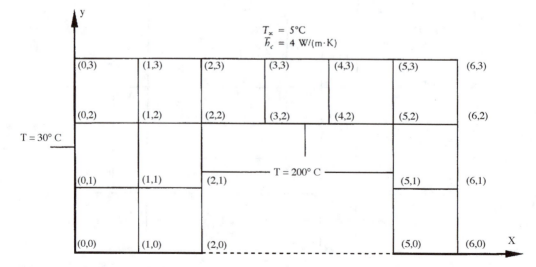

FIGURE 3.27 One half of the chimney of Figure 3.26 divided into a number of nodes, using $\Delta x = \Delta y = 10$ cm.

node 1, 3 (insulated boundary, item 3)

$$30 + 2T_{1,2} + T_{2,3} - 4T_{1,3} = 0$$

node 2, 3 (convective boundary, item 2)

$$T_{3,3} + 2(200) + T_{1,3} + \frac{2(4)(0.1)(5)}{0.658} - 2T_{2,3}\left[2 + \frac{4(0.1)}{0.658}\right] = 0$$

node 3, 3 (convective boundary, item 2)

$$T_{4,3} + 2(200) + T_{2,3} + \frac{2(4)(0.1)(5)}{0.658} - 2T_{3,3}\left[2 + \frac{4(0.1)}{0.658}\right] = 0$$

node 4, 3 (convective boundary, item 2)

$$T_{5,3} + 2(200) + T_{3,3} + \frac{2(4)(0.1)(5)}{0.658} - 2T_{4,3}\left[2 + \frac{4(0.1)}{0.658}\right] = 0$$

node 5, 3 (convective boundary, item 2)

$$T_{6,3} + 2(200) + T_{4,3} + \frac{2(4)(0.1)(5)}{0.658} - 2T_{5,3}\left[2 + \frac{4(0.1)}{0.658}\right] = 0$$

node 6, 3 (exterior corner, item 5)

$$T_{6,2} + T_{5,3} + \frac{2(4)(0.1)(5)}{0.658} - 2T_{6,3}\left[1 + \frac{4(0.1)}{0.658}\right] = 0$$

node 6, 2 (convective boundary, item 2)

$$T_{6,1} + 2(200) + T_{6,3} + \frac{2(4)(0.1)(5)}{0.658} - 2T_{6,2}\left[2 + \frac{4(0.1)}{0.658}\right] = 0$$

node 6, 1 (convective boundary, item 2)

$$T_{6,0} + 2(200) + T_{6,2} + \frac{2(4)(0.1)(5)}{0.658} - 2T_{6,1}\left[2 + \frac{4(0.1)}{0.658}\right] = 0$$

Node 6, 0 is on a line of symmetry and on a convective boundary. In this case, we use item 2 for the convective boundary with $T_{6,1}$, the neighboring node, appearing as $T_{m,n+1}$ and $T_{m,n-1}$. Thus

$$T_{6,1} + 2(200) + T_{6,1} + \frac{2(4)(0.1)(5)}{0.658} - 2T_{6,0}\left[2 + \frac{4(0.1)}{0.658}\right] = 0$$

These finite-difference equations must now be solved simultaneously to obtain the temperature at each point. Note that the boundary conditions appear as part of the equations. ❑

3.7 METHODS OF SOLVING SIMULTANEOUS EQUATIONS

As seen in the preceding section, a set of simultaneous algebraic equations results for a system divided up into a nodal network. Methods for solving simultaneous equations are presented here, even though you will probably have access to a computing facility capable of solving simultaneous equations. The purpose here is to present the mechanics involved in making the calculations.

The Gaussian elimination method was discussed in the preceding section. Solution by this method is appropriate for systems in which the equations can be set up in matrix form. Matrix algebra is then used to obtain desired results.

Another method of solution involving matrix representation is the matrix-inversion method. The simultaneous equations can be written in matrix form as per the following notation:

$$C_{11}T_1 + C_{12}T_2 + \ldots + C_{1n}T_n = A_1$$
$$C_{21}T_1 + C_{22}T_2 + \ldots + C_{2n}T_n = A_2$$
$$\vdots$$
$$C_{n1}T_1 + C_{n2}T_2 + \ldots + C_{nn}T_n = A_n$$

The coefficients C_{11}, C_{12}, etc., and the constants A_1, A_2, etc., are known numerically. They involve the physical parameters of the problem such as thermal conductivity k, convection coefficient $\bar{h}_c$, and increment Δx. In matrix form the simultaneous equations become

$$[C][T] = [A] \tag{3.63}$$

The solution for temperature is obtained when the inverse of the coefficient matrix $[C]^{-1}$ is found:

$$[C][C]^{-1} [T] = [C]^{-1} [A]$$

Because the product $[C][C]^{-1}$ gives a diagonal matrix with elements equal to unity (the *unity matrix*), the above equation becomes

$$[T] = [C]^{-1} [A] \tag{3.64}$$

As an example of how to determine the inverse of a matrix, consider the 0.5-m-square slab of Figure 3.28. The upper surface is at a temperature of 200°C; the left, right, and lower faces are at 100°C. We assume a unit depth and divide the region up into nine squares. There are four nodes, and we wish to determine the temperature at each node. Each node in the figure is labeled with a double subscript. To simplify the notation, we will use a single subscript:

$$T_{1,1} = T_1$$
$$T_{1,2} = T_2$$
$$T_{2,1} = T_3$$
$$T_{2,2} = T_4$$

Referring to Table 3.3, the finite-difference equation for each node is

$$\left. \begin{array}{l} T_{1,1}\colon\ 100 + T_2 + T_3 + 100 - 4T_1 = 0 \\ T_{1,2}\colon\ T_1 + 100 + 200 + T_4 - 4T_2 = 0 \\ T_{2,2}\colon\ T_3 + T_2 + 200 + 100 - 4T_4 = 0 \\ T_{2,1}\colon\ 100 + T_1 + T_4 + 100 - 4T_3 = 0 \end{array} \right\} \tag{3.65}$$

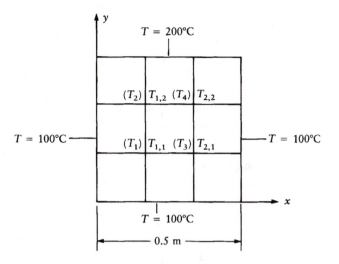

FIGURE 3.28 A square slab.

Rewriting gives

$$
\begin{aligned}
4T_1 - T_2 - T_3 \qquad &= 200 \\
T_1 - 4T_2 \qquad + T_4 &= -300 \\
T_2 + T_3 - 4T_4 &= -300 \\
T_1 \qquad - 4T_3 + T_4 &= -200
\end{aligned}
$$

In matrix form we have

$$
\begin{bmatrix}
4 & -1 & -1 & 0 \\
1 & -4 & 0 & 1 \\
0 & 1 & 1 & -4 \\
1 & 0 & -4 & 1
\end{bmatrix}
\begin{bmatrix}
T_1 \\
T_2 \\
T_3 \\
T_4
\end{bmatrix}
=
\begin{bmatrix}
200 \\
-300 \\
-300 \\
-200
\end{bmatrix}
\tag{3.66}
$$

It is possible to find the inverse of the coefficient matrix in a number of ways. Here, for purposes of illustration, we select a method that lends itself to hand calculations. We first write the coefficient matrix next to the unity matrix:

$$
[C][1] =
\begin{bmatrix}
4 & -1 & -1 & 0 \\
1 & -4 & 0 & 1 \\
0 & 1 & 1 & -4 \\
1 & 0 & -4 & 1
\end{bmatrix}
\begin{bmatrix}
1 & 0 & 0 & 0 \\
0 & 1 & 0 & 0 \\
0 & 0 & 1 & 0 \\
0 & 0 & 0 & 1
\end{bmatrix}
\tag{3.67}
$$

Next, matrix algebra computations are made to the coefficient matrix to transform it into the unity matrix. However, all calculations made on $[C]$ in Equation 3.67 are made also on $[1]$. At the conclusion of the transformation operation, the unity matrix becomes the inverse of $[C]$. The mechanics of some of the calculations are as follows:

1. Divide row 1 by 4:

$$\begin{bmatrix} 1 & -\frac{1}{4} & -\frac{1}{4} & 0 \\ 1 & -4 & 0 & 1 \\ 0 & 1 & 1 & -4 \\ 1 & 0 & -4 & 1 \end{bmatrix} \begin{bmatrix} \frac{1}{4} & 0 & 0 & 0 \\ 0 & 1 & 0 & 0 \\ 0 & 0 & 1 & 0 \\ 0 & 0 & 0 & 1 \end{bmatrix}$$

2. Multiply row 1 by (−1), add to row 2, and replace row 2:

$$\begin{bmatrix} 1 & -\frac{1}{4} & -\frac{1}{4} & 0 \\ 0 & -3\frac{3}{4} & \frac{1}{4} & 1 \\ 0 & 1 & 1 & -4 \\ 1 & 0 & -4 & 1 \end{bmatrix} \begin{bmatrix} \frac{1}{4} & 0 & 0 & 0 \\ -\frac{1}{4} & 1 & 0 & 0 \\ 0 & 0 & 1 & 0 \\ 0 & 0 & 0 & 1 \end{bmatrix}$$

3. Multiply row 1 by (−1), add to row 4, and replace row 4:

$$\begin{bmatrix} 1 & -\frac{1}{4} & -\frac{1}{4} & 0 \\ 0 & -3\frac{3}{4} & \frac{1}{4} & 1 \\ 0 & 1 & 1 & -4 \\ 0 & \frac{1}{4} & -3\frac{3}{4} & 1 \end{bmatrix} \begin{bmatrix} \frac{1}{4} & 0 & 0 & 0 \\ -\frac{1}{4} & 1 & 0 & 0 \\ 0 & 0 & 1 & 0 \\ -\frac{1}{4} & 0 & 0 & 1 \end{bmatrix}$$

We proceed systematically in this fashion until the following results:

$$[1][C]^{-1} = \begin{bmatrix} 1 & 0 & 0 & 0 \\ 0 & 1 & 0 & 0 \\ 0 & 0 & 1 & 0 \\ 0 & 0 & 0 & 1 \end{bmatrix} \begin{bmatrix} 0.2916 & -0.0833 & -0.0416 & -0.0833 \\ 0.0833 & -0.2916 & -0.0833 & -0.0416 \\ 0.0833 & -0.0416 & -0.0833 & -0.2916 \\ 0.0416 & -0.0833 & -0.2916 & -0.0833 \end{bmatrix} \qquad (3.68)$$

It is easy to verify that $[C][C]^{-1} = [1]$. The solution, then, for temperature is

$$[T] = [C]^{-1}[A]$$

$$\begin{bmatrix} T_1 \\ T_2 \\ T_3 \\ T_4 \end{bmatrix} = \begin{bmatrix} 0.2916 & -0.0833 & -0.0416 & -0.0833 \\ 0.0833 & -0.2916 & -0.0833 & -0.0416 \\ 0.0833 & -0.0416 & -0.0833 & -0.2916 \\ 0.0416 & -0.0833 & -0.2916 & -0.0833 \end{bmatrix} \begin{bmatrix} 200 \\ -300 \\ -300 \\ -200 \end{bmatrix}$$

For T_1 we obtain

$$T_1 = \quad 0.291\ 6(200)\ -0.083\ 3(-300)$$
$$-0.041\ 6(-300)\ -0.083\ 3(-200)$$

or

$$T_1 = 112.5°C$$

Likewise,

$$T_2 = 137.5°C$$

$$T_3 = 112.5°C$$

$$T_4 = 137.5°C$$

A computer can perform these calculations far more quickly than they can be done by hand.

Another method of solving simultaneous equations is by Gauss-Seidel iteration method. In this method, the finite-difference equations are written explicitly for each node. An initial value for each temperature is estimated. Using the explicit equations, new values for each temperature are determined. The iteration procedure is continued until convergence within some tolerable error is achieved. The method is illustrated with Equations 3.65, which are rewritten to solve for each node explicitly:

$$
\begin{aligned}
T_1 &= 50 \quad + T_2/4 + T_3/4 \\
T_2 &= T_1/4 + T_4/4 + 75 \\
T_3 &= 50 \quad + T_1/4 + T_4/4 \\
T_4 &= T_{3/4} \quad + T_2/4 + 75
\end{aligned}
\tag{3.65}
$$

The calculations are conveniently summarized in Table 3.4. The first row are estimates of each temperature. Substituting into Equations 3.65 yields the results displayed in the second row, and so on. The calculations were stopped at the ninth row. The results are seen to have excellent agreement with those obtained by the matrix-inversion technique.

The final method of solution that we will consider is the relaxation method, in which the equations for the nodal temperatures are rewritten in terms of what are known as *residuals*. To illustrate, we again use Equations 3.65:

$$50 - T_1 + \frac{T_2}{4} + \frac{T_3}{4} = R_1 \tag{3.65a}$$

$$75 - T_2 + \frac{T_1}{4} + \frac{T_4}{4} = R_2 \tag{3.65b}$$

$$50 - T_3 + \frac{T_1}{4} + \frac{T_4}{4} = R_3 \tag{3.65c}$$

$$75 - T_4 + \frac{T_3}{4} + \frac{T_2}{4} = R_4 \tag{3.65d}$$

TABLE 3.4
Summary of Calculations Made by Using the Gauss-Seidel Iteration Method

Trial	T_1	T_2	T_3	T_4
1 (estimates)	125	150	125	150
2	118.75	143.75	118.75	143.75
3	115.63	140.63	115.63	140.63
4	114.06	139.06	114.06	139.06
5	113.28	138.28	113.28	138.28
6	112.89	137.89	112.89	137.89
7	112.70	137.66	112.70	137.66
8	112.60	137.60	112.60	137.60
9	112.55	137.55	112.55	137.55

The R_i values are the residuals. The objective is to find the temperatures T_1 through T_4 that reduce the residuals to 0. Initial estimates are made for each temperature, and the residuals are calculated. The largest residual is identified, and the estimate for the corresponding temperature is changed to reduce the largest residual to zero. This process is continued until all residuals are made zero or a value close to zero that is within a tolerable limit. The calculations are summarized in a relaxation table (Table 3.5). The first row contains the estimated temperatures and the residuals that result from using Equations 3.65. The residual R_3 is the farthest from 0, so we modify T_3 (referring to Equation 3.65c for R_3) by adding -22.5 ($= R_3$) to 140 ($= T_3$), obtaining 117.5. New calculations are then made as displayed in row 2.

TABLE 3.5
Relaxation Table Summarizing Calculations Made with Equations 3.65

Trial	T_1	R_1	T_2	R_2	T_3	R_3	T_4	R_4
1 (estimate temperatures)	120	−2.5	130	12.5	140	−22.5	150	−7.5
2	120	−8.1	130	12.5	117.5	0	150	−13.1
3	120	−8.1	130	9.2	117.5	−3.3	136.8	0
4	120	−5.8	139.2	0	117.5	−3.3	136.8	2.3
5	114.2	0	139.2	−1.4	117.5	−4.7	136.8	2.3
6	114.2	−1.2	139.2	−1.4	112.8	0	136.8	1.1
7	114.2	−1.6	137.8	0	112.8	0	136.8	0.8
8	112.6	0	137.8	−0.4	112.8	−0.4	136.8	0.8
9	112.6	0	137.8	−0.2	112.8	−0.2	137.6	0

Next, relaxing R_4 to zero involves subtracting 13.1 ($= R_4$) from 150 ($= T_4$), giving a new T_4 of 136.8. Calculations are continued in this way until all residuals are less than, say, 0.25 (selected arbitrarily). Notice that exact values are not obtained by this method unless a large number of iterations are used.

There are variations of this technique. For example, it is possible to *overrelax* the residuals, wherein they are reduced past zero. This is done to achieve convergence more quickly.

Once temperature is known, the heat transferred can be calculated. To illustrate the calculation, we refer again to the solution of Equations 3.65 written for the system of Figure 3.28. Figure 3.29 is of the same system, sketched with nodal temperatures. Heat is transferred from the 200°C face to each of the 100°C faces. The heat leaving the 200°C surface is

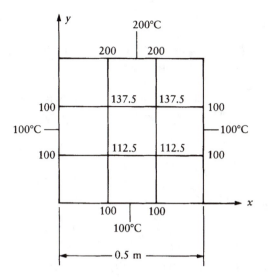

FIGURE 3.29   The solution to the square-slab problem of Figure 3.28 and Equations 3.65.

$$q = -k\Delta x(L)\frac{(137.5 - 200)}{\Delta y} - k\Delta x(L)\frac{(137.5 - 200)}{\Delta y}$$

with

$$\Delta x = \Delta y,$$

$$\frac{q}{Lk} = 125 \text{ m·K}$$

The heat transferred to each 100° C surface is

$$q = -k\Delta y(L)\frac{(137.5 - 100)}{\Delta x} - k\Delta y(L)\frac{(112.5 - 100)}{\Delta x}$$

$$- k\Delta x(L)\frac{(112.5 - 100)}{\Delta y} - k\Delta x(L)\frac{(112.5 - 100)}{\Delta y}$$

$$- k\Delta y(L)\frac{(112.5 - 100)}{\Delta x} - k\Delta y(L)\frac{(137.5 - 100)}{\Delta x}$$

Solving,

$$\frac{q}{Lk} = -125 \text{ m·K}$$

3.8   SUMMARY

In this chapter, we investigated conduction heat transfer in more than one dimension. The analytical method was presented for a relatively simple two-dimensional geometry, in which an explicit equation for temperature was obtained. Heat transferred could then be determined. It was concluded that the analytical method was difficult to apply, especially for an irregular geometry. We then resorted to alternative methods.

The graphical method of flux plotting was presented. In using this method to determine heat transferred, the conduction shape factor was defined. A table of shape factors for various systems was presented.

Numerical methods were outlined. It was shown that the differential equation could be rewritten in terms of simultaneous algebraic equations. Techniques for solving such equations were also discussed.

3.9 PROBLEMS

3.9.1 ANALYTICAL METHODS

1. Equation 3.21 is graphed in Figure 3.21 for various temperatures. Plot the isotherm on this graph for $T = 105°C$.
2. Derive Equation 3.26, beginning with Equation 3.23.
3. Graph the adiabats given by Equation 3.26 on a likeness of Figure 3.4.
4. The two-dimensional system shown in Figure 3.2 has three surfaces at T_1 and the fourth surface at $T_1 + T_2 \sin(\pi x/W)$. If the fourth surface is at constant temperature (T_2) instead, the solution to Equation 3.2 becomes

$$\frac{T - T_1}{T_2 - T_1} = \frac{2}{\pi} \sum_{n=1}^{\infty} \frac{(-1)^n + 1}{n} \sin(n\pi x/W) \frac{\sinh(n\pi y/W)}{\sinh(n\pi H/W)}$$

   (a) Write the first eight terms ($n = 1, 2, 3,..., 8$) of this series.
   (b) Given that $H = 1.5 W$, determine the temperature at the midpoint of the system. Let $T_2 = 200°C$ and $T_1 = 100°C$.
5. Verify that the temperature profile given in Problem 4 is indeed a solution of Equation 3.7.
6. For the geometry and temperature distribution described in Problem 4, determine the heat-flow equations for q_x and q_y.
7. In the section about the analytical method of solution, a rectangular plate was analyzed, and the two-dimensional conduction equation (3.7) was solved subject to four boundary conditions (Equations 3.8). Suppose we again solve the two-dimensional conduction equation for the rectangular plate of Figure 3.2, but with different boundary conditions. Obtain the temperature profile for the following boundary conditions:
   (a) $x = 0$ $\theta = 0$
   (b) $x = W$ $\theta = T_2 \sin \pi(y/H)$
   (c) $y = 0$ $\theta = 0$
   (d) $y = H$ $\theta = 0$
8. Obtain an equation for the heat flow rate q'' (similar to Equation 3.26) for the temperature profile derived in Problem 7.
9. In the section about the analytical method of solution, a rectangular plate was analyzed, and the two-dimensional conduction equation (3.7) was solved subject to four boundary conditions (Equations 3.8). Suppose we again solve the two-dimensional conduction equation for the rectangular plate of Figure 3.2, but with different boundary conditions. Obtain the temperature profile for the following boundary conditions:
   (a) $x = 0$ $\theta = 0$
   (b) $x = W$ $\theta = 0$
   (c) $y = 0$ $\theta = 0$
   (d) $y = H$ $\theta = T_2 \sinh(\pi x/W)$ for $x \leq W/2$

For $x \geq W/2$, the temperature profile is symmetric about $x/W = 0.5$. A graph of the fourth boundary condition is as shown in Figure P3.1.

10. Obtain an equation for the heat flow rate q'' (similar to Equation 3.26) for the temperature profile derived in Problem 9.

3.9.2 FIELD PLOTTING—GRAPHICAL METHOD

11. Determine the shape factor per unit depth for the square chimney shown in Figure P3.2. Calculate the heat transferred if $T_1 = 150°C$, $T_2 = 20°C$, and masonry brick is used.

12. Figure P3.3 is a square chimney having a square hole. Compared to Figure P3.2, both have the same external dimensions and the same amount of material or area removed from the center. Determine and compare the shape factors per unit depth for both chimneys.

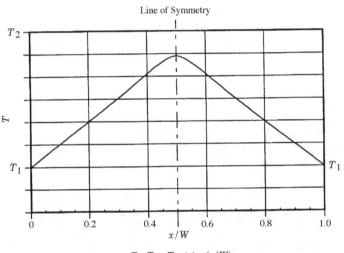

$$T = T_1 + T_2 \sinh \pi (x/W)$$

FIGURE P3.1

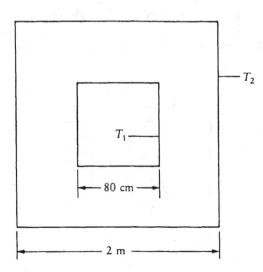

FIGURE P3.2

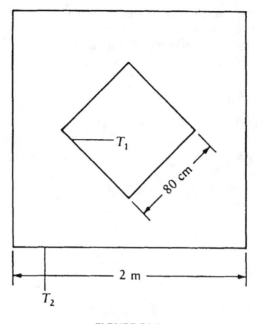

FIGURE P3.3

13. Figure P3.4 is a square chimney having a round hole. Compared to Figure P3.2, both have the same external dimensions and the same amount of material or area removed from the center. Determine and compare the shape factors per unit depth for both chimneys.

14. Figure P3.5 is of a wall safe built along part of the wall of a house. The wall itself is insulated. The safe is made of stainless steel (Type 304). The design of the safe includes consideration of fire protection. That is, if a fire breaks out in the room, the heat-transfer rate through the walls of the safe must be retarded enough so that the safe contents can survive for a certain length of time. The worst case exists when the external-wall temperature of the safe is 1000°F and the inside-wall temperature is 70°F. Determine the heat-flow rate per unit depth through the safe wall for the dimensions given in the figure.

15. Figure P3.6 is of a brick wall 60 cm tall, 30 cm wide, made of masonry brick. During daylight hours the sun heats the brick surface to 52°C while the shaded rear of the wall remains at 16°C. Determine the heat transferred per unit depth through the wall.

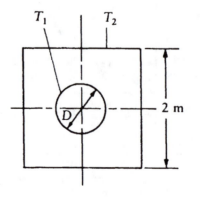

FIGURE P3.4

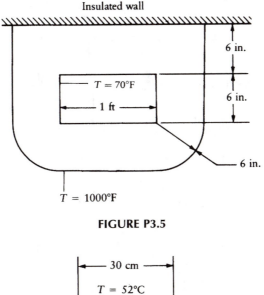

FIGURE P3.5

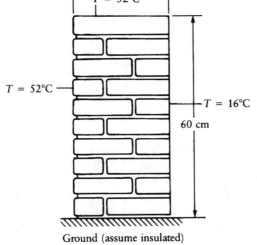

Ground (assume insulated)

FIGURE P3.6

16. Figure P3.7 is of a wall supported by concrete columns, which in turn are reinforced with (1C) steel I-beams. The columns are 0.67 m apart point to point. For the dimensions and temperatures shown in the figure, determine the heat transferred per unit depth through the wall.

17. Figure P3.8 shows the cross section of a 6-in concrete ceiling supported by trapezoidal concrete beams. Whether fire sprinklers are to be installed will depend on the heat-transfer rate through the ceiling under the worst-case conditions. For the dimensions and temperatures shown in the figure, calculate the heat transferred per unit depth through the ceiling.

18. Figure P3.9 shows a metal slab that is to be formed or rolled into sheet metal. The metal slab is at an elevated temperature and allowed to cool somewhat on a bed made of concrete and steel channels. The metal slab heats the low-carbon (1C) steel channels uniformly to 200°C. The bottom of the concrete is at 17°C. Determine the heat transferred per unit depth through the bed.

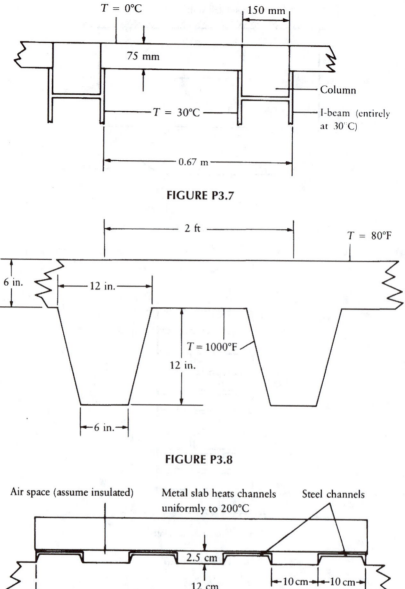

FIGURE P3.7

FIGURE P3.8

FIGURE P3.9

19. Figure P3.10 shows a concrete stack of internal dimensions 3 × 3 ft. The stack is part of the ventilation system of a heated sulfuric acid bath used in metal treatment. Sulfuric acid fumes exit through the stack, heating the inside wall to 200°F. The stack wall contains copper tubes through which cold (40°F) Freon® is piped to cool the inside wall and condense as much sulfuric acid as possible. Condensed acid finds its way back to the tank. The stack external-wall temperature is 40°F. Estimate the heat transferred per unit length to the Freon.

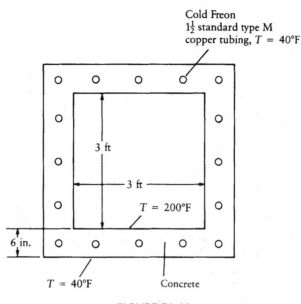

FIGURE P3.10

20. Figure P3.11 shows the casting of a plastic into a beam. The molten plastic heats the surrounding (1C) steel mold (k_1) to a temperature of 250°F. The steel mold fits inside a brass cooling jacket (k_2) that contains water-cooling tubes that are at 70°F. Determine the heat transferred per unit length from the steel-plastic interface to the water tubes.

21. Figure P3.12 shows the casting of a PVC plastic sphere that is to be machined and become part of a ball valve. The plastic is at 200°C, and the exterior wall of the steel (1C) is at 40°C. Determine the heat-flow rate from the plastic-steel interface to the exterior steel surface.

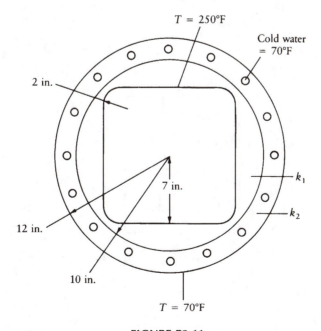

FIGURE P3.11

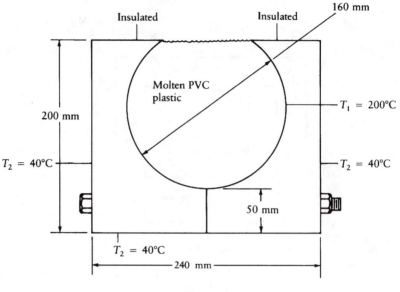

FIGURE P3.12

3.9.3 Shape Factor Method

22. Figure P3.13 is of a brick masonry chimney built into the wall of a house. For temperatures of $T_1 = 400°F$, $T_2 = 70°F$, and T_3 30°F, determine the net heat transferred per unit depth to the surface at T_2 between the lines labeled A-A.

23. Figure P3.14 is a cross section of a piping system and surrounding brickwork. The two pipes are for conveying acid to a plating operation. The 4 nominal pipe reaches a wall temperature of 92°C as it conveys heated liquid to the bath. The 6 nominal return line reaches a wall temperature of 22°C. The pipes are surrounded by clay earth (28%

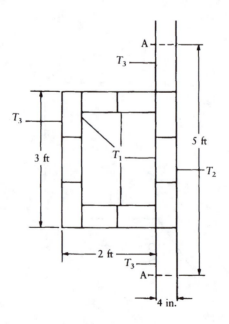

FIGURE P3.13

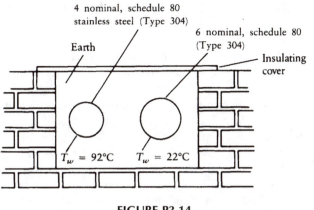

4 nominal, schedule 80
stainless steel (Type 304)

6 nominal, schedule 80
(Type 304)

Earth

Insulating
cover

$T_w = 92°C$ $T_w = 22°C$

FIGURE P3.14

moisture) in a fully insulated, brick-lined channel. Any heat lost by the heated pipe is gained only by the other pipe. Determine the heat-transfer rate per unit length between the pipes if they are 0.4 m apart center to center.

24. An underground cold-storage room where films are kept is shown schematically in Figure P3.15. The ceiling of the room is 20 ft below the ground surface. Four vent pipes, made of 6 nominal, schedule 40 cast iron, lead to the room. The entire room and the vent pipes are at a temperature of about 38°F. The isothermal surface above is at 70°F. Determine the heat transferred, which would be the first step in sizing the cooling unit. Assume the earth to be sandy with 8% moisture.

25. Two steam pipes are placed alongside one another as shown in Figure P3.16. One pipe carries superheated steam at 300°C. The other pipe carries wet steam at 100°C. Both

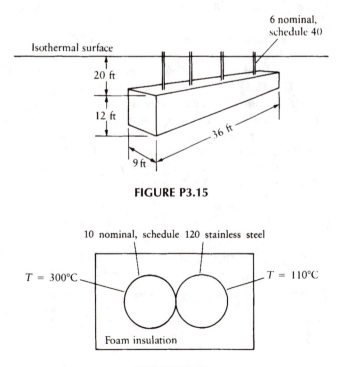

6 nominal,
schedule 40

Isothermal surface

20 ft

12 ft

9 ft

36 ft

FIGURE P3.15

10 nominal, schedule 120 stainless steel

$T = 300°C$

$T = 110°C$

Foam insulation

FIGURE P3.16

pipes are 10 nominal, schedule 120 stainless steel and are surrounded by a foam insulation having a thermal conductivity of 0.07 W/(m·K). Determine the heat transferred per unit length from one pipe to the other.

26. Systems 7 and 12 of Table 3.1 are both of two-cylinder configurations. By using the identities of Appendix Table A.4, show that, if the eccentricity of system 12 is 0, the corresponding (absolute value of the) shape factor becomes equal to that of system 7.

27. System 14 of Table 3.1 is of a horizontal disk buried in a semi-infinite medium. The shape factor is 8 times the radius. System 5 of Table 3.1 is of a horizontal rectangular plate buried in a semi-infinite medium. If the rectangular plate is really square and has the same area as the disk, which will have the larger shape factor?

28. System 14 of Table 3.1 is of a horizontal disk buried in a semi-infinite medium. The shape factor is 8 times the radius. System 5 of Table 3.1 is of a horizontal rectangular plate buried in a semi-infinite medium. If the rectangular plate has a width equal to half its length but with the same area as the disk, which will have the larger shape factor?

29. It is proposed to use a corrugated material as a solar collector surface. A cross section of the shape is shown in Figure P3.17. The material is a plastic having a thermal conductivity of 0.233 W/(m·K) and a thickness of 7 mm. The concern is whether or not to place insulation between the corrugated surface and the plywood backing to retard heat loss through the collector to the roof. The decision rests on how much heat is transferred through the collector. During operation the upper surface of the corrugated material reaches a temperature of 90°C. Calculate the heat transferred per unit length through the corrugated material for lower-surface temperatures ranging from 30° to 90°C. Graph the results (heat transferred vs. temperature of lower surface). Assume a length of 3 m into the page.

30. An underground power cable is buried 4 ft below the surface. The cable is 3 inches in diameter (OD), and its outer surface reaches a temperature of 175°F. The ground-surface temperature is 70°F. How much heat is lost by the cable per unit length? Take the cable wire to be copper, insulated with spongy rubber, and the earth to be sandy with 8% moisture.

31. Use the graphical method to determine the shape factor of the geometry of system 6 in Table 3.1. Take $W = 10$ cm and $K = 3.5$ cm. Compare the result to that shown in the table.

32. A barbell is cast in a sand mold as shown in Figure P3.18. The sand has a thermal conductivity of 0.7 BTU/(hr·ft·°R), and its surface temperature is 90°F. If the barbell material is at 1800°F, estimate the heat transferred.

33. Shape factor 4 of Table 3.1 is of a rectangular parallelepiped buried in a semi-infinite medium having an isothermal surface. Shape factor 18 is similar except the shape is spherical. If the sphere and parallelepiped have the same volume, at what depth must the sphere be buried to ensure that the shape factors are equal? Take the parallelepiped dimensions to be $W = 1$, $H = 1/2$, $L = 2$, and $d = 0.4$.

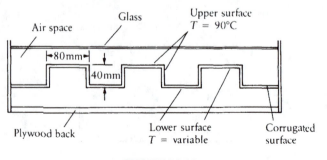

FIGURE P3.17

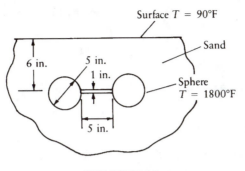

FIGURE P3.18

34. Would using a different pipe in the piping system of Example 3.4 reduce the heat transferred? Calculate the heat transferred for 2 nominal, 6 nominal, and 8 nominal pipe, keeping all other dimensions the same. Graph nominal diameter (not actual diameter) vs. shape factor.

3.9.4 NORMALIZATION/TRANSFORMATION OF EQUATIONS

35. Normalize the following differential equation and boundary conditions for the infinitely long pin fin:

$$\frac{d^2T}{dz^2} - \frac{\bar{h}_cP}{kA}(T - T_\infty) = 0$$

$$\text{B.C.1.} \quad z = 0; \quad T = T_w$$

$$\text{B.C.1.} \quad z = \infty; \quad T = T_\infty$$

36. Transform the differential equation and boundary conditions for the pin fin with an uninsulated end using Equations 3.35 and 3.36:

$$\frac{d^2T}{dz^2} - \frac{\bar{h}_cP}{kA}(T - T_\infty) = 0$$

$$\text{B.C.1.} \quad z = 0; \quad T = T_w$$

$$\text{B.C.1.} \quad z = L; \quad -k\frac{dT}{dz} = \bar{h}_c(T - T_\infty)$$

37. Show that Equation 2.53 can be transformed into

$$\frac{d^2\theta}{d\xi^2} + \frac{1}{A}\frac{dA}{d\xi}\frac{d\theta}{d\xi} - \frac{dA_s}{d\xi}\theta = 0$$

by using

$$\theta = T - T_\infty$$

and

$$\xi = \bar{h}_cz/kA$$

38. The parabolic spine of Figure P3.19 is described by the following differential equation:

$$z^2\frac{d^2T}{dz^2} + z\frac{dT}{dz} - \frac{2\bar{h}_cP}{kA}\sqrt{L}z^{3/2}(T - T_\infty) = 0$$

Transform this equation by using

$$\theta = T - T_\infty$$

$$\xi = z/z_0$$

Express the result in terms of z_0, then let

$$z_0^{3/2} = \frac{k\delta}{2\bar{h}_cL^{1/2}}$$

and simplify accordingly.

39. The pin-fin problem solved numerically is shown schematically in Figure 3.16, in which the fin is divided into four sections. Repeat the development for such a problem by dividing the region into three sections. Solve for θ in any convenient way, using $mL = 2$. Graph θ vs. ξ for the three-section formulation, the four-section formulation, and the exact solution.

3.9.5 NUMERICAL METHODS FOR ONE-DIMENSIONAL PROBLEMS

40. Repeat Problem 39 for a five-section formulation. Graph θ vs. ξ for the five-section formulation and compare to the exact solution. Let $mL = 2$.

41. Figure P3.20 shows a cup of coffee with a spoon partly submerged. The coffee is at 90°C, which is the temperature of the spoon at the liquid surface. The spoon handle

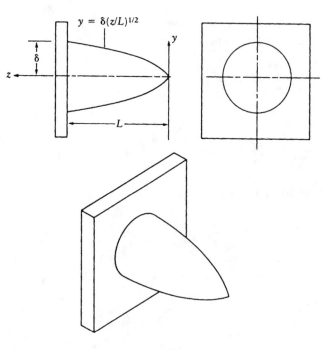

FIGURE P3.19

FIGURE P3.20

extends 80 mm above the coffee surface and can be approximated as a rod 7 mm in diameter. The ambient temperature is 22°C, and the convection coefficient is 3 W/(m²·K). The spoon is made of stainless steel (assume Type 304). Graph temperature vs. length from the free surface. Use the exact solution and the four-section numerical formulation. Compare the results.

42. A glass stirring rod is used in Example 3.5. Repeat the (numerical formulation) calculations for two metal rods:
    (a) An aluminum rod.
    (b) A stainless steel (Type 347) rod.
    (c) Graphically compare the temperature profile with that of the glass rod.

43. Repeat Problem 39 for a six-section formulation. Compare the results to the exact solution.

3.9.6 NUMERICAL METHODS FOR TWO-DIMENSIONAL PROBLEMS

44. Figure 3.23 is the sketch used in performing a heat balance on node *m, n*. Equation 3.57 resulted. Perform a heat balance on the cylindrical geometry of Figure P3.21 to obtain a finite-difference equation for an interior node in a cylindrical system.

45. Derive item #4 of Table 3.3.

46. Derive item #5 of Table 3.3.

47. A square slab 0.6 m on a side is shown in Figure P3.22. The top surface is at 200°C. The two vertical sides are at 50°C while the lower surface is at 0°C. Divide up the region into nine squares containing four internal nodes. Write the finite-difference equations for each node.

48. Solve the system of equations for temperature in Problem 47. If the material is concrete, determine the heat flow per unit length to or from each surface.

49. A rectangular plate is shown in Figure P3.23. The top surface is at 150°F, the left surface is at 100°F, and the lower surface is at 50°F. The right surface is a convective boundary

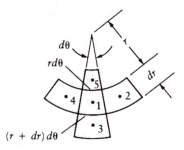

FIGURE P3.21

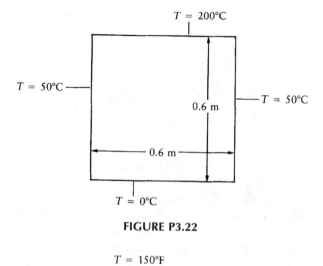

FIGURE P3.22

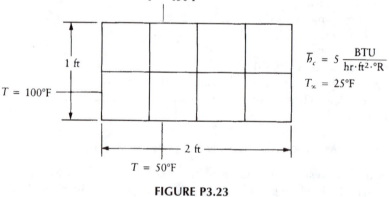

FIGURE P3.23

in which $\bar{h}_c = 5$ BTU/(hr·ft²·°R) and $T_\infty = 25°F$. The plate has dimensions of 2 ft by 1 ft. Divide the region into eight squares and write the finite-difference equations for the nodes.

50. Solve the system of equations for temperature in Problem 49 if the material is type 304 stainless steel. Determine also the heat flow per unit length to or from each surface.

51. A rectangular slab 50 by 75 cm is shown in Figure P3.24. The top surface is maintained at 100°C; the left and lower surfaces are at 50°C. The right surface is insulated. Divide the region into six squares and write the finite-difference equations for each node.

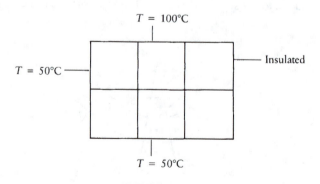

FIGURE P3.24

52. Solve the system of equations for temperature in Problem 51. If the material is cast iron, determine the heat flow per unit length to or from each surface.

53. The cross section of a corner of a wall is shown in Figure P3.25. For the boundary conditions and dimensions shown in the figure, divide the region into a number of squares containing nine internal nodes. Write the finite-difference equations.

54. Solve the system of equations for temperature in Problem 53. If the wall is of masonry brick, determine the heat flow per unit length to or from each isothermal surface.

55. A portion of the wall of a vacuum chamber is shown in Figure P3.26. The upper surface is at 100°C, while the left surface is at 200°C. The lower surface is a line of symmetry. The right surface radiates heat away from the wall. Divide the region into six squares and write the finite-difference equations for each node.

56. Solve the system of equations for temperature in Problem 55. If the wall is made of silica brick, determine the heat flow per unit length from the isothermal surface.

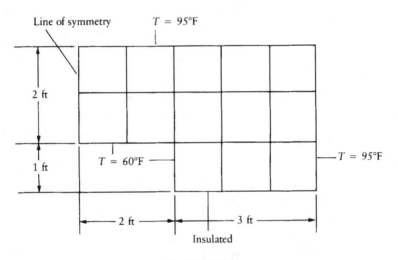

FIGURE P3.25

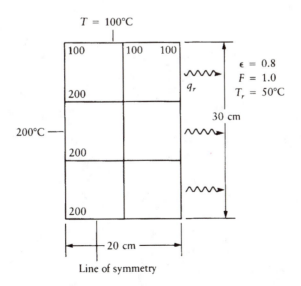

FIGURE P3.26

57. Item 2 of Table 3.3 is for convection from a surface. Item 7 is for radiation from a surface. Derive a finite-difference expression for a surface that transfers heat away by convection and radiation.

4 Unsteady-State Heat Conduction

CONTENTS

4.1 INTRODUCTION

In this chapter, we consider transient-conduction problems in which there is no internal heat generation. Temperature will therefore vary with location within the system and with time. Examples of heat transfer in transient systems include quenching of materials to attain a desired microstructure, cooling of canned foods as a preservative measure, heating tubes to be joined together by brazing in an oven, cooking foods and then combining them into what will become frozen dinners, etc. The list of applications is long.

In some problems, such as in predicting the required size of an oven, finding the heat-transfer rate is important. For other systems, however, the temperature history is the desired quantity of interest, as in the material-treatment process. In still other problems, such as in roasting a turkey, the time to attain a certain temperature is important.

We will approach transient-conduction problems in much the same way as was done in previous chapters. The general equation for the geometry is written. Appropriate functional dependencies—such as $T = T(x, t)$—are identified and unimportant terms are canceled. The resulting differential equation must then be solved subject to the boundary and initial conditions. For relatively simple geometries, the above method works well, but for more complicated systems, we will resort to other methods of solution. These include numerical and graphical techniques.

We can identify three types of problems that arise for a transient problem. To illustrate, consider a slab of thickness $2L$ that extends to infinity into the page, as shown in Figure 4.1. Also shown are the temperature vs. x axes. Initially, the plate is at temperature T_i; then at some time zero, fluid streams at T_∞ are directed at both left and right faces.

Heat is transferred by convection at each surface. As the surface temperature decreases, heat is transferred from the center of the slab to the surface, then to the fluid. Now, if the material itself is copper, the temperature response within the plate is considerably different from that if the material is glass. The response has to do with what is called the *internal thermal resistance* of the material. Furthermore, if the convection coefficient is very high, then the surface temperature almost becomes identical to the free-stream temperature quickly. Alternatively, for a low convection coefficient, a large temperature difference exists between the surface and the free stream. The value of the

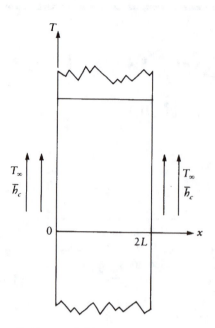

FIGURE 4.1   An infinite plate cooled by two fluid streams.

convection coefficient controls what is known as the *surface resistance* to heat transfer. For some problems, the internal resistance is negligible compared to the surface resistance. For other problems, the surface resistance is negligible. For still others, both effects must be taken into account. In this chapter, we will examine problems that fit into each of these categories.

4.2   SYSTEMS WITH NEGLIGIBLE INTERNAL RESISTANCE

Consider an object at an initial temperature T_i that is being cooled (or heated) by a fluid stream at temperature T_∞. If the object has a high thermal conductivity, then its internal thermal resistance to heat flow is small. This means that the temperature profile within the object is uniform and everywhere equal (or almost equal) to the surface temperature. At any instant in time, there are no temperature gradients within the object—only a temperature difference between the surface and the free stream. The heat transferred between the object and the fluid is controlled almost exclusively by the surface resistance, i.e., the value of the convection coefficient $\bar{h}_c$. The heat-transfer process in this case is known as Newtonian cooling (or heating).

Because temperature does not change (significantly) within the solid, we can formulate a model by performing an energy balance on the object. Figure 4.2 shows a heated object immersed in a

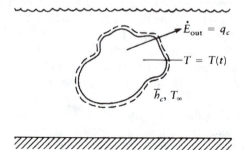

FIGURE 4.2   An object being cooled in a fluid.

cold fluid. The object itself is the system we select for analysis. An energy balance for the system states that the decrease in stored energy equals the heat transferred by convection:

$$-\dot{E}_{st} = \dot{E}_{out} \tag{4.1}$$

The rate of decrease in stored energy $(\dot{E}_{st})$ is the decrease in internal energy of the system $(\Delta u/\Delta t)$. We therefore write

$$\dot{E}_{st} = \frac{\Delta u}{\Delta t} = mc\frac{dT}{dt} = \rho_s V_s c\frac{dT}{dt} \tag{4.2}$$

where ρ_s is the density of the solid, V_s is the volume of the solid, and c is its specific heat. The temperature of the solid changes only with time (dT/dt). The heat transfer by convection is

$$q_c = \bar{h}_c A_s(T - T_\infty) \tag{4.3}$$

in which A_s is the surface area of the solid. Equating 4.2 and 4.3 gives

$$\rho_s V_s c\frac{dT}{dt} = -\bar{h}_c A_s(T - T_\infty) \tag{4.4}$$

with the initial condition that

$$\text{I.C.} \quad t = 0, T = T_i \tag{4.5}$$

To simplify the differential equation, we introduce the following variable:

$$\theta = \frac{T - T_\infty}{T_i - T_\infty} \tag{4.6}$$

The time derivative becomes

$$\frac{dT}{dt} = (T_i - T_\infty)\frac{d\theta}{dt}$$

Substituting into Equation 4.4 gives

$$\rho_s V_s c(T_i - T_\infty)\frac{d\theta}{dt} = -\bar{h}_c A_s(T_i - T_\infty)$$

or

$$-\frac{\rho_s V_s c}{\bar{h}_c A_s}\frac{d\theta}{dt} = \theta$$

Rearranging into a form more suitable for integration brings

$$-\frac{d\theta}{\theta} = \frac{\bar{h}_c A_s}{\rho_s V_s c}dt \tag{4.7}$$

Now the temperature within the solid will vary from T_i at $t = 0$ to a temperature T at some future time t. Therefore, θ varies from 1 to 0. Integrating Equation 4.7 yields

$$-\int_1^\theta \frac{d\theta}{\theta} = \frac{\bar{h}_c A_s}{\rho_s \mathcal{V}_s c} \int_0^t dt$$

$$-\ln(\theta) = \frac{\bar{h}_c A_s}{\rho_s \mathcal{V}_s c} t$$

Solving,

$$\theta = \frac{T - T_\infty}{T_i - T_\infty} = \exp\left[-\left(\frac{\bar{h}_c A_s}{\rho_s \mathcal{V}_s c}\right)t\right] \tag{4.8}$$

The quantity $(\rho_s \mathcal{V}_s c / \bar{h}_c A_s)$ is called the *thermal time constant* for the geometry and has dimensions of time T (s). The numerator of the time constant is called the *lumped thermal capacitance* of the solid, and $1/\bar{h}_c A_s$ is recognized as the convective resistance.

Example 4.1

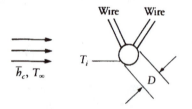

Wire   Wire

T_i

$\bar{h}_c, T_\infty$

D

FIGURE 4.3 A spherical thermo-couple junction in a fluid stream.

A thermocouple is used to measure the temperature of a gas stream. The thermocouple junction is shown schematically in Figure 4.3. Take the junction to be spherical and to have the following properties:

$k = 12$ BTU/(hr·ft·°R)
$c = 0.1$ BTU/(lbm·°R)
$D = 0.025$ in
$\rho_s = 525$ lbm/ft³

The initial temperature of the sphere is 65°F. The fluid temperature is 140°F, and the velocity is such that the convection coefficient is 80 BTU/(hr·ft²·°R). Determine the response time of the junction.

Solution

The thermocouple junction is a metal having a relatively high thermal conductivity and a small diameter. When the sphere is immersed in a high-velocity stream of different temperature, heat is transferred at the upstream side at a different rate than at the downstream side.

Assumptions

1. Metal properties are constant during the process.
2. The convection coefficient is an average that applies over the entire surface area of the sphere.
3. The controlling parameter during heat transfer is the convective resistance.

Equation 4.8 thus applies:

$$\frac{T - T_\infty}{T_i - T_\infty} = \exp\left[-\left(\frac{\bar{h}_c A_s}{\rho_s \mathcal{V}_s c}\right)t\right]$$

Density, specific heat, diameter, and the convection coefficient are given, as are the temperatures. The surface area of the sphere is

$$A_s = \pi D^2$$

The volume is

$$V_s = \frac{\pi D^3}{6}$$

Therefore, $A_s/V_s = 6/D$. Evaluating the reciprocal of the time constant gives

$$\frac{\bar{h}_c A_s}{\rho_s V_s c} = \frac{6\bar{h}_c}{\rho_s c D} = \frac{6(80)}{525(0.1)(0.025/12)} = 4388.6/hr$$

Substituting yields

$$\frac{T - 140}{65 - 140} = \exp(-4388.6t)$$

Now if $T = 140°F$ is substituted into the above equation, the time required becomes infinite—a peculiarity of the model. We therefore select $139°F$ arbitrarily to estimate the time. Thus,

$$\ln \frac{139 - 140}{65 - 140} = -4388.6t$$

Solving,

$$t = 9.84 \times 10^{-4} \text{ hr} = 3.5 \ s \qquad \square$$

The lumped-capacitance equation was used to solve this example. It was remarked that the size of the sphere was small, and for other various reasons it was permissible to use Equation 4.8 and obtain accurate results. Obviously, it is necessary to have a more quantitative basis for using the lumped-capacitance equation. Let us examine the exponent of Equation 4.8 and rearrange it to obtain

$$\frac{\bar{h}_c A_s t}{\rho_s V_s c} = \frac{\bar{h}_c V_s}{k} \frac{k}{A_s \rho_s c} \frac{t A_s^2}{V_s^2} = \frac{\bar{h}_c V_s}{k A_s} \frac{\alpha t A_s^2}{V_s^2}$$

Notice that the time constant itself does not contain thermal conductivity k but that k is introduced into the parameter. The ratio of volume to surface area for an object is often used as a characteristic length; i.e.,

$$L_c = \frac{V_s}{A_s}$$

The argument of the exponent in the lumped-capacitance equation now becomes

$$\frac{\bar{h}_c A_s t}{\rho_s V_s c} = \frac{\bar{h}_c L_c}{k} \frac{\alpha t}{L_c^2} \qquad (4.9)$$

The term $\bar{h}_c L_c / k$ $(= \bar{h}_c V_s / k A_s)$ is defined as the Biot number for the object. The term $\alpha t / L_c^2$ $(= \alpha t A_s^2 / V_s^2)$ is known as the Fourier number.

The Biot number is a dimensionless ratio of convection to conduction resistance to heat transfer. Thus, the Biot number gives an indication of the temperature drop within the solid compared to the temperature difference between the solid surface and the fluid. It has been found that, if the following condition is satisfied for the Biot number,

$$\mathrm{Bi} = \frac{\bar{h}_c L_c}{k} = \frac{\bar{h}_c V_s}{k A_s} \le 0.1 \tag{4.10}$$

then the lumped-capacitance equation can be used with little error. In solving transient-conduction problems, it is therefore prudent to *first* calculate the Biot number.

As discussed, the criterion for use of the lumped-capacitance method is that the Biot number should be less than 0.1. The influence of the Biot number on the transient temperature distribution is shown in Figure 4.4. Shown are three cases of interest; each is of a plane wall initially at temperature T_i exposed to a fluid at $T_\infty < T_i$. Figure 4.4a shows the case where the Biot number is less than 0.1. The temperature profile at any time is (nearly) a constant within the wall. The controlling resistance to heat flow is the convective resistance. Figure 4.4c shows the case where the controlling resistance to heat transfer is due to conduction. The convective resistance is negligible. Figure 4.4b shows an example in which both conduction and convection resistances must be accounted for.

The Fourier number is a dimensionless time parameter. It represents the ratio of heat transfer by conduction to the energy-storage rate within the material. In terms of the dimensionless ratios just discussed, Equation 4.8 becomes

$$\frac{T - T_\infty}{T_i - T_\infty} = \exp[-(\mathrm{Bi})(\mathrm{Fo})] \tag{4.11}$$

For certain types of problems, the ability to calculate a heat-transfer rate is required. Two separate calculations can be made: the *instantaneous heat rate q* and a *cumulative heat rate* $\tilde{Q}$. The instantaneous heat rate at any time t is found with

$$q = \rho_s V_s c \frac{dT}{dt} \tag{4.12}$$

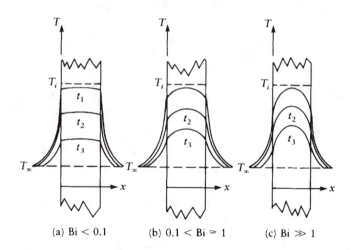

(a) Bi < 0.1   (b) 0.1 < Bi ≈ 1   (c) Bi ≫ 1

FIGURE 4.4 Relationship between the Biot number and the temperature profile.

Substituting for dT/dt from Equation 4.8 gives

$$q = \bar{h}_c A_s (T_\infty - T_i) \exp\left[\left(-\frac{\bar{h}_c A_s}{\rho_s V_s c}\right) t\right] \tag{4.13}$$

The instantaneous heat rate has dimensions of energy per unit time (W or BTU/hr). If the instantaneous heat rate is integrated over the entire time interval during which heat is transferred, the cumulative heat rate results:

$$\tilde{Q} = \int_0^t q \, dt \tag{4.14}$$

Combining with Equation 4.13 gives

$$\tilde{Q} = \rho_s V_s c (T_\infty - T_i) \left\{ 1 - \exp\left[\left(-\frac{\bar{h}_c A_s}{\rho_s V_s c}\right) t\right] \right\} \tag{4.15}$$

The cumulative heat rate has dimensions of energy (J or BTU).

Example 4.2

A 5-cm diameter, 60-cm long aluminum cylinder initially at 50°C is submerged in an ice-water bath at 2°C. The unit surface conductance between the metal and the bath is 550 W/(m²·K). (a) Determine the temperature of the aluminum after 1 minute. (b) Calculate the cumulative heat rate for the 1 minute.

Solution

When the aluminum is submerged, natural convection effects cause a flow of liquid past the cylinder, because liquid near the surface becomes slightly warmer than that in the surrounding bath. Aluminum properties change somewhat during cooling.

Assumptions

1. The surface conductance $\bar{h}_c$ is an average value that is constant and applicable to the entire surface.
2. Aluminum properties are constant.

From Appendix Table B.1 for aluminum

$$k = 236 \text{ W/(m·K)}$$
$$c_p = 896 \text{ J/(kg·K)}$$
$$\text{sp.gr.} = 2.702; \; \rho_s = 2\,702 \text{ kg/m}^3$$

The cylinder volume and surface area are

$$V_s = \frac{\pi D^2 L}{4} = \frac{\pi (0.05)^2 (0.6)}{4} = 0.001\,18 \text{ m}^2$$

$$A_s = \frac{2\pi D^2}{4} + \pi D L = \frac{\pi (0.05)^2}{2} + \pi (0.05)(0.6) = 0.098 \text{ m}^2$$

The Biot number is found with Equation 4.10:

$$\text{Bi} = \frac{\bar{h}_c V_s}{k A_s} = \frac{[550 \text{ W/(m}^2 \cdot \text{K})](0.001\ 18 \text{ m}^2)}{[236 \text{ W/(m} \cdot \text{K})](0.102 \text{ m}^2)} = 0.028 < 0.1$$

With a Biot number less than 0.1, the lumped-capacitance equations apply—Equations 4.8 for temperature and 4.15 for cumulative heat rate. Equation 4.8 is used to find the temperature of the aluminum:

$$\frac{T - T_\infty}{T_i - T_\infty} = \exp\left[-\left(\frac{\bar{h}_c A_s}{\rho_s V_s c}\right)t\right]$$

After rearranging and substituting, the temperature after one minute becomes

$$T = T_\infty + (T_i - T_\infty)\exp\left[-\left(\frac{\bar{h}_c A_s}{\rho_s V_s c}\right)t\right] = 2 + (50 - 2)\exp\left\{\frac{-[550(0.102)](1)(60)}{(2\ 702)(0.001\ 18)(896)}\right\} = 2 + 48(0.308)$$

Solving,

(a) $T = 16.8°C$

Equation 4.15 gives the cumulative heat transferred as

$$\tilde{Q} = \rho_s V_s c (T_\infty - T_i)\left\{1 - \exp\left[\left(-\frac{\bar{h}_c A_s}{\rho_s V_s c}\right)t\right]\right\} = 2\ 702(0.001\ 18)(896)(2 - 50)(1 - 0.308)$$

Solving,

(b) $$\tilde{Q} = 94\ 890 \text{ J} = 94.9 \text{ kJ} \qquad \square$$

4.3 SYSTEMS WITH FINITE INTERNAL AND SURFACE RESISTANCES

The previous section dealt with a system in which the internal resistance was negligible. This section deals with the case in which internal and surface resistances are both significant. A variety of problems that fall into this category have been solved.

Results appear in the literature and in several texts on conduction. In general, such solutions are complex in form, often involving series solutions. Calculations made from such solutions are presented here in graphical form (known as Heisler charts and Grober charts) for three common geometries: the finite slab or plane wall, the infinite cylinder, and the sphere. We will also consider the solution for a semi-infinite slab.

Let us first consider a plate of width 2L that extends to infinity into the page, as shown in Figure 4.5. Also shown are the temperature vs. x axes. Initially, the plate is at a uniform temperature T_i. At some time zero, the left and right surfaces are cooled by a fluid at temperature T_∞. Our interest is in predicting the temperature profile with time and the heat-transfer rate.

Figure 4.6a is a Heisler chart for an infinite plate of thickness 2L. Conduction is in the x direction. The vertical axis of the graph is of a dimensionless temperature $(T_c - T_\infty)/(T_i - T_\infty)$ where T_c is the temperature at the centerplane of the plate. The dimensionless temperature varies from

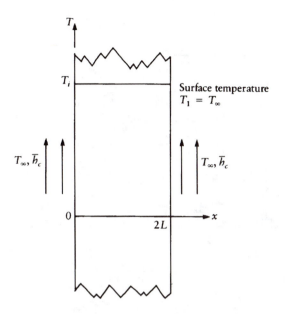

FIGURE 4.5   An infinite plate of thickness $2L$.

0.001 to 1.0. The horizontal axis is the Fourier number, which varies from 0 to 700. Lines that appear on the chart are for various values of the reciprocal of the Biot number, ranging from 0.05 to 100. In the Biot and Fourier numbers the characteristic length is one-half the plate thickness. A wide variety of problems are represented.

The Heisler chart of Figure 4.8b also applies to the infinite-plate problem. The dimensionless temperature $(T - T_\infty)/(T_c - T_\infty)$ appearing on the vertical axis ranges from 0 to 1.0. The horizontal axis is of the reciprocal of the Biot number, which ranges from 0.01 to 100. Lines on the chart are for various values of x/L. Figure 4.8a can be used to calculate the midplane temperature, which in turn can be used with Figure 4.8b to determine the temperature at any other x location within the plate.

Figure 4.8c is a Grober chart for the infinite plate. The horizontal axis is Bi^2Fo, ranging from 10^{-5} to 10^4. The vertical axis is of a dimensionless heat-flow ratio $\tilde{Q}/\tilde{Q}_0$, where $\tilde{Q}_0$ is the initial internal energy of the wall relative to the fluid temperature T_∞. By definition,

$$\tilde{Q}_0 = \rho_s V_s c (T_i - T_\infty)$$

The parameter $\tilde{Q}$ is the total amount of energy that has passed through the wall up to any time t of interest. The maximum internal energy change that can occur is $\tilde{Q}_0$, and $\tilde{Q}/\tilde{Q}_0$ is a fraction of that total. $\tilde{Q}/\tilde{Q}_0$ varies from 0 to 1 on the Grober chart. Lines on the chart are of the Biot number, which varies from 0.001 to 50.

The infinite-cylinder problem we consider is one in which the cylinder is at an initial temperature T_i, and it is suddenly exposed to a fluid of temperature T_∞ with surface coefficient $\bar{h}_c$. The cylinder is long enough (at least having a length-to-radius ratio ≥ 10) such that heat flow occurs only or primarily in the radial direction.

Figure 4.7a is a Heisler chart for an infinite cylinder. Again the vertical axis is of a dimensionless centerline temperature ranging from 0.001 to 1.0. The horizontal axis is of the Fourier number, which ranges from 0 to over 300. Various lines on the chart are of the reciprocal of the Biot number, ranging from 0 to 100. The characteristic length in the Fourier and Biot numbers is the cylinder radius.

Figure 4.7b, also for an infinite cylinder, is of a dimensionless temperature vs. the reciprocal of the Biot number. The dimensionless temperature $(T - T_\infty)/(T_c - T_\infty)$ varies from 0 to 1.0 while

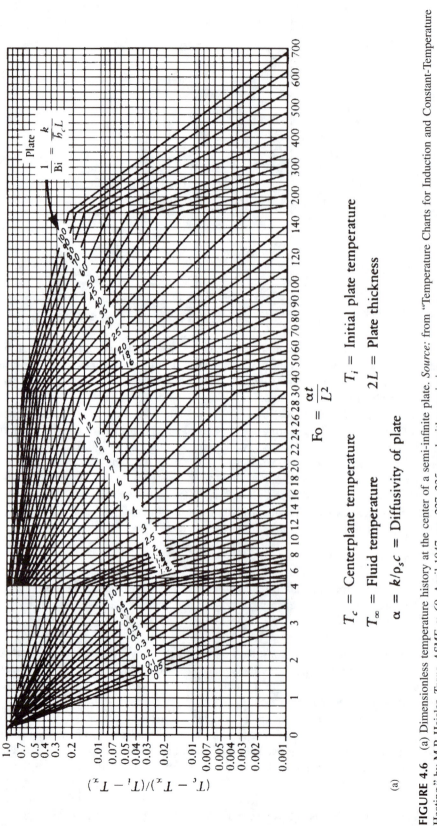

(a)

FIGURE 4.6 (a) Dimensionless temperature history at the center of a semi-infinite plate. *Source:* from "Temperature Charts for Induction and Constant-Temperature Heating," by M.P. Heisler, *Trans. ASME*, v. 69, April 1947, pp. 227–235, used with permission.

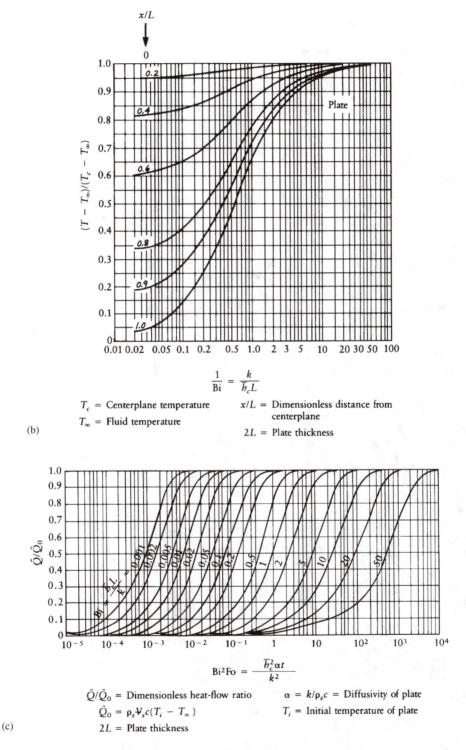

$$\frac{1}{\text{Bi}} = \frac{k}{h_c L}$$

T_c = Centerplane temperature x/L = Dimensionless distance from
T_∞ = Fluid temperature   centerplane
                                                        $2L$ = Plate thickness

(b)

$$\text{Bi}^2\text{Fo} = \frac{h_c^2 \alpha t}{k^2}$$

$\tilde{Q}/\tilde{Q}_0$ = Dimensionless heat-flow ratio $\alpha = k/\rho_s c$ = Diffusivity of plate
$\tilde{Q}_0 = \rho_s V_s c(T_i - T_\infty)$ T_i = Initial temperature of plate
$2L$ = Plate thickness

(c)

FIGURE 4.6   (b) Dimensionless temperature distribution in a semi-infinite plate, and (c) change in internal-energy ratio as a function of dimensionless time for a semi-infinite plate. *Sources:* (b) from "Temperature Charts for Induction and Constant-Temperature Heating," by M.P. Heisler, *Trans. ASME,* v. 69, April 1947, pp. 227–235, used with permission; (c) from *Fundamentals of Heat Transfer,* by H. Grober and S. Erk, McGraw-Hill, 1961, Chapter 3, used with permission.

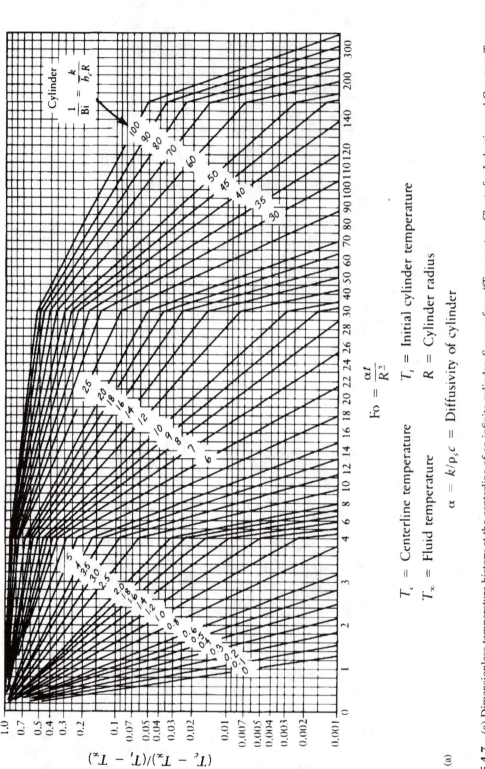

$$Fo = \frac{\alpha t}{R^2}$$

T_c = Centerline temperature T_i = Initial cylinder temperature

T_∞ = Fluid temperature R = Cylinder radius

$\alpha = k/\rho_s c$ = Diffusivity of cylinder

FIGURE 4.7 (a) Dimensionless temperature history at the centerline of an infinite cylinder. *Source:* from "Temperature Charts for Induction and Constant-Temperature Heating," by M.P. Heisler, *Trans. ASME,* v. 69, April 1947, pp. 227–235, used with permission.

(a)

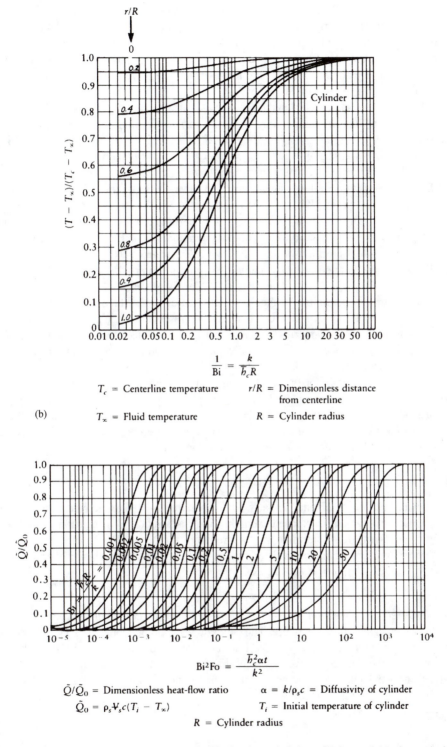

$$\frac{1}{\text{Bi}} = \frac{k}{\bar{h}_c R}$$

T_c = Centerline temperature r/R = Dimensionless distance from centerline

(b) T_∞ = Fluid temperature R = Cylinder radius

$$\text{Bi}^2\text{Fo} = \frac{\bar{h}_c^2 \alpha t}{k^2}$$

$\bar{Q}/\bar{Q}_0$ = Dimensionless heat-flow ratio $\alpha = k/\rho_s c$ = Diffusivity of cylinder

$\bar{Q}_0 = \rho_s \mathcal{V}_s c(T_i - T_\infty)$ T_i = Initial temperature of cylinder

(c) R = Cylinder radius

FIGURE 4.7   (b) Dimensionless temperature distribution in an infinite cylinder, and (c) change in internal-energy ratio as a function of dimensionless time for an infinite cylinder. *Sources:* (b) from "Temperature Charts for Induction and Constant-Temperature Heating," by M.P. Heisler, *Trans. ASME,* v. 69, April 1947, pp. 227–235, used with permission; (c) from *Fundamentals of Heat Transfer,* by H. Grober and S. Erk, McGraw-Hill, 1961, Chapter 3, used with permission.

1/Bi ranges from 0.01 to 100. Lines on the chart are of various values of r/R. Figure 4.7a can be used to find the centerline temperature of the cylinder. Figure 4.7b can be used with the centerline temperature to determine the temperature at any radial location within the cylinder.

Figure 4.7c is a Grober chart of dimensionless internal energy change vs. Bi^2Fo for an infinite cylinder. $\tilde{Q}/\tilde{Q}_0$ varies from 0 to 1 and Bi^2Fo ranges from 10^{-4} to 10^5. Lines on the chart are of the Biot number, which ranges from 0.001 to 50.

The sphere problem is one in which the sphere exists at some initial temperature T_i, and suddenly it is immersed in a fluid of temperature T_∞ with surface coefficient $\bar{h}_c$. Figure 4.8a is of a dimensionless center temperature (ranging from 0.001 to 1.0) vs. the Fourier number (ranging from 1 to 250). Various lines on the graph are of the reciprocal of the Biot number (ranging from 0 to 100).

Figure 4.8b is a dimensionless temperature $(T - T_\infty)/(T_c - T_\infty)$ vs. the reciprocal of the Biot number. Dimensionless temperature ranges from 0 to 1.0 while 1/Bi ranges from 0.01 to 100. Lines on the graph are of r/R, where R is the sphere radius.

Figure 4.8c is the Grober chart for a sphere. The vertical axis is of a dimensionless internal-energy-change ratio ranging from 0 to 1. The horizontal axis is of Bi^2Fo, which varies from 10^{-5} to 10^4. Lines on the chart are for Biot numbers, which range from 0.001 to 50.

Recall that for the lumped-capacity analysis (negligible internal resistance) the Biot number should be less than 0.1. If $Bi = 0.1$, then $1/Bi = 10$. This is represented on the Heisler charts. On the other hand, for the negligible-surface-resistance case, the surface coefficient is large; i.e., $\bar{h}_c \to \infty$. For this case, $1/Bi = k/\bar{h}_cL = 0$. This too is represented on the charts. The Heisler charts are general solutions for the geometries to which each applies. (Note, however, that the characteristic length used in the Biot number for the lumped-capacity method is different from the characteristic length used in the Heisler and Grober charts.)

Other charts are available for the infinite plate, infinite cylinder, and sphere problems. These are known as Gurney-Lurie Charts,[*] and they are graphs of dimensionless temperature vs. dimensionless distance for each geometry. They provide temperature at any location, whereas the Heisler Charts require the use of two figures (the first two in each geometry) to determine temperature anywhere (other than the center temperature) within the solid.

Example 4.3

A concrete wall is 24 cm thick and, initially, its temperature is 50°C. One side of the wall is well insulated, while the other side is exposed to combustion gases at 600°C. The heat transfer coefficient on this side is 30 W/(m²·K). Determine the time required for the insulated surface to reach 150°C, and the temperature profile in the wall at that instant. Use the following properties: $k = 1.25$ W/(m²·K), $\rho_s = 550$ kg/m³, $c = 890$ J/(kg·K).

Solution

Heat is transferred by convection to one side of the wall and by conduction through it. Properties of the wall will change with temperature. The concrete wall can be modeled using the infinite plate solution, making suitable assumptions. The infinite plate solution will have a temperature profile that is symmetric about its centerline. A plane of symmetry is treated the same mathematically as an insulated surface, so the half thickness of the infinite plate problem $(1/2 \times 2L)$ equals the actual thickness of the wall in this example.

Assumptions

1. The convection coefficient is constant over the exposed side of the wall.
2. Material properties are constant.
3. One dimensional heat flow exists in the wall, and the infinite plate solution charts apply.

---

[*] H.P. Gurney and J. Lurie, *Ind. Eng. Chem.* 15, 1170 (1923).

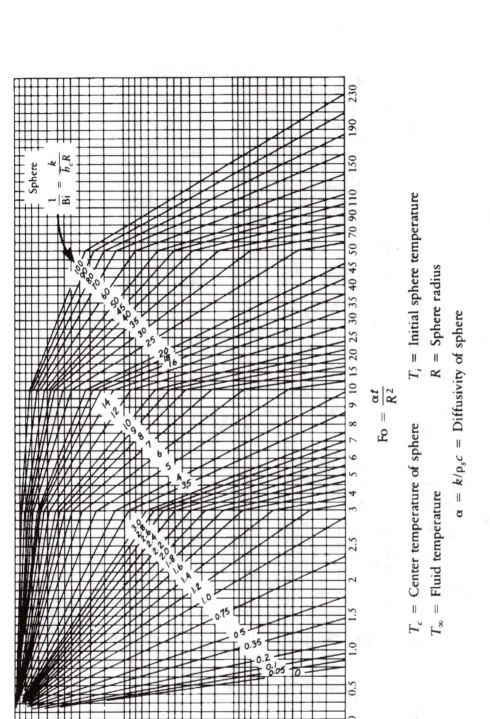
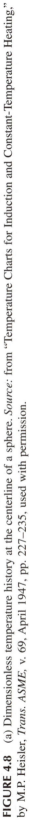

FIGURE 4.8 (a) Dimensionless temperature history at the centerline of a sphere. *Source:* from "Temperature Charts for Induction and Constant-Temperature Heating," by M.P. Heisler, *Trans. ASME*, v. 69, April 1947, pp. 227–235, used with permission.

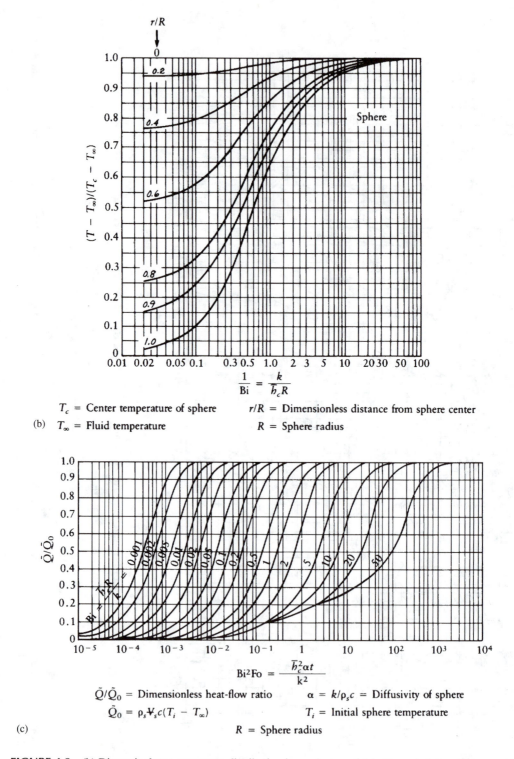

T_c = Center temperature of sphere r/R = Dimensionless distance from sphere center

(b) T_∞ = Fluid temperature R = Sphere radius

$\tilde{Q}/\tilde{Q}_0$ = Dimensionless heat-flow ratio $\alpha = k/\rho_s c$ = Diffusivity of sphere

$\tilde{Q}_0 = \rho_s V_s c(T_i - T_\infty)$ T_i = Initial sphere temperature

(c) R = Sphere radius

FIGURE 4.8   (b) Dimensionless temperature distribution in a sphere, and (c) change in internal-energy ratio as a function of dimensionless time for a sphere. *Sources:* (b) from "Temperature Charts for Induction and Constant-Temperature Heating," by M.P. Heisler, *Trans. ASME,* v. 69, April 1947, pp. 227–235, used with permission; (c) from *Fundamentals of Heat Transfer,* by H. Grober and S. Erk, McGraw-Hill, 1961, Chapter 3, used with permission.

The first step in this (or any unsteady) problem is to calculate the Biot number to determine if a lumped capacity analysis can be used. For this concrete wall, we have:

$$\mathrm{Bi} = \frac{\bar{h}_c L}{k}$$

where L in this equation is the thickness of the wall. Substituting,

$$\mathrm{Bi} = \frac{30(0.24)}{1.25} = 5.76$$

Because this value is greater than 0.1, the lumped capacity approach will not give an accurate result. We can, however, use Figure 4.6 to obtain a solution. Note that to use that figure, we must identify a wall thickness $2L$. Recall that, for this problem, we are modeling a wall whose thickness is $2(0.24\ \mathrm{m})$, and whose temperature profile is symmetric about its centerline. Therefore, $L = 0.24$ m.
   The reciprocal of the Biot number (required to use the Figure 4.6) is

$$\frac{1}{\mathrm{Bi}} = \frac{k}{\bar{h}_c L} = \frac{1}{5.76} = 0.174$$

The dimensionless temperature is $(T_c - T_\infty)/(T_i - T_\infty)$ where T_c is the temperature at the plate centerline, which, due to symmetry, is at the surface of the insulated side. We therefore write

$$\frac{T_c - T_\infty}{T_i - T_\infty} = \frac{150°\mathrm{C} - 600°\mathrm{C}}{50°\mathrm{C} - 600°\mathrm{C}} = 0.82$$

Using these values, Figure 4.6a gives at

$$\mathrm{Fo} = \frac{\alpha t}{L^2} \approx 0.4$$

It is difficult to read the graph accurately at these values. Nevertheless, we calculate the time using the Fourier number

$$t = \frac{L^2 \mathrm{Fo}}{\alpha}$$

The diffusivity is calculated from the given properties:

$$\alpha = \frac{k}{\rho_s c} = \frac{1.25}{550(890)} = 2.55 \times 10^{-6}\ \mathrm{m^2/s}$$

and the time is

$$t = \frac{(0.24)^2(0.4)}{2.55 \times 10^{-6}}$$

or

$$t = 9\ 022\ \mathrm{s} = 2.5\ \mathrm{hr}$$

To determine the temperature profile, we use Figure 4.6b. At a value of $1/Bi = 0.174$, we read the following values:

x/L	0.0	0.2	0.4	0.6	0.8	0.9	1.0
$\dfrac{T_c - T_\infty}{T_i - T_\infty}$	1.0	0.97	0.86	0.68	0.48	0.36	0.24

We use these values with the dimensionless centerline temperature we calculated earlier (= 0.82). So, for example, at $x/L = 0.4$, we have

$$\frac{T - T_\infty}{T_i - T_\infty} = \frac{T - T_\infty}{T_c - T_\infty} \cdot \frac{T_c - T_\infty}{T_i - T_\infty} = 0.86(0.82) = 0.71$$

The temperature then is

$$T = 0.71(T_i - T_\infty) + T_\infty = 0.71(50 - 600) + 600 = 210°C$$

This temperature exists at $x/L = 0.4$, which corresponds to

$$x = 0.4(0.24) = 0.096 \text{ m} = 9.6 \text{ cm}$$

from the insulated side (i.e., the centerline). Repeating such calculations for the other positions yields the following results:

X/L	0.0	0.2	0.4	0.6	0.8	0.9	1.0
x, cm	0	4.8	9.6	14.4	19.2	21.6	24.0
T, °C	149	163	210	293	384	438	492

The wall temperature at the exposed side is 492°C, and the temperature at the insulated side is 149°C. The temperature profile is graphed in Figure 4.9. Had we used the lumped-capacitance approach, the equation would have predicted that the temperature within the wall is a constant, and clearly, Figure 4.9 shows that it is not. Admittedly, these results are approximate, due primarily to the inaccuracies associated with reading the figures. ❑

Example 4.4

Oranges are usually refrigerated as a preservative measure. However, some people prefer to eat oranges that are a little cooler than room temperature but not as cold as the refrigerator makes them. Determine the time it takes for an orange removed from a refrigerator to reach 20°C. Use the following conditions:

Refrigerated temperature = 4°C
Ambient room temperature = 23°C
Surface conductance = 6 W/(m²·K)
Thermal conductivity of an orange = 0.431 W/(m·K)[*]
Density of orange = 998 kg/m³
Specific heat of orange = 2 kJ/(kg·K)
Orange diameter = 105 mm

---

[*] Geonkoplis, Christie J., *Transport Processes and Unit Operations,* Allyn & Bacon, Boston, 1978, p. 629.

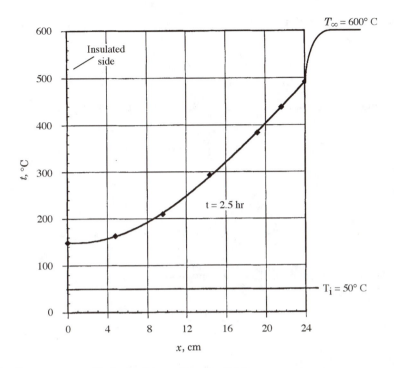

FIGURE 4.9   Temperature profile in the 24-cm slab after 2.5 hr.

Also, estimate the heat transferred to the ambient temperature during this time.

Solution

The time it takes for the entire orange to reach 20°C is difficult to calculate because, under the conditions specified, a temperature gradient would probably exist within the orange. So the entire orange might never be at 20°C unless 20°C is the ambient temperature. The rind of the orange has properties that differ from those of the meat within.

Assumptions

1. The time asked for in the problem statement is the time required for the temperature at the center to reach 20°C.
2. The convection coefficient is a constant over the surface area of the orange.
3. The orange is a sphere.
4. Orange properties are uniform and constant.

The first step is to calculate the Biot number for a lumped-capacitance approach:

$$\text{Bi} = \frac{\bar{h}_c V_s}{k A_s} = \frac{\bar{h}_c \pi D^3 / 6}{k \pi D^2} = \frac{\bar{h}_c D}{6k} = \frac{6(0.105)}{6(0.431)} = 0.243$$

The lumped-capacitance analysis does not apply, because Bi > 0.1. We therefore resort to using Figure 4.8. The Biot number for that figure (*different from that calculated above*) is

$$\text{Bi} = \frac{\bar{h}_c R}{k} = \frac{6(0.105/2)}{0.431} = 0.731$$

$$1/\text{Bi} = 1.36$$

The dimensionless temperature is found as

$$\frac{T_c - T_\infty}{T_i - T_\infty} = \frac{20 - 23}{4 - 23} = 0.158$$

For a dimensionless temperature of 0.158 and $1/Bi = 1.36$, Figure 4.8a gives

$$Fo \cong \frac{\alpha t}{R^2} = 1.05$$

The diffusivity is

$$\alpha = \frac{k}{\rho_s c} = \frac{0.431}{998(2\ 000)} = 2.159 \times 10^{-7}\ m^2/s$$

Rearranging the Fourier equation and substituting yields

$$t = \frac{1.05(0.105/2)^2}{2.159 \times 10^{-7}}$$

Solving,

$$t = 1.34 \times 10^4\ s = 3.7\ hr$$

The heat transferred during 3.7 hr is found by using Figure 4.8c. First, we calculate

$$Bi^2 Fo = (0.731)^2(1.05) = 5.61 \times 10^{-1}$$

Figure 4.8c shows

$$\frac{\tilde{Q}}{\tilde{Q}_0} = \frac{\tilde{Q}}{\rho_s V_s c(T_i - T_\infty)} = 0.7$$

The heat transferred is

$$\tilde{Q} = 0.7(998)(2\ 000)\left[\frac{\pi}{6}(0.105)^3\right](4 - 23) = -1.609 \times 10^4\ J$$

where the negative sign indicates that heat is transferred into the orange.   ❑

The last system we consider in this section is the semi-infinite solid. Figure 4.10 shows a semi-infinite solid with one of three types of possible boundary conditions. We consider first the condition where the surface temperature is known. The differential equation for the system is

$$\frac{\partial^2 T}{\partial x^2} = \frac{1}{\alpha}\frac{\partial T}{\partial t} \tag{4.16}$$

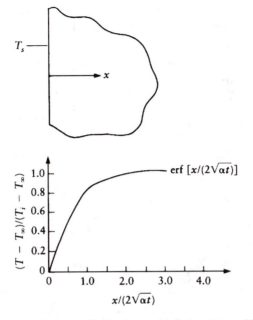

FIGURE 4.10 Solution for temperature profile for the semi-infinite-plate problem having constant surface temperature.

The boundary and initial conditions are

B.C.1. $x = 0$, $T = T_s$
B.C.2. $x = \infty$, $T = T_i$
I.C. $t = 0$, $T = T_i$ for all x

The differential equation can be solved subject to the boundary conditions by a number of methods. The final solution is

$$\frac{T - T_\infty}{T_i - T_\infty} = \frac{2}{\sqrt{\pi}} \int_0^{x/(2\sqrt{\alpha t})} \exp(-\lambda^2)\,d\lambda \tag{4.17}$$

It is not possible to evaluate the integral unless a specific value for its upper limit $x/(2\sqrt{\alpha t})$ is selected. So it is customary to present as a solution a chart where values of the upper limit are listed and the corresponding values of the right-hand side of Equation 4.17 are given. The right-hand side of Equation 4.17 is called the *error function* or probability integral and denoted as erf$[x/2\sqrt{\alpha t}]$. An error-function chart is provided in Appendix Table A.5 and graphed in Figure 4.10 as a solution to Equation 4.17. The heat flux at the surface is found as

$$q_s'' = -k\frac{\partial T}{\partial x}\bigg|_{x=0} = \frac{k(T_s - T_i)}{\sqrt{\pi\alpha t}} \tag{4.18}$$

Figures 4.11a and 4.11b show two other cases of heat transfer into a semi-infinite body. Figure 4.11a is for the case where, at the left boundary, the heat added is a constant. Figure 4.11b is for the case where convection at the surface is taken into account. Resultant temperature profiles for these cases are also shown. The solutions to these problems involve exponential functions and the *complementary error function* (erfc) defined as

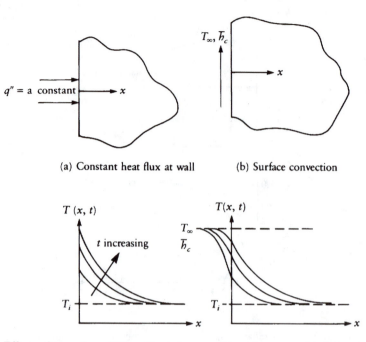

(a) Constant heat flux at wall   (b) Surface convection

FIGURE 4.11 Effects of two other additional boundary conditions for the semi-infinite-plate problem.

$$\text{erfc}[x/(2\sqrt{\alpha t})] = 1 - \text{erf}[x/(2\sqrt{\alpha t})]$$

Rather than write the solutions here, results are presented in graphical form in Figure 4.12. The vertical axis is of a dimensionless temperature $(T - T_i)/(T_\infty - T_i)$, which ranges from 0.01 to 1.0. The horizontal axis is the argument of the error function discussed earlier, $x/(2\sqrt{\alpha t})$, which varies from 0 to 1.5. Lines on the graph are of various values of the parameter $\text{Bi}\sqrt{\text{Fo}} = \bar{h}_c\sqrt{\alpha t}/k$. This parameter ranges from 0.05 to infinity. When the surface resistance is negligible (as in Figure 4.10), $\bar{h}_c \to \infty$. This case is also represented in the solution chart of Figure 4.12.

Example 4.5

Water pipes are run from a city water main to individual dwellings. During the winter months, the soil may be exposed to freezing temperatures for several months, and the soil depth above which the temperature stays at or below freezing is called the *freeze line*. To ensure that water in the buried pipe does not freeze, expand, and rupture the pipe, it is important that water pipes be located beneath the freeze line. Estimate the depth of the freeze line in soil having properties of sp.gr. = 2.1, $c = 0.41$ BTU/(lbm·°R), $k = 0.3$ BTU/(hr·ft·°R). Assume a surface temperature of 10°F and that this temperature exists for 3 months (≈30 days/month). Take the initial soil temperature to be 70°F.

Solution

Soil varies in composition and properties from region to region and even with depth, depending on water content and percentage of rock, etc. As shown in Figure 4.13, we are trying to calculate the depth x below which a water-filled pipe can be located without fear that the water will freeze during winter months. The water itself may be under pressure, which affects the freezing temperature.

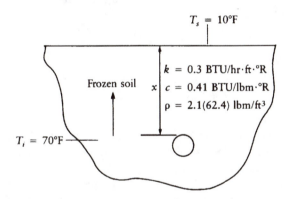

$x =$ Depth into plate $\alpha = k/\rho_s c =$ Diffusivity of plate

$T_i =$ Initial plate temperature $t =$ Time

FIGURE 4.12   Temperature variation with time for the semi-infinite solid.

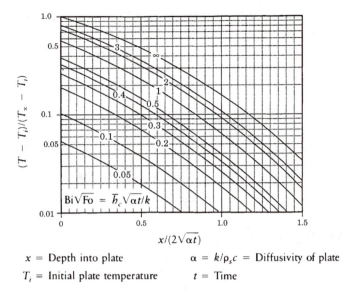

FIGURE 4.13   The soil of Example 4.4 treated as a semi-infinite body.

Assumptions

1. Soil properties are uniform and constant.
2. Freezing refers to the temperature at which liquid water changes to ice, taken to be 32°F.
3. The soil can be treated as a semi-infinite body because the depth of the freeze line is small compared to the soil depth.
4. The surface temperature of 10°F is constant and uniform, and lasts for 90 days.

The diffusivity of the soil is

$$\alpha = \frac{k}{\rho_s c} = \frac{0.3}{2.1(62.4)(0.41)} = 5.58 \times 10^{-3} \text{ ft}^2/\text{hr}$$

The time in hours is

$$t = 3 \text{ months } (30 \text{ days/month}) \ (24 \text{ hr/day}) = 2160 \text{ hr}$$

Figure 4.12 applies for semi-infinite geometries. Thus, the argument of the error function is

$$\frac{x}{2\sqrt{\alpha t}} = \frac{x}{2\sqrt{5.58 \times 10^{-3}(2160)}} = 0.144x$$

Also,

$$Bi\sqrt{Fo} = \frac{\bar{h}_c\sqrt{\alpha t}}{k} = \infty$$

because the surface temperature was prescribed in the problem. Thus,

$$T_\infty = T_s$$

The dimensionless temperature ratio is

$$\frac{T - T_i}{T_\infty - T_i} = \frac{32 - 70}{10 - 70} = 0.6333$$

At this value of dimensionless temperature and infinite $Bi\sqrt{Fo}$ Figure 4.12 shows

$$\frac{x}{2\sqrt{\alpha t}} = 0.144x = 0.38$$

Solving,

$$x = 2.64 \text{ ft} \qquad \square$$

One important application of the semi-infinite-body solution is in determining contact temperature. Consider two blocks of different materials and at different temperatures brought into contact, as shown in Figure 4.14. The left block is initially at T_{Li}, while the right block is initially at T_{Ri}. For short times after contact, or for small temperature differences $T_{Li} - T_{Ri}$, both blocks act as semi-infinite bodies. If the contact resistance is neglected, then it can be assumed that the contact plane suddenly attains some intermediate temperature T_c. Thus, the semi-infinite-body solution applies to each body. The heat flow into the right block is given by Equation 4.18 as follows:

$$q_R'' = \frac{k_R(T_c - T_{Ri})}{(\sqrt{\pi \alpha t})_R}$$

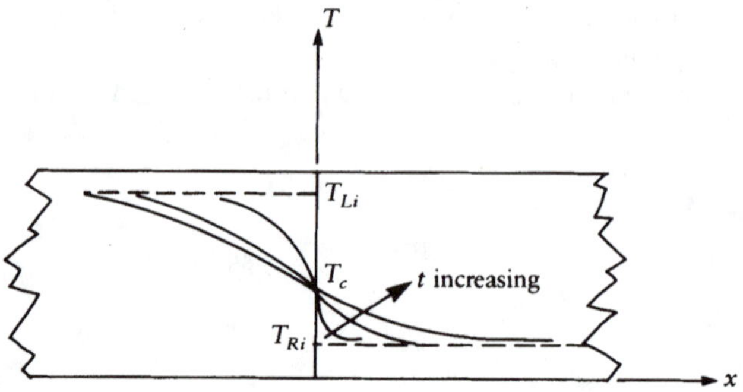

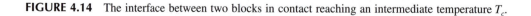

FIGURE 4.14 The interface between two blocks in contact reaching an intermediate temperature T_c.

Likewise, the heat leaving the left block is

$$q_L'' = -\frac{k_L(T_c - T_{Li})}{(\sqrt{\pi \alpha t})_L}$$

At any time, these two heat-flow equations must be equal. Therefore,

$$\frac{k_R(T_c - T_{Ri})}{(\sqrt{\pi \alpha t})_R} = \frac{-k_L(T_c - T_{Li})}{(\sqrt{\pi \alpha t})_L}$$

Rearranging,

$$\frac{T_c - T_{Ri}}{T_{Li} - T_c} = \frac{(\sqrt{k\rho c})_L}{(\sqrt{k\rho c})_R}$$

Solving for the contact temperature gives

$$T_c = \frac{T_{Li}(\sqrt{k\rho c})_L + T_{Ri}(\sqrt{k\rho c})_R}{(\sqrt{k\rho c})_L + (\sqrt{k\rho c})_R} \tag{4.19}$$

Example 4.6

An air-conditioned laboratory is maintained at 20°C; all objects in the lab are at this temperature. You enter the lab and, in the course of performing an experiment, you touch a block of aluminum and a piece of oak furniture. Would both objects feel as if they were at the same temperature? Assume the $k\rho c$ product of your skin is 4 W$^2\cdot$s/(m$^4\cdot$K^2) and that your body temperature is 37.3°C.

Solution

When objects first come into contact, the heat transferred during the initial few moments affects the temperature of each object only to a small depth. If this affected depth is small compared to the dimensions of the object, then each object is in essence a semi-infinite body. For oak furniture, glass, etc., this is the case. For skin, however, the penetration depth must be small, because convective effects associated with blood flow influence the heat exchanged.

Assumptions

1. The skin properties are constant.
2. The skin can be treated as a semi-infinite medium.
3. Properties of the other objects are constant.
4. One-dimensional transient heat flow exists.

We apply Equation 4.19 between the skin and the oak and again between the skin and the aluminum. The greater the contact temperature, the warmer the object will feel. From Table B.1 for aluminum,

$k = 236$ W/(m·K)
$\rho = 2.7(1\,000)$ kg/m^3
$c = 896$ J/(kg·K)

From Table B.3 for oak,

$k = 0.19$ W/(m·K)

$p = 0.705(1\ 000)$ kg/m^3
$c = 2\ 390$ J/(kg·K)

For the aluminum,

$$\sqrt{k\rho c} = [(236)(2\ 700)(896)]^{1/2} = 2.39 \times 10^4 \ \text{W}^2\text{·s/(m}^4\text{·K}^2)$$

Substituting into Equation 4.19 gives

$$T_c = \frac{T_{Li}(k\rho c)_L^{1/2} + T_{Ri}(k\rho c)_R^{1/2}}{(k\rho c)_L^{1/2} + (k\rho c)_R^{1/2}} = \frac{20(2.39 \times 10^4) + 37.3(2)}{2.39 \times 10^4 + 2}$$

Solving,

$$T_c = 20.0°\text{C} \quad (\text{aluminum–skin})$$

Similarly, for the oak,

$$\sqrt{k\rho c} = [(0.19)(705)(2\ 390)]^{1/2} = 5.66 \times 10^2$$

Substituting into Equation 4.19, we get

$$T_c = \frac{20(566) + 37.3(2)}{566 + 2}$$

Solving,

$$T_c = 20.1°\text{C} \quad (\text{oak–skin})$$

The oak will therefore feel warmer to the touch than will the aluminum.   ❑

4.4   SOLUTIONS TO MULTIDIMENSIONAL GEOMETRY SYSTEMS

We have considered a number of problems in which solutions were not written explicitly in equation form. Charts have been presented based on the solutions. The plane and cylindrical geometries discussed in Section 4.3 were for infinite systems. In this section, we will see how solutions to such one-dimensional problems can be combined to obtain solutions to problems having finite geometries.

Consider that we are trying to predict the temperature at some point $(r,\ \theta,\ z,\ t)$ within a cylinder of length $2L$ and radius R, as shown in Figure 4.15. The cylinder is initially at temperature T_i. At time zero, it is exposed to a fluid at temperature T_∞ with convection coefficient of $\bar{h}_c$. Temperature within the cylinder is a function of r, z, and t. The θ dependence vanishes, because we assume that temperature is symmetric with respect to θ. For a system with no internal heat generation, constant thermal conductivity, and $T = T(r,\ z,\ t)$, the general conduction equation in two dimensions is (without derivation):

$$\frac{1}{r}\frac{\partial}{\partial r}\left(r\frac{\partial T}{\partial r}\right) + \frac{\partial^2 T}{\partial z^2} = \frac{1}{\alpha}\frac{\partial T}{\partial t} \qquad (4.20)$$

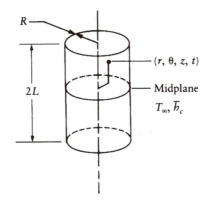

FIGURE 4.15 A finite cylinder.

The boundary and initial conditions are

B.C.1. $\quad r = 0; \quad \dfrac{\partial T}{\partial r} = 0$ $\hspace{5cm}$ (4.21a)

B.C.2. $\quad r = R; \quad -k\dfrac{\partial T}{\partial r} = \bar{h}_c(T_s - T_\infty)$ $\hspace{3cm}$ (4.21b)

B.C.3. $\quad z = 0; \quad \dfrac{\partial T}{\partial z} = 0$ $\hspace{5cm}$ (4.21c)

B.C.4. $\quad z = \pm L; \quad -k\dfrac{\partial T}{\partial z} = \bar{h}_c(T_s - T_\infty)$ $\hspace{2.5cm}$ (4.21d)

I.C. $\quad t = 0; \quad T = T_i$

$\quad\quad\quad t > 0; \quad T_s = $ surface temperature $\hspace{3.5cm}$ (4.22)

The differential equation can be solved by the separation-of-variables method, where we assume a solution of the form

$$T(r, z, t) = T_1(r, t) \cdot T_2(z, t) \tag{4.23}$$

The dependence on r and on z (two perpendicular coordinates) has been separated such that T_1 satisfies

$$\frac{1}{r}\frac{\partial}{\partial r}\left(r\frac{\partial T_1}{\partial r}\right) = \frac{1}{\alpha}\frac{\partial T_1}{\partial t} \tag{4.24}$$

subject to the associated boundary and initial conditions (Equations 4.21a, 4.21b, and 4.22). Furthermore, T_2 must satisfy

$$\frac{\partial^2 T_2}{\partial z^2} = \frac{1}{\alpha}\frac{\partial T_2}{\partial t} \tag{4.25}$$

subject to the associated boundary and initial conditions (Equations 4.21c, 4.21d, and 4.22). Before substituting Equation 4.23 into the differential equation (4.20), we first evaluate the derivatives, beginning with 4.23:

$$\frac{\partial T}{\partial r} = T_2 \frac{\partial T_1}{\partial r}$$

$$\frac{\partial T}{\partial z} = T_1 \frac{\partial T_2}{\partial z}$$

$$\frac{\partial^2 T}{\partial z^2} = T_1 \frac{\partial^2 T_2}{\partial z^2}$$

$$\frac{\partial T}{\partial t} = T_1 \frac{\partial T_2}{\partial t} + T_2 \frac{\partial T_1}{\partial t}$$

Combining with Equations 4.24 and 4.25 gives

$$\frac{\partial T}{\partial t} = T_1 \alpha \frac{\partial^2 T_2}{\partial z^2} + T_2 \frac{\alpha}{r} \frac{\partial}{\partial r}\left(r \frac{\partial T_1}{\partial r}\right)$$

Substituting into Equation 4.20 yields

$$\frac{1}{r}\frac{\partial}{\partial r}\left(r T_2 \frac{\partial T_1}{\partial r}\right) + T_1 \frac{\partial^2 T_2}{\partial z^2} = T_1 \frac{\partial^2 T_2}{\partial z^2} + \frac{T_2}{r}\frac{\partial}{\partial r}\left(r \frac{\partial T_1}{\partial r}\right)$$

which is an identity. Therefore, if T_1 and T_2 are known solutions to two separate one-dimensional problems, their product gives the solution to the combined problem. Based on information given in this chapter, we recognize T_1 as being the solution to the infinite-cylinder problem and T_2 as being the solution to the infinite-plane problem. Thus, the solution to the finite-cylinder problem of Figure 4.15 can be obtained by using the product of two one-dimensional problems. This is illustrated in Figure 4.16. Moreover, this analysis can be extended to three dimensions appropriately.

Figure 4.16 shows how the solution to two one-dimensional problems are combined to obtain the solution to a two-dimensional problem. Other combinations can be made. For purposes of simplifying the notation, we shall denote the solution to the semi-infinite problem as

$$S(x, t) = S\left(\frac{x}{2\sqrt{\alpha t}}\right) = \text{semi-infinite solid} \qquad (4.26)$$

Likewise, for other one-dimensional solutions, we have

$$P(x, t) = P(\alpha t/L^2) = P(\text{Fo}) = \text{infinite plate} \qquad (4.27)$$

$$C(r, t) = C(\alpha t/R^2) = C(\text{Fo}) = \text{infinite cylinder} \qquad (4.28)$$

Appropriate combinations of the above solutions yield solutions to a quarter-infinite solid, an eighth-infinite solid, a semi-infinite plate, a quarter-infinite plate, an infinite rectangular bar, a semi-infinite rectangular bar, a rectangular parallelepiped, an infinite cylinder, a semi-infinite cylinder, and a short cylinder. These are summarized in Table 4.1. In each case, the solid exists at an initial

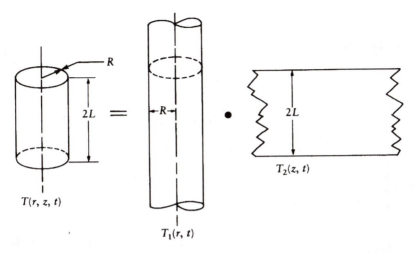

FIGURE 4.16 Solution to the finite-cylinder problem obtained by multiplication of the solutions to two one-dimensional problems: $T(r, z, r) = T_1(r, t) \cdot T_2(z, t)$.

temperature T_i, and at time zero it is subjected to a fluid at temperature T_∞ with convection coefficient $\bar{h}_c$.

Example 4.7

On short notice, a friend has been invited to a party and desires to take along some beer as liquid refreshment. The only beer available is at room temperature $(T = 72°F)$, and to cool beer cans quickly, they are placed in the freezer compartment of a refrigerator $(T = 30°F)$. A typical beer can measures 2-1/2 inches in diameter by 4-3/4 inches tall. What will the beer temperature at the center of the can become after four hours in the freezer? What is the beer temperature near the metal of the can? Take the average convection coefficient between the can and the freezer to be 1.7 BTU/(hr·ft²·°R).

Solution

The can of beer is placed in an environment where the temperature is low. The can is probably placed so that its bottom is in contact with the freezer shelf; thus, heat is transferred by conduction through the can bottom. Beer in the vicinity of the can metal cools before the beer in the can center does. As the beer near the wall cools, its properties change, and natural convective effects begin to occur. The beer experiences some mixing, which tends to distribute energy more nearly evenly, resulting in a more nearly uniform temperature profile.

Assumptions

1. Lacking better information, we assume the properties of beer are the same as those of water at 72°F (or thereabout).
2. Properties of beer are constant.
3. Natural convective effects occurring within the beer are negligible, and the beer can be treated as if it were a solid.
4. The convection coefficient is constant and applies to the entire surface of the can.

From Appendix Table C.11 for water at 68°F we find

$\rho = 62.46$ lbm/ft³ $c_p = 0.9988$ BTU(lbm·°R)
$k = 0.345$ BTU/(hr·ft·°R) $\alpha = 5.54 \times 10^{-3}$ ft²/hr

TABLE 4.1
Product Solutions to Various Problems by Combination of One-Dimensional Transient Systems

S(X) Figure 4.12   P(Fo) Figure 4.6   C(Fo) Figure 4.7

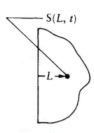

S(L, t)

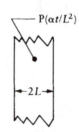

P($\alpha t/L^2$)

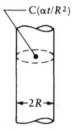

C($\alpha t/R^2$)

Semi-infinite solid   Infinite plate   Infinite cylinder

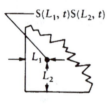

S(L_1, t)S(L_2, t)

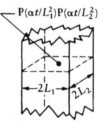

P($\alpha t/L_1^2$)P($\alpha t/L_2^2$)

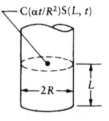

C($\alpha t/R^2$)S(L, t)

Quarter-infinite solid   Infinite rectangular bar   Semi-infinite cylinder

S(L_1, t)S(L_2, t)S(L_3, t)

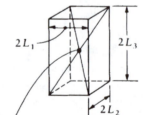

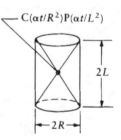

C($\alpha t/R^2$)P($\alpha t/L^2$)

P($\alpha t/L_1^2$)P($\alpha t/L_2^2$)P($\alpha t/L_3^2$)

Eighth-infinite solid   Rectangular parallelepiped   Short cylinder

P($\alpha t/L_1^2$)S(L_2, t)

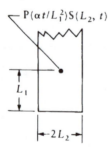

P($\alpha t/L_1^2$)P($\alpha t/L_2^2$)S(L_3, t)

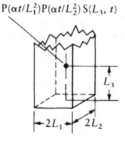

P($\alpha t/L_1^2$)S(L_2, t)S(L_3, t)

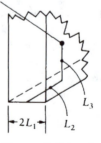

Semi-infinite plate   Semi-infinite rectangular bar   Quarter-infinite plate

These properties vary within the beer during the time that the temperature falls from 72°F to, say, 32°F. For example, at 32°F, Table C.11 shows the following properties:

$\rho = 62.57$ lbm/ft^3
$k = 0.319$ BTU/(hr·ft·°R)

$c_p = 1.0074$ BTU/(lbm·°R)
$\alpha = 5.07 \times 10^{-3}$ ft^2/hr

The diffusivity changes by as much as 9%. For the calculations here, it is assumed that the properties at 68°F are representative.

Table 4.1, for the short-cylinder problem, indicates that the solution is the product of the infinite-cylinder problem (Figure 4.7) and the infinite-plate problem (Figure 4.6). However, we must determine if the lumped-capacitance formulation applies. We first calculate the Biot number,

$$\mathrm{Bi} = \frac{\bar{h}_c L_c}{k} = \frac{\bar{h}_c V_s}{k A_s}$$

The can volume is

$$V_s = \frac{\pi D^2 L}{r} = \frac{\pi \left(\frac{2.5}{12}\right)^2 \left(\frac{4.75}{12}\right)}{4} = 0.0135 \text{ ft}^3$$

The surface area of the can is

$$A_s = \pi D L + 2\left(\frac{\pi D^2}{4}\right) = \pi\left(\frac{2.5}{12}\right)\left(\frac{4.75}{12}\right) + 2\left[\pi\left(\frac{2.5}{12}\right)^2/4\right] = 0.280 \text{ ft}^2$$

The characteristic length for the can dimensions is

$$L_c = \frac{V_s}{A_s} = 0.054 \text{ ft}$$

The Biot number then becomes

$$\mathrm{Bi} = \frac{1.7(0.054)}{0.345} = 0.266 \text{ (lumped capacitance)}$$

A lumped-capacitance analysis is not warranted because Bi > 0.1. The Biot number equation above is appropriate for the lumped-capacitance method but not for the chart solutions of Figures 4.6 and 4.7.

For the cylinder solution, we find

$$\mathrm{Fo} = \frac{\alpha t}{R^2} = \frac{(5.54 \times 10^{-3} \text{ ft}^2/\text{hr})(4 \text{ hr})}{\left[\left(\frac{2.5}{2}\right)\left(\frac{1}{12}\right)\right]^2 \text{ ft}} = 2.04$$

and

$$\mathrm{Bi} = \frac{\bar{h}_c R}{k} = \frac{1.7\left(\frac{2.5}{2}\right)\left(\frac{1}{12}\right)}{0.345} = 0.513$$

$$1/\mathrm{Bi} = 1.95 \text{ (chart solution)}$$

From Figure 4.7a at Fo = 2.04 and 1/Bi = 1.95 we read

$$\left.\frac{T_c - T_\infty}{T_i - T_\infty}\right|_{cylinder} = 0.175$$

For the infinite-plate solution,

$$Bi = \frac{\bar{h}_c L}{k} = \frac{1.7\left[\frac{4.75}{2(12)}\right]}{0.345} = 0.98$$

$$1/Bi = 1.0$$

$$Fo = \frac{\alpha t}{L^2} = \frac{5.54 \times 10^{-3}(4)}{\left[\frac{4.75}{2(12)}\right]^2} = 0.566$$

From Figure 4.6a at Fo = 0.566 and 1/Bi = 1.0, we read

$$\left.\frac{T_c - T_\infty}{T_i - T_\infty}\right|_{plate} = 0.55$$

Therefore,

$$\left.\frac{T_c - T_\infty}{T_i - T_\infty}\right|_{short\ cylinder} = 0.175(0.55) = 0.096$$

Substituting and solving gives

$$T_c = 0.096(72 - 30) + 30 = 34.0°F$$

This is the temperature at the center of the can after four hours in the freezer.

To obtain an estimate of the temperature at the can wall, we refer to Figures 4.6b and 4.7b. For 1/Bi = 1.95 and r/R = 1, we read from Figure 4.7b:

$$\left.\frac{T - T_\infty}{T_c - T_\infty}\right|_{cylinder} = 0.77$$

From Figure 4.6b, with x/L = 1 and 1/Bi = 1.0,

$$\left.\frac{T - T_\infty}{T_c - T_\infty}\right|_{plate} = 0.65$$

At the wall then

$$\left.\frac{T_w - T_\infty}{T_c - T_\infty}\right|_{short\ cylinder} = 0.77(0.65) = 0.50$$

Substituting and solving,

$$T_w = 0.50(34.0 - 30) + 30$$

or

$$T_w = 32.0°F \qquad \Box$$

It is prudent to note that there is a difference in the semi-infinite solid graph (Figure 4.12) and the graphs of the other solutions (infinite slab, infinite cylinder, and sphere); specifically, a difference in the dimensionless temperature. For the semi-infinite slab, the vertical axis is of dimensionless temperature at some depth x; namely: $(T - T_i)/(T_\infty - T_i)$ in which the reference temperature is T_i. For the other graphs, the vertical axis is of dimensionless center temperature: $(T_c - T_\infty)/(T_i - T_\infty)$. The reference temperature is T_∞.

As mentioned in the preceding paragraph, the reference temperature in the semi-infinite solid is T_i. If, instead, the graph is produced so that the reference is T_∞, the lines do not "spread" out as they do in Figure 4.12, and so they are difficult to read and obtain an accurate result. To change the reference temperature in the semi-infinite solid solution, we merely subtract the dimensionless temperature from 1 to obtain

$$1 - \frac{T - T_i}{T_\infty - T_i} = \frac{T_\infty - T_i - T + T_i}{T_\infty - T_i} = \frac{T_\infty - T}{T_\infty - T_i} = \frac{T - T_\infty}{T_i - T_\infty}$$

This is a necessary step when combining the results of the various problems, as shown in the following example.

Example 4.8

A long rectangular bar 3×3 cm made of stainless steel (type 304) is initially at 95°C. It is submerged in water whose temperature is 17°C. Determine the time it takes for the center temperature 4 cm from the bottom to reach 50°C. Assume a convection coefficient of 50 W/(m²·K).

Solution

A "long" bar suggests that the length is infinite. While the bar in reality is not infinite in length, we can obtain a solution (an estimate of the temperature) by modeling it as such. Referring to Table 4.1, we apply the semi-infinite rectangular bar solution with $2L_1 = 2L_2 = 2L$, because the bar is square in cross section.

Assumptions

1. The convection coefficient is constant.
2. Material properties are constant.
3. The semi-infinite bar model applies. For stainless steel, we read from Appendix Table B.2,

$$\rho = 7\,817 \text{ kg/m}^3 \qquad\qquad c = 461 \text{ J/(kg·K)}$$
$$k = 14.4 \text{ W/(m·K)} \qquad\qquad \alpha = 0.387 \times 10^{-5} \text{ m}^2/\text{s}$$

We will use the infinite plate solution $P(\alpha t/L^2)$, and the semi-infinite slab solution $S(L_3, t)$ in which $L_3 = 4$ cm. We thus calculate:

infinite plate

$$L = \frac{0.03}{2} = 0.015 \text{ m}$$

$$\frac{1}{\text{Bi}} = \frac{k}{\bar{h}_c L} = \frac{14.4}{50(0.015)} = 19.2$$

$$\text{Fo} = \frac{\alpha t}{L^2} = \frac{0.387 \times 10^{-5} t}{(0.015)^2} = 0.017 t$$

semi-infinite solid

$$\text{Bi}\sqrt{\text{Fo}} = \frac{\bar{h}_c \sqrt{\alpha t}}{k} = \frac{50\sqrt{(0.387 \times 10^{-5} t)}}{14.4} = 6.83 \times 10^{-3} t^{1/2}$$

$$\frac{x}{2\sqrt{\alpha t}} = \frac{0.04}{2\sqrt{(0.387 \times 10^{-5} t)}} = \frac{10.16}{t^{1/2}}$$

The time at this point is unknown. Consequently, we will have to resort to a trial-and-error procedure that sounds unappealing but is immensely rewarding when a solution is obtained. The following serves as an example of the procedure:

First trial: assume a time of 3000 s, then,

$$\text{infinite plate} \quad \left. \begin{array}{l} 1/\text{Bi} = 19.2 \\ \text{Fo} = 0.017 t = 51.6 \end{array} \right\} \frac{T_c - T_\infty}{T_i - T_\infty} \approx 0.085 \text{ (Figure 4.6a)}$$

$$\text{semi-infinite solid} \quad \left. \begin{array}{l} \text{Bi}\sqrt{\text{Fo}} = 0.374 \\ \dfrac{x}{2\sqrt{\alpha t}} = 0.186 \end{array} \right| \frac{T - T_i}{T_\infty - T_i} \approx 0.225 \text{ (Figure 4.12)}$$

The temperature at a depth of 4 cm then is calculated as:

$$(P_{L=3\text{ cm}}) \cdot (P_{L=3\text{ cm}}) \cdot (S_{L=4\text{ cm}}) = \frac{T_c - T_\infty}{T_i - T_\infty} \cdot \frac{T_c - T_\infty}{T_i - T_\infty} \cdot \left(1 - \frac{T - T_i}{T_\infty - T_i}\right)$$

or

$$\frac{T - T_\infty}{T_i - T_\infty} = (0.085)(0.085)(1 - 0.225) = 0.0056$$

Solving for temperature, we find

$$T_{L=4\text{ cm}} = 0.0056(95 - 17) + 17 = 17.4°\text{C}$$

Thus, 3000 s is too long a time; we are seeking the time it takes for the point of interest to reach 50°C. The result for other times is provided in the following table:

time t, s	Infinite plate			Semi-infinite solid			Dimensionless temperature	Temperature
	1/Bi	Fo	$\dfrac{T_c - T_\infty}{T_i - T_\infty}$	$Bi\sqrt{Fo}$	$\dfrac{x}{2\sqrt{\alpha t}}$	$\dfrac{T - T_i}{T_\infty - T_i}$	$\dfrac{T - T_\infty}{T_i - T_\infty}$	T
3000	19.2	51.6	0.085	0.374	0.186	0.225	0.0056	17.4
1500	19.2	25.8	0.34	0.265	0.262	0.14	0.099	24.8
700	19.2	12.0	0.55	0.181	0.384	0.075	0.28	38.8
400	19.2	6.88	0.7	0.137	0.508	0.046	0.26	37.6
200	19.2	3.44	0.8	0.097	0.719	0.02	0.63	65.9
300	19.2	5.16	0.8	0.118	0.587	0.035	0.62	65.2
350	19.2	6.02	0.7	0.128	0.543	0.042	0.47	53.6

In using the graphs, it quickly becomes apparent that obtaining an accurate result from Figure 4.6a is extremely difficult because of the region we are working in. In view of this shortcoming, we remember that 350 s is an *estimate* of the time required for the temperature at the point of interest to be within 10% of 50°C.

Had we used the lumped capacitance approach (assuming a length of 4 m), the result would have been a time of 465 s.

4.5 APPROXIMATE METHODS OF SOLUTION TO TRANSIENT-CONDUCTION PROBLEMS

Several approximate methods of solution to transient-heat-conduction problems exist. We will consider two of them: the numerical method and the graphical method. The advantages that each has will become evident in the following paragraphs.

4.5.1 NUMERICAL METHODS

Numerical methods were introduced earlier, and it was stated that the technique involves rewriting the differential equation in terms of a set of simultaneous algebraic equations. The method is readily extended to transient problems.

Consider the system sketched in Figure 4.17, which is of a two-dimensional region with a sudden temperature difference imposed. In this case, we are concerned with the temperature response of the material itself, specifically, how temperature at any point within the region varies with time. Also shown in the figure are the x-y axes and a nodal network. The region has been divided into a grid containing a number of nodal points. Five interior nodes are labeled with subscripts m and n to denote spatial position. The superscript p denotes a time. Under steady conditions, for a region having constant properties with no internal heat generation, Equation 3.4 reduces to

$$\frac{\partial^2 T}{\partial x^2} + \frac{\partial^2 T}{\partial y^2} = \frac{1}{\alpha}\frac{\partial T}{\partial t} \tag{4.29}$$

This equation can be rewritten in finite-difference form. The second derivatives of temperature for node m, n (see Chapter 3 for a detailed derivation) become

$$\left.\frac{\partial^2 T}{\partial x^2}\right|_{m,n} = \frac{T^p_{m+1,n} + T^p_{m-1,n} - 2T^p_{m,n}}{(\Delta x)^2} \tag{4.30a}$$

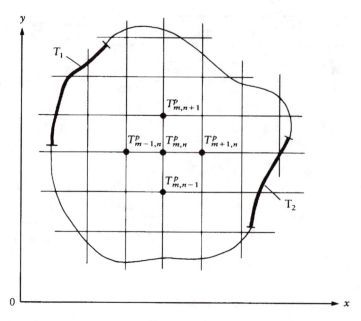

FIGURE 4.17   A two-dimensional region divided into a number of nodal points.

$$\left.\frac{\partial^2 T}{\partial y^2}\right|_{m,n} = \frac{T^p_{m,n+1} + T^p_{m,n-1} - 2T^p_{m,n}}{(\Delta y)^2} \tag{4.30b}$$

Just as the space variables are divided into Δx by Δy increments, time will also be divided into discrete time increments Δt. The parameter p is an integer used to denote the time t that has elapsed according to

$$t = p \, \Delta t \tag{4.31}$$

The finite-difference solution provides values of temperature at discrete points in the region for various values of time that are apart by the time interval Δt.

The first derivative of temperature with respect to time can be written in several ways, one of which is

$$\left.\frac{\partial T}{\partial t}\right|_{m,n} = \frac{T^{p+1}_{m,n} - T^p_{m,n}}{\Delta t} \tag{4.32}$$

The numerator of the right-hand side of Equation 4.55 contains the temperature of node m, n at some future time Δt away ($T^{p+1}_{m,n}$), and the temperature of node m, n at the present time. The expression is thus referred to as a forward-difference scheme. Substituting Equations 4.30 and 4.32 into 4.29 gives

$$\frac{T^p_{m+1,n} + T^p_{m-1,n} - 2T^p_{m,n}}{(\Delta x)^2} + \frac{T^p_{m,n+1} + T^p_{m,n-1} - 2T^p_{m,n}}{(\Delta y)^2} = \frac{T^{p+1}_{m,n} - T^p_{m,n}}{\Delta t} \frac{1}{\alpha}$$

Setting Δx equal to Δy and solving for ($T^{p+1}_{m,n}$), we get

$$T_{m,n}^{p+1} = \frac{\alpha\,\Delta t}{(\Delta x)^2}(T_{m+1,n}^p + T_{m-1,n}^p + T_{m,n+1}^p + T_{m,n-1}^p) + \left[1 - \frac{4\alpha\,\Delta t}{(\Delta x)^2}\right]T_{m,n}^p \qquad (4.33)$$

This expression gives the temperature at any interior node m, n at a future time $t + \Delta t$ explicitly in terms of the temperature at node m, n and its surrounding nodes at the preceding time. The equation is thus referred to as an *explicit formulation*. The temperature at every interior node might be known from the initial condition $(p = 0, t = 0)$. Equation 4.33 is then used to calculate the temperature at every interior node after one time interval $[p = 1, t = (1)\Delta t]$. Once these are known, Equation 4.33 is used again to calculate the temperature at every interior node after another time interval $[p = 2, t = (2)\Delta t]$. The process is repeated until a desired time or condition is reached.

Equation 4.33 can be reduced to a simpler form for a one-dimensional problem where temperature varies only with x and t; Equation 4.33 becomes

$$T_m^{p+1} = \frac{\alpha\,\Delta t}{(\Delta x)^2}(T_{m+1}^p + T_{m-1}^p) + \left[1 - \frac{2\alpha\,\Delta t}{(\Delta x)^2}\right]T_m^p \qquad (4.34)$$

Equations 4.33 and 4.34 both contain a finite-difference version of the Fourier number:

$$\text{Fo}_\Delta = \frac{\alpha\,\Delta t}{(\Delta x)^2} \qquad (4.35)$$

The value of Fo_Δ must be specified when solving a problem. When the space increment Δx is chosen and Fo_Δ is specified, then Δt is automatically determined for a given material through Equation 4.35. A larger Δx and a larger Δt help to effect a solution more rapidly (the fewer nodal temperatures to calculate, the fewer Δt values to reach the end condition). Alternatively, smaller Δx and Δt values yield more accurate results. Limits on Fo_Δ are required for convenience as well as for guidance.

Now if $\text{Fo}_\Delta[= \alpha\,\Delta t/(\Delta x^2)] > 1/2$ in Equation 4.34, then the coefficient of T_m^p becomes negative. Furthermore, if the adjoining nodes are all equal but less than T_m^p, then at the conclusion of the time interval, T_m^{p+1} might not be equal to or less than T_{m+1}^p and T_{m-1}^p. Thus, Equation 4.34 would predict that heat has been transferred from a node at some temperature to a node at a higher temperature. This violates what actually happens in a system, so Equation 4.34 ceases to be an effective model. We therefore conclude that

$$\left.\begin{array}{ll} \text{Fo}_\Delta = \dfrac{\alpha\,\Delta t}{(\Delta x)^2} \le \dfrac{1}{2} & \text{(one-dimensional system)} \\[4mm] \text{Fo}_\Delta = \dfrac{\alpha\,\Delta t}{(\Delta x)^2} \le \dfrac{1}{4} & \text{(two-dimensional system)} \end{array}\right\} \qquad (4.36)$$

Although these restrictions were based on physical grounds, results of calculations made with a computer are in agreement. For example, if Equation 4.34 is used with $\alpha\,\Delta t/(\Delta x)^2 > 1/2$, a solution is not obtained. Calculations would show that temperatures at successive times for any node diverge rather than converge. So, for example, the temperature of one node at time t might be 30°C. After $t + \Delta t$, temperature jumps to 32°C. After $t + 2\Delta t$, temperature falls to 25°C. After $t + 3\Delta t$, temperature rises to 40°C, and so on.

Equations 4.33 and 4.34 can be further simplified by selecting a specific value of Fo_Δ. For example, if

$$\text{Fo}_\Delta = \frac{\alpha\,\Delta t}{(\Delta x)^2} = \frac{1}{2}$$

then Equation 4.34 becomes

$$T_m^{p+1} = \frac{T_{m+1}^p + T_{m-1}^p}{2} \tag{4.37}$$

The temperature at the future time $t + \Delta t$ is the arithmetic average of the temperature of the surrounding nodes during the preceding time t.

As mentioned earlier, the above formulation is known as an explicit formulation in which the equations are developed on what is called a forward-difference technique. A corresponding analysis exists for an arrangement called the *backward-difference technique.* Instead of writing second derivatives with a superscript of p, as in Equations 4.30, we write, for any interior node m, n

$$\left.\frac{\partial^2 T}{\partial x^2}\right|_{m,n} = \frac{T_{m+1,n}^{p+1} + T_{m-1,n}^{p+1} - 2T_{m,n}^{p+1}}{(\Delta x)^2} \tag{3.38}$$

The two-dimensional unsteady differential equation (4.29) then becomes

$$\frac{T_{m+1,n}^{p+1} + T_{m-1,n}^{p+1} - 2T_{m,n}^{p+1}}{(\Delta x)^2} + \frac{T_{m,n+1}^{p+1} + T_{m,n-1}^{p+1} - 2T_{m,n}^{p+1}}{(\Delta y)^2} = \frac{1}{\alpha}\frac{T_{m,n}^{p+1} - T_{m,n}^p}{\Delta t}$$

Rearranging and solving for $T_{m,n}^p$ gives

$$T_{m,n}^p = -\frac{\alpha \Delta t}{(\Delta x)^2}(T_{m+1,n}^{p+1} + T_{m-1,n}^{p+1} + T_{m,n+1}^{p+1} + T_{m,n-1}^{p+1}) + \left(1 + \frac{4\alpha \Delta t}{(\Delta x)^2}\right)T_{m,n}^{p+1} \tag{3.39}$$

This equation does not provide for explicitly calculating the temperature at an interior node at the future time $p + 1$. A whole system of equations must he written for the entire nodal system and solved simultaneously. This scheme is thus referred to as an *implicit formulation.*

The advantage of the explicit formulation is that future temperature can be calculated directly. The disadvantage is that the stability of the calculations is fixed by selection of Δx and Δt. The advantage of the implicit method is that no such stability criterion must be satisfied. Thus, any time increment can be selected. A larger increment of time would reduce the number of calculations required. The disadvantage of the implicit method is that a large number of calculations are required for each time increment to solve the simultaneous equations.

The above explicit and implicit difference equations apply only to regions in which boundary temperatures have been specified. Let us now consider a system where a convective boundary exists. As an alternative to merely substituting into the descriptive differential equation, an energy balance will be formulated for a one-dimensional system.

Figure 4.18 shows a plane wall exposed on its left surface to a fluid at temperature T_∞ with a convection coefficient of $\bar{h}_c$. The region is divided into increments Δx wide with the surface node only $\Delta x/2$ wide. Such a division for the surface node, although not necessary, is made for the sake of more accurately determining temperature near the surface. The energy convected into the surface node equals the energy stored in that increment plus the energy transferred to the adjacent increment by conduction. In equation form,

$$\bar{h}_c A(T_\infty - T_0^p) = \rho_s c(A \Delta x/2)\frac{T_0^{p+1} - T_0^p}{\Delta t} + \frac{kA}{\Delta x}(T_0^p - T_1^p)$$

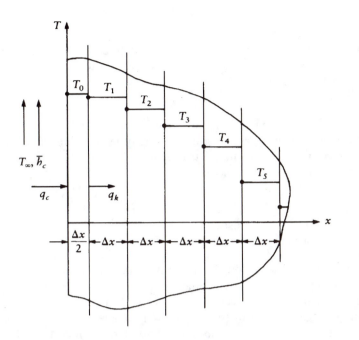

FIGURE 4.18 A one-dimensional system with a convective boundary.

where A is the area normal to the direction of heat flow, and ρ_s and c are density and specific heat of the solid, respectively. Simplifying and rearranging to solve for T_0^{p+1} gives

$$T_0^{p+1} = \frac{2\bar{h}_c \Delta t}{\rho_s c \Delta x}T_\infty + \frac{2k\,\Delta t}{\rho_s c (\Delta x)^2}T_1^p + \left(1 - \frac{2k\,\Delta t}{\rho_s c (\Delta x)^2} - \frac{2\bar{h}_c\,\Delta t}{\rho_s c \Delta x}\right)T_0^p \tag{4.40a}$$

The coefficients of temperature on the right-hand side can be manipulated in the following way:

$$\frac{\bar{h}_c \Delta t}{\rho_s c\,\Delta x} = \frac{\bar{h}_c \Delta x}{k}\frac{k\,\Delta t}{\rho_s c (\Delta x)^2} = \frac{\bar{h}_c \Delta x}{k}\frac{\alpha\,\Delta t}{(\Delta x)^2} = \mathrm{Bi}_\Delta \mathrm{Fo}_\Delta$$

where a finite-difference form of the Biot and Fourier number has emerged. Continuing,

$$\frac{k\,\Delta T}{\rho_s c (\Delta x)^2} = \frac{\alpha\,\Delta t}{(\Delta x)^2} = \mathrm{Fo}_\Delta$$

Equation 4.40a now becomes

$$T_0^{p+1} = 2\mathrm{Fo}_\Delta(\mathrm{Bi}_\Delta T_\infty + T_1^p) + (1 - 2\mathrm{Fo}_\Delta - 2\mathrm{Bi}_\Delta \mathrm{Fo}_\Delta)T_0^p \tag{4.40b}$$

This explicit formulation yields a stable solution only if

$$2\mathrm{Fo}_\Delta(\mathrm{Bi}_\Delta + 1) \le 1$$

Example 4.9

A 30-cm-thick concrete slab is initially at 20°C. Suddenly the temperature of the left exposed surface is raised to 200°C. Determine the time that will pass before the heat added will be felt at the opposite face, which is insulated.

Solution

Heat added to the slab goes into increasing its internal energy, which is manifested as an increase in the slab temperature. Figure 4.19 is a scale drawing of temperature vs. distance at time zero. We are looking for the time required for the temperature of the right face to become elevated over 20°C. This problem will be solved numerically.

Assumptions

1. One-dimensional heat flow exists.
2. Material properties are constant.

As a first step in the numerical method, the region is divided into a number of smaller regions Δx wide. The dimension of Δx can be made arbitrarily. The decision will be influenced by the desired accuracy and by the computer time that can be devoted to the problem. So we arbitrarily select Δx = 5 cm. The divided region is shown also in Figure 4.19. Each line is labeled with a value for m. The numerical solution will provide a temperature for each of the lines labeled 0 through 6.

Appendix Table B.3 for concrete shows

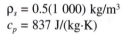

$$\rho_s = 0.5(1\ 000) \text{ kg/m}^3$$
$$c_p = 837 \text{ J/(kg·K)}$$

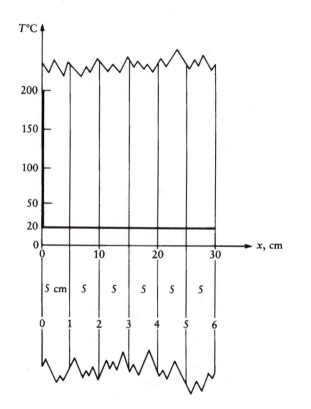

FIGURE 4.19   Initial temperature distribution in the 30-cm concrete slab of Example 4.9.

$$k = 0.128 \text{ W/(m·K)}$$
$$\alpha = 0.049 \times 10^{-5} \text{ m}^2/\text{s}$$

For stability of the numerical method, we have from Equation 4.36

$$\text{Fo}_\Delta = \frac{\alpha \, \Delta t}{(\Delta x)^2} \leq \frac{1}{2} \tag{4.36}$$

Arbitrarily, let $\text{Fo}_\Delta = 1/2$. With this value and the above equation, the time increment is automatically specified; i.e.,

$$\frac{1}{2} = \frac{\alpha \, \Delta t}{(\Delta x)^2} = \frac{0.049 \times 10^{-5} \Delta t}{(0.05)^2}$$

and the time increment we must use is

$$\Delta t = 2.551 \times 10^5 \text{ s} = 0.709 \text{ hr}$$

With the Fourier number equal to 1/2 the resulting equation for temperature is

$$T_m^{p+1} = \frac{T_{m+1}^p + T_{m-1}^p}{2} \tag{4.37}$$

The parameter m ranges from 1 to 6 while p starts at 1 and increases until $T_6 > 20°\text{C}$. At $m = 0$, the temperature is a constant at 200°C. The distribution at time zero ($p = 0$) is shown in Figure 4.19. So after one time interval, we write

$p = 0$

$m = 1$ $\qquad\qquad T_1^1 = \dfrac{T_2^0 + T_0^0}{2} = \dfrac{20 + 200}{2} = 110$

$m = 2$ $\qquad\qquad T_2^1 = \dfrac{T_3^0 + T_1^0}{2} = \dfrac{20 + 20}{2} = 20$

$m = 3$ $\qquad\qquad T_3^1 = \dfrac{T_4^0 + T_2^0}{2} = \dfrac{20 + 20}{2} = 20$

$m = 4$ $\qquad\qquad T_4^1 = \dfrac{T_5^0 + T_3^0}{2} = \dfrac{20 + 20}{2} = 20$

$m = 5$ $\qquad\qquad T_5^1 = \dfrac{T_6^0 + T_4^0}{2} = \dfrac{20 + 20}{2} = 20$

$m = 6$ $\qquad\qquad T_6^1 = \dfrac{T_5^0 + T_5^0}{2} = \dfrac{20 + 20}{2} = 20$

(insulated)

(An insulated boundary was discussed in Chapter 3.) The above calculated temperatures exist after 0.709 hr has elapsed and are shown graphically in Figure 4.20a. Continuing for the next time interval,

$$p = 1$$

$m = 1$
$$T_1^2 = \frac{T_2^1 + T_0^1}{2} = \frac{20 + 200}{2} = 110$$

$m = 2$
$$T_2^2 = \frac{T_3^1 + T_1^1}{2} = \frac{20 + 110}{2} = 65$$

$m = 3$
$$T_3^2 = \frac{T_4^1 + T_2^1}{2} = \frac{20 + 20}{2} = 20$$

$m = 4$
$$T_4^2 = \frac{T_5^1 + T_3^1}{2} = \frac{20 + 20}{2} = 20$$

$m = 5$
$$T_5^2 = \frac{T_6^1 + T_4^1}{2} = \frac{20 + 20}{2} = 20$$

$m = 6$
$$T_6^2 = \frac{T_5^1 + T_5^1}{2} = \frac{20 + 20}{2} = 20$$

The results are shown in Figure 4.20b. Further calculations yield the results shown in Figures 4.20c, d, e, and f. It is evident that after 7(0.709) hr, the temperature at T_6 will begin to rise. Thus, the solution is

$$t = 4.97 \text{ hr} \qquad \qquad ❑$$

4.5.2 THE GRAPHICAL METHOD

At least two graphical methods have been developed to solve unsteady conduction problems: the Schmidt plot and the Saul'ev plot.

The Schmidt plot is a graphical equivalent of the one-dimensional explicit numerical method of solution. Because of the confusing nature of the completed graph, it is advantageous to spend more time on the Saul'ev method, because it is easier to draw and interpret. It applies to one-dimensional systems, as does the Schmidt method, but the Saul'ev method produces solutions more quickly. For a one-dimensional unsteady-heat-conduction problem we have

$$\frac{\partial^2 T}{\partial x^2} = \frac{1}{\alpha} \frac{\partial T}{\partial t}$$

In finite-difference form, however, the above equation is written somewhat differently from that obtained earlier for explicit and implicit formulations. Here we write

$$\frac{T_{m+1}^p - T_m^p - T_m^{p+1} + T_{m-1}^{p+1}}{(\Delta x)^2} = \frac{1}{\alpha} \frac{T_m^{p+1} - T_m^p}{\Delta t} \qquad \text{(Saul'ev method)}$$

Note that a combination of implicit and explicit terms appears in the numerator of the left-hand side. The time differentiation is unchanged. Rearranging and simplifying, we get

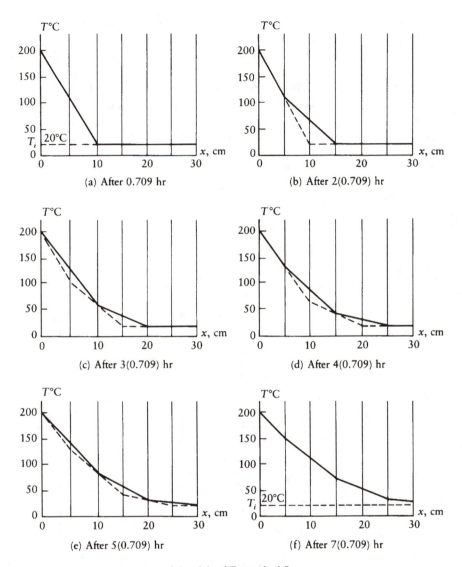

FIGURE 4.20   Temperature response of the slab of Example 4.8.

$$T_m^{p+1} = \frac{\alpha\,\Delta t}{(\Delta x)^2}(T_{m+1}^p - T_m^p - T_m^{p+1} + T_{m-1}^{p+1} + T_m^p) \tag{4.41}$$

Equation 4.41 is unconditionally stable, so, regardless of what is selected for $\alpha\,\Delta t/(\Delta x)^2$, the solution will converge if solved numerically. The equation simplifies greatly if

$$\mathrm{Fo}_\Delta = \frac{\alpha\,\Delta t}{(\Delta x)^2} = 1$$

Equation 4.41 becomes

$$T_m^{p+1} = \frac{T_{m+1}^p + T_{m-1}^{p+1}}{2}$$

Here again, we use temperatures at adjacent nodes $(m + 1)$ and $(m - 1)$ to determine the temperature at node m. However, the temperature at the $(m - 1)$ node is at the future time. So, once a new temperature at $(m - 1)$ is found, it is used immediately to find the temperature at m. In addition, the Fourier number here is 1, in contrast to 1/2 for the explicit formulation. The Saul'ev method will therefore converge to the solution in half the time required in the explicit method.

The graphical interpretation of Equation 4.43 is illustrated in Figure 4.21. Shown is a one-dimensional system divided up into intervals that are all Δx wide. The initial temperature profile is labeled at the grid locations with T_i^o, where the subscript i refers to nodal locations $(m - 2 \leq i \leq m + 3)$, and the superscript 0 refers to time. The initial condition, or initial profile, exists at time $t = 0$. To find the temperature profile at time $t = 1\Delta t$, we use Equation 4.43 as follows:

Align T_{m-2}^0 with T_m^0 and locate T_{m-1}^1.

Align newly found T_{m-1}^1 with T_{m+1}^0 and locate T_m^1.

Align newly found T_m^1 with T_{m+2}^0 and locate T_{m+1}^1.

Align newly found T_{m+1}^1 with T_{m+0}^0 and locate T_{m+2}^1.

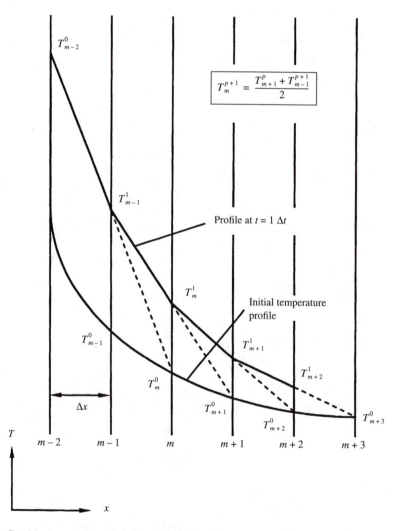

$$T_m^{p+1} = \frac{T_{m+1}^p + T_{m-1}^{p+1}}{2}$$

Profile at $t = 1\,\Delta t$

Initial temperature profile

FIGURE 4.21 Graphical equivalent of the one-dimensional Saul'ev numerical method.

Eventually, the profile after 1 time interval emerges. The point-location process is continued until a desired condition is reached. This is exactly what is done in the numerical method; the Saul'ev plot is merely a graphical way of solving a one-dimensional transient problem with a numerical technique. Note that as time increases (p increases), the temperature profiles approach the steady-state distribution.

Many problems encountered have a convective boundary that can be modeled graphically. At the surface, we write

$$\frac{q}{A} = -k\frac{\partial T}{\partial x}\bigg|_m = \bar{h}_c(T_\infty - T_w)$$

This boundary condition can be written as

$$\frac{\partial T}{\partial x}\bigg|_m^p = \frac{T_w^p - T_\infty^p}{k/\bar{h}_c}$$

Thus, for any time $t = p\,\Delta t$ the temperature gradient at the wall (location m) equals the difference between wall and fluid temperatures divided by $k/\bar{h}_c$. In other words, the definition of the slope $\partial T/\partial x$ equals the rise $(T_w - T_\infty)$ over the run $k/\bar{h}_c$. This is shown graphically in Figure 4.22. Note that $k/\bar{h}_c$ has dimensions of length.

We can also model an insulated boundary, for which we write

$$q'' = k\frac{dT}{dx} = 0$$

Thus, the slope of the temperature profile at an insulated boundary is 0; or, the temperature profile intersects the insulated boundary at a right angle. This is illustrated also in Figure 4.22.

Example 4.9

A stainless steel (type 304) wall is 8 in thick and is part of the chimney section of a furnace. When the furnace is in operation, the temperature of the inside-wall surface is 600°F, while the outside-wall temperature is 95°F. The profile within the wall is at steady state. The chimney section is to be cleaned, so the furnace is shut down and cooling fans are used to move 75°F air upward through the chimney. The right face stays at 95°F during the cooling process. Determine the time it takes for the stainless steel slab to cool to the point where the temperature at any point within the wall is less than 200°F. Use a convection coefficient of 84 BTU/(hr·ft²·°R).

Solution

As the cooling air moves upward through the chimney, the air becomes warmer, absorbing heat as it rises. Air at the chimney bottom has a greater capacity for absorbing heat, because that is where the greatest temperature difference (chimney minus air) exists. Temperature of the chimney outside surface will probably vary with height and will be 95°F only over a certain portion of the stack. Further, the air and chimney temperatures vary with height and so will the film conductance.

Assumptions

1. The chimney surface temperature of 95°F for the outside wall is uniform and constant at the section of interest.
2. One-dimensional heat flow exists.
3. Convection coefficient is constant and applies at the section of interest.

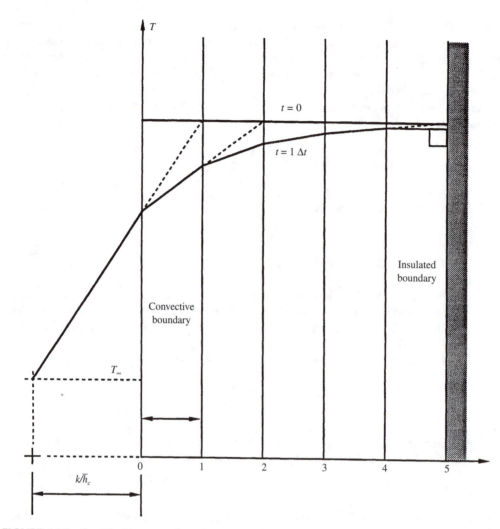

FIGURE 4.22 Graphical interpretation of convective and insulated boundaries.

4. Material properties are constant.
5. The warm chimney loses energy only to the air within by convection.

The problem is solved graphically, and the Saul'ev plot is shown in Figure 4.23. The plot shown is the completed picture. When drawn, the temperature profiles develop one by one, and the number of elapsed time intervals is determined.

Arbitrarily, Δx is selected as 1 in. From Appendix Table B.2 for type 304 stainless steel,

$\rho_s = 7.817(62.4)$ lbm/ft^3
$c = 0.110$ BTU/(lbm·°R)
$k = 8.32$ BTU/(hr·ft·°R)
$\alpha = 0.417 \times 10^{-4}$ ft^2/s

For the convective boundary,

$$\frac{k}{\bar{h}_c} = \frac{8.32}{84} = 0.099 \text{ ft} = 1.19 \text{ in}$$

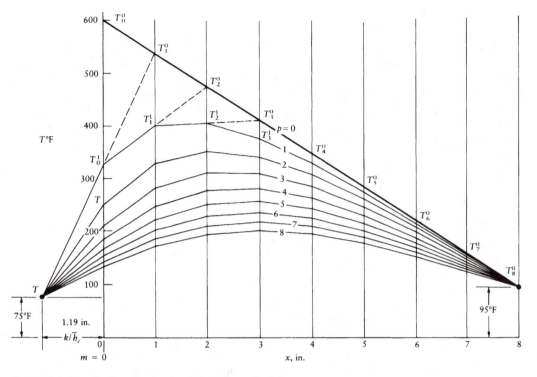

FIGURE 4.23 Saul'ev plot for the slab of Example 4.9.

The method is as follows:

Align T_∞ and T_1^0 to locate T_0^1.

Align newly found T_0^1 and T_2^0 to locate T_1^1.

Align newly found T_1^1 and T_3^0 to locate T_2^1.

and so forth. The solution clearly illustrates temperature profiles at each time interval. Here, eight time intervals have elapsed before the condition that any point within the wall is cooler than 200°F. We calculate

$$\mathrm{Fo}_\Delta = \frac{\alpha\,\Delta t}{(\Delta x)^2} = \frac{0.417 \times 10^{-4}\Delta t}{(1/12)^2} = 1$$

$$\Delta t = 166.5 \text{ s}$$

With eight intervals, the time required is

$$t = 8\,\Delta t = 1332 \text{ s} = 0.37 \text{ hr}$$

Some conditions that might be encountered could lead to simplifications or difficulties. For example, if silica brick is the material, $k = 0.618$ BTU/(hr·ft·°R), and $k/\bar{h}_c = 0.618/84 = 0.0074$ ft $= 0.088$ in. Thus, the surface temperature would be approximately equal to the free-stream temperature for all time. On the other hand, if the convective coefficient is 8.4, then, for stainless steel, $k/\bar{h}_c = 8.32/8.4 = 0.99$ ft $= 11.9$ in. This is large compared to the wall thickness, so some accuracy may be lost. ❑

4.6 SUMMARY

It was remarked in the beginning of this chapter that three types of transient-conduction problems exist—systems with negligible internal resistance, systems with negligible surface resistance, and systems with finite internal and surface resistances. Various problems that fit into these categories were discussed. Analytical methods were not expressly emphasized, although the solution charts presented were constructed from analytical results. Multidimensional geometries were also considered. The approximate methods reviewed included the numerical method and the graphical method. Transient problems abound, and it is hoped that the few examples discussed in this chapter provided an insight into the physics of what happens in such problems and how solutions may be formulated.

4.7 PROBLEMS

4.7.1 LUMPED CAPACITANCE METHOD

1. Calculate the Biot number for the thermocouple junction of Example 4.1. Is use of the lumped-capacitance method justified?
2. Beginning with Equation 4.12, derive Equation 4.15.
3. Brass rod is used in a brazing operation. The brass must first be heat treated to make it into a suitable filler metal. The rods are 1/8 in diameter and 3 ft long. They are heated to a temperature of 200°F, then cooled individually in air. The ambient air temperature is 70°F with a convection coefficient of 2 BTU/(hr·ft²·°R). Determine the time required for one rod to cool to 75°F.
4. Springs used in the suspension system of an automobile are made of steel rod (1.5C), 2 cm in diameter, 1.2 m long. The rods are heated individually and, while ductile, they are wound into the shape of a coil. The rods are heated from an initial temperature of 20°C to 250°C in an oven where the ambient temperature is 500°C, and the convection coefficient is 5 W/(m²·K). How much energy must be added to each rod, and how much time is required to heat one rod?
5. A sheet of glass 1/4 in thick, 3 ft wide, and 4 ft tall is at an initial temperature of 50°F. It is heated to 75°F by two air streams, one on either side of the glass. The air temperature is 90°F, and the convection coefficient between the air and the glass is 2 BTU/(hr·ft²·°R). Determine the time required for the process to occur. Graph the instantaneous heat transferred as a function of time.
6. Ornamental iron gates consist of various types of square cast-iron bars. Consider one such casting that is $1 \times 1 \times 1$ cm long. It is cast in a sand mold and released when it has solidified to some extent. It then cools from an initial temperature of 1 000°C in air whose temperature is 28°C with a convection coefficient of 12 W/(m²·K). How much time elapses before the temperature at the rod center is 30°C?
7. A half-gallon ice cream container having dimensions of 6-7/8 × 5-3/8 × 3-1/2 in, taken from a freezer, is at an initial temperature of 25°F. It is exposed to air at a temperature of 95°F, $\bar{h}_c$ = 0.8 BTU/(hr·ft²·°R). Determine and graph the variation of center temperature with time until the center temperature reaches 35°F. Use the following properties for ice cream:

   $k = 2.2$ W/(m·K)

   $\rho = 1\ 030$ kg/m³

   $c_p = 2.42$ kJ/(kg·K)

4.7.2 CHART SOLUTIONS: SLABS, CYLINDERS, SPHERES

8. A 2-cm diameter sphere is made of plastic and cast in a mold. The plastic has the same properties as Plexiglas®, with a specific heat of 2 000 J/(kg·K). When the sphere is removed from the mold, the sphere temperature is uniform at 40°C. Determine the time required for the center temperature of the sphere to reach 25°C. Use a convection coefficient of 4 W/(m²·K) and $T_\infty = 21°C$.

9. A spongy rubber ball $[c_p = 0.4$ BTU/(lbm·°R)] used as a toy for pets has a 3-in diameter. During manufacture, it is cast in a mold, removed, and submerged in an oil bath. The initial ball temperature is uniform and equal to 150°F. The oil in the bath circulates at a high velocity such that the ball cools quickly. Is 20 min sufficient time for the temperature at the ball center to reach 72°F? Take the convection coefficient to be 10 BTU/(hr·ft²·°R) and $T_\infty = 60°F$.

10. An asphalt $[c_{p=}800$ J/(kg·K)] driveway is 10 cm thick. On a hot summer day, the asphalt reaches a uniform temperature of 54°C. Water is then applied to the surface, suddenly reducing the surface temperature to 17°C. How long will it take for the temperature 5 cm below the surface to reach one-half the initial temperature? How much heat per unit area is transferred to the water? (Hint: the convection coefficient is infinite.)

11. A sheet of rubber is 3/4 in thick. It is to be cured at a temperature greater than 200°F for a certain length of time. Heat is applied at both surfaces, and it can be assumed that these surfaces are at 300°F. The initial temperature of the sheet is 72°F, and the thermal diffusivity α is 0.003 ft²/hr. Graph the temperature at the center of the sheet vs. time and determine how long it will take for the center temperature to reach 200°F.

12. There is sometimes a controversy over whether cold water or hot water freezes faster in ice-cube trays placed in a freezer. Consider a cube of water 2.5 × 2.5 × 2.5 cm whose surfaces are maintained at −5°C. Determine the time required for the center of the cube to reach 0°C. Assume that Figure 4.8a applies with $V/A = R/3$.

    (a) Perform the calculations for an initial temperature of 16°C.

    (b) Perform the calculations for an initial temperature of 50°C.

    Use diffusivity values at these temperatures from Table C.11.

13. The opening doors of an elevator have to meet certain standards regarding fire protection. One such door is constructed of two sheets of 16-gage stainless steel with a layer of asbestos between them. Assume the temperature of one of the sheet-metal sides of the door is elevated from 70°to 1000°F, and the other sheet-metal side is to become no warmer than 200°F after 45 min.

    (a) Estimate the thickness of asbestos required.

    (b) What is the instantaneous heat-transfer rate through the door after the 45 min have elapsed?

14. Potatoes $[k = 0.63$ BTU/(hr·ft·°R)]* are baked in an oven where the ambient temperature is 400°F and the convection coefficient is 1.2 BTU/(hr·ft²·°R). The initial temperature of a potato to be baked is 70°F. After one hour of baking, the temperature at the potato center is 160°F. The potato is then removed from the oven; the ambient temperature is now 70°F, and $\bar{h}_c$ is again 1.2 BTU/(hr·ft²·°R). How long before the temperature at the potato center is 140°F? Assume the potato is spherical with a diameter of 3-1/2 in.

15. An egg $[k \approx 0.65$ W/(m·K)]* taken from a refrigerator is initially at 8°C. It is submerged in boiling water at 100°C for 3 min. Estimate the temperature at the center of the egg. Assume the egg to be spherical in shape with a diameter of 5 cm and having a thermal

---

* Geonkoplis, Christie J., *Transport Properties and Unit Operations,* Allyn & Bacon, Inc., Boston, 1978, p. 629.

diffusivity of 6×10^{-7} m²/s. Take the surface convection coefficient to be 3 500 W/(m²·K).

16. An apple (assumed spherical) is approximately 2-1/2 inches in diameter. It is initially at a temperature of 75°F; then it is placed in a refrigerator where the temperature is 40°F, and the convection coefficient is 1.15 BTU/(hr·ft²·°R). Estimate the time required for the center of the apple to reach 45°F. How much heat must be removed by the refrigerator to effect this cooling? The apple properties are as follows:

   c_p =3.8 kJ/(kg·K)[*]
   k = 0.7 W/(m·K)
   ρ = 990 kg/m³

17. According to a cookbook,[†] the center temperature of a 4-lbm prime rib roast (assume spherical) reaches 140°F after 2-1/4 hr in an oven where the temperature is 325°F. If the thermal conductivity of beef is 0.4 W/(m·K), the specific heat is 3.4 kJ/(kg·K), and the density is 62.1 lbm/ft³, determine the average convection coefficient in the oven. Take the initial temperature of the roast to be 60°F.

18. A 4-in dia., very long, copper cylinder initially at 70°F is heated in a bath whose temperature is 250°F and where the convection coefficient is 200 BTU/(hr·ft²·°R).
    (a) What is the centerline temperature after 3 minutes have elapsed?
    (b) Suppose that a copper cylinder only 3 inches in diameter is heated in the same way. What is the convection coefficient required in order that the smaller cylinder center-line temperature equal that of the larger cylinder after the same amount of time has elapsed?
    (c) Repeat part (b) for a cylinder diameter of 5 in.

19. A stainless steel (type 304) slab is initially at a temperature of 20°C. It is heated in a heat-treating oven where the ambient temperature is 100°C. How much time is required for the slab centerplane temperature to double? Perform the calculations for slab thicknesses of 1, 3, 5, 10, and 15 cm. Determine also the corresponding ratio of cumulative heat transferred to total heat ($\tilde{Q}/\tilde{Q}_0$). Take the convection coefficient to be 50 W (m²·K).

20. An aluminum slab is 2 inches thick, and it is air-cooled from an initial temperature of 200°F until its centerline temperature is 100°F. The air temperature is 75°F, and the convection coefficient is 10 BTU/(hr·ft²·°R).
    (a) How long does this cooling process take?
    (b) Repeat the calculations for copper, iron, magnesium, and tin.
    (c) Graph diffusivity vs. time.

21. Example 4.4 deals with an orange taken from a refrigerator and allowed to warm up. In this problem, the reverse of that process is investigated. How long does it take an orange initially at 20°C to cool so its center temperature is 5°C in a refrigerator where the ambient temperature is 4°C? Use the same physical dimensions and thermal properties as in the example.

4.7.3 Semi-infinite Slabs

22. The temperature of the left face of a semi-infinite nickel slab is elevated from an initial temperature of 50°F to 200°F. Graph the temperature profile existing within the slab after 1, 5, and 10 minutes have elapsed. Use the error function chart in the appendix.

23. A particular soil has a diffusivity of approximately 0.006 ft²/hr. It exists initially at 58°F. Its surface is exposed to heat and reaches a temperature of 95°F. Graph temperature vs. depth for an elapsed time of 7 days, assuming these conditions are constant. Use the error function chart in the appendix.

---

[*] Geonkoplis, Christie J., *Transport Properties and Unit Operations,* Allyn & Bacon, Inc., Boston, 1978, p. 629.
[†] Better Homes & Gardens, *New Cook Book,* 1976.

24. The asphalt ($\alpha = 0.05 \times 10^{-5}$ m²/s) of a street is 8 cm thick. During daylight hours, the asphalt surface is exposed to warm air that raises the asphalt surface temperature to 35°C. If the initial temperature of the asphalt is 23°C, determine and graph the temperature vs. time response of a point 1 cm below the surface of the asphalt up to a time when the temperature 8 cm below the surface begins to receive energy.

25. Repeat the calculations of Example 4.6 for other objects in the laboratory, such as a cork board, concrete, and window-pane glass.

26. A hot aluminum block with a temperature T_{al} of 200°F is brought into contact with another block whose temperature is 100°F. Determine the contact temperature T_c if the second block is
    (a) copper
    (b) lead
    (c) tin

4.7.4 MULTIDIMENSIONAL PROBLEMS

27. Lenses for telescopes are made from glass that must be heat treated. One such piece of glass is 32 cm in diameter and 6 cm thick. The glass is initially at 25°C, and it must be heated to 400°C. The oven to be used has a controllable ambient temperature. Determine the time required to heat-treat the glass for an ambient temperature of 425°C. Take the convection coefficient to be 6 W/(m²·K) and the material to have the same properties as window-pane glass.

28. Common bricks are made of an earth-and-clay mixture that is shaped, cut, and oven-cured. One brick has dimensions of $8 \times 4 \times 4$ in. The bricks can be oven treated in one of two ways—either in a batch type of process or in a continuous process. The batch process is similar to placing each brick individually into the oven. The continuous process is one where the bricks are placed lengthwise one behind the other on a conveyor, forming a 4×4-in infinite rectangular bar that travels slowly through the oven. The oven in either heating method is the same; the ambient temperature is 1200°F, and the convection coefficient is 2 BTU/(hr·ft²·°R). How much time is required in each method for the center temperature of a brick to reach 1000°F from an initial temperature of 60°F?

29. A thin-walled casserole dish contains a mixture of vegetables with a uniform initial temperature of 22°C. The dish itself has an external diameter of 23 cm and a height of 10 cm. The entire mix is heated, and after 20 min in a 165°C oven, the center temperature of the food reaches 65°C.
    (a) Determine the average diffusivity of the dish and vegetable composite.
    (b) The casserole is removed from the oven and allowed to cool in air (temperature of air is 22°C). How long must it be allowed to cool before its center temperature reaches 50°C? Assume the convection coefficient in the oven to be 6.5 W/(m²·K).
    For all cases, the thermal conductivity is 1 W/(m·K).

30. Some people buy loaves of bread that are unsliced, because they believe such loaves do not need (and therefore do not contain) preservatives. Often, several loaves are purchased and frozen. When necessary, a loaf is removed from the freezer (where the ambient temperature is 0°C) and set out on a rack to thaw. One loaf measures $13 \times 13 \times 21$ cm long and weighs 1 lbm. After 3 hr at room temperature [25°C ambient temperature and $\bar{h}_c = 5$ W/(m²·K)], the loaf center reaches a temperature of 23°C. Determine the thermal conductivity of the bread if its specific heat is 2 800 J/(kg·K).

4.7.5 NUMERICAL METHODS

31. Solve the problem described in Example 4.9, but use $Fo_\Delta = 1.4$. Compare the results to those of the example.

32. A large slab of cast iron 12 in thick is initially at 400°F. Suddenly its left face is reduced to 40°F while its right face is kept at 400°F. Using the numerical method, determine the temperature at the midplane after 15 minutes have elapsed.

33. A large slab of copper 10 cm wide is initially at a uniform temperature of 25°C. Suddenly the temperature of its left and right faces in increased to 200°C. How much time elapses before the midplane temperature reaches 140°C? Solve numerically.

34. It is desired to construct a fireproof wall for a safe. The wall consists of two 12-gage pieces of sheet metal with asbestos insulation sandwiched between them. The criterion for protection is that, if the temperature of the outside piece of sheet metal is suddenly raised to 1000°F, then no less than 2 hr should elapse before the temperature of the inside piece of sheet metal uses to 200°F from an initial and uniform temperature of 75°F. Determine the required thickness of the asbestos. Solve numerically.

35. A stainless steel (type 347) wall serves as part of a steam-conveying line. It is 120 mm thick and at an initial temperature of 18°C. Its left face is exposed to a high-speed stream of steam at a temperature of 250°C, with a convection coefficient of 500 W/(m²·K). The right face is exposed to a stream of water at 15°C with a convection coefficient of 250 W/(m²·K). Determine the left- and right-surface temperatures when the midplane temperature reaches 75°C. Solve numerically.

4.7.6 GRAPHICAL METHODS

36. Solve Problem 32 graphically.
37. Solve Problem 33 graphically.
38. Solve Problem 34 graphically.
39. Solve Problem 35 graphically.

4.7.7 PROJECT PROBLEMS

40. Data from a conventional cookbook[*] indicate the following roasting times for chicken (all at constant oven temperatures):

Time for roasting, hours	Weight of chicken, lbm
0.75–1	1.5–2
1–1.25	2–2.5
1.25–1.5	2.5–3
1.5–2	3–4

Given these data, it is reasonable to assume that, for any chicken from 1.5 to 4 lbm, the cooking time should result in a center temperature that is the same. We investigate this expectation in this problem. For chicken, use a thermal conductivity of 0.502 W(m·K),[†] density of 995 kg/m³, and a specific heat (c_p) of 3.2 kJ/(kg·K). Complete the steps in the following paragraphs to solve this problem.

Discussion

The chicken is probably cooked in a pan, so the surface in contact with the metal is receiving heat by conduction. The oven temperature varies within a few degrees of the set point. As the chicken cooks, its properties will change considerably, depending on the water content of the meat and how it is prepared (e.g., type of stuffing). To obtain a solution, however, we can assume that:

---

[*] Better Homes & Gardens, *New Cook Book,* 1976.

[†] Geonkoplis, Christie J., *Transport Properties and Unit Operations,* Allyn & Bacon, Inc., Boston, 1978, p. 629.

- The chicken receives energy solely by convection heat transfer.
- The film coefficient $\bar{h}_c$ is an average that applies to the entire surface.
- Properties of the meat are constant during the process.
- Oven temperature is constant.
- Chicken is spherical in shape.
- The convection coefficient is 2.0 BTU/(hr·ft²·°R).

Solution

Time t is given by the data table. The volume is calculated from the weight data and the density. By definition,

$$V_s = \frac{\text{weight}}{\rho_s} = \frac{\text{wt}}{995 \text{ kg/m}^3} = \frac{\text{wt}}{62.1 \text{ lbm/ft}^3}$$

(a) Determine the volume for all eight time-weight entries in the table. To determine the surface area, we use the assumption that the chicken is spherical, so a diameter can be calculated from the volume:

$$D = (6V_s/\pi)^{1/3}$$

The surface area becomes

$$A_s = \pi D^2$$

(b) Calculate diameter and surface area for all data in the table. The Fourier number is given by

$$\text{Fo} = \frac{\alpha t}{L_c^2} = \frac{\alpha t A_s^2}{V_2^2} = \frac{k t A_s^2}{\rho c V_s^2}$$

(c) Determine the Fourier number for the data.
(d) Calculate the Biot number for the data using D as the characteristic dimension.
(e) Use the Biot and Fourier numbers to obtain numbers from the appropriate Heisler and Grober charts.
(f) Calculate the center temperature for all the data.
(g) Are the results reasonable?

41. Integral Solution Method
    The integral method of solution, although approximate, is a powerful tool in the solution of one-dimensional unsteady problems. We shall apply the method to unsteady heat flow in a semi-infinite slab, as shown in Figure P4.1. The problem considered is one in which the slab is initially at temperature T_i. Then, at time zero, the left face is raised to some temperature T_s. The differential equation for this system is Equation 4.16:

$$\frac{\partial^2 T}{\partial x^2} = \frac{1}{\alpha}\frac{\partial T}{\partial t} \tag{4.16}$$

    with the following boundary and initial conditions:

    B.C. 1. $x = 0$, $T = T_s$
    B.C.2. $x = \infty$, $T = T_i$
    IC. $t = 0$, $T = T_i$

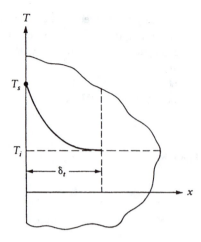

FIGURE P4.1 A semi-infinite body.

The solution to this problem is expressed in Equation 4.17 and graphed in Figure 4.10:

$$\frac{T - T_s}{T_i - T_s} = \mathrm{erf}\left[\frac{x}{(2\sqrt{\alpha t})}\right] \tag{4.17}$$

In the integral method, we integrate the differential equation over a distance δ_t (see Figure P4.1), called the *thermal boundary layer*, which is a distance that varies with time. When the temperature of the left face of the slab is raised to T_s, the heat added will penetrate the material only to the edge of the thermal boundary layer. The slab temperature beyond δ_t is still at T_i, although B.C. 2 states that, for this to occur, x must be infinite. The solution we seek in the integral method is the temperature variation within δ_t and the variation of δ_t with time.

Our next step would be to integrate the differential equation over the thermal boundary layer to obtain what is known as the *heat balance integral* for the problem. This step involves some mathematics beyond the scope of what we wish to cover, so we will merely state the results when necessary.

The next step in the integral method is to assume a solution for temperature in terms of x/δ_t. A polynomial profile is usually assumed, although trigonometric or exponential approximations can also be written. The polynomial equation selected will have unknown coefficients that we will evaluate through the use of the boundary and initial conditions, Thus, we assume

$$T = C_1 + C_2\frac{x}{\delta_t} + C_3\frac{x^2}{\delta_t^2} + C_4\frac{x^3}{\delta_t^3} + C_5\frac{x^4}{\delta_t^4} \tag{i}$$

where $0 \le \dfrac{x}{\delta_t} \le 1$

We must apply conditions to Equation i to evaluate the unknown C_i coefficients. We could write only two conditions:

1. $x = 0$, $T = T_s$
2. $x = \delta_t$, $T = T_i$

The coefficients in Equation i would become

$$C_3 = C_4 = C_5 = 0$$

1. $T_s = C_1 + C_2(0)$ or $C_1 = T_s$
2. $T_i = T_s + C_2$ or $C_2 = T_1 - T_s$

Equation i then reduces to

$$\frac{T - T_i}{T_s - T_i} = \frac{x}{\delta_t}$$

This equation is a linear profile where T varies from T_s to T_i with increasing x/δ_t. A better approximation (to the exact solution) will result if more coefficients are retained. This requires more conditions. In addition to 1 and 2 above, we might also write the condition that, at the edge of the boundary layer, the slope of the temperature profile is zero:

3. $x = \delta_t$, $\dfrac{\partial T}{\partial x} = 0$

(a) When conditions 1, 2, and 3 are applied to Equation i, the profile becomes

$$\frac{T - Ti}{T_s - T_i} = 1 - 2(x/\delta_t) + (x/\delta_t)^2$$

(b) A fourth condition can also be written by referring to the original differential equation. At $x = 0$, the left face, the temperature is T_s, which does not vary with time. So

$$\left.\frac{\partial T}{\partial t}\right|_{x=0} = 0$$

Combining with Equation 4.17 gives a fourth condition:

4. $\left.\dfrac{\partial^2 T}{\partial x^2}\right|_{x=0} = 0$

When the above four conditions are applied to Equation i, show that the result is

$$\frac{T - T_i}{T_s - T_i} = 1 - \frac{3}{2}(x/\delta_t) + \frac{1}{2}(x/\delta_t)^3$$

(c) Now, at the edge of the boundary layer ($x = \delta_t$), $T = T_i$, which is a constant and not a function of time. Thus,

$$\left.\frac{\partial T}{\partial t}\right|_{x=\delta_t} = 0$$

Combining with the original differential equation (4.17) gives yet a fifth condition:

5. $\left.\dfrac{\partial^2 T}{\partial x^2}\right|_{x=\delta_t} = 0$

When all five conditions are used, show that Equation i gives the following fourth-degree approximation for temperature:

$$\frac{T-T_i}{T_s-T_i} = 1 - 2(x/\delta_t)^2 + 2(x/\delta_t)^3 - (x/\delta_t)^4$$

(d) All of the above derived profiles satisfy the boundary conditions. It is necessary, however, to find the depth of the thermal boundary layer δ_t. This is done by using the heat balance integral, a step that we skipped. The result would be

$$\delta_t = \sqrt{8\alpha t} = 2\sqrt{2}\sqrt{\alpha t}$$

Construct a graph of the dimensionless temperature $(T = T_i)/(T_s = T_i)$ as a function of $x/\sqrt{4\alpha t}$ using all four of the profiles derived in this problem.

(e) It would be instructive to also graph the error function on the same set of axes as in part (d) so that a comparison can be made. The dimensionless temperature in Equation 4.17, however, is not the same as that in the profiles derived here. Note that

$$1 - \frac{T-T_s}{T_i-T_s} = \frac{T_i-T_s-T+T_s}{T_i-T_s} = \frac{T_i-T}{T_i-T_s} = \frac{T-T_i}{T_s-T_i}$$

Use the preceding equation to help graph the error function with the other profiles, and compare the results.

5 Introduction to Convection

CONTENTS

5.1 INTRODUCTION

The preceding chapters dealt primarily with conduction heat transfer. Convective effects were discussed briefly and considered almost exclusively as boundary conditions. In this and the following chapters, we turn our attention to the convection problem.

There are two categories of heat transfer involving fluids: forced convection and free or natural convection. In forced convection, the fluid flow is established by an external driving force provided by, say, a fan or pump. In natural convection, the fluid flow is caused by the buoyant effect resulting from heating or cooling the fluid.

The method of solution in forced- and natural-convection problems differs. Generally, for the forced-convection problem, the velocity profile is found first, then used to derive the temperature profile. For the natural-convection problem, the velocity and temperature profiles are intimately related and are therefore determined together. The final objective of an analysis is, again, to determine the heat-transfer rate. In either case, it is important to have information on the fluid properties that influence fluid flow and on the equations of motion for a fluid. These topics are discussed in this chapter.

We will derive continuity, momentum, and energy equations that will be of use to us in formulating convection problems. The equations will be applied to some laminar-flow problems to show how the equations are reduced or simplified according to assumptions made in specific problems. Simplifying the general equations to obtain the applicable expression is the method recommended here rather than developing equations from an energy balance for each problem of interest. The recommended method is safer and provides a unifying overview of convection problems.

In chapters that follow, we will consider laminar- and turbulent-flow problems and natural-convection problems in greater detail. This chapter thus serves as an introduction to convection by providing general equations of fluid mechanics and convection heat transfer.

5.2   FLUID PROPERTIES

Our objective is to determine equations of fluid mechanics and use them to solve for velocity profiles in a system. We will need information on the important properties of fluids. Those properties discussed here are dynamic or absolute viscosity, pressure, density, kinematic viscosity, surface tension, internal energy, enthalpy, specific heat, compressibility factor, and volumetric thermal expansion coefficient.

5.2.1   ABSOLUTE VISCOSITY

A fluid (liquid or gas) is a substance that deforms continuously under the action of an applied shear stress. The rate of deformation is defined as a strain rate. The applied shear stress is proportional to the strain rate, and the proportionality factor is called the fluid *viscosity.*

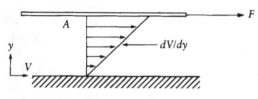

To illustrate, consider the experimental setup shown in Figure 5.1. A liquid is placed in the space between two parallel plates. The lower plate is stationary, and the upper plate is pulled to the right at some velocity V with some force F_1. The area of contact between the upper plate and the liquid is A. Thus the plate is exerting a shear stress equal to $\tau = F_1/A$. A velocity vs. y coordinate system is established. If we could measure the velocity at each point within the fluid, a velocity distribution like that shown would result (in the absence of a pressure gradient in the flow direction). The fluid velocity at the lower plate is zero and at the upper plate is V, due to the nonslip condition. The slope of the velocity distribution is dV_1/dy (the strain rate corresponding to τ_1), and it is linear under laminar-flow conditions.

FIGURE 5.1   Shear stress applied to a fluid.

Suppose that we repeat the experiment, using a different force F_2. We would measure a different slope dV_2/dy. Likewise, if the force is F_3, dV_3/dy would result. If these data are graphed for a fluid like water as τ vs. dV/dy, a plot such as Figure 5.2 would result. The data points lie on a straight line that passes through the origin. The slope of the resulting line is the absolute viscosity of the fluid, because it is a measure of the fluid's resistance to the applied shear stress. Thus, the fluid viscosity gives an indication of how the fluid behaves *(dV/dy)* under the action of an applied shear stress (τ).

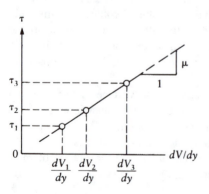

The graph of Figure 5.2 is a straight line that passes through the origin. This is characteristic of what is known as a *Newtonian fluid,* which can be described by

$$\tau_{yx} = \mu \frac{dV_x}{dy} \tag{5.1}$$

where τ_{yx} is the x-directed force (denoted with second subscript) acting normal to the y direction (first subscript), μ is the absolute or dynamic viscosity of the fluid, and dV_x/dy is the strain rate. The shear stress has dimensions of F/L^2 ($N/m^2 = Pa$ or lbf/ft^2), the viscosity has dimensions of $F \cdot T/L^2$ ($N \cdot s/m^2$ or $lbf \cdot s/ft^2$), and the strain rate has dimensions of $1/T$ (rad/s).

FIGURE 5.2   Graph of shear stress vs. strain rate for a Newtonian fluid.

From our experience with fluids, we know that, if we invert a glass of water, the flow pattern is vastly different from that of an inverted jar of mayonnaise. The mayonnaise behaves quite differently under the action of an applied shear stress than does water, and the mayonnaise cannot be accurately described by Equation 5.1. Therefore, mayonnaise is an example of what is known as a *non-Newtonian fluid*. Other examples include paint, printer's ink, chocolate mixtures, and flour dough. Examples of Newtonian fluids are water, oil, air, and glycerine. The absolute or dynamic viscosity of various substances can be obtained from data in the Appendix tables (C for liquids and D for gases). In this text, we will work exclusively with Newtonian fluids.

To get an intuitive notion regarding shear stresses and strain rates, consider the simple exercise of spreading butter with a knife over a slice of toasted bread. Suppose we move the knife across the bread at a velocity of 3 in/s (7.6 cm/s), and that the butter between the knife and bread is on the average 1/8 in (0.32 cm) thick. The strain rate is then calculated as

$$\frac{dV}{dx} \approx \frac{\Delta V}{\Delta y} \approx \frac{3 \text{ in/s}}{\frac{1}{8} \text{ in}} = 24 \text{ rad/s}$$

Assume next that the force with which we move the knife is roughly 2 oz (0.56 N) and that the knife area is rectangular with dimensions of 2-1/2 × 1/2 inches in contact with the butter. The shear stress applied then is

$$\tau = \frac{F}{A} = \frac{2}{16} \text{ lbf} \frac{144}{2.5 \times 0.5 \text{ ft}^2} = 14.4 \text{ lbf/ft}^2 \text{ (689 Pa)}$$

5.2.2 PRESSURE

Fluids at rest or moving experience some form of pressure variation, either with height or with horizontal distance. *Pressure* is defined as a normal force per unit area acting on or existing within the fluid. Pressure p has dimensions of F/L^2 (Pa = N/m^2 or lbf/ft^2).

Pressure units depend on whether atmospheric pressure is included. A pressure gage, for example, is calibrated to give a zero reading when exposed to the pressure of the atmosphere. Readings from such gages are known as gage pressures and denoted in Engineering units as psig (lbf/in^2–gage). An absolute reading is zero only when a perfect vacuum exists and is denoted in Engineering units as psia (lbf/in^2–absolute). All readings in SI units are of absolute pressure unless stated otherwise.

5.2.3 DENSITY

The *density* ρ of a fluid is its mass per unit volume. Density has dimensions of M/L^3 (kg/m^3 or lbm/ft^3). The density of various gases is provided in Appendix D tables. A property related to density is specific weight, *SW*. Specific weight is a weight expressed in force units per unit volume and is related to density by

$$SW = \frac{\rho g}{g_c} \tag{5.2}$$

The dimensions of specific weight are F/L^3 (N/m^3 or lbf/ft^3).

Another useful quantity related to density is the specific gravity of a substance. *Specific gravity* s is the ratio of the density of a substance to that of water:

$$s = \frac{\rho}{\rho_w} \tag{5.3}$$

where ρ_w is the density of water at a particular condition so that this reference property remains unchanged. Values of specific gravity (rather than density) are provided in the Appendix tables (B for solids and C for liquids) for various substances. For our purposes, we assume that

$$\rho_w = 1\,000 \text{ kg/m}^3 = 62.4 \text{ lbm/ft}^3 \tag{5.4}$$

5.2.4 KINEMATIC VISCOSITY

In many situations, the ratio of absolute viscosity to density arises. This ratio is defined as the *kinematic viscosity,* given by

$$\nu = \frac{\mu g_c}{\rho} \tag{5.5}$$

The dimensions of kinematic viscosity are L² (m²/s or ft²/s). The property tables of the Appendix contain kinematic viscosity and density, from which absolute viscosity can be calculated. For example, from Table C.11 for water at 60°C,

$$\rho = 0.985(1\,000) \text{ kg/m}^3 = 0.985(62.4) \text{ lbm/ft}^3$$

and

$$\nu = 0.478 \times 10^{-6} \text{ m}^2\text{/s} = 0.514 \times 10^{-5} \text{ ft}^2\text{/s}$$

The absolute viscosity is

$$\mu = \frac{\rho\nu}{g_c} = 0.985(1\,000)\frac{\text{kg}}{\text{m}^3}\left(0.478 \times 10^{-6}\,\frac{\text{m}^2}{s}\right)$$

$$= 4.71 \times 10^{-4} \text{ kg/(m·s)} = 4.71 \times 10^{-4} \text{ kg·m·s/(m}^2\text{·s}^2)$$

$$= 4.71 \times 10^{-4} \text{ N·s/m}^2$$

In English units

$$\mu = \frac{0.985(62.4)\dfrac{\text{lbm}}{\text{ft}^3}\left(0.514 \times 10^{-5}\,\dfrac{\text{ft}^2}{s}\right)}{32.2\dfrac{\text{lbm·ft}}{\text{lbf·s}^2}} = 9.81 \times 10^{-6} \text{ lbf·s/ft}^2$$

It is important to remember that density and viscosity change with temperature. For example, from Table C.1 for mercury, when temperature varies from 0°C (32°F) to 315°C (600°F), specific gravity varies from 13.628 to 12.847. Over this same temperature range, kinematic viscosity varies from 0.124×10^{-6} m²/s (0.133×10^{-5} ft²/s) to $0.067\,3 \times 10^{-6}$ m²/s (0.0724×10^{-5} ft²/s).

5.2.5 SURFACE TENSION

Surface tension is a measure of the energy required to reach below the surface of a liquid to bring molecules to the surface and form a new area. Surface tension thus arises from molecular considerations and has meaning only for liquids. The dimensions of surface tension σ are energy/area, $F \cdot L/L^2 = F/L$ (N/m or lbf/ft). Surface tension is important in the study of droplet flows or where liquid-liquid and liquid-vapor interfaces exist. In the majority of problems we discuss, surface-tension forces are negligible compared to pressure, gravity, and viscous forces.

Before discussing internal energy, enthalpy, and specific heat, it is necessary to mention the ideal gas law. This approximation relates density to pressure and temperature:

$$\rho = \frac{p}{RT}$$

where R^* is the gas constant for the particular gas of interest (values of which are provided in Appendix D tables for various gases), p is pressure in absolute units, and T is temperature in absolute units. The gas constant for any gas can be obtained by dividing the *universal gas constant* by the molecular weight (called molecular mass in SI) of the gas. The universal gas constant is

$$\bar{R} = 1545 \text{ ft} \cdot \text{lbf/(lbmole} \cdot {}^\circ\text{R}) = 49709 \text{ ft} \cdot \text{lbf/(slug} \cdot \text{mol} \cdot {}^\circ\text{R})$$

$$\bar{R} = 8\,312 \text{ N} \cdot \text{m/(mol} \cdot \text{K})$$

The molecular mass has units of kg/mol or lbm/lbmole. The ideal gas closely approximates the behavior of most real gases at low to moderate pressure.

5.2.6 INTERNAL ENERGY

Internal energy (U or U/m = u) is energy associated with the molecular motions of a substance, distinct from the energy of position (*potential energy*) and energy of motion (*kinetic energy*). The effect of an increase in internal energy is manifested macroscopically as an increase in temperature. Internal energy has dimensions of energy, $F \cdot L$ (J or BTU). Specific internal energy has dimensions of energy per unit mass, $F \cdot L/M$ (J/kg or BTU/lbm).

5.2.7 ENTHALPY

In the course of simplifying the First Law of Thermodynamics for steady-flow systems, it is found that the work term is more conveniently expressed when shaft work is separated from flow work. When this is accomplished, the sum of internal energy and flow work arises $(u + p/\rho)$. This quantity is given the special name of *enthalpy (H or H/m = $h^\dagger$)*. Enthalpy has the same dimensions as internal energy, $F \cdot L$ (J or BTU) or, on a per-unit-mass basis, $F \cdot L/M$ (J/kg or BTU/lbm).

5.2.8 SPECIFIC HEAT

Specific heat, a property first encountered in Chapter 3, is defined as the heat required to raise the temperature of a unit mass of a substance by one degree. How this heat is added will make a difference, especially for gases; a constant-pressure process of heat addition involves a different specific heat than under constant-volume conditions. The dimensions of specific heat are energy/mass·temperature: $F \cdot L/(M \cdot t)$ (J/(kg·K) or BTU/(lbm·°R). The BTU is the unit of energy measurement in the Engineering system, with 1 BTU defined (roughly) as the energy required to raise the temperature of 1 lbm of water at 68°F by 1°F.

---

* Not to be confused with degrees Rankine, °R.
† Not to be confused with the convection coefficient, $\bar{h}_c$.

The specific heat at constant pressure is c_p and at constant volume is c_v. Both vary with temperature for real substances. Values of specific heat are provided in the Appendix tables (B for solids, C for liquids, and D for gases). For ideal gases, it can be shown that a change in enthalpy equals the product of specific heat at constant pressure and temperature difference:

$$dh = c_p dT$$

Likewise, for internal energy change of an ideal gas,

$$du = c_v dT$$

Example 5.1

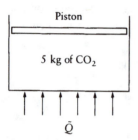

Piston

5 kg of CO$_2$

$\tilde{Q}$

FIGURE 5.3 5 kg of CO$_2$ being heated.

Five kilograms of carbon dioxide are placed in the chamber formed by a piston and cylinder, as shown in Figure 5.3. The piston is free to move vertically. (a) How much heat must be added to the carbon dioxide to raise its temperature by 20 K? (b) How much heat must be added for the same temperature increase if the piston is not free to move?

Solution

As heat is added, the CO$_2$ expands and forces the piston to move upward. The expansion during heat addition essentially occurs at constant pressure if the friction between the piston and cylinder is neglected. On the other hand, for a constrained piston heat addition occurs at constant volume. As heat is added, however, CO$_2$ nearest the walls heats first; after mixing, due to natural convection effects, the heat is distributed uniformly.

Assumptions

1. Heat addition is uniform so that any increase in temperature is experienced by the entire volume of CO$_2$ at once.
2. Piston movement is frictionless.
3. The CO$_2$ has constant properties evaluated at room temperature.

(a) From the definition of specific heat at constant pressure, the heat added equals the product of mass, specific heat, and temperature difference:

$$\tilde{Q} = mc_p(T_2 - T_1) \tag{i}$$

From Appendix Table D.2 for CO$_2$ at 300 K

$$c_p = 871 \text{ J/(kg·K)}, \quad \gamma = 1.3$$

By definition,

$$c_v = \frac{c_p}{\gamma} = \frac{871}{1.3} = 670 \text{ J/(kg·K)}$$

With mass given as 5 kg and ΔT as 20 K, we get

$$\tilde{Q} = 5(871)(20) = 87\,100 \text{ J} = 87.2 \text{ kJ}$$

(b) With the piston held in place, heat addition occurs at constant volume:

$$\tilde{Q} = mc_v(T_2 - T_1) = 5(670)(20) \tag{ii}$$

Solving,

$$\tilde{Q} = 67 \text{ kJ}$$

We see that more heat is required in the constant-pressure process because some of the heat added goes into performing work by raising the piston. In the constant-volume process, heat added goes directly into increasing the CO_2 temperature.

Note that this problem might be approached by writing the First Law of Thermodynamics, where for part (a) heat transferred equals a change in enthalpy. Then, by using the ideal gas assumption, Equation (i) above is derived from the First Law. The ideal gas assumption is a good approximation only under certain conditions, such as for a gas whose state is in the superheated range far from saturation.

5.2.9 COMPRESSIBILITY FACTOR

One difference between liquids and gases is that a liquid will take the *shape* of its container, but a gas will take the *volume* and the *shape* of its container. Another difference is in how each is affected by an external pressure. If the external pressure acting on a volume of water, for example, is increased isothermally by a factor of 5, the density or volume of the water decreases by less than 1%. For a gas, we can use the ideal gas law to estimate the effect of pressure. Isothermally increasing the external pressure acting on a gas by a factor of 5 decreases its volume by a factor of 5 and increases its density by a factor of 5.

The property that describes the behavior of the fluid under the action of external pressure is called the isothermal *compressibility factor,* defined as

$$\beta' = -\frac{1}{\Psi}\left(\frac{\partial \Psi}{\partial p}\right)_T \tag{5.6}$$

where Ψ is the volume, $\partial \Psi / \partial p$ is the change in volume with respect to pressure, and the temperature T subscript denotes that the process is to occur isothermally. The dimension of the compressibility factor in Equation 5.6 is the reciprocal of pressure, L^2/F (1/Pa or ft²/lbf).

5.2.10 VOLUMETRIC THERMAL-EXPANSION COEFFICIENT

It was mentioned that fluid density and volume change with pressure, and the compressibility factor was introduced to account for such behavior. Density and volume of a fluid change also with temperature. The factor that describes this phenomenon is called the *volumetric thermal-expansion coefficient:*

$$\beta' = -\frac{1}{\Psi}\left(\frac{\partial \Psi}{\partial T}\right)_p \tag{5.7}$$

where Ψ is the volume, $\partial \Psi / \partial T$ is the change in volume with respect to temperature, and the notation indicates that constant-pressure conditions are maintained. In terms of density,

$$\beta = -\frac{1}{\rho}\left(\frac{\partial \rho}{\partial T}\right)_p \tag{5.8}$$

The dimension of β is the reciprocal of temperature, $1/T$ ($1/K$ or $1/°R$). Values of the volumetric thermal-expansion coefficient are provided for various fluids in the Appendix tables (C for liquids and D for gases).

Example 5.2

Dichlorodifluoromethane is used in refrigeration and air-conditioning systems and is more commonly known as Freon-12. Using the data in Appendix Table C.3 for Freon-12, verify the value shown for the volumetric thermal-expansion coefficient. Repeat the calculations for helium at 400 K.

Solution

Equation 5.8 is the defining equation. Before it can be used, however, it must be rewritten in difference, rather than differential, form. Note that the data on Freon-12 in the appendix are for the saturated liquid state. The saturation pressure changes with temperature, so the information in Appendix Table C.3 is not constant-pressure data. It is all that is available in this text, however. The helium data are for constant-pressure conditions.

Assumptions

1. $\partial\rho/\partial T$ can be rewritten as $\Delta\rho/\Delta T$.
2. β is calculated on the basis of data available at $-50°C$ and $-40°C$ for Freon-12, even though the corresponding saturation pressures are not equal as required by the definition of β (Equation 5.8).
3. β is calculated on the basis of data available at 366 K and 477 K for helium, which are at constant pressure.

Equation 5.8 is rewritten as

$$\beta = -\frac{1}{\rho}\left(\frac{\partial\rho}{\partial T}\right)_p \cong -\frac{1}{\rho_1}\frac{(\rho_1-\rho_2)}{(T_1-T_2)}$$

From Appendix Table C.3 for Freon-12 we find

$$T_1 = -50°\,C, \quad \rho_1 = 1.546(1\,000)\ \text{kg/m}^3$$

$$T_2 = -40°\,C, \quad \rho_2 = 1.518(1\,000)\ \text{kg/m}^3$$

Substituting,

$$\beta = -\frac{1}{1.546}\frac{1.546-1.518}{-50-(-40)} = 1.811\times10^{-3}/K$$

The accurate value shown in the table is $2.63\times10^{-3}/K$. The error introduced by using data at different pressures is

$$\frac{2.63-1.811}{2.63}\times100\% = 31\%$$

From Appendix Table D.3 for helium,

$$T_1 = 366 \text{ K}, \quad \rho_1 = 0.132\ 80 \text{ kg/m}^3$$

$$T_2 = 477 \text{ K}, \quad \rho_2 = 0.102\ 04 \text{ kg/m}^3$$

Substituting,

$$\beta = -\frac{1}{0.132\ 80}\ \frac{0.132\ 80 - 0.102\ 04}{366 - 477} = 2.087 \times 10^{-3}/\text{K} \qquad \qquad \square$$

5.3 CHARACTERISTICS OF FLUID FLOW

It is important to review the types of flows that exist. In general, there are three types of flow situations:

- *Closed-conduit flows* are enclosed completely by solid restraining surfaces. Flow through a pipe or in an air-conditioning duct are examples.
- *External flows* are not enclosed but do have contact with a solid surface. Air flowing past a flat surface such as a roof is an example.
- *Unbounded flows* do not contact any solid surface. Flow from a spray-paint can or from a faucet are examples.

In most flow situations, effecting a solution requires that a velocity be found. However, the fluid velocity may vary in the flow field, so it is advantageous to classify the flow according to how velocity varies. A *one-dimensional flow* exists if the flow parameters and the fluid properties are constant at any cross section normal to the flow, or if they are represented by average values over the cross section. Velocity distributions for a one-dimensional flow in a circular geometry are illustrated in Figure 5.4. The flow velocity changes from section 1 to 2, but at either section it is a constant. A pressure difference from point A to point B is imposed on the fluid, which causes the fluid to flow. The pressure thus changes with z, and a gradient dp/dz is the only gradient that is recognized.

A *two-dimensional flow* exists if two gradients are accounted for in the flow field. Figure 5.5 illustrates a one-directional, two-dimensional flow in a circular pipe. A pressure gradient dp/dz exists. A velocity gradient also exists in which the z-directed instantaneous velocity V_z varies only with the r coordinate. With pressure a function of one space variable and velocity a function of one space variable, the flow is said to be two-dimensional.

A *three-dimensional flow* is one where three gradients are accounted for in the flow field. For example, if pressure varies with axial distance in a pipe, and velocity varies with the radial and tangential coordinates (a vortex), a three-dimensional flow exists.

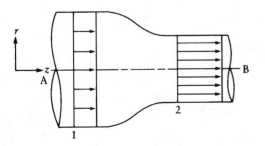

FIGURE 5.4 One-dimensional velocity profiles at sections 1 and 2.

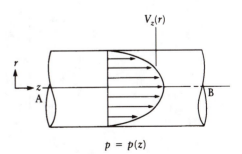

FIGURE 5.5 One-dimensional, two-dimensional flow.

In a number of flow situations, one-dimensional flow can be assumed where a two-dimensional flow exists. This is done to simplify the calculations required to obtain a solution. An average constant velocity (Figure 5.4) might be used in place of a parabolic velocity profile (Figure 5.5). The parabolic profile gives a better description of a real flow, because it shows a zero velocity at the wall and the maximum velocity at the centerline.

A *steady flow* is one where conditions do not vary with time, or where fluctuations are small with respect to mean-flow conditions, and mean-flow conditions do not vary with time. An unsteady flow is one where conditions do vary with time. A quasi-steady flow is one where an unsteady flow exists but, because the time dependence is weak, a steady flow is assumed. Thus, in a quasi-steady flow, steady-flow equations can be applied to an unsteady flow for selected instants in time. The acceptability of the inaccuracy involved depends on whether the solution is within a tolerable limit and on how well the simplified model describes the phenomena in question.

A *streamline* is an aid in flow visualization. Streamlines are everywhere tangent to the velocity vector, and, for steady flow, a particle of fluid follows a path coincident with a streamline.

With the above review of the definitions and concepts of flow behavior, we can now proceed with the development of the equations of fluid mechanics and of convection heat transfer.

5.4   EQUATIONS OF FLUID MECHANICS

The equations we consider in this section include the continuity and momentum equations written for a fluid volume.

5.4.1   THE CONTINUITY EQUATION

Consider a stationary volume element Δx by Δy by Δz through which fluid is flowing (see Figure 5.6 for the x-directed flow). Performing a mass balance for the element gives

Rate of mass in = Rate of mass out + Rate of mass accumulation

The rate of mass flow into the volume element in the x direction is equal to the product of density, area, and velocity at location x, which is $(\rho V_x)|_x\, \Delta y\, \Delta z$. The rate of mass flow out of the volume element in the x direction is given by $(\rho V_x)|_{x+\Delta x}\, \Delta y\, \Delta z$. Similar expressions can be written for the y and z directions as well. The rate of mass accumulation within the element is $\Delta x\, \Delta y\, \Delta z(\partial \rho/\partial t)$. Substituting into the mass balance gives

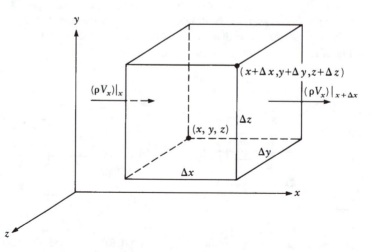

FIGURE 5.6   One-dimensional velocity profiles at sections 1 and 2.

$$\Delta y \Delta z (\rho V_x)|_x + \Delta x \Delta z (\rho V_y)|_y + \Delta x \Delta y (\rho V_z)|_z$$

$$= \Delta y \Delta z (\rho V_x)|_{x+\Delta x} + \Delta x \Delta z (\rho V_y)|_{y+\Delta y}$$

$$+ \Delta x \Delta y (\rho V_z)|_{z+\Delta z} + \Delta x \Delta y \Delta z \left(\frac{\partial \rho}{\partial t}\right)$$

(5.9)

Rearranging and dividing by $\Delta x \, \Delta y \, \Delta z$, we get

$$\frac{(\rho V_x)|_x - (\rho V_x)|_{x+\Delta x}}{\Delta x} + \frac{(\rho V_y)|_y - (\rho V_y)|_{y+\Delta y}}{\Delta y} + \frac{(\rho V_z)|_z - (\rho V_z)|_{z+\Delta z}}{\Delta z} = \frac{\partial \rho}{\partial t}$$

(5.10)

Taking the limit of the above expression as the volume is allowed to shrink to zero, we obtain

$$\frac{\partial \rho}{\partial t} + \frac{\partial (\rho V_x)}{\partial x} + \frac{\partial (\rho V_y)}{\partial y} + \frac{\partial (\rho V_z)}{\partial z} = 0$$

(5.11)

Equation 5.11 is the continuity equation for any fluid in Cartesian coordinates. An important form of the continuity equation results for an incompressible fluid (density is constant)

$$\frac{\partial V_x}{\partial x} + \frac{\partial V_y}{\partial y} + \frac{\partial V_z}{\partial z} = 0$$

(5.12)

For flow of an incompressible fluid in two dimensions, Equation 5.11 reduces to

$$\frac{\partial V_x}{\partial x} + \frac{\partial V_y}{\partial y} = 0$$

(5.13)

The continuity equation in cylindrical coordinates can be derived from a mass balance performed on a cylindrical element. The result is

$$\frac{\partial \rho}{\partial t} + \frac{1}{r} \frac{\partial}{\partial r}(\rho r V_r) + \frac{1}{r} \frac{\partial}{\partial \theta}(\rho V_\theta) + \frac{\partial}{\partial z}(\rho V_z) = 0$$

(5.14)

5.4.2   THE MOMENTUM EQUATION (OR THE EQUATION OF MOTION)

We begin again with a control volume Δx by Δy by Δz, as shown in Figure 5.7. A momentum balance written for the element is

$$\begin{pmatrix} \text{Sum of external} \\ \text{forces acting} \\ \text{on element} \end{pmatrix} = \begin{pmatrix} \text{Rate of momentum} \\ \text{accumulation} \\ \text{within element} \end{pmatrix} + \begin{pmatrix} \text{Rate of} \\ \text{momentum out} \\ \text{of element} \end{pmatrix} - \begin{pmatrix} \text{Rate of} \\ \text{momentum} \\ \text{into element} \end{pmatrix} \quad (5.15)$$

It is known from Newton's Law that Equation 5.15 can be written for a particular direction—x, y, or z. Equation 5.15 is thus a vector equation having three component equations. For our development, we will work only with the x-directed momentum components and forces; similar developments for y and z directions can be handled analogously.

Figure 5.7 shows shear stresses acting on the fluid element selected as our system. The shear stresses are related to the x component of momentum and show how x-directed momentum is transported through the element. Momentum flows into and out of the element by convection and by molecular transfer. The convective momentum flow is due to bulk-fluid flow. The molecular transfer is due to the existence of velocity gradients. The *rate* at which the x component of momentum enters the element at x is the product of mass flow (density times area times velocity) and x-directed velocity, that is, $\rho V_x V_x|_x \Delta y \Delta z$. Similarly, the rate at which the x component enters at y is $\rho V_y V_x|_y \Delta x \Delta z$, and for the z face, $\rho V_z V_x|_z \Delta y \Delta x$. The rate at which the x component of momentum leaves the element at $x + \Delta x$ is $\rho V_x V_x|_{x+\Delta x} \Delta y \Delta z$, with similar expressions for the faces at $y + \Delta y$ and $z + \Delta z$. The convective flow of the x component of momentum is thus evaluated across all faces of the element. The out-minus-in flow of x momentum then is

$$(\rho V_x V_x|_{x+\Delta x} - \rho V_x V_x|_x)\Delta y \Delta z + (\rho V_y V_x|_{y+\Delta y} - \rho V_y V_x|_y) \times \Delta y \Delta z + (\rho V_z V_x|_{z+\Delta z} - \rho V_z V_x|_z)\Delta x \Delta y \quad (5.16)$$

Expression 5.16 represents the *net-rate-out* (out-minus-in) flow of the x component of momentum from the volume element by bulk-fluid flow.

Velocity gradients exist in the fluid due to forces exerted on the fluid. As illustrated in Figure 5.1, where the definition of viscosity was developed, an external force F in the x direction is transferred to the fluid in the form of a shear stress, τ_{yx}. A strain rate dV_x/dy results. Fluid layers not in contact with either plate move in the x direction. The momentum of the x-directed force is "transported" through the fluid by molecular effects arising from the fact that the fluid has a finite or nonzero viscosity. Thus, momentum is transported (perpendicular to the x direction) through the fluid by molecular effects. We identify the viscous forces with surface stresses.

Returning now to Figure 5.7, the rate at which the x component of momentum enters the element at x due to molecular transport is $\tau_{xx}|_x \Delta y \Delta z$. The rate at which the x component of momentum enters the element at y is $\tau_{yx}|_y \Delta x \Delta z$. The rate at which the x component of momentum

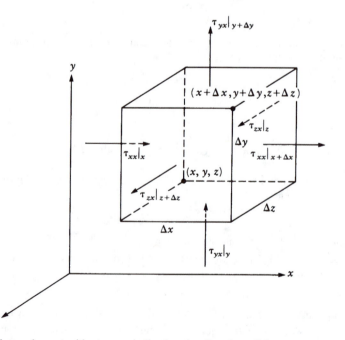

FIGURE 5.7 A volume element with stresses indicating the direction of the x component of momentum transported through the element.

leaves the element at $x + \Delta x$ is $\tau_{xx}|_{x+\Delta x} \Delta y \Delta z$. Similar expressions can be written for the other faces. The net rate out of the x component of momentum from the element due to molecular transport then is

$$(\tau_{xx}|_{x+\Delta x} - \tau_{xx}|_x)\Delta y \Delta z + (\tau_{yx}|_{y+\Delta y} - \tau_{yx}|_y)\Delta x \Delta z + (\tau_{zx}|_{z+\Delta z} - \tau_{zx}|_z)\Delta x \Delta y \qquad (5.17)$$

The forces acting on the fluid element can be due to a number of effects, such as pressure, gravity, surface tension, and magnetic effects. In fluid mechanics, it is usual to consider only pressure and gravity forces. For the x direction these are

$$(p|_x - p|_{x+\Delta x})\Delta y \Delta z + \frac{\rho g_x}{g_c}\Delta x \Delta y \Delta z \qquad (5.18)$$

Note that pressure is a scalar quantity, while gravity is a vector quantity of which g_x is the x-directed component.

The rate of accumulation of x momentum within the element is

$$\frac{\partial(\rho V_x)}{\partial t} \frac{\Delta x \Delta y \Delta z}{g_c} \qquad (5.19)$$

Substituting Expressions 5.16 through 5.19 into Equation 5.15, dividing by $\Delta x \, \Delta y \, \Delta z$, and taking the limit as Δx, Δy, and Δz approach zero, we obtain

$$\frac{1}{g_c}\frac{\partial(\rho V_x)}{\partial t} = -\left[\frac{\partial(\rho V_x V_x)}{\partial x} + \frac{\partial(\rho V_y V_x)}{\partial y} + \frac{\partial(\rho V_z V_x)}{\partial z}\right]\frac{1}{g_c} + \left(\frac{\partial \tau_{xx}}{\partial x} + \frac{\partial \tau_{yx}}{\partial y} + \frac{\partial \tau_{zx}}{\partial z}\right) - \frac{\partial p}{\partial x} + \frac{\rho g_x}{g_c} \qquad (5.20)$$

Equation 5.20 is the x component of the equation of motion. The dimensions of 5.20 are F/L^3 (lbf/ft^3 or N/m^3). Performing the indicated differentiation, substituting from the equation of continuity, and rearranging gives

$$\frac{\rho}{g_c}\left(\frac{\partial V_x}{\partial t} + V_x\frac{\partial V_x}{\partial x} + V_y\frac{\partial V_x}{\partial y} + V_z\frac{\partial V_x}{\partial z}\right) = -\frac{\partial p}{\partial x} + \left(\frac{\partial \tau_{xx}}{\partial x} + \frac{\partial \tau_{yx}}{\partial y} + \frac{\partial \tau_{zx}}{\partial z}\right) + \frac{\rho g_x}{g_c} \qquad (5.21)$$

We can physically interpret each term of the above equation as follows:

- The left-hand side represents a mass per unit volume times acceleration.
- The pressure term represents the force due to pressure per unit volume acting on the element.
- The shear-stress terms represent a viscous force per unit volume acting on the element.
- The gravity term represents the gravitational force per unit volume acting on the element.

Essentially, then, Equation 5.21 is in the form of mass times acceleration equals sum of forces, all on a per-unit-volume basis. Similar expressions can be developed for the y and z directions.

As indicated in Equation 5.1, the shear stresses are related to strain rates. For Newtonian fluids with constant properties of density and viscosity, we have

$$\tau_{xx} = 2\mu\frac{\partial V_x}{\partial x} + \frac{2}{3}\mu\left(\frac{\partial V_x}{\partial x} + \frac{\partial V_y}{\partial y} + \frac{\partial V_z}{\partial z}\right) \qquad (5.22)$$

$$\tau_{yy} = 2\mu\frac{\partial V_y}{\partial y} + \frac{2}{3}\mu\left(\frac{\partial V_x}{\partial x} + \frac{\partial V_y}{\partial y} + \frac{\partial V_z}{\partial z}\right) \tag{5.23}$$

$$\tau_{zz} = 2\mu\frac{\partial V_z}{\partial z} + \frac{2}{3}\mu\left(\frac{\partial V_x}{\partial x} + \frac{\partial V_y}{\partial y} + \frac{\partial V_z}{\partial z}\right) \tag{5.24}$$

$$\tau_{xy} = \tau_{yx} = \mu\left(\frac{\partial V_x}{\partial y} + \frac{\partial V_y}{\partial x}\right) \tag{5.25}$$

$$\tau_{yz} = \tau_{zy} = \mu\left(\frac{\partial V_y}{\partial z} + \frac{\partial V_z}{\partial y}\right) \tag{5.26}$$

$$\tau_{zx} = \tau_{xz} = \mu\left(\frac{\partial V_z}{\partial x} + \frac{\partial V_x}{\partial z}\right) \tag{5.27}$$

Equations 5.22 through 5.27 are written here without proof, and you are asked to accept them as presented.[*] When Equations 5.22 through 5.27 are combined with 5.21, the result is

$$\frac{\rho}{g_c}\left(\frac{\partial V_x}{\partial t} + V_x\frac{\partial V_x}{\partial x} + V_y\frac{\partial V_x}{\partial y} + V_z\frac{\partial V_x}{\partial z}\right) = -\frac{\partial p}{\partial x} + \mu\left(\frac{\partial^2 V_x}{\partial x^2} + \frac{\partial^2 V_x}{\partial y^2} + \frac{\partial^2 V_x}{\partial z^2}\right) + \frac{\rho g_x}{g_c} \tag{5.28a}$$

Collectively, Equation 5.28a and the analogous equations for the y and z directions are called the *Navier-Stokes Equations*. The continuity equation and the Navier-Stokes Equations in both Cartesian and cylindrical coordinates are tabulated in the Summary section of this chapter.

The continuity and Navier-Stokes Equations can be used to set up descriptive equations for various problems. While it is possible to formulate an equation for each flow situation, it is far safer to begin with the general equations and cancel terms that do not apply. That is the method recommended here. The solution technique involves obtaining the descriptive equation for the system, and solving (if possible) subject to the boundary conditions. For a majority of problems, especially in turbulent flow, the equations cannot be solved exactly, however. This is due to the presence of the nonlinear acceleration terms. A number of solutions exist for laminar-flow (also called viscous-flow) problems in which the acceleration terms vanish from the differential equation of motion. Use of the continuity and Navier-Stokes Equations to develop equations for common fluid-mechanics problems is illustrated in the following examples.

Example 5.3

Liquid flows down an incline, as illustrated in Figure 5.8. Derive an equation for the velocity distribution, assuming laminar flow exists.

Solution

As shown in the figure, x vs. z axes have been imposed. The velocity distribution we are seeking is labeled as V_z in the figure.

Assumptions

1. Laminar flow exists.
2. Liquid properties of density and viscosity are constant.

---

[*] For further discussion, see *Transport Phenomena*, by R. B. Bird, W. E. Stewart, and F. N. Lightfoot, John Wiley & Sons, Inc., 1960, pp. 79–89.

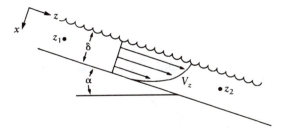

FIGURE 5.8 Steady laminar flow of a Newtonian liquid down an incline.

3. The liquid is Newtonian.
4. The flow is steady.

Under these conditions, we begin by determining what velocities exist in the flow and on what variables the velocities may depend. There are at most three velocities to consider:

$$\tilde{V} = \tilde{e}_x V_x + \tilde{e}_y V_y + \tilde{e}_z V_z \qquad (5.29)$$

where the velocity vector $\tilde{V}$ is seen to have three components—V_x, V_y, and V_z—which are the instantaneous velocities in the indicated directions, and the $\tilde{e}_i$ variables are corresponding unit vectors. For laminar flow,

$$V_x = V_y = 0$$

Velocity in the z direction is nonzero. Furthermore, V_z can depend on at most four variables; that is,

$$V_z = V_z(x, y, z, t)$$

The time dependence is rejected for steady-flow conditions. The velocity V_z will not vary in the y direction (out of the page). Because the velocity profile will be the same at z_1 and z_2, we conclude that V_z is not dependent on z. We are thus left with

$$V_z = V_z(x)$$

With density constant, continuity Equation 5.11 reduces to

$$\frac{\partial V_z}{\partial z} = 0$$

which means that V_z is independent of z, a conclusion already drawn. The Navier-Stokes Equations (5.28a, b, and c of Table 5.2 in the Summary section) reduce to

x component $\quad 0 = -\dfrac{\partial p}{\partial x} + \dfrac{\rho g_x}{g_c}$

y component $\quad 0 = -\dfrac{\partial p}{\partial y} + \dfrac{\rho g_y}{g_c}$

z component $0 = -\dfrac{\partial p}{\partial z} + \mu\dfrac{\partial^2 V_z}{\partial x^2} + \dfrac{\rho g_z}{g_c}$

Pressure variations in the flow are not significant. The x- and y-component equations thus vanish, and we are left with

$$\mu\frac{\partial^2 V_z}{\partial x^2} = -\frac{\rho g_z}{g_c} \tag{i}$$

The vertical component of g_z is $g \sin \alpha$. Equation (i) states that viscous forces oppose the gravity force, which causes liquid to flow. The boundary conditions are

B.C.1. $x = \delta,$ $V_z = 0$ (nonslip condition)
B.C.2. $x = 0,$ $\tau_{xz} = 0$ (no viscous interaction between the air and the liquid)

Equation (i) can be integrated directly and the boundary conditions applied. The result is

$$V_z = \frac{\rho g \delta^2 \sin\alpha}{2\mu g_c}\left(1 - \frac{x^2}{\delta^2}\right) \tag{3.30}$$

Thus, the profile is parabolic.   ❏

Example 5.4

Liquid flows under the action of a pressure drop within a horizontal tube. Under laminar-flow conditions, derive an equation for the velocity profile.

Solution

Figure 5.9 is a sketch of the system, with an expected velocity profile. The pressure drop imposed can be written as $\Delta p/L$ or, more generally, as will be seen in the equations, $\partial p/\partial z$. The r, θ, and z axes are indicated in the figure.

Assumptions

1. The flow is laminar.
2. Liquid properties are constant.
3. The liquid is Newtonian.
4. Steady flow exists.
5. Gravitational forces are neglected.

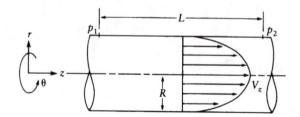

FIGURE 5.9   Steady laminar flow of a Newtonian fluid in a tube induced by a pressure drop $(p_1 - p_2)/L = \Delta p/L$.

As before, we start by writing

$$\tilde{V} = \tilde{e}_r V_r + \tilde{e}_\theta V_\theta + \tilde{e}_z V_z$$

We conclude that $V_r = V_\theta = 0$, with V_z the only nonzero velocity to consider. Next, we write

$$V_z = V_z(r, \theta, z, t)$$

Based on the assumptions and an analysis of the flow, we get

$$V_z = V_z(r)$$

The continuity equation in cylindrical coordinates reduces to

$$\frac{\partial V_z}{\partial z} = 0$$

which states that V_z is independent of z. The Navier-Stokes Equations (5.29a, b, and c of Table 5.2 in the Summary section) become

r component $\qquad 0 = -\dfrac{\partial p}{\partial r}$ (p is not a function of r)

θ component $\qquad 0 = -\dfrac{\partial p}{\partial \theta}$ (p is not a function of θ)

z component $\qquad 0 = -\dfrac{\partial p}{\partial z} + \dfrac{\mu}{r}\dfrac{\partial}{\partial r}\left(r\dfrac{\partial V_z}{\partial r}\right)$

The boundary conditions are

B.C.1. $\qquad r = R, \qquad V_z = 0$ (nonslip condition)

B.C.2. $\qquad r = 0, \qquad \dfrac{dV_z}{dr} = 0$ (velocity is a maximum at the centerline due to symmetry)

The solution to the differential equation as found by direct integration is

$$V_z = \frac{R^2}{4\mu}\left(-\frac{dp}{dz}\right)\left(1 - \frac{r^2}{R^2}\right) \qquad (5.31)$$

The velocity profile is thus parabolic. It is important to note that as z increases, pressure p decreases. Therefore the pressure gradient ($-dp/dz$) is actually a positive quantity. ❑

Let us next consider flow in an annulus, as illustrated in Figure 5.10. The outer tube is of radius R, and the inner tube is of radius κR, where $0 < \kappa < 1$. If flow is induced by a pressure drop, if the fluid is Newtonian with constant properties, and if laminar, steady conditions exist, then the Navier-Stokes Equations can be applied. The equation to be solved is Equation (i) of Example 5.4, which deals with flow in a tube. What will differentiate these formulations is the boundary conditions. For the annulus, we write

B.C.1. $\qquad r = R, \qquad V_z = 0$
B.C.2. $\qquad r = \kappa R, \qquad V_z = 0$

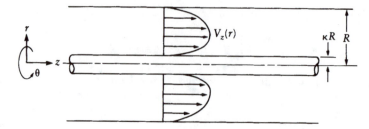

FIGURE 5.10 Steady laminar flow of a Newtonian liquid in an annulus induced by a pressure drop.

When solved, the velocity distribution for laminar flow in an annulus becomes

$$V_z = \frac{R^2}{4\mu}\left(-\frac{dp}{dz}\right)\left[1 - \frac{r^2}{R^2} - \frac{1-\kappa^2}{\ln(\kappa)}\ln\left(\frac{r}{R}\right)\right]\tag{5.32}$$

Once the velocity distribution is known, other flow parameters can be determined. To illustrate the definitions that follow, we use the parabolic profile for laminar flow in a tube, Equation 5.31:

$$V_z = \frac{R^2}{4\mu}\left(-\frac{dp}{dz}\right)\left(1 - \frac{r^2}{R^2}\right)\tag{5.31}$$

The *volume flow rate* is obtained by integration of the velocity profile over the flow area normal to the flow direction. For flow in a tube

$$Q = \iint V_z dA = \int_0^{2\pi}\int_0^R \frac{R^2}{4\mu}\left(-\frac{dp}{dz}\right)\left(1 - \frac{r^2}{R^2}\right)r\ dr\ d\theta$$

or

$$Q = \left(-\frac{dp}{dz}\right)\frac{\pi R^4}{8\mu}\quad\text{(flow in a tube)}\tag{5.33}$$

The *average velocity* is obtained by dividing the volume flow rate by the area. For flow in a tube,

$$V = \frac{Q}{A} = \left(-\frac{dp}{dz}\right)\frac{\pi R^4}{8\mu}\frac{1}{\pi R^2}$$

$$V = \left(-\frac{dp}{dz}\right)\frac{R^2}{8\mu}\quad\text{(flow in a tube)}\tag{5.34}$$

The significance of the average velocity is that it can be used in place of the instantaneous velocity profile as a one-dimensional approximation of V_z.

The *maximum velocity* is determined by differentiating the velocity-profile equation and setting the result equal to zero unless, of course, obtaining the maximum velocity can be done in a more obvious fashion. For example, for flow in a tube, the maximum velocity can be determined by setting $r = 0$ in Equation 5.31:

$$V_{z\,max} = \left(-\frac{dp}{dz}\right)\frac{R^2}{4\mu}\quad\text{(flow in a tube)}\tag{5.35}$$

The *shear stress* in the fluid is obtained from Newton's Law of Viscosity. For flow in a tube,

$$\tau_{rz} = -\mu \frac{dV_z}{dr} = \left(-\frac{dp}{dz}\right)\frac{r}{2} \quad \text{(flow in a tube)} \tag{5.36}$$

The *wall shear stress* is found by setting $r = R$ in the shear-stress equation for the case of flow in a tube:

$$\tau_w = \left(-\frac{dp}{dz}\right)\frac{R}{2} \quad \text{(flow in a tube)} \tag{5.37}$$

5.5   THE THERMAL-ENERGY EQUATION

The energy equation can be derived for an arbitrary control volume. To be general, a total energy conservation equation would include mechanical as well as thermal energy. It is sufficient here, however, to derive the thermal energy equation and neglect mechanical energy. (Recall from thermodynamics that mechanical and potential energy changes are small compared with thermal-energy changes for many problems.)

Consider a stationary volume element through which fluid is flowing, as shown in Figure 5.11. The conservation-of-energy equation written for the volume element is

$$\begin{pmatrix} \text{Net rate of} \\ \text{heat addition} \\ \text{by conduction} \end{pmatrix} + \begin{pmatrix} \text{Rate of} \\ \text{internal energy} \\ \text{in by convection} \end{pmatrix}$$

$$= \begin{pmatrix} \text{Rate of} \\ \text{accumulation of} \\ \text{internal energy} \end{pmatrix} + \begin{pmatrix} \text{Rate of} \\ \text{internal energy out} \\ \text{by convection} \end{pmatrix} + \begin{pmatrix} \text{Net rate of work} \\ \text{done by system} \\ \text{on surroundings} \end{pmatrix} \tag{5.38}$$

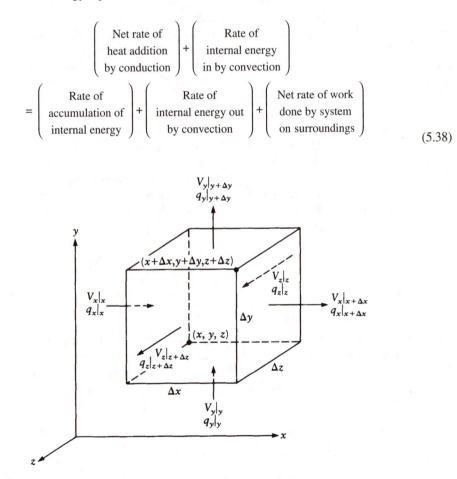

FIGURE 5.11   A volume element for deriving the thermal-energy equation.

Equation 5.38 is recognized as the First Law of Thermodynamics written for a system that may or may not be at steady state.

The net rate of heat addition by conduction in the x direction is obtained by subtracting heat conducted out of the element from heat conducted into it:

$$(q_x''|_x - q_x''|_{x+\Delta x})\Delta y \Delta z \tag{5.39}$$

where q'' is the heat conducted per unit area. Analogous expressions can be written for the y and z directions.

Equation 5.38 shows two convective terms; one is a rate of internal energy in, and the other is a rate out. The *net* rate of internal energy convected into (in minus out) the volume element in the x direction is

$$[(V_x \rho u)|_x - (V_x \rho u)|_{x+\Delta x}]\Delta y \Delta z \tag{5.40}$$

where u is the internal energy per unit mass of the fluid element.

The rate of accumulation of internal energy within the element is

$$\Delta x \Delta y \Delta z \frac{\partial}{\partial t}(\rho u) \tag{5.41}$$

The work done by the fluid within the volume element can include several terms. Work done against gravity will yield expressions recognized as potential-energy terms, which are neglected here. Work done against static-pressure forces is known as flow work. Noting that work = force × distance and that rate of work done = force × velocity, we write for the x-directed flow work:

$$\Delta y \Delta z[(pV_x)|_{x+\Delta x} - (pV_x)|_x] \tag{5.42}$$

Another work term that we could include is the work done against the viscous forces. Viscous dissipation is an irreversible rate of internal energy increase due to viscous effects. Calculations made with viscous dissipation terms show that they are negligible compared to the magnitude of the other terms in the energy equation. We will not include viscous dissipation in our derivation.

We now combine the foregoing expressions with corresponding y- and z-directed formulations and substitute into Equation 5.38. Dividing by the volume $\Delta x \Delta y \Delta z$ and taking the limit as the volume approaches zero gives

$$\begin{aligned}
\frac{\partial \rho u}{\partial t} = &-\frac{1}{g_c}\left[\frac{\partial}{\partial x}(V_x \rho u) + \frac{\partial}{\partial y}(V_y \rho u) + \frac{\partial}{\partial z}(V_z \rho u)\right] \\
&-\left(\frac{\partial q_x''}{\partial x} + \frac{\partial q_y''}{\partial y} + \frac{\partial q_z''}{\partial z}\right) - \left[\frac{\partial}{\partial x}(pV_x) + \frac{\partial}{\partial y}(pV_y) + \frac{\partial}{\partial z}(pV_z)\right]
\end{aligned} \tag{5.43}$$

For most applications of the energy equation, it is more convenient to have the equation in terms of viscosity, temperature, and heat capacity rather than internal energy and shear stresses. Temperature is more easily measured than internal energy. To effect various simplifications on Equation 5.44, let us examine the first group of terms in parentheses on the right-hand side. By performing the indicated differentiation, we get

$$\frac{\partial}{\partial x}(V_x \rho u) + \frac{\partial}{\partial y}(V_y \rho u) + \frac{\partial}{\partial z}(V_z \rho u) = V_x \rho \frac{\partial u}{\partial x} + u\frac{\partial}{\partial x}(\rho V_x) + V_y \rho \frac{\partial u}{\partial y} + u\frac{\partial}{\partial y}(\rho V_y) + V_z \rho \frac{\partial u}{\partial z} + u\frac{\partial}{\partial z}(\rho V_z)$$

$$\tag{5.44}$$

From continuity Equation 5.11 it is seen that the right-hand side of Equation 5.44 reduces to

$$\rho\left(V_x\frac{\partial u}{\partial x} + V_y\frac{\partial u}{\partial y} + V_z\frac{\partial u}{\partial z}\right) - u\frac{\partial \rho}{\partial t} \qquad (5.45)$$

Furthermore, if density is assumed constant as with an incompressible fluid, $\partial\rho/\partial t = 0$.

Returning again to Equation 5.43, we expand the term representing work done against pressure forces to obtain

$$\frac{\partial}{\partial x}(pV_x) + \frac{\partial}{\partial y}(pV_y) + \frac{\partial}{\partial z}(pV_z) = V_x\frac{\partial p}{\partial x} + p\frac{\partial V_x}{\partial x} + V_y\frac{\partial p}{\partial y} + p\frac{\partial V_y}{\partial y} + V_z\frac{\partial p}{\partial z} + p\frac{\partial V_z}{\partial z} \qquad (5.46)$$

Again, for constant density, we obtain from the continuity equation:

$$\frac{\partial V_x}{\partial x} + \frac{\partial V_y}{\partial y} + \frac{\partial V_z}{\partial z} = 0$$

Substituting the above simplifications into Equation 5.43 gives

$$\frac{\partial u}{\partial t} = -\left(V_x\frac{\partial u}{\partial x} + V_y\frac{\partial u}{\partial y} + V_z\frac{\partial u}{\partial z}\right) - \left(\frac{\partial q_x''}{\partial x} + \frac{\partial q_y''}{\partial y} + \frac{\partial q_z''}{\partial z}\right)\frac{1}{\rho} - \left(V_x\frac{\partial p}{\partial x} + V_y\frac{\partial p}{\partial y} + V_z\frac{\partial p}{\partial z}\right)\frac{1}{\rho} \qquad (5.47)$$

The internal energy and pressure terms on the right-hand side of Equation 5.47 can be combined to give

$$\frac{\partial u}{\partial t} + \left[V_x\frac{\partial}{\partial x}\left(u + \frac{p}{\rho}\right) + V_y\frac{\partial}{\partial y}\left(u + \frac{p}{\rho}\right) + V_z\frac{\partial}{\partial z}\left(u + \frac{p}{\rho}\right)\right] \qquad (5.48)$$

Assuming that pressure does not vary with time, $\partial p/\partial t = 0$. Therefore, we can write

$$\frac{\partial u}{\partial t} = \frac{\partial}{\partial t}\left(u + \frac{p}{\rho}\right)$$

and so the argument of each derivative in Expression 5.48 contains $u + p/\rho$. With enthalpy defined as

$$h = u + \frac{p}{\rho}$$

Equation 5.47 can be simplified to give

$$\rho\left(\frac{\partial h}{\partial t} + V_x\frac{\partial h}{\partial x} + V_y\frac{\partial h}{\partial y} + V_z\frac{\partial h}{\partial z}\right) = -\left(\frac{\partial q''}{\partial x} + \frac{\partial q''}{\partial y} + \frac{\partial q''}{\partial z}\right) \qquad (5.49)$$

Further substitutions can be made from various definitions. For enthalpy, we can write for a constant-pressure process:

$$\left(\frac{\partial h}{\partial T}\right)_p = c_p$$

From Fourier's Law of Heat Conduction, for a fluid with constant thermal conductivity k_f,

$$q_x'' = -k_f \frac{\partial T}{\partial x}$$

with similar equations for the y and z components. When the above equations for enthalpy and heat conducted are incorporated into Equation 5.49, we obtain the energy equation for Newtonian fluids having constant density and viscosity:

$$\rho c_p\left(\frac{\partial T}{\partial t} + V_x\frac{\partial T}{\partial x} + V_y\frac{\partial T}{\partial y} + V_z\frac{\partial T}{\partial z}\right) = k_f\left(\frac{\partial^2 T}{\partial x^2} + \frac{\partial^2 T}{\partial y^2} + \frac{\partial^2 T}{\partial z^2}\right) \tag{5.50}$$

See the summary table at the end of this chapter for the energy equation in cylindrical coordinates, which is not derived here. Thus, Equation 5.50 states (on a per-unit-volume basis) that the temperature of a moving fluid changes due to heat conduction and expansion effects. Viscous dissipation (not included in Equation 5.50) usually has a negligible effect on the fluid temperature and may be omitted except for highly viscous fluids or problems having velocities near the sonic velocity.

5.6 APPLICATIONS TO LAMINAR FLOWS

Equation 5.50 can be simplified for various problems. For a solid, all velocities vanish, and we are left with the conduction equation first introduced in Chapter 2. Equation 5.50 is the form of the energy equation that we will use to set up convection problems. The energy equation in Cartesian and in cylindrical coordinates appears in Table 5.3 in the Summary section of the chapter. It should be mentioned that energy Equation 5.43 is most general and that many forms of the energy equation exist, depending on the assumptions made in the course of simplifying it. How the equations are used in setting up a heat-transfer problem is illustrated in the following example.

Example 5.5

Consider laminar flow down an incline in which the free surface of the liquid is maintained at temperature T_s and the solid surface is maintained at temperature T_w. Develop the descriptive differential equations for velocity and temperature.

Solution

Figure 5.12 illustrates the situation, with x vs. z axes imposed. Also shown is the velocity distribution and an expected temperature profile. For the majority of real fluids, the viscosity, density, and thermal conductivity all change with temperature. So, if the viscosity near the wall is different from

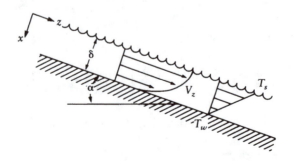

FIGURE 5.12 Steady laminar flow of a Newtonian liquid down an incline in which a temperature difference has been imposed.

that near the surface, the velocity profile becomes affected. This is a forced-convection problem where fluid motion is caused by gravity forces overcoming viscous forces rather than by thermal effects.

Assumptions

1. Laminar flow exists.
2. Velocity and temperature profiles are fully developed and will not change with increasing z.
3. Properties are constant.
4. The fluid is Newtonian.
5. The flow is steady.

The continuity, momentum, and energy equations of the Summary tables apply. By determining which terms are not important, we obtain the appropriate equations. Having worked through this analysis in Example 5.3, here we merely write the results.

The velocity profile:

$$V_z = \frac{\rho g \delta^2 \sin\alpha}{2\mu g_c}\left(1 - \frac{x^2}{\delta^2}\right)$$

Temperature can depend on, at most, three space variables and on time:

$$T = T(x, y, z, t)$$

The time dependence is rejected for steady conditions. The y dependence is not significant for this two-dimensional flow, and temperature does not vary with z. Thus, $T = T(x)$ only. Energy Equation 5.50 reduces to

$$0 = k_f \frac{\partial^2 T}{\partial x^2}$$

with boundary conditions

B.C.1. $x = 0$, $T = T_s$
B.C.2. $x = \delta$, $T = T_w$

By direct integration, we find

$$\frac{dT}{dx} = C_1$$

$$T = C_1 x + C_2$$

where C_1 and C_2 are constants of integration. Applying the boundary conditions and evaluating the constants gives, for temperature

$$\frac{T - T_s}{T_w - T_s} = \frac{x}{\delta} \tag{5.51}$$

❏

Once the temperature profile is known, a variety of other parameters can be calculated. The *heat flux* is obtained by differentiating the temperature profile by Fourier's Law of Heat Conduction or by Newton's Law of Cooling, whichever is appropriate. The *maximum temperature rise* can be obtained by differentiating the temperature profile and setting the result equal to zero. The *average temperature rise* is obtained in a manner analogous to finding average velocity. The temperature profile is integrated over the cross-sectional area, and the result is divided by the area itself:

$$T_{avg} - T_r = \frac{\iint [T(x) - T_r] dA}{\iint dA} \tag{5.52}$$

where T_r is a reference temperature. As an example, consider the profile equation of 5.51. The average temperature rise is found with

$$\frac{T_{avg} - T_s}{T_w - T_s} = \frac{\int_0^b \int_0^\delta \left(\frac{x}{\delta}\right) dx\, dy}{bh} = \frac{b\left(\frac{x^2}{2\delta}\right)_0^\delta}{bh}$$

Solving,

$$\frac{T_{avg} - T_s}{T_w - T_s} = \frac{1}{2}$$

The *bulk temperature* at any z location is the temperature obtained if all the fluid issuing forth at z were collected in a container and mixed thoroughly. This temperature is often called the *cup mixing temperature* or the *flow-average temperature*. Mathematically, the bulk temperature T_b is defined as

$$T_b - T_r = \frac{\iint V_z(x)[T(x) - T_r] dA}{\iint V_z(x) dA} \tag{5.53}$$

The denominator of Equation 5.53 is recognized as the volume flow rate. To show how bulk temperature is calculated, consider again the temperature profile of Equation 5.51 and the velocity profile of Equation 5.30. Substituting into Equation 5.53 gives

$$\frac{T_b - T_s}{T_w - T_s} = \frac{\int_0^b \int_0^\delta \frac{\rho g \delta^2 \sin\alpha}{2\mu}\left(1 - \frac{x^2}{\delta^2}\right)\left(\frac{x}{\delta}\right) dx\, dy}{\int_0^b \int_0^\delta \frac{\rho g \delta^2 \sin\alpha}{2\mu}\left(1 - \frac{x^2}{\delta^2}\right) dx\, dy} = \frac{b\left(\frac{x}{2\delta} - \frac{x^4}{4\delta^3}\right)_0^\delta}{b\left(x - \frac{x^3}{3\delta^2}\right)_0^\delta} = \frac{\left(\frac{\delta}{2} - \frac{\delta}{4}\right)}{\left(\delta - \frac{\delta}{3}\right)}$$

Solving,

$$\frac{T_b - T_s}{T_w - T_s} = \frac{3}{8}$$

Example 5.6

Consider laminar flow of a fluid in a tube of radius R. Heat is applied at a constant rate of q_w'' at the tube surface. Develop the differential equation for temperature of the fluid.

Solution

Figure 5.13 shows flow in a tube and a constant heat rate being applied. Various forms of this problem exist. For example, if the flow is near an entrance, we must contend with entrance effects. Alternatively, the velocity profile could be fully developed and the temperature profile starting to develop, or both velocity and temperature profiles could be fully developed.

Assumptions

1. One-dimensional flow exists.
2. The flow is steady.
3. The velocity profile is fully developed. The temperature profile begins developing at $z = 0$, where the bulk-fluid temperature is uniform and equal to T_i.
4. Fluid properties are constant.
5. The fluid is Newtonian.

The velocity profile for this problem is derived in Example 5.4 and given in Equation 5.54 as

$$V_z = \frac{R^2}{4\mu}\left(-\frac{dp}{dz}\right)\left(1 - \frac{r^2}{R^2}\right) \tag{5.54}$$

The maximum velocity is obtained from V_z when $r = 0$:

$$V_{z\,max} = (-dp/dz)R^2/4\mu$$

In terms of $V_{z\,max}$, Equation 5.31 becomes

$$\frac{V_z}{V_{z\,max}} = \left(1 - \frac{r^2}{R^2}\right)$$

Temperature can vary with at most three space variables and time:

$$T = T(r, \theta, z, t)$$

Under steady conditions, the time dependence vanishes. Due to symmetry, the θ dependence vanishes. Thus,

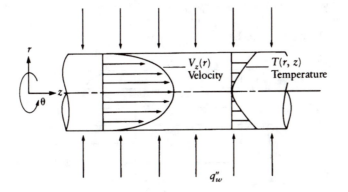

FIGURE 5.13   Steady laminar flow of a Newtonian fluid in a tube where a constant heat is being applied at the wall.

$$T = T(r, z)$$

The energy equation (Table 5.3 of the Summary section) therefore reduces to

$$\rho c_p\left(V_z\frac{\partial T}{\partial z}\right) = \frac{k_f}{r}\frac{\partial}{\partial r}\left(r\frac{\partial T}{\partial r}\right) + k_f\frac{\partial^2 T}{\partial z^2}$$

Usually, heat conducted in the axial direction (represented by $k_f\ \partial^2 T/\partial z^2$) is small compared to the convective terms (i.e., the terms containing a velocity). Neglecting axial conduction and substituting the velocity profile gives

$$\rho c_p V_{z\,max}\left(1 - \frac{r^2}{R^2}\right)\frac{\partial T}{\partial z} = \frac{k_f}{r}\frac{\partial}{\partial r}\left(r\frac{\partial T}{\partial r}\right) \tag{5.55}$$

The boundary conditions are

B.C.1. $r = 0$, $\dfrac{\partial T}{\partial r} = 0$

B.C.2. $r = R$, $-k_f\dfrac{\partial T}{\partial r} = q_w''$ (a constant)

B.C.3. $z = 0$, $T = T_i$ (for all r)   ❑

5.7   APPLICATIONS TO TURBULENT FLOWS

As seen in the previous section, when the equations for laminar flows are developed, the nonlinear terms in the Navier-Stokes Equations were set equal to zero for the problems considered. The remaining expression, if not too complex, can then be solved. On the other hand, for turbulent flows none of the velocities vanish, and all of the nonlinear terms remain. However, if the Navier-Stokes Equations could be solved, the instantaneous values of the velocities would result. The instantaneous velocities (pressure and temperature also) fluctuate randomly about mean values. To illustrate this behavior, consider Figure 5.14, which is of a turbulent flow in a tube. If the velocity at each point in the cross section is measured with a relatively insensitive device, such as a pitot-static tube, the velocity profile in Figure 5.14a results. If the velocity is instead measured with a more sensitive device, such as a hot-wire anemometer, the profile in Figure 5.14b is obtained. Figure 5.14b shows the instantaneous velocity V_z profile. This profile changes with all variables, including time. Thus if velocity is measured again at this same location, but at a different time, a different instantaneous velocity V_z profile results. Figure 5.14b shows that the instantaneous velocity V_z fluctuates randomly about the mean z-directed velocity $\bar{V}_z$.

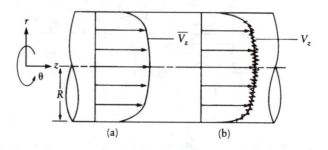

(a)   (b)

FIGURE 5.14   Turbulent flow in a tube of radius R.

The differences between laminar and turbulent flow are worth noting. Laminar flow exists at relatively low velocities. A fluid particle in one layer stays in that layer. Streamlines are parallel and smooth. Turbulent flow is characterized by random variations or fluctuations of the instantaneous velocity about a mean value. This random nature gives rise to an effective mixing phenomenon. Fluid particles usually do not stay in one layer but move about tortuously throughout the flow. Turbulent flow exists at velocities that are much higher than in laminar flow. The difference in velocity profiles is also significant, as shown in Figure 5.15. Figure 5.15a is of a laminar profile for flow in a tube, and Figure 5.15b is of a turbulent profile (mean velocity is shown rather than instantaneous). The mixing effect in the turbulent profile tends to distribute more evenly the kinetic energy of the flow, which causes the flatter profile. The momentum of the flow is transported closer to the tube surface when the flow is turbulent.

When working with parameters in turbulent flow, it is customary to *time smooth* the instantaneous velocity, pressure, and temperature to obtain mean values. The time-smoothing operation involves integrating the time dependence out of each parameter and rewriting the continuity, momentum, and energy equations in terms of mean, rather than instantaneous, values. Once this is accomplished, the equations become easier to use but still represent sometimes insurmountable difficulties in the course of trying to solve them. Consequently, alternative solution methods have been devised. The remainder of this chapter is devoted to a discussion of such methods.

5.8 BOUNDARY-LAYER FLOW

The majority of convection problems encountered will involve turbulent flow. The analysis of such problems would require a simultaneous solution to the continuity, momentum, and energy equations. Because of the nonlinearity of the momentum equations, they remain insoluble for most turbulent-flow problems, and some simplifications must be made. In that regard, we divide the flow region into two portions. One portion is a nonviscous region away from any solid surfaces. Near a boundary, however, the fluid adheres to the surface due to viscous or frictional effects. The velocity at the wall is zero relative to the wall. So near a boundary, viscosity is important. Away from a boundary, viscous effects can be neglected. The flow near a boundary is known as *boundary-layer flow*.

Consider flow past a flat plate, as shown in Figure 5.16. Upstream the flow velocity is uniform and equal to V_∞. Anywhere along the plate the profile changes from the uniform free-stream profile. The velocity increases from zero at the surface to nearly the free-stream value at some distance away. The *momentum-boundary-layer thickness* δ varies in thickness or height with x. Within the

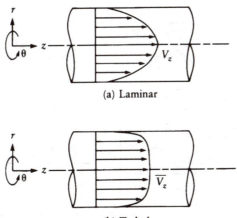

(a) Laminar

(b) Turbulent

FIGURE 5.15   Laminar- and turbulent-velocity profiles for flow in a tube.

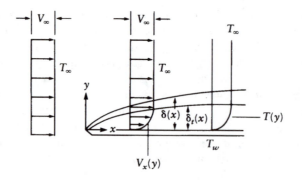

FIGURE 5.16 Steady uniform flow of a Newtonian fluid past a heated flat plate.

boundary layer, the velocity in the x direction V_x varies only with y. In the free stream beyond δ, the velocity V_x (nearly) equals the free-stream value V_∞ (actually, V_x asymptotically approaches V_∞).

Similar behavior occurs with temperature. Assuming the free-stream temperature T_∞ is different from the (constant) wall temperature T_w, a profile exists that ranges from T_w at the plate to T_∞ at some distance δ_t away. (T approaches T_∞ asymptotically.) The *thermal-boundary-layer thickness* δ_t also varies in thickness with x.

Typically, the momentum- and thermal-boundary-layer thicknesses are a few thousandths of a millimeter thick. Despite their size, both are important in determining drag forces (such as for a ship moving through water) and heat-transfer rates (such as cold winds blowing past a window). For a gas the momentum and thermal boundary layers have about the same thickness. For liquid metals, $\delta_t \gg \delta$, whereas for oils, $\delta_t \ll \delta$.

Laminar- and turbulent-flow regimes for flow over a flat plate are illustrated in Figure 5.17. As shown, the boundary layer grows from the leading edge and continues growing with distance. Near the leading edge the boundary layer is laminar. Thus the viscous forces are large in this region compared to the inertia of the flow. At some location downstream the flow enters a transition region where purely laminar conditions cease to exist. After transition, the boundary layer becomes turbulent. In the turbulent region, however, there is still a small layer of flow near the surface (where velocity is zero) that is laminar. This layer is called the *laminar sublayer* δ_l. Transition usually occurs over a range and not at a single point. Transition varies over the range

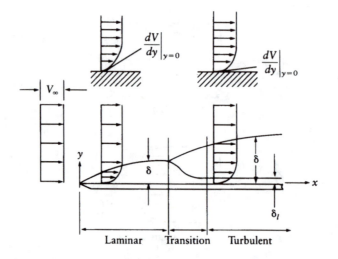

FIGURE 5.17 Laminar- and turbulent-flow regimes for flow over a flat plate.

$$2\times10^5 \le \mathrm{Re}_{cr} \le 3\times10^6$$

where $\mathrm{Re}_{cr}\ (= V_\infty\, x_{cr}/\nu)$ is called the critical Reynolds number based on x_{cr}, which is the x distance where transition exists. For purposes of calculation, it is customarily assumed that transition occurs at

$$\mathrm{Re}_{cr} = \frac{V_\infty x_{cr}}{\nu} = 5\times10^5$$

The velocity profiles in laminar and turbulent flows are different, as shown in Figure 5.17. The turbulent profile is flatter over a greater portion of the boundary layer due primarily to the mixing effect.

An important class of problems associated with boundary-layer flow has been developed. To solve such problems, we must first formulate the applicable equations. Again, we begin with the continuity, momentum, and energy equations. Consider that the problem we are investigating is the two-dimensional flow of an incompressible Newtonian fluid with constant properties over a flat plate. The flow is steady, and gravity is neglected. Pressure does not vary in the direction normal to the surface because the boundary layer is thin. Under these conditions, the continuity, momentum, and energy equations in Cartesian coordinates reduce to

$$\frac{\partial V_x}{\partial x} + \frac{\partial V_y}{\partial y} = 0$$

$$\left(V_x\frac{\partial V_x}{\partial x} + V_y\frac{\partial V_x}{\partial y}\right) = -\frac{g_c}{\rho}\frac{dp}{dx} + \frac{\mu g_c}{\rho}\left(\frac{\partial^2 V_x}{\partial x^2} + \frac{\partial^2 V_x}{\partial y^2}\right)$$

$$\left(V_x\frac{\partial V_y}{\partial x} + V_y\frac{\partial V_y}{\partial y}\right) = \frac{\mu g_c}{\rho}\left(\frac{\partial^2 V_y}{\partial x^2} + \frac{\partial^2 V_y}{\partial y^2}\right)$$

$$\left(V_x\frac{\partial T}{\partial x} + V_y\frac{\partial T}{\partial y}\right) = \alpha\left(\frac{\partial^2 T}{\partial x^2} + \frac{\partial^2 T}{\partial y^2}\right)$$

These equations are still not solvable due to the nonlinear terms. There is a method, known as an *order-of-magnitude analysis,* that has been used traditionally to simplify the equations so they can be solved. The essence of the method is to identify terms in the equations that are negligibly small and cancel them. Without going into all the mathematical details here, we merely summarize the method.

The boundary layer is thin, so velocities and thicknesses in the y direction have small orders of magnitude. On the other hand, velocities and thicknesses in the x direction have larger orders of magnitude. The relative orders of magnitude are important in a comparative rather than an absolute sense. We can now evaluate the relative importance of each term in the continuity, momentum and energy equations. Those terms that have relatively small magnitudes can be discarded.

Performing an order-of-magnitude analysis for the flat-plate problem and discarding terms that are comparatively small, we are left with

$$\frac{\partial V_x}{\partial x} + \frac{\partial V_y}{\partial y} = 0$$

$$V_x\frac{\partial V_x}{\partial x} + V_y\frac{\partial V_x}{\partial y} = \nu\frac{\partial^2 V_x}{\partial y^2}$$

$$V_x \frac{\partial T}{\partial x} + V_y \frac{\partial T}{\partial y} = \alpha \frac{\partial^2 T}{\partial y^2}$$

The entire y-component momentum equation is negligible. The continuity and momentum equations have been solved in a number of ways for the velocity V_x. It is sufficient here to summarize what is known as the Blasius solution:

Plate width	b
Plate length in flow direction	L
Reynolds number at x	$\mathrm{Re}_x = V_\infty x / v$
Reynolds number at plate end	$\mathrm{Re}_L = V_\infty L / v$
Laminar-boundary-layer thickness	$\delta = 5.0 x / \sqrt{\mathrm{Re}_x}$
Wall shear stress, laminar flow	$\tau_w = 0.332 V_\infty \mu \sqrt{\mathrm{Re}_x} / x$
Local drag coefficient skin-friction coefficient for laminar flow	$C_d = \dfrac{2\tau_w g_c}{\rho V_\infty^2} = 0.664 / \sqrt{\mathrm{Re}_x}$
Total drag coefficient for laminar flow (C_d integrated over the plate length)	$C_D = 1.328 / \sqrt{\mathrm{Re}_L}$
Total drag force exerted (laminar flow)	$D_f = 0.664 V_\infty \mu b \sqrt{\mathrm{Re}_L}$

These equations are valid only if laminar flow exists over the entire plate.

Solutions for temperature profiles depend on the boundary conditions specified. For example, the plate could be heated and transferring energy to the fluid. Heating could begin at the leading edge of the plate or somewhere downstream. The heat flux could be a constant or allowed to vary so that the plate-surface temperature remained constant. On the other hand, for velocity, velocity is zero along the wall and increases with height to the free-stream value. So although the momentum equations are more complex than the energy equation, the velocities have more clearly defined boundary conditions than does the temperature in a convection system. Solutions for temperature are provided in Chapter 7.

5.9 THE NATURAL-CONVECTION PROBLEM

The equations of motion derived earlier are applicable to fluids with constant properties. That is, density and viscosity are assumed constant in the equations. In a natural-convection problem, fluid near a warmed surface experiences a decrease in density, making the fluid "lighter." The fluid near the warmed surface thus rises. Flow then is caused by the fact that density is not a constant in the fluid. Clearly, the Navier-Stokes Equations, in which density is constant, do not apply. There are at least two possible approaches that we can take to obtain descriptive equations for the natural-convection problem. One approach is to rederive equations analogous to the Navier-Stokes Equations from an energy balance. The other approach is to modify the density term in the Navier-Stokes Equations to make them applicable. Here we will adopt the second approach.

For a natural-convection problem, heat added to the fluid causes the density change and induces flow. Thus, we seek a term or property that relates density change to temperature change. Defined earlier in this chapter (Equations 5.7 and 5.8) is the volumetric thermal-expansion coefficient,

$$\beta = -\frac{1}{\rho}\left(\frac{\partial \rho}{\partial T}\right)_p \tag{5.8}$$

This coefficient can be approximated by

$$\beta = -\frac{1}{\rho}\frac{(\Delta\rho)}{(\Delta T)} \tag{5.56}$$

Before substituting specific values for $\Delta\rho$ and ΔT, we must first define the value of density ρ. Density will vary with temperature throughout the flow field. Arbitrarily, we select a mean temperature T_m as our reference. The mean temperature must be specified for each individual problem. The density at temperature T_m is ρ_m, and Equation 5.56 is used to obtain the deviation of ρ from ρ_m for a given temperature difference:

$$\beta = -\frac{1}{\rho_m}\frac{\rho-\rho_m}{T-T_m} \tag{5.57}$$

Thus, the density term in the Navier-Stokes Equations becomes

$$\rho - \rho_m = -\beta\rho_m(T-T_m)$$

or

$$\rho = \rho_m - \beta\rho_m(T-T_m) \tag{5.58}$$

How this equation is used in natural convection problems is illustrated in the following example.

Example 5.7

A fluid is placed between two vertical walls of different temperature, as shown in Figure 5.18. Determine equations for velocity and temperature of the fluid.

Solution

Expected velocity and temperature profiles are sketched in the figure. The left wall is heated to a temperature T_1, and the right wall is at T_2 with $T_2 < T_1$. Fluid near the heated wall rises, and fluid near the cold wall descends.

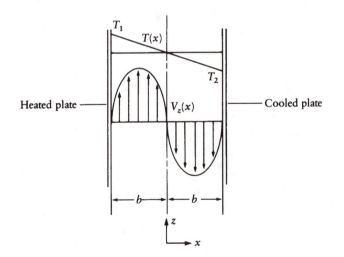

FIGURE 5.18 Fluid placed between two vertical plates of different temperatures.

Assumptions

1. The system is at steady state.
2. The fluid is Newtonian with constant properties.
3. The system is set up such that volume flow upward equals volume flow downward.
4. Viscous dissipation is neglected.
5. Flow is two dimensional.
6. Edge effects are neglected.
7. Temperature varies only with the x coordinate.

The velocity vector has three components—V_x, V_y, and V_z. Only V_z is nonzero. Based on the assumptions made, V_z varies only with x. The continuity and momentum equations become

continuity

$$\frac{\partial V_z}{\partial z} = 0$$

z component, Navier-Stokes Equations

$$0 = -\frac{\partial p}{\partial z} + \mu \frac{\partial^2 V_z}{\partial x^2} + \frac{\rho g_z}{g_c}$$

The boundary conditions are

B.C.1. $x = \pm b,$ $V_z = 0$
B.C.2. $x = 0,$ $V_z = 0$

Temperature can depend on at most three space variables and time. For this problem $T = T(x)$, and the energy equation reduces to

$$0 = k_f \frac{d^2 T}{dx^2}$$

with boundary conditions

B.C.1. $x = -b,$ $T = T_1$
B.C.2. $x = b,$ $T = T_2$

The energy equation is integrated directly giving

$$\frac{dT}{dx} = C_1$$

$$T = C_1 x + C_2$$

Applying the boundary conditions, we obtain for the temperature

$$T = \frac{T_1 + T_2}{2} - \frac{x}{2b}(T_1 - T_2)$$

It is convenient to define the mean temperature as

$$T_m = \frac{T_1 + T_2}{2}$$

In terms of mean temperature, the temperature profile becomes

$$\frac{T - T_m}{T_1 - T_2} = -\frac{x}{2b}$$

With mean temperature defined, we can now modify the z-component equation to make it applicable to this problem. We begin with Equation 5.58:

$$\rho = \rho_m - \beta\rho_m(T - T_m) \tag{5.58}$$

Substituting into the equation of motion and rearranging gives

$$\mu\frac{d^2 V_z}{dx^2} = \frac{dp}{dz} - \frac{\rho_m g_z}{g_c} + \frac{\beta\rho_m g_z}{g_c}(T - T_m)$$

A pressure gradient dp/dz in the system exists because the fluid has a nonzero density. Therefore, from hydrostatics,

$$\frac{dp}{dz} = \frac{\rho_m g_z}{g_c}$$

The equation of motion now becomes

$$\frac{d^2 V_z}{dx^2} = \frac{\beta\rho_m g_z}{\mu g_c}(T - T_m)$$

Substituting from the temperature profile, we get

$$\frac{d^2 V_z}{dx^2} = \frac{\beta\rho_m g_z x(T_2 - T_1)}{2\mu g_c b}$$

The equation is directly integrable; the result is

$$V_z = \frac{\beta\rho_m g_z x^3}{12\mu g_c b}(T_2 - T_1) + C_3 x + C_4$$

Applying the boundary conditions, we obtain

$$0 = \frac{\beta \rho_m g_z b^3}{12 \mu g_c b}(T_2 - T_1) + C_3 b + C_4$$

$$0 = -\frac{\beta \rho_m g_z b^3}{12 \mu g_c b}(T_2 - T_1) - C_3 b + C_4$$

Solving,

$$C_4 = 0$$

and

$$C_3 = -\frac{\beta \rho_m g_z b}{12 \mu g_c}(T_2 - T_1)$$

The velocity distribution becomes

$$V_z = \frac{\beta \rho_m g_z x^3}{12 \mu g_c b}(T_2 - T_1) - \frac{\beta \rho_m g_z b x}{12 \mu g_c}(T_2 - T_1)$$

$$V_z = \frac{\beta \rho_m g_z b^2 (T_2 - T_1)}{12 \mu g_c}\left(\frac{x}{b} - \frac{x^3}{b^3}\right)$$

The buoyancy forces exist due to a temperature difference that induces fluid movement. Thus, if $T_1 = T_2$, then $V_z = 0$. ❏

5.10 DIMENSIONAL ANALYSIS

In the preceding sections considerable time and space are devoted to presentation of continuity, momentum, and energy equations, and to boundary-layer equations. For a given problem in which the descriptive equations are written, if they can be solved, velocity and temperature profiles are obtained. In most cases, however, the final objective is to predict a heat-transfer rate. Furthermore, it is desirable to use Newton's Law of Cooling because of its simplicity:

$$q_c'' = \bar{h}_c(T_w - T_\infty)$$

where $T_w - T_\infty$ is the temperature difference causing the heat flow. In addition to the temperature difference, the convection coefficient $\bar{h}_c$ must also be known to calculate the heat flux q_c''.

In a majority of fluid-mechanics and heat-transfer problems, turbulent-flow conditions exist. The equations of motion cannot be solved explicitly for turbulent-flow problems, unless some significant and limiting assumptions are made. Consequently, for such problems the method of *dimensional analysis* is used. Traditionally, dimensional analysis is utilized to obtain a functional relationship between the convection coefficient $\bar{h}_c$ and the relevant physical properties and kinematic parameters of the flow situation. Once the convective coefficient is known, the heat-transfer rate can be determined.

As an illustration of the mechanics of dimensional analysis, consider that we are trying to predict the pressure drop Δp within a pipe. The pressure-drop measurement is made between two points a distance L apart. We start by theorizing as to which flow parameters influence the pressure drop. Based on experience or physical intuition, we could write

$$\Delta p = \Delta p(\rho, V, D, L, \mu, \varepsilon, g_c) \tag{5.59}$$

Thus, pressure drop is *assumed* to depend on

ρ = density of fluid
V = average velocity of the flow
D = characteristic length representing the flow-geometry cross section
L = distance between points where pressure drop is measured
μ = dynamic viscosity of the fluid
ε = a parameter (with dimensions of length) that characterizes the roughness of the pipe wall
g_c = conversion factor between force and mass units

The dimensions of each of these and other variables are provided in Appendix Table A.6. It is further assumed that the functional dependence in Equation 5.59 can be rewritten as

$$\Delta p = C_1 \rho^{a1} V^{a2} D^{a3} L^{a4} \mu^{a5} \varepsilon^{a6} g_c^{a7} \tag{5.60}$$

where C_1 and the a exponents are constants yet to be determined. Equation 5.60 must be dimensionally homogeneous. Therefore units on the left-hand side must equate to those on the right-hand side. Substituting the dimensions of each term into 5.60 gives

$$\frac{F}{L^2} = C_1 \left(\frac{M}{L^3}\right)^{a1} \left(\frac{L}{T}\right)^{a2} (L)^{a3} (L)^{a4} \left(\frac{F \cdot T}{L^2}\right)^{a5} (L)^{a6} \left(\frac{M \cdot L}{F \cdot T^2}\right)^{a7}$$

The force units must balance. Thus, the exponents of force on both sides must be related according to

F: $1 = a_5 - a_7$

Likewise, for mass, time, and length, we get

M: $0 = a_1 + a_7$

T: $0 = -a_2 + a_5 - 2a_7$

L: $-2 = -3a_1 + a_2 + a_3 + a_4 - 2a_5 + a_6 + a_7$

We have four equations and seven unknowns (a_1 through a_7). To continue, we must solve for four of the unknowns in terms of the remaining three. Proceeding,

$$a_7 = a_5 - 1$$

$$a_1 = -a_7 = 1 - a_5 = a_1$$

$$a_2 = a_5 - 2a_7 = a_5 - 2a_5 + 2$$

$$a_2 = 2 - a_5$$

and after a lengthy algebraic exercise,

$$a_3 = -a_5 - a_4 - a_6$$

Substituting into Equation 5.60 yields

$$\Delta p = C_1 \frac{\rho}{\rho^{a5}} \frac{V^2}{V^{a5}} \frac{1}{D^{a5} D^{a4} D^{a6}} L^{a4} \mu^{a5} \varepsilon^{a6} \frac{g_c^{a5}}{g_c}$$

Grouping terms,

$$\frac{\Delta p g_c}{\rho V^2} = C_1 \left(\frac{\mu g_c}{\rho VD}\right)^{a5} \left(\frac{L}{D}\right)^{a4} \left(\frac{\varepsilon}{D}\right)^{a6}$$

or in functional-dependence form,

$$\frac{\Delta p g_c}{\rho V^2} = \Delta p\left(\frac{\mu g_c}{\rho VD}, \frac{L}{D}, \frac{\varepsilon}{D}\right) \tag{5.61}$$

All terms are dimensionless. The constant C_1 and the exponents a_4, a_5, and a_6 must now be found by experiment. If indeed there is a relationship between the dimensionless groups of Equation 5.61, the only way to find it is to obtain data and calculate the values of C_1 and the exponents. If they cannot be evaluated satisfactorily, it is necessary to go back to Equation 5.59 and reevaluate our thinking regarding which variables we originally assumed to be significant.

The exponents left to evaluate are a_4, a_5, and a_6. Suppose, however, that we solved the algebraic equations a little differently and obtained

$a_5 = 1 + a_7$

$a_1 = -a_7$

$a_2 = 1 - a_7$

and

$a_4 = -a_3 - a_7 - a_6 - 1$

The final functional dependence would become

$$\frac{\Delta p L}{V \mu} = \Delta p\left(\frac{\mu g_c}{\rho VL}, \frac{\varepsilon}{L}, \frac{D}{L}\right) \tag{5.62}$$

The above form at this point is just as valid as Equation 5.61; both are essentially correct. It is not necessary, however, to go back to the four simultaneous equations and re-solve them each time a permutation of the basic groups is desired. In other words, it is permissible to manipulate Equation 5.61 algebraically and obtain another dependence. We can multiply $\Delta p g_c/\rho V^2$ by $\rho VD/\mu g_c$ to obtain $\Delta p D/\mu V$, which can then be used to replace either $\Delta p g_c/\rho V^2$ or $\rho VD/\mu g_c$ in Equation 5.61. It is convenient to express the dependencies in certain forms, because various dimensionless groups are traditionally recognized in the literature. For example, the dependent variable in Equation 5.61, when written as $\Delta p g_c/\frac{1}{2}(\rho V^2)$, is known as the *pressure coefficient*, the 1/2 being added as a matter of custom to make the denominator a kinetic-energy term. Also, the term $\rho VD/\mu g_c$ is recognized as the Reynolds number.

We now apply the method of dimensional analysis to the general classes of convection heat-transfer problems. It is desired to let the convection coefficient $\bar{h}_c$ be the dependent variable, which can depend on a multitude of variables. We consider three types of problems: internal flows, external flows, and natural convection. The following analyses serve as an introduction to each class of problems.

5.10.1 INTERNAL FLOWS

The typical problem in internal flows is a forced-convection system. Fluid under the action of an external force travels through a conduit, which can be round, annular, square, or some other shape. A temperature difference between the conduit wall and the fluid causes energy to be transferred. The amount of energy transfer can be calculated once the convection coefficient is known. We therefore write

$$\bar{h}_c = \bar{h}_c(\rho, V, D, \mu, c_p, \Delta T, k_f, g_c) \tag{5.63}$$

where

$\bar{h}_c$ = a convection coefficient [F·L/(T·L²·t)]
ρ = the fluid density (M/L³)
V = the average fluid velocity (L/T)
D = a characteristic dimension of the cross section (L)
p = the dynamic viscosity of the fluid (F·T/L²)
c_p = the specific heat of the fluid [F·L/(M·t)]
ΔT = temperature difference between the wall and fluid (bulk) (t)
k_f = the thermal conductivity of the fluid [F/(t·T)]
g_c = conversion factor between mass and force units [M·L/(F·T²)]

The convection coefficient $\bar{h}_c$ could be one of several that exist in the conduit. There is a local coefficient that varies with axial distance; an average coefficient that results when the local coefficient is integrated over the conduit length; etc. The characteristic dimension D of the conduit cross section is often referred to as the *hydraulic diameter* and defined as

$$D = \frac{4 \times \text{cross-sectional area}}{\text{perimeter}} \tag{5.64}$$

The hydraulic diameter is sometimes denoted as D_h. Using the hydraulic diameter allows us to represent noncircular cross sections—such as squares, rectangles, and triangles—with a single linear dimension. Performing a dimensional analysis on Equation 5.63, we get

$$\frac{\bar{h}_c D}{k_f} = \bar{h}_c\left(\frac{\rho V D}{\mu g_c}, \frac{c_p \mu g_c}{k_f}, \frac{V^2}{c_p \Delta T g_c}\right) \tag{5.65}$$

The dimensionless groups in the above equation are known by specific names:

Nusselt number $\qquad = \mathrm{Nu}_D = \dfrac{\bar{h}_c D}{k_f}$

Reynolds number $\qquad = \mathrm{Re}_D = \dfrac{\rho V D}{\mu g_c}$

Prandtl number $\qquad = \mathrm{Pr} = \dfrac{c_p \mu g_c}{k_f} = \dfrac{c_p \rho}{k_f}\,\dfrac{\mu g_c}{\rho} = \dfrac{\nu}{\alpha}$

Eckert number $\qquad = \mathrm{Ec} = \dfrac{V^2}{c_p \Delta T g_c}$

The Eckert number is related to what is known as the Mach number of the flow, which is significant in high-speed compressible flows. Equation 5.65 can be rewritten as

$$\text{Nu}_D = \text{Nu}_D(\text{Re}_D, \text{Pr}, \text{Ec}) \tag{5.66}$$

For an internal-flow heat-transfer problem, it remains to determine the relationship between these dimensionless groups, which can be done at this point only by experiment.

Another dimensionless group that could be derived instead of the Eckert number is the Brinkman number, where

$$\text{Br} = \frac{\mu V^2}{k_f \Delta T} = \frac{c_p \mu g_c}{k_f} \frac{V^2}{c_p \Delta T g_c} = (\text{Pr})(\text{Ec})$$

Similarly, the combination of the Reynolds and Prandtl numbers gives yet another dimensionless group called the Peclet number:

$$\text{Pe} = (\text{Re})(\text{Pr}) = \frac{\rho V D}{\mu g_c} \frac{c_p \mu g_c}{k_f} = \frac{VD}{\alpha}$$

The choice of which ratios to use is arbitrary.

Recall, from Chapter 4 on transient conduction, the Biot number:

$$\text{Bi} = \frac{\bar{h}_c L}{k}$$

Notice the similarity between the Nusselt number and the Biot number. The difference between these ratios is the characteristic length and the thermal conductivity. The conductivity in the Nusselt number is for the fluid, while the Biot number applies to solids.

Each ratio mentioned above has a physical significance or interpretation. For example, the Nusselt number is a ratio of convection heat transfer to conduction in a fluid. In contrast, the Biot number is a ratio of internal thermal resistance of a solid to the convective resistance at the surface. The significance of the various ratios is presented in Table 5.4 in the Summary section of this chapter. Internal flows are the subject of Chapter 6.

5.10.2 EXTERNAL FLOWS

The external-flow problem we consider is a forced-convection problem. An external source causes fluid to move past a flat or curved surface. The surface temperature is different from the fluid temperature, so heat transfer takes place. An example of a flat-surface external flow is wind moving past a warmed roof. Flow past a heated or cooled cylinder is a curved-surface example. A dimensional analysis for the external-flow problem might begin with

$$\bar{h}_c = \bar{h}_c(\rho, V, x, \mu, c_p, \Delta T, k_f, g_c) \tag{5.67}$$

which is similar to Equation 5.63 for internal flows. One difference is that in external flows the characteristic dimension could be distance from the leading edge in the flat-plate problem. Because the temperatures of the fluid and the surface probably vary with distance from the leading edge, the coefficient will also vary. Performing the dimensional analysis, we obtain

$$\frac{\bar{h}_c x}{k_f} = \bar{h}_c\left(\frac{\rho V x}{\mu g_c}, \frac{c_p \mu g_c}{k_f}, \frac{V^2}{c_p \Delta T g_c}\right) \tag{5.68}$$

or

$$\text{Nu}_x = \text{Nu}_x(\text{Re}_x, \text{Pr}, \text{Ec}) \tag{5.69}$$

External flows are discussed in greater detail in Chapter 7.

5.10.3 NATURAL CONVECTION

The representative natural-convection problem is one where a fluid is in contact with a warmer plate or curved surface (cylinder, for example). As fluid near the warmed surface begins to increase in temperature, local fluid density decreases. The buoyant force causes the less dense fluid to rise, and more cooler fluid is brought into contact with the surface. Fluid-motion and heat-transfer effects are thus intimately related. For the general natural-convection problem, we write

$$\bar{h}_c = \bar{h}_c(\rho, V, L, \mu, c_p, \Delta T, k_f, \beta, g_c) \tag{5.70}$$

where L is a characteristic length and β is the volumetric expansion coefficient. Performing a dimensional analysis, we obtain

$$\frac{\bar{h}_c L}{k_f} = \bar{h}_c\left(\frac{\rho V L}{\mu g_c}, \frac{c_p \mu g_c}{k_f}, \frac{\rho^2 g \beta \Delta T L^3}{\mu^2 g_c^2}, \frac{V^2}{c_p \Delta T g_c}\right) \tag{5.71}$$

The group containing the volumetric expansion coefficient is called the Grashof number:

$$\text{Gr} = \frac{\rho^2 g \beta \Delta T L^3}{\mu^2 g_c^2} = \frac{\beta \Delta T L^3}{v^2} \tag{5.72}$$

The Grashof number arises in natural-convection problems where β is significant. Other dimensionless groups of importance that could arise are the Rayleigh number (Ra) and the Stanton number (St), defined as

$$\text{Ra} = (\text{Gr})(\text{Pr}) = \frac{g \beta \Delta T L^3}{v^2} \frac{v}{\alpha} = \frac{g \beta \Delta T L^3}{v \alpha}$$

$$\text{St} = \frac{\text{Nu}}{(\text{Re})(\text{Pr})} = \frac{\bar{h}_c L}{k_f} \frac{v}{VL} \frac{\alpha}{v} = \frac{h_c L}{k_f} \frac{v}{VL} \frac{\alpha}{v} \frac{c_p \rho}{c_p \rho}$$

$$\text{St} = \frac{\bar{h}_c}{\rho c_p V}$$

Equation 5.71 can therefore be rewritten as

$$\text{Nu}_L = \text{Nu}_L(\text{Re}_L, \text{Pr}, \text{Gr}_L, \text{Ec}) \tag{5.73}$$

Natural convection is the subject of Chapter 8.

5.11 SUMMARY

In this chapter, we have discussed fluid properties and the equations of fluid mechanics. Continuity and momentum equations were derived. The thermal-energy equation was also derived, and selected laminar-flow problems were solved. The continuity, momentum, and energy equations are provided in the Tables 5.1 through 5.4.

For turbulent-flow problems, the equations of motion cannot be solved exactly, so alternative methods were mentioned. Boundary-layer equations were discussed. The method of dimensional analysis was presented. Dimensionless groups were derived for internal-flow heat-transfer problems, external-flow heat-transfer problems, and natural-convection problems. A number of dimensionless groups are listed and defined in the tables of this section. This chapter thus serves as an introduction to the convection chapters that follow.

TABLE 5.1
The Equation of Continuity

Cartesian coordinates

$$\frac{\partial \rho}{\partial t} + \frac{\partial}{\partial x}(\rho V_x) + \frac{\partial}{\partial y}(\rho V_y) + \frac{\partial}{\partial z}(\rho V_z) = 0 \qquad (5.11)$$

Cylindrical coordinates

$$\frac{\partial \rho}{\partial t} + \frac{1}{r}\frac{\partial}{\partial r}(\rho r V_r) + \frac{1}{r}\frac{\partial}{\partial \theta}(\rho V_\theta) + \frac{\partial}{\partial z}(\rho V_z) = 0 \qquad (5.14)$$

5.12 PROBLEMS

5.12.1 FLUID PROPERTIES

1. Regarding the "bread and butter" example on viscosity in Section 5.2, use the data there to calculate the dynamic viscosity of the butter.
2. Calculate the absolute viscosity of Freon-12 at 0°C.
3. Verify that the ideal gas equation is used to calculate density in the air table (Appendix D) at 200, 400, 600, and 1 000 K.
4. Verify that the ideal gas equation is used to calculate density in the carbon dioxide table (Appendix D) for temperatures of 200, 300, 400, and 500 K.
5. Calculate the absolute viscosity of helium at 500 K.
6. A gas (mass = 0.2 lbm) is placed in a chamber formed by a piston and cylinder, as shown in Figure P5.1. The piston is free to move vertically. The gas is heated with 10 BTU of energy. Determine the increase in temperature realized by the gas if it is
   (a) Hydrogen
   (b) Helium
   (c) Carbon dioxide
   Take the initial gas temperature to be 60°F and assume constant properties.

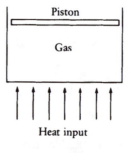

FIGURE P5.1

TABLE 5.2
The Equation of Motion for a Newtonian Fluid Having Constant Density and Viscosity (Navier-Stokes Equations)

Cartesian coordinates

x component

$$\frac{\rho}{g_c}\left(\frac{\partial V_x}{\partial t} + V_x\frac{\partial V_x}{\partial x} + V_y\frac{\partial V_x}{\partial y} + V_z\frac{\partial V_x}{\partial z}\right) = -\frac{\partial p}{\partial x} + \mu\left(\frac{\partial^2 V_x}{\partial x^2} + \frac{\partial^2 V_x}{\partial y^2} + \frac{\partial^2 V_x}{\partial z^2}\right) + \frac{\rho g_x}{g_c} \qquad (5.28a)$$

y component

$$\frac{\rho}{g_c}\left(\frac{\partial V_y}{\partial t} + V_x\frac{\partial V_y}{\partial x} + V_y\frac{\partial V_y}{\partial y} + V_z\frac{\partial V_y}{\partial z}\right) = -\frac{\partial p}{\partial y} + \mu\left(\frac{\partial^2 V_y}{\partial x^2} + \frac{\partial^2 V_y}{\partial y^2} + \frac{\partial^2 V_y}{\partial z^2}\right) + \frac{\rho g_y}{g_c} \qquad (5.28b)$$

z component

$$\frac{\rho}{g_c}\left(\frac{\partial V_z}{\partial t} + V_x\frac{\partial V_z}{\partial x} + V_y\frac{\partial V_z}{\partial y} + V_z\frac{\partial V_z}{\partial z}\right) = -\frac{\partial p}{\partial z} + \mu\left(\frac{\partial^2 V_z}{\partial x^2} + \frac{\partial^2 V_z}{\partial y^2} + \frac{\partial^2 V_z}{\partial z^2}\right) + \frac{\rho g_z}{g_c} \qquad (5.28c)$$

Cylindrical coordinates

r component

$$\frac{\rho}{g_c}\left(\frac{\partial V_r}{\partial t} + V_r\frac{\partial V_r}{\partial r} + \frac{V_\theta}{r}\frac{\partial V_r}{\partial \theta} + V_z\frac{\partial V_r}{\partial z} - \frac{V_\theta^2}{r}\right)$$

$$= -\frac{\partial p}{\partial r} + \mu\left\{\frac{\partial}{\partial r}\left[\frac{1}{r}\frac{\partial}{\partial r}(rV_r)\right] + \frac{1}{r^2}\frac{\partial^2 V_r}{\partial \theta^2} + \frac{\partial^2 V_r}{\partial z^2} - \frac{2}{r^2}\frac{\partial V_\theta}{\partial \theta}\right\} + \frac{\rho g_r}{g_c} \qquad (5.29a)$$

θ component

$$\frac{\rho}{g_c}\left(\frac{\partial V_\theta}{\partial t} + V_r\frac{\partial V_\theta}{\partial r} + \frac{V_\theta}{r}\frac{\partial V_\theta}{\partial \theta} + \frac{V_r V_\theta}{r} + V_z\frac{\partial V_\theta}{\partial z}\right)$$

$$= -\frac{1}{r}\frac{\partial p}{\partial \theta} + \mu\left\{\frac{\partial}{\partial r}\left[\frac{1}{r}\frac{\partial}{\partial r}(rV_\theta)\right] + \frac{1}{r^2}\frac{\partial^2 V_\theta}{\partial \theta^2} + \frac{2}{r^2}\frac{\partial V_r}{\partial \theta} + \frac{\partial^2 V_\theta}{\partial z^2}\right\} + \frac{\rho g_\theta}{g_c} \qquad (5.29b)$$

z component

$$\frac{\rho}{g_c}\left(\frac{\partial V_z}{\partial t} + V_r\frac{\partial V_z}{\partial r} + \frac{V_\theta}{r}\frac{\partial V_z}{\partial \theta} + V_z\frac{\partial V_z}{\partial z}\right)$$

$$= -\frac{\partial p}{\partial z} + \mu\left[\frac{1}{r}\frac{\partial}{\partial r}\left(r\frac{\partial V_z}{\partial r}\right) + \frac{1}{r^2}\frac{\partial^2 V_z}{\partial \theta^2} + \frac{\partial^2 V_z}{\partial z^2}\right] + \frac{\rho g_z}{g_c} \qquad (5.29c)$$

In cylindrical coordinates, the term $\rho V_\theta^2/r$ is the centrifugal force that gives the effective *r*-directed force resulting from fluid motion in the θ direction. When transforming equations from Cartesian to cylindrical coordinates, this term arises automatically. The term $\rho V_r V_\theta/r$ is the Coriolis force, which is the effective θ-directed force resulting from flow in both the *r* and θ directions. The term also arises automatically when transforming from Cartesian to cylindrical coordinates.

7. A gas having a mass of 2 kg is placed in a constant-volume chamber and heated with 2 W of energy. Determine the temperature increase experienced by the gas if it is
   (a) Air
   (b) Helium
   (c) Carbon dioxide
   The initial temperature of the gas is 20°C. Assume constant gas properties.
8. Calculate the value of the volumetric thermal-expansion coefficient for air at 30°C.

TABLE 5.3
The Thermal-Energy Equation for Newtonian Fluids Having Constant Density and Viscosity, with Viscous Dissipation Terms Included

Cartesian coordinates

$$\rho c_p\left(\frac{\partial T}{\partial t} + V_x\frac{\partial T}{\partial x} + V_y\frac{\partial T}{\partial y} + V_z\frac{\partial T}{\partial z}\right) = k\left(\frac{\partial^2 T}{\partial x^2} + \frac{\partial^2 T}{\partial y^2} + \frac{\partial^2 T}{\partial z^2}\right) \tag{5.51}$$

Cylindrical coordinates

$$\rho c_p\left(\frac{\partial T}{\partial t} + V_r\frac{\partial T}{\partial r} + \frac{V_\theta}{r}\frac{\partial T}{\partial \theta} + V_z\frac{\partial T}{\partial z}\right) = k\left[\frac{1}{r}\frac{\partial}{\partial r}\left(r\frac{\partial T}{\partial r}\right) + \frac{1}{r^2}\frac{\partial^2 T}{\partial \theta^2} + \frac{\partial^2 T}{\partial z^2}\right]$$

Viscous dissipation terms have been neglected in these equations.

9. Calculate the value of the volumetric thermal-expansion coefficient for carbon dioxide at 200°F.
10. Which of the following gases has the highest value of the volumetric thermal-expansion coefficient at 0°C?
    (a) Air
    (b) Helium
    (c) Carbon dioxide

5.12.2 MOMENTUM EQUATION

11. The x-directed instantaneous velocity of a two-dimensional system is given by

$$V_x = 2x$$

Use the continuity equation to find V_y, given that at $y = 0$, $V_y = 0$.
12. Starting with Equation 5.21, derive Equation 5.28a.
13. Work through Example 5.3, and verify that Equation 5.30 is correct.
14. Graph Equation 5.30 (V_z vs. x) for a system where the fluid is glycerine at 20°C, $\alpha = 5°$, and $\delta = 1$ cm.
15. Use Equation 5.30 for velocity of flow down an incline to derive equations for
    (a) Volume flow rate
    (b) Average velocity
    (c) Maximum velocity
    (d) Shear stress
    (e) Wall shear stress
16. Work through Example 5.4 and verify that Equation 5.31 is correct.
17. Combine Equation 5.31 with Equation 5.35 and graph $V_z/V_{z\,max}$ vs. r/R.
18. Starting with Equation (i) of Example 5.4, derive Equation 5.32.
19. Show that the average velocity for flow in an annulus is

$$V = \frac{R^2}{8\mu}\left(-\frac{dp}{dz}\right)\left[1 + \kappa^2 + \frac{1-\kappa^2}{\ln(\kappa)}\right]$$

20. Average velocity for flow in a tube is defined as

$$V = \frac{\iint V_z dA}{\iint dA}$$

TABLE 5.4
Some Common Dimensionless Ratios Encountered in Fluid Mechanics and Heat Transfer

Ratio	Name	Symbol	Interpretation
$\bar{h}_c L/k$	Biot number	Bi	Internal thermal resistance/ External resistance for a solid
$\mu V^2/k_f(T_w - T_\infty)$	Brinkman number	Br = (Pr)(Ec)	Viscosity × Inertia/ Heat-transfer rate
$\tau_w g_c/\frac{1}{2}\rho V^2$	Coefficient of friction	C_f	Wall shear stress/Kinetic energy of the flow
$D_f g_c/\frac{1}{2}\rho V^2 D^2$	Drag coefficient	C_D	Drag force/Kinetic energy or inertia of the flow
$V^2/c_p(T_w - T_\infty)g_c$	Eckert number	Ec	Kinetic energy of flow/ Enthalpy difference
$\alpha t/L^2$	Fourier number	Fo	Heat conduction rate/Thermal-energy storage
$\Delta p D/\frac{1}{2}\rho V^2 L$	Friction factor	f	Pressure drop/Kinetic energy of the flow
$F g_c/\frac{1}{2}\rho V^2 D^2$	Force coefficient	C_F	Force (any)/Kinetic energy or inertia of the flow
V^2/gL	Froude number	Fr	Inertia of flow/Gravity force
$g\beta(T_w - T_\infty)L^3/v^2$	Grashof number	Gr	Buoyancy force/Viscous force
$L_f g_c/\frac{1}{2}\rho V^2 D^2$	Lift coefficient	C_L	Lift force/Kinetic energy of the flow
V/a	Mach number	M	Inertia of flow at V/Inertia of flow at α
$\bar{h}_c L/k_f$	Nusselt number	Nu	Convection heat transfer/ Conduction heat transfer in a fluid
VL/α	Peclet number	Pe = (Re)(Pr)	Inertia of flow/Diffusivity of fluid
$c_p \mu/k = v/\alpha$	Prandtl number	Pr	Kinematic viscosity/Diffusivity for a fluid
$\Delta p g_c/\frac{1}{2}\rho V^2$	Pressure coefficient	C_p	Pressure forces/Inertia of flow
$g\beta(T_w - T_\infty)L^3/v\alpha$	Rayleigh number	Ra = (Gr)(Pr)	Buoyant forces/Viscosity × Diffusivity of the fluid
$\rho V D/\mu g_c$	Reynolds number	Re	Inertia/Viscous forces
$\bar{h}_c/\rho V c_p$	Stanton number	St = Nu/(Re)(Pr)	Convection coefficient/ Inertia of flow
$\rho V^2 L/\sigma g_c$	Weber number	We	Inertia of flow/Surface-tension forces

Using the following velocity profile for turbulent flow of a Newtonian fluid in a tube,

$$\frac{V_z}{V_{z\,max}} = (1 - r/R)^{1/7}$$

derive an equation for the average velocity. Graph the above equation for $V_z/V_{z\,max}$ vs. r/R.

21. For laminar flow down an incline, the velocity distribution is

$$V_z = \frac{\rho g \delta^2 \sin\alpha}{2\mu g_c}\left(1 - \frac{x^2}{\delta^2}\right)$$

(a) Determine the average velocity of the flow for a width of b.
(b) Express V_z in terms of the average velocity.

22. Integrate the velocity profile derived in Example 5.7 and determine if the assumption that volume flow upward between the plates equals volume flow downward as assumed.

5.12.3 THE THERMAL ENERGY EQUATION

23. Example 5.7 deals with a natural-convection problem in which fluid between parallel plates moves under the action of buoyant forces caused by a temperature difference between the plates and the fluid. Graph temperature vs. x and velocity vs. x. Use the following conditions:
    Fluid is water
    $T_1 = 40°C$
    $T_2 = 0°C$
    $b = 4$ cm

24. Repeat Problem 23 for air with $T_1 = 40°C$, $T_2 = 0°C$, and $b = 4$ cm.

25. Consider a cooled Newtonian liquid film falling down a vertical wall, as shown in Figure P5.2. The wall is at an elevated temperature T_s. Derive the appropriate continuity, momentum, and energy equations for the film (do not solve). Write also all the boundary conditions. List all assumptions.

26. Consider a fluid placed in a chamber similar in shape to a coffee percolator, as shown in Figure P5.3. The center tube is of temperature T_1 while the outer wall is at T_2, with $T_1 > T_2$. Fluid moves due to convective effects. Assuming two-dimensional, steady, laminar flow of a Newtonian fluid,
    (a) Show that the applicable continuity and momentum equations are
        continuity

$$\frac{\partial V_z}{\partial z} = 0$$

z component, Navier-Stokes equations

$$0 = -\frac{\partial p}{\partial z} + \mu\left[\frac{1}{r}\frac{\partial}{\partial r}\left(r\frac{\partial V_z}{\partial r}\right)\right] + \frac{\rho g_z}{g_c}$$

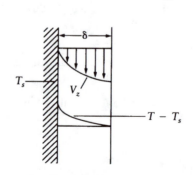

FIGURE P5.2

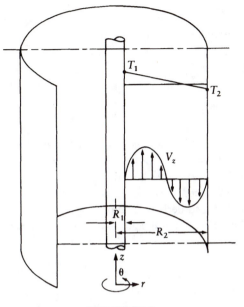

FIGURE P5.3

(b) Write the boundary conditions for the above equations.
(c) Neglecting axial conduction and viscous dissipation, and assuming temperature varies only with r, show that the energy equation for the system is

$$0 = k_f \left[\frac{1}{r} \frac{\partial}{\partial r} \left(r \frac{\partial T}{\partial r} \right) \right]$$

(d) Write the boundary conditions for the above equation.
(e) Solve the energy equation subject to the boundary conditions and show that

$$\frac{T - T_2}{T_1 - T_2} = \frac{\ln(r/R_2)}{\ln(R_1/R_2)}$$

(f) Selecting T_2 as the temperature at which fluid properties are evaluated, show that the z component of the momentum equation becomes

$$\frac{\rho_2 \beta g_z}{g_c}(T - T_2) = \frac{\mu}{r} \frac{\partial}{\partial r} \left(r \frac{\partial V_z}{\partial r} \right)$$

(g) Substitute from the temperature profile to obtain

$$\frac{r \rho_2 \beta g_z}{\mu g_c}(T_1 - T_2) \frac{\ln(r/R_2)}{\ln(R_1/R_2)} = \frac{\partial}{\partial r} \left(r \frac{\partial V_z}{\partial r} \right)$$

Integrate twice and show that the result is

$$V_z = \frac{\rho_2 \beta g_z (T_1 - T_2) R_2^2}{4 \mu g_c \ln(R_2/R_1)} \left\{ \left(\frac{R_1^2}{R_2^2} - 1 \right) \frac{\ln(r/R_2)}{\ln(R_1/R_2)} + \frac{R_1^2}{R_2^2} \left[\left(\frac{r^2}{R_1^2} - 1 \right) \ln(r/R_2) - \left(1 + \frac{r^2}{R_2^2} \right) \right] + 1 - \frac{r^2}{R_2^2} \right\}$$

5.12.4 Dimensional Analysis

27. Determine the hydraulic diameter for the following cross sections:
    (a) Circular tube of diameter D.
    (b) Square tube of dimensions $s \times s$.
    (c) Rectangular tube of dimensions $H \times W$, where H is a multiple of W:

$$H = nW$$

    (d) Rectangular tube of width W and height H such that $W \ll H$.
    (e) An annulus with OD $= D$ and ID $= \kappa D$, where $0 < \kappa < 1$.
28. Start with Equation 5.63 and derive Equation 5.66.
29. Start with Equation 5.67 and derive Equation 5.69.
30. Start with Equation 5.70 and derive Equation 5.73.
31. The velocity distribution for natural convection between two parallel plates is

$$V_z = \frac{\beta \rho_m g b^2 (T_1 - T_2)}{12 \mu g_c} \left(\frac{x}{b} - \frac{x^3}{b^3} \right)$$

    Express this equation in terms of the following dimensionless groups:

$$\mathrm{Re} = \frac{\rho_m V_z b}{\mu g_c}$$

$$\mathrm{Gr} = \frac{g \beta \Delta T b^3}{v^2}$$

32. The temperature profile for laminar flow of a Newtonian fluid down an incline is

$$\frac{T - T_s}{T_w - T_s} = \frac{\rho^2 g^2 \delta^4 \sin^2 \alpha}{12 \mu g_c^2 k_f (T_w - T_s)} \left(\frac{x^4}{\delta^4} - \frac{x}{\delta} \right) + \frac{x}{\delta}$$

    where viscous dissipation is not neglected. Express this equation in terms of applicable dimensionless groups. (*Hint:* Use the velocity profile

$$V_z = \frac{\rho g \delta^2}{2 \mu g_c} \left(1 - \frac{x^2}{\delta^2} \right)$$

    expressed in terms of dimensionless groups.)

5.12.5 Miscellaneous Problems

33. A Newtonian fluid having constant properties is flowing under laminar conditions through a narrow slit, as shown in Figure P5.4. Both plates are maintained at temperature T_s, which is different from the fluid temperature. Neglecting gravity but including viscous dissipation,
    (a) Show that the z-directed instantaneous velocity is

$$V_z = \frac{\delta^2}{2\mu} \left(-\frac{dp}{dz} \right) \left(1 - \frac{x^2}{\delta^2} \right)$$

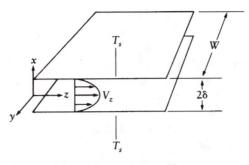

FIGURE P5.4

When viscous dissipation is included in the energy equation for this problem, the energy equation reduces to

$$0 = k_f \frac{\partial^2 T}{\partial x^2} + \mu \left(\frac{\partial V_z}{\partial x} \right)^2$$

Starting with this equation,

(b) Show that the temperature profile is

$$T - T_s = \frac{\mu V_{z\,max}^2}{3 k_f} \left(1 - \frac{x^4}{\delta^4} \right)$$

or

$$\frac{1}{\text{Br}} = \frac{1}{3} \left(1 - \frac{x^4}{\delta^4} \right)$$

(c) Graph $V_z/V_{z\,max}$ vs. x/δ.

(d) Graph $(T - T_s) k_f / \mu V_{z\,max}^2$ vs. x/δ.

34. Couette flow between parallel plates is illustrated in Figure P5.5. Fluid is placed between two plates. The upper plate at temperature T_2 moves at velocity U while the lower plate at temperature T_1 is stationary. Include viscous dissipation, the energy equation reduces to

$$0 = k_f \frac{\partial^2 T}{\partial y^2} + \mu \left(\frac{\partial V_z}{\partial y} \right)^2$$

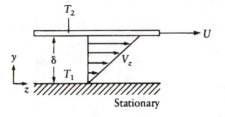

FIGURE P5.5

Derive equations for the velocity and temperature profiles. Do not neglect viscous dissipation and assume steady laminar flow of a Newtonian fluid. Neglect gravity and take the pressure gradients to be zero.

35. Couette flow between cylinders is illustrated in Figure P5.6. A fluid is placed in the annulus formed by two concentric cylinders. The inner cylinder is maintained at temperature T_1 and rotates at an angular velocity of ω. The outer cylinder is maintained at T_2 and is stationary. Assuming steady laminar flow of a Newtonian liquid, derive equations for velocity and temperature of the fluid. Neglect viscous dissipation. Gravity and all pressure gradients can be neglected.

36. Consider laminar flow down an incline in which the free surface of the liquid is maintained at temperature T_s and the solid surface at T_w, as shown in Figure P5.7. Including viscous dissipation, the energy equation is:

$$0 = k_f \frac{\partial^2 T}{\partial x^2} + \mu \left(\frac{\partial V_z}{dx} \right)^2$$

(a) For a steady laminar flow of a liquid with constant properties and with velocity and temperature profiles fully developed, show that the temperature profile is given by

$$\frac{T - T_s}{T_w - T_s} = \frac{\rho^2 g^2 \delta^2 \sin^2 \alpha}{12 \mu k_f (T_w - T_s) g_c} \left(\frac{x}{\delta} - \frac{x^4}{\delta^4} \right) + \frac{x}{\delta}$$

for the case where viscous dissipation is not neglected.

(b) Differentiate the temperature profile given above to determine the heat flux.

(c) Show that the maximum temperature rise exists for a given value of x given by

$$x = \left[\frac{(T_s - T_w) 3 \mu k_f g_c^2}{\rho^2 g^2 \delta \sin^2 \alpha} + \frac{\delta^3}{4} \right]^{1/3}$$

(d) Show that the average temperature rise T_{avg} is given by

$$\frac{T_{avg} - T_s}{T_w - T_s} = \frac{1}{2} + \frac{\rho^2 g^2 \delta^4 \sin^2 \alpha}{40 \mu k_f (T_w - T_s) g_c^2}$$

(e) Show that the bulk-fluid temperature T_b is given by

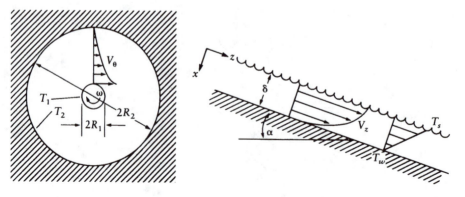

FIGURE P5.6   **FIGURE P5.7**

$$\frac{T_b - T_s}{T_w - T_s} = \frac{3}{8} + \frac{27}{1120} \frac{\rho^2 g^2 \delta^4 \sin^2 \alpha}{\mu k_f (T_w - T_s) g_c^2}$$

(f) Graph T vs. x for the case where the fluid is glycerine and fluid properties are evaluated at 25°C. Let $\alpha = 5°$ and $\delta = 1$ cm. Calculate also T_{avg} and T_b. Assume a wall temperature of 30°C and a liquid surface temperature of 20°C.

(g) Based on calculations made above, does neglecting viscous dissipation have a significant effect?

6 Convection Heat Transfer in a Closed Conduit

CONTENTS

6.1   INTRODUCTION

In this chapter we examine flow of and heat transfer to or from a fluid flowing in a conduit. The problems that can be investigated are numerous and varied. We therefore examine only a few representative cases.

In Chapter 5, considerable attention is devoted to a derivation of general equations, applications to specific problems, and solutions. The result obtained is usually an explicit equation for temperature. The relationship between temperature profile and heat transfer at a bounding surface is defined in terms of the convection coefficient introduced in Newton's Law of Cooling. As we will see, though, the convection coefficient depends on the temperature difference selected. For flow in a conduit, there are several temperature differences of importance, and each has an associated convection coefficient. Consequently, it is necessary to properly define the problem of flow in a closed conduit.

In this chapter, the general problem of flow and heat transfer in a circular conduit is defined. Hydrodynamic and heat-transfer solutions are presented for laminar, and then again for turbulent, flow. Finally, solutions for other cross sections are discussed. There are two objectives in the analysis of heat transfer in closed conduits. One is to relate the heat transferred to the temperature change experienced by the fluid. The second is to relate the heat transferred to the fluid to the difference in temperature between the wall and the fluid. These objectives are realized for numerous problems, a sampling of which are included in this chapter.

6.2   HEAT TRANSFER TO AND FROM LAMINAR FLOW IN A CIRCULAR CONDUIT

Consider a reservoir from which fluid flows under laminar conditions into a circular duct, as shown in Figure 6.1. The duct entrance is well rounded. At the tube entrance the velocity is constant and uniform. As the fluid moves along the duct, the velocity profile changes from being a constant at inlet to being what is called *fully developed* downstream. The axial distance required for the profile to become fully developed or nonchanging is called the *hydrodynamic entrance length* z_e. The

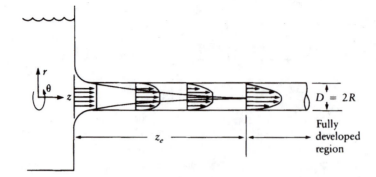

FIGURE 6.1   Laminar flow in a circular duct from constant to fully developed velocity profiles.

entrance length for fully developed flow is a somewhat controversial issue. In terms of the Reynolds number, the entrance length can be written as

$$\frac{z_e}{D} \cong 0.05 \mathrm{Re}_D = 0.05\frac{\rho VD}{\mu g_c} = 0.05\frac{VD}{\nu} \quad \text{(laminar flow)} \tag{6.1}$$

where D is the duct diameter, V is the average flow velocity, and ρ and μ are the fluid density and absolute viscosity, respectively. The constant 0.05 indicates that the shear stress on the pipe wall is within 1.4% of the value at $z = \infty$. If the constant is 0.03, the wall shear stress is within 5% of the fully developed value at $z = \infty$. We shall use 0.05 in this text. From Chapter 5, the velocity profile for fully developed laminar flow of a Newtonian fluid in a circular duct is given by

$$V_z = \left(-\frac{dp}{dz}\right)\left(\frac{R^2}{4\mu}\right)\left(1 - \frac{r^2}{R^2}\right) \tag{6.2}$$

or

$$V_z = V_{z\,\mathrm{max}}\left(1 - \frac{r^2}{R^2}\right)$$

The wall shear stress is

$$\tau_w = \tau_{rz}\big|_{r=R} = +\mu\frac{dV_z}{dr}\bigg|_{r=R} = \left(-\frac{dp}{dz}\right)\left(\frac{R}{2}\right) \tag{6.3}$$

The average velocity is given by

$$V = \frac{R^2}{8\mu}\left(-\frac{dp}{dz}\right) \tag{6.4}$$

Laminar flow exists when the Reynolds number is less than 2100:

$$\mathrm{Re}_D = \frac{VD}{\nu} < 2100$$

Although transition can occur over a range of Reynolds numbers, we will use 2100 as a specific value with which to work.

6.2.1 CONSTANT HEAT FLUX AT THE WALL

The heat-transfer problem we consider first is one in which the heat flux q/A_s at the duct wall is a constant. This can be achieved experimentally by passing an electric current through a metal pipe, for example, or by wrapping the duct with a material through which an electric current is passed (see Figure 6.2). The resulting temperature-profile variation with axial distance is illustrated in Figure 6.3. In this case, the velocity profile and the temperature profile both begin changing at the tube entrance. The fluid temperature at inlet is T_∞, the tank-fluid temperature. As fluid travels downstream, the temperature begins to vary within the tube. At any axial location, T varies from T_w, the wall temperature, to some value at the centerline. Just as the velocity profile eventually becomes fully developed, so does the temperature profile. The temperature profile itself seems eventually to reach a nonchanging shape, but with the constant heat input that exists, the difference $T - T_\infty$ increases with increasing axial distance.

The term *fully developed,* when applied to the velocity, referred to a nonchanging profile that was attained by the fluid. Mathematically, we define a fully developed velocity profile as

$$\frac{d(V_z/V_{z\,max})}{dz} = 0 \tag{6.5}$$

This equation is valid at each radial location in the cross section. For the heat-transfer problem a fully developed profile is one in which

$$\frac{d[(T_w - T)/(T_w - T_b)]}{dz} = 0 \tag{6.6}$$

where T_w is the wall temperature and T_b is the bulk-fluid temperature defined in Chapter 5 for a circular conduit as

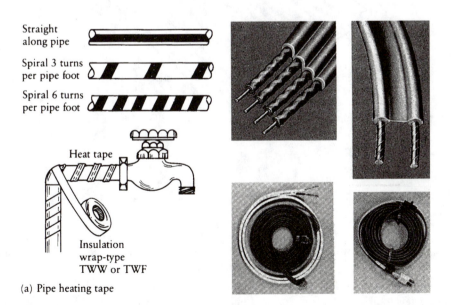

(a) Pipe heating tape

FIGURE 6.2a Commercial devices used in producing a constant wall flux. *Source:* courtesy of Emerson-Chromolox Div. of Emerson Electric Co.

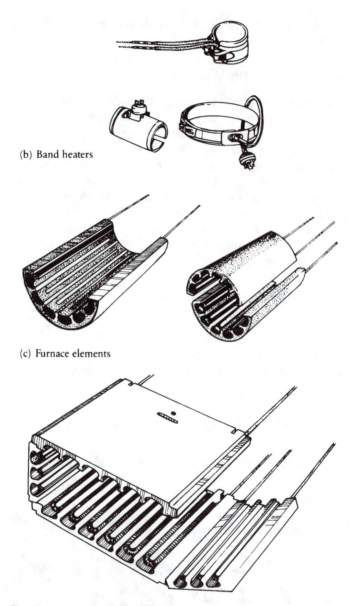

(b) Band heaters

(c) Furnace elements

FIGURE 6.2b and c   *Source:* courtesy of Thermal Corp., Huntsville, Alabama.

$$T_b = \frac{\iint \rho V_z c_p T \ dA}{\iint \rho V_z c_p \ dA} = \frac{\int_0^{2\pi} \int_0^R \rho V_z c_p T r \ dr \ d\theta}{\int_0^{2\pi} \int_0^R \rho V_z c_p r \ dr \ d\theta} \tag{6.7}$$

Note that the definition of the bulk temperature applies to either laminar or turbulent flow. The thermal entry length z_t is shown in Figure 6.3. The wall and bulk temperatures T_w and T_b vary only with z, while the fluid temperature T varies with z and r.

Figure 6.3 shows also the variation of the wall temperature and the bulk temperature with increasing z. The bulk temperature increases linearly with distance. The wall temperature changes over the thermal entry length z_t. Beyond z_t the wall- and bulk-temperature lines are parallel. The entry length required for the flow to be thermally developed is given by

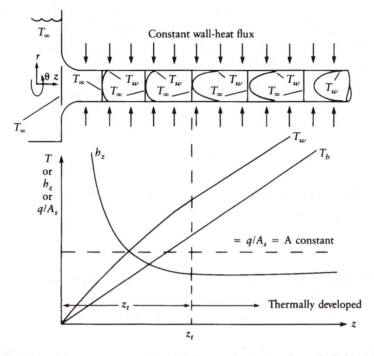

FIGURE 6.3 Variation of flow parameters with axial distance for laminar flow of a fluid with constant wall-heat flux, q_w'' .

$$\frac{z_t}{D} = 0.05 \mathrm{Re}_D \, \mathrm{Pr} = 0.05 \frac{VD}{v} \frac{v}{\alpha} = 0.05 \frac{VD}{\alpha}$$

$$z_t = 0.05 \frac{VD^2}{\alpha} \qquad \text{(laminar flow)} \qquad (6.8)$$

Comparing Equations 6.1 and 6.8 reveals that, for fluids with Prandtl numbers greater than 1, the flow develops hydrodynamically more rapidly than it does thermally. If Pr < 1, the equations predict that the flow develops thermally first. For substances with high Prandtl numbers, such as oils (Pr > 100), the flow develops hydrodynamically so much more quickly that a fully developed velocity profile can be reasonably assumed throughout. For some oils, the Prandtl number is over 1 000 and, in this case, the temperature profile might never become fully thermally developed. The flow must be fully developed hydrodynamically before it can be fully developed thermally.

As evident from Equation 6.6, the temperature difference between the wall and the fluid $(T_w - T_\infty)$ is an important quantity. It is possible to use this difference in Newton's Law of Cooling to define what is called a local heat-transfer coefficient h_z:

$$\frac{q}{A_s} = h_z(T_w - T_b)$$

The local coefficient h_z is somewhat different from what we have been using thus far: $\bar{h}_c$. The subscript c is present to indicate that convection is taking place, and the overbar indicates that the coefficient is an average of some kind. If we integrate the local coefficient h_z over a definite area (such as the surface area A_s), the integrated result would be written with an overbar.

If the wall-minus-bulk temperature difference varies with z, then the local coefficient h_z for constant wall flux will also vary with axial distance. This is the case in the entrance region, where the flow is not fully developed. In the thermally developed region, however, where the T_w vs. z and

T_b vs. z lines (Figure 6.3) are parallel, the difference $(T_w - T_b)$ is a constant. With the heat flux q/A_s constant, we conclude from Newton's Law that the local coefficient h_z is constant in the thermally developed region. The variation of h_z with z is shown in Figure 6.3. Note that at the entrance of the tube the local coefficient h_z is infinite.

To obtain a relationship between the fluid bulk-temperature change and the heat flux, consider the incremental volume element shown in Figure 6.4. We perform a heat balance for the element, which is dz wide. The wall-heat flux is

$$q_w'' = \frac{dq}{dA_s} = \frac{dq}{P\,dz}$$

where q_w'' has dimensions of energy/(area·time) (W/m² or BTU/hr·ft²) and P is the duct perimeter. The heat added then is

$$dq = q_w'' P\,dz$$

The energy added to the fluid goes into increasing the enthalpy h by an amount given as

$$dq = \dot{m}\,dh = \rho A V c_p dT_b$$

where A is the cross-sectional area and $\dot{m}$ is the mass flow rate of fluid. For a circular tube, $P = 2\pi R$ and $A = \pi R^2$. Substituting and setting the two above equations for dq equal gives, for the axial variation of bulk temperature,

$$2\pi R q_w''\,dz = \rho\pi R^2 V c_p\,dT_b$$

$$\frac{dT_b}{dz} = \frac{2q_w''}{\rho c_p V R} = \frac{2q_w'' k_f}{\rho c_p V R k_f}$$

or

$$\frac{dT_b}{dz} = \frac{2q_w'' \alpha}{V k_f R} \qquad (6.9a)$$

where k_f is the thermal conductivity of the fluid. The slope of the bulk-temperature line (T_b vs. z in Figure 6.3) is thus a constant. Therefore, T_b varies linearly with z, a result that is independent

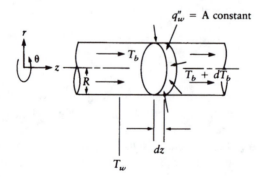

FIGURE 6.4   A control volume for determining axial variation of bulk temperature.

of whether fully developed conditions exist. Integrating Equation 6.9a from inlet T_{bi} to outlet T_{bo} as limits on temperature and 0 to L as limits on z, we obtain

$$T_{bo} - T_{bi} = \frac{2q_w'' \alpha L}{V k_f R} \tag{6.9b}$$

This equation applies to laminar- or turbulent-flow regimes.

To obtain an equation for temperature in the thermally developed region, we apply the energy equation (Table 5.3). Canceling terms, we are left with

$$V_z \frac{\partial T}{\partial z} = \frac{\alpha}{r} \frac{\partial}{\partial r}\left(r \frac{\partial T}{\partial r}\right) \tag{6.10}$$

We use the criterion for fully developed flow to evaluate $\partial T / \partial z$. From Equation 6.6

$$0 = \frac{\partial[(T_w - T)/(T_w - T_b)]}{\partial z}$$

$$0 = \frac{(T_w - T_b)\left(\dfrac{\partial T_w}{\partial z} - \dfrac{\partial T}{\partial z}\right) - (T_w - T)\left(\dfrac{\partial T_w}{\partial z} - \dfrac{\partial T_b}{\partial z}\right)}{(T_w - T_b)^2}$$

Simplifying and rearranging, we obtain

$$\frac{\partial T}{\partial z} = \frac{\partial T_w}{\partial z} + \frac{T_w - T}{T_w - T_b} \frac{\partial T_b}{\partial z} - \frac{T_w - T}{T_w - T_b} \frac{\partial T_w}{\partial z} \tag{6.11}$$

From Figure 6.3, it is seen that the wall- and bulk-temperature lines are parallel in the thermally developed region. Another way of stating this is that the slopes of both lines are equal; that is,

$$\frac{\partial T_w}{\partial z} = \frac{\partial T_b}{\partial z} \tag{6.12}$$

in the thermally developed region. Combining with Equation 6.11, we conclude that

$$\frac{\partial T}{\partial z} = \frac{\partial T_w}{\partial z} \tag{6.13}$$

Furthermore, from Equations 6.12 and 6.9a

$$\frac{\partial T}{\partial z} = \frac{\partial T_w}{\partial z} = \frac{\partial T_b}{\partial z} = \frac{2q_w'' \alpha}{V k_f R} \tag{6.14}$$

With all parameters in the above equation constant, it is seen that $\partial T / \partial z$ is constant in the thermally developed region.

We now return to the differential equation and substitute for velocity from Equation 6.2 and for $\partial T / \partial z$ from Equation 6.14 to obtain

$$\frac{\alpha}{r} \frac{\partial}{\partial r}\left(r \frac{\partial T}{\partial r}\right) = 2V\left(1 - \frac{r^2}{R^2}\right)\frac{2q_w'' \alpha}{V k_f R} \tag{6.15}$$

Simplifying, rearranging, and integrating once gives

$$\frac{d}{dr}\left(r\frac{dT}{dr}\right) = \frac{4q_w''}{k_fR}\left(r - \frac{r^3}{R^2}\right)$$

$$r\frac{dT}{dr} = \frac{4q_w''}{k_fR}\left(\frac{r^2}{2} - \frac{r^4}{4R^2}\right) + C_1$$

Dividing by r and integrating again gives

$$\frac{dT}{dr} = \frac{4q_w''}{k_fR}\left(\frac{r}{2} - \frac{r^3}{4R^2}\right) + \frac{C_1}{r}$$

$$T = \frac{4q_w''}{k_fR}\left(\frac{r^2}{4} - \frac{r^4}{16R^2}\right) + C_1\ln(r) + C_2 \tag{6.16}$$

To evaluate the constants of integration, C_1 and C_2, we apply the boundary conditions:

B.C.1. $r = 0$, T = finite or $dT/dr = 0$
B.C.2. $r = R$, $T = T_w$

The first boundary condition, when applied, yields

$$C_1 = 0$$

The natural logarithm function becomes infinite at $r = 0$. The second boundary condition gives

$$T_w = \frac{4q_w''}{k_fR}\left(\frac{R^2}{4} - \frac{R^4}{16R^2}\right) + C_2$$

or

$$C_2 = T_w - \frac{4q_w''}{k_fR}\left(\frac{3R^2}{16}\right)$$

Substituting into Equation 6.16 gives

$$T - T_w = \frac{4q_w''}{k_fR}\left(\frac{r^2}{4} - \frac{r^2}{16R^2} - \frac{3R^2}{16}\right)$$

Simplifying,

$$T - T_w = \frac{4q_w''}{k_fR}\frac{R^2}{4}\left(\frac{r^2}{R^2} - \frac{r^4}{4R^4} - \frac{3}{4}\right)$$

$$T_w - T = \frac{q_w''R}{k_f}\left(\frac{3}{4} - \frac{r^2}{R^2} - \frac{1}{4}\frac{r^4}{R^4}\right) \tag{6.17}$$

The bulk temperature may now be determined by substituting into Equation 6.7 and performing the indicated integration:

$$T_b = \frac{2\pi \int_0^R \left(\rho(2V)\left[1 - \left(\frac{r}{R}\right)^2\right]\left[\frac{q_w'' R}{k_f}\left(\frac{r^2}{R^2} - \frac{r^4}{4R^4} - \frac{3}{4}\right)\right] - T_w\right) r \, dr}{2\pi \int_0^R \rho(2V)\left[1 - \left(\frac{r}{R}\right)^2\right] r \, dr}$$

Assuming constant properties, we obtain

$$T_b = T_w - \frac{11}{24}\frac{q_w'' R}{k_f} \tag{6.18}$$

In terms of duct diameter $D = 2R$, we get

$$T_w - T_b = \frac{11}{48}\frac{q_w'' D}{k_f} \tag{6.19}$$

To check on the original assumption of a thermally developed temperature profile, we find that the ratio $(T_w - T)/(T_w - T_b)$—which is Equation 6.17/Equation 6.19—is indeed not a function of z.

It is possible to further simplify Equation 6.19 by expressing the result in terms of the local coefficient h_z. Recall that the definition of h_z is found in Newton's Law of Cooling as

$$q_w'' = \frac{q}{A_s} = h_z(T_w - T_b)$$

Combining with Equation 6.19 gives

$$T_w - T_b = \frac{11}{48}\frac{h_z(T_w - T_b)D}{k_f}$$

or

$$Nu_D = \frac{h_z D}{k_f} = \frac{48}{11} = 4.364 \tag{6.20}$$

(constant wall flux; hydrodynamically and thermally developed laminar flow)

The Nusselt number based on the local coefficient and the duct diameter is thus a constant in the thermally developed region of the flow. Remember that the heat flux is related to the bulk-fluid temperature change and the fluid properties. The heat-transfer coefficient relates this same heat flux to the difference between the bulk-fluid temperature and the wall temperature.

Example 6.1

A flat-plate solar collector contains a number of copper tubes that are painted black to enhance the absorptance of thermal radiation from the sun. Consider one such copper tube that is 1/2 standard type M and is 3 m long. It conveys ethylene glycol at a volume flow rate of 3.25×10^{-6} m³/s. The liquid enters the tube at a (bulk) temperature of 20°C. Determine the fluid outlet temperature and the tube-wall temperature at the outlet. Take the incident heat flux to be 2 200 W/m².

Solution

The heat flux is incident only on the upper surface of the tube. However, because of various effects such as reflection to the tube from the bottom of the collector, assuming a uniform heat flux over the entire tube is not too bad an approximation. Ethylene glycol or an ethylene glycol-water mix is often used as a working fluid because of its low freezing point.

Assumptions

1. Steady-state flow exists.
2. The fluid has constant properties.
3. The temperature of the tube does not vary in the radial direction.
4. The tube is uniformly heated with a flux of 2 200 W/m².

From Appendix Table C.5 for ethylene glycol at 20°C we read

$$c_p = 2\ 382 \text{ J/(kg·K)} \qquad p = 1.116(1\ 000) \text{ kg/m}^3$$
$$v = 19.18 \times 10^{-6} \text{ m}^2\text{/s} \qquad k_f = 0.249 \text{ W/(m·K)}$$
$$\alpha = 0.939 \times 10^{-7} \text{ m}^2\text{/s} \qquad \text{Pr} = 204$$

We will use these values as a first approximation. It is customary to evaluate properties at a mean temperature such as the average of the inlet and outlet bulk fluid temperatures. This cannot be done yet, because the outlet temperature here is unknown. From Appendix Table F.2 for 1/2 standard type M seamless copper water tubing

$$\text{OD} = 1.588 \text{ cm} \qquad\qquad \text{ID} = 1.446 \text{ cm}$$
$$A = 1.642 \text{ cm}^2$$

The volume flow rate Q is given, and with area known, the average flow velocity can be calculated as

$$V = \frac{Q}{A} = \frac{3.25 \times 10^{-6}}{1.642 \times 10^{-4}} = 0.02 \text{ m/s} = 2 \text{ cm/s}$$

Before we can apply the equations of this section, we must first determine if the flow is laminar. The Reynolds number is

$$\text{Re}_D = \frac{VD}{v} = \frac{0.02(0.014\ 46)}{19.18 \times 10^{-6}} = 15.1$$

The flow is laminar because $\text{Re}_D < 2100$. The hydrodynamic entry length is next calculated to be

$$z_e = 0.05D\ \text{Re}_D = 0.05(0.014\ 46)(15.1)$$
$$z_e = 0.011 \text{ m} = 1.1 \text{ cm}$$

Equation 6.9 relates the bulk-temperature difference to the pertinent variables

$$T_{bo} - T_{bi} = \frac{2q_w'' \alpha L}{V k_f R}$$

All parameters are known. Rearranging and substituting gives the bulk-fluid outlet temperature as

$$T_{bo} = T_{bi} + \frac{2q_w'' \alpha L}{V k_f R} = 20 + \frac{2(2\,200)(0.939 \times 10^{-7})(3)}{(0.02)(0.249)\left(\frac{0.014\,46}{2}\right)}$$

Solving,

$$T_{bo} = 54.4°C$$

This result is based on fluid properties evaluated at 20°C. As a second approximation the properties can be reevaluated at $(54.4 + 20)/2 = 37.2°C$ and the calculations repeated. This is reserved as an exercise.

The wall temperature at the outlet can be calculated with Equation 6.19 applied at the outlet provided thermally developed flow exists there, which is the case if Equation 6.8 is satisfied. The thermal entry length is given by

$$z_t = 0.05D \, \mathrm{Re}_D \, \mathrm{Pr} \tag{6.8}$$

Substituting,

$$z_t = 0.05(0.014\,46)(15.1)(204) = 2.2 \text{ m}$$

The thermal entry length is greater than the 1.1-cm hydrodynamic entry length. The tube in this example is 3 m long, so thermally developed flow exists at the outlet. Equation 6.19 applied at the tube outlet is

$$T_{wo} - T_{bo} = \frac{11}{48} \frac{q_w'' D}{k_f}$$

Rearranging and substituting gives

$$T_{wo} = 54.4 + \frac{11}{48} \frac{(2\,200)(0.014\,46)}{0.249}$$

Solving,

$$T_{wo} = 83.7°C$$

Remember that these results are a first approximation because they are based on flow properties evaluated at the fluid inlet temperature. ❏

6.2.2 CONSTANT WALL TEMPERATURE

The next heat-transfer problem we consider is one in which the temperature of the duct wall is constant, and heat is transferred to or from a fluid flowing within the tube to the tube wall. The hydrodynamic details of the flow are the same as those discussed earlier for the constant-wall-flux case. Equations 6.1 through 6.4 are for entrance length, velocity, and shear stress for laminar flow of a Newtonian fluid. Note that it is not possible to maintain both a constant heat flux and a constant wall temperature.

A constant wall temperature can be achieved experimentally, for example, by allowing a fluid to condense on the outside surface of the tube. The process of condensation (and of vaporization) is isothermal for pure substances. The problem we consider is one where fluid at temperature T_∞ is allowed to flow from a reservoir through a circular duct of wall temperature T_w ($> T_\infty$). The resulting variation of temperature profile with axial distance is illustrated in Figure 6.5. As shown, the bulk temperature T_b changes from inlet and approaches a constant value equal to T_w. A fully developed profile is not easily discerned, so once again we rely on the definition given in Equation 6.6:

$$\frac{\partial}{\partial z}\left(\frac{T_w - T}{T_w - T_b}\right) = 0 \tag{6.6}$$

This equation is satisfied only when the flow is fully developed thermally.

The bulk temperature variation with axial distance is shown also in Figure 6.5, as is the thermal entry length. Experiments have shown that the thermal entry length for this case is given by Equation 6.8:

$$z_t = 0.05 D\, \mathrm{Re}_D\, \mathrm{Pr} = 0.05 \frac{VD}{\nu}\frac{\nu}{\alpha} = 0.5\frac{VD}{\alpha}$$

or

$$z_t = 0.05 \frac{VD^2}{\alpha} \quad \text{(laminar flow)} \tag{6.8}$$

Note that the bulk temperature continues to approach the wall temperature with increasing z. The heat flux decreases with increasing z.

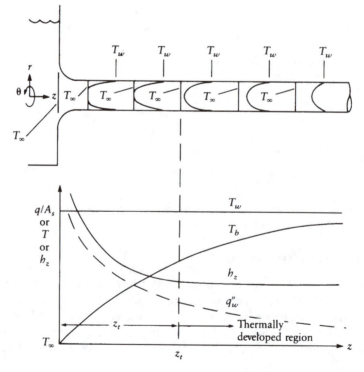

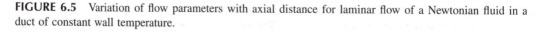

FIGURE 6.5 Variation of flow parameters with axial distance for laminar flow of a Newtonian fluid in a duct of constant wall temperature.

We can define a local convective heat-transfer coefficient h_z with Newton's Law:

$$q_w'' = \frac{q}{A_s} = h_z(T_w - T_b) \tag{6.21}$$

The wall-minus-bulk temperature difference decreases with z, but so does the heat flux q/A_s. The combined effect is that the local coefficient h_z becomes constant in the thermally developed region. The variation of h_z with z is shown also in Figure 6.5.

Consider the control volume sketched in Figure 6.6. For a circular duct the heat convected into the fluid is given by

$$dq = q_w'' P \, dz = q_w'' (2\pi R) dz$$

The added heat causes the bulk-fluid temperature to increase by dT_b. Thus, for a circular tube,

$$dq = \dot{m} \, dh = \rho A V c_p \, dT_b = \rho \pi R^2 V c_p \, dT_b \tag{6.22}$$

Equating both expressions for dq gives

$$\rho \pi R^2 V c_p dT_b = q_w'' (2\pi R) dz$$

or

$$\frac{dT_b}{dz} = \frac{2q_w'' \alpha}{V k_f R} \tag{6.23}$$

When the wall flux is constant, Equation 6.23 is directly integrable. In the case of constant wall temperature, however, the equation must be modified before we can integrate it. Substituting for q_w'' from Equation 6.21 gives

$$\frac{dT_b}{dz} = \frac{2\alpha h_z}{V k_f R}(T_w - T_b)$$

With wall temperature constant, $dT_w/dz = 0$, so we can write

$$\frac{d(T_w - T_b)}{dz} = \frac{dT_w}{dz} - \frac{dT_b}{dz} = -\frac{dT_b}{dz}$$

This result, combined with the previous equation, gives

$$-\frac{d(T_w - T_b)}{dz} = \frac{2\alpha h_z}{V k_f R}(T_w - T_b)$$

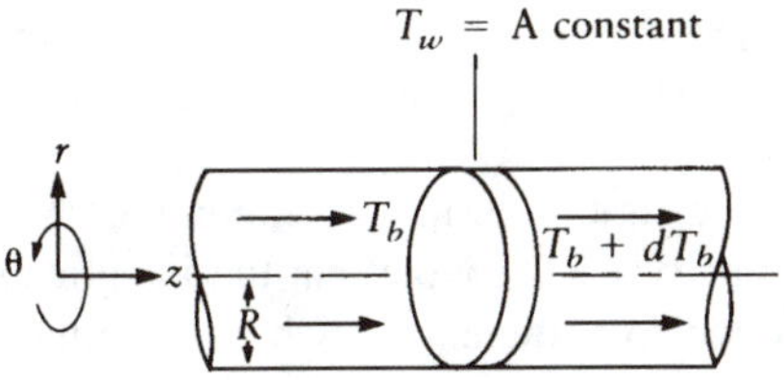

$T_w = $ A constant

FIGURE 6.6 Control volume for flow in a tube of constant wall temperature.

Separating variables and integrating gives

$$\int_i^0 -\frac{d(T_w - T_b)}{(T_w - T_b)} = \frac{2\alpha}{V k_f R} \int_0^L h_z dz$$

$$\ln\left[\frac{(T_w - T_b)_o}{(T_w - T_b)_i}\right] = -\frac{2\alpha L \bar{h}_L}{V k_f R}$$

The limits on the temperature integral indicate an inlet i and outlet o temperature difference between the wall and bulk-fluid temperatures. The convection coefficient $\bar{h}_L$ is the average for the entire tube of length L. The above equation can be rewritten as

$$\frac{T_w - T_{bo}}{T_w - T_{bi}} = \exp\left(-\frac{2\alpha L \bar{h}_L}{V k_f R}\right) \qquad (6.24a)$$

where T_w is constant $(T_w$ at inlet equals that at outlet) and the bulk-fluid temperature at the inlet is the tank-fluid temperature. More generally, we can write

$$\frac{T_w - T_{bz}}{T_w - T_{bi}} = \exp\left(-\frac{2\alpha z \bar{h}_z}{V k_f R}\right) \qquad (6.24b)$$

where T_{bz} is the bulk-fluid temperature at any axial location and $\bar{h}_z$ is the average convection coefficient from inlet to location z.

The total heat transferred to the fluid can be calculated by integrating Equation 6.22 from inlet to outlet to obtain

$$q = \dot{m} c_p \Delta T = \rho A V c_p (T_{bo} - T_{bi})$$

Adding and subtracting T_w gives

$$q = \rho A V c_p[(T_w - T_{bo}) - (T_w - T_{bi})]$$

Solving Equation 6.24 for the average velocity V and substituting, we get

$$V = \frac{2\alpha L \bar{h}_L}{k_f R} \frac{1}{\ln[(T_w - T_{bo})/(T_w - T_{bi})]}$$

$$q = \rho \pi R^2 c_p \frac{2\alpha L \bar{h}_L}{k_f R} \frac{[(T_w - T_{bo}) - (T_w - T_{bi})]}{\ln[(T_w - T_{bo})/(T_w - T_{bi})]}$$

or, finally,

$$q = \bar{h}_L A_s \frac{[(T_w - T_{bo}) - (T_w - T_{bi})]}{\ln[(T_w - T_{bo})/(T_w - T_{bi})]} \qquad (6.25)$$

where the surface area $A_s = 2\pi R L$ for the circular duct. The temperature term is commonly known as the *log-mean temperature difference* and is often abbreviated as ΔT_{lm}. Note that the integrated coefficient $\bar{h}_L$ is defined in terms of ΔT_{lm}, in contrast to the local coefficient $\bar{h}_z$, which is defined in terms of $(T_w - T_b)$. Equation 6.25 applies to laminar flow of a fluid in a duct having a constant wall temperature. The equations and results obtained thus far do not require that fully developed conditions exist within the duct.

To obtain an equation for temperature in the thermally developed region, we apply the energy equation (Table 5.3). Canceling terms, we are left with

$$V_z \frac{\partial T}{\partial z} = \frac{\alpha}{r} \frac{\partial}{\partial r}\left(r \frac{\partial T}{\partial r}\right) \tag{6.10}$$

We have laminar fully developed flow with constant wall temperature. The condition to be satisfied for thermally developed flow is

$$\frac{\partial}{\partial z}[(T_w - T)/(T_w - T_b)] = 0$$

Evaluating the derivative for the axial variation of temperature (developed earlier), we get

$$\frac{\partial T}{\partial z} = \frac{\partial T_w}{\partial z} + \frac{T_w - T}{T_w - T_b} \frac{\partial T_b}{\partial z} - \frac{T_w - T}{T_w - T_b} \frac{\partial T_w}{\partial z} \tag{6.11}$$

For constant wall temperature $\partial T_w/\partial z = 0$. Equation 6.11 then reduces to

$$\frac{\partial T}{\partial z} = \frac{T_w - T}{T_w - T_b} \frac{\partial T_b}{\partial z} \tag{6.26}$$

Substituting Equations 6.26, 6.2, and 6.4 for V_z into Equation 6.10 gives

$$2V\left(1 - \frac{r^2}{R^2}\right)\frac{T_w - T}{T_w - T_b} \frac{\partial T_b}{\partial z} = \frac{\alpha}{r} \frac{\partial}{\partial r}\left(r \frac{\partial T}{\partial r}\right) \tag{6.27}$$

This equation cannot be integrated directly because temperature T varies with r. The other parameters either are constant or are independent of r. The boundary conditions are

B.C.1. $r = R$, $T = T_w$
B.C.2. $r = 0$, $\partial T/\partial r = 0$ or T is finite

One way to solve Equation 6.27 is by using a successive approximation method. We could first substitute a polynomial equation (such as that obtained for constant wall flux) for the dimensionless temperature $(T_w - T)/(T_w - T_b)$. The result is directly integrable to yield a new temperature profile, and the Nusselt number is calculated numerically. The new profile is then substituted back into Equation 6.27 for the dimensionless temperature. The result is then integrated to obtain another profile and another numerical value for the Nusselt number. This procedure is repeated until the Nusselt number approaches a limiting value. For the case of constant wall temperature in the thermally developed region, the result for the Nusselt number is

$$\mathrm{Nu}_D = \frac{h_z D}{k_f} = 3.658 \tag{6.28}$$

(constant wall temperature; hydrodynamically and thermally developed laminar flow)

The solution for temperature (not given here) is an infinite series.

Example 6.2

Figure 6.7 illustrates a device customarily known as a double-pipe heat exchanger. It consists of a tube within a tube. Usually, copper tubing is used and soldered together at the fittings. One fluid flows through the annular space and a second through the inner tube. The outside of the outer tube is well insulated and, with the fluids at different temperatures, there is a net heat transfer through the wall of the inner tube. A double-pipe heat exchanger could be used to recover heat in a warmed fluid, for example. If the warmed fluid is to be discarded, it can first be passed through an exchanger to transfer heat to another fluid.

Water is heated in a double-pipe heat exchanger. Water passes through the inner tube, and steam passes through the annulus. The steam is saturated and condenses at a temperature of 240°F on the outer surface of the inner tube. The water is to be heated from an inlet temperature of 70°F to an outlet temperature of 140°F. The inner tube is 1 standard type M copper tubing. The exchanger is 20 ft long. Calculate the average convection coefficient if the mass flow rate of water is 1.5 lbm/s.

Solution

The condensing steam elevates the outside wall temperature of the inner tube to a constant, but the inside surface temperature might be somewhat lower.

Assumptions

1. The inner-wall temperature of the inner tube is at 240°F.
2. Water properties evaluated at the average bulk temperature are constant.

From Appendix Table C.11 for water at an average bulk temperature of $(140 + 70)/2 = 105 \approx 104°F$:

$$\rho = 0.994(62.4) \text{ lbm/ft}^3 \qquad k_f = 0.363 \text{ BTU/(hr·ft·°R)}$$
$$c_p = 0.9980 \text{ BTU/(lbm·°R)} \qquad \alpha = 5.86 \times 10^{-3} \text{ ft}^2/\text{hr}$$
$$v = 0.708 \times 10^{-5} \text{ ft}^2/\text{s} \qquad Pr = 4.34$$

From Appendix Table F.2 for 1 standard type M copper tube,

$$OD = 1.125 \text{ in} \qquad ID = 0.8792 \text{ ft}$$
$$A = 0.006071 \text{ ft}^2$$

The total heat transferred to the water is given by the heat-balance equation:

$$q = \dot{m} c_p (T_{bo} - T_{bi})$$

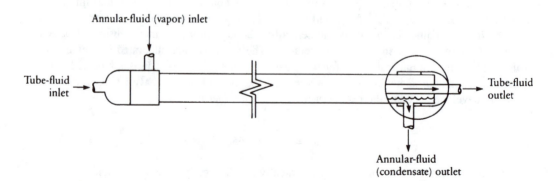

Annular-fluid (vapor) inlet

Tube-fluid inlet

Tube-fluid outlet

Annular-fluid (condensate) outlet

FIGURE 6.7   Control volume for flow in a tube of constant wall temperature.

In terms of the average convection coefficient,

$$q = \bar{h}_L A_s \frac{[(T_w - T_{bo}) - (T_w - T_{bi})]}{\ln[(T_w - T_{bo})/(T_w - T_{bi})]}$$

Equating both expressions for q and solving for the convection coefficient gives

$$\bar{h}_L \pi D L \frac{(T_{bi} - T_{bo})}{\ln[(T_w - T_{bo})/(T_w - T_{bi})]} = \rho \frac{\pi D^2}{4} V c_p (T_{bo} - T_{bi})$$

or

$$\bar{h}_L = -\frac{\rho V D c_p}{4L} \ln[(T_w - T_{bo})/(T_w - T_{bi})] \qquad (i)$$

The mass flow rate is given as 1.5 lbm/s. The velocity then is

$$V = \frac{\dot{m}}{\rho A} = \frac{1.5}{0.994(62.4)(0.006071)} = 3.98 \text{ ft/s} = 14300 \text{ ft/hr}$$

Substituting into Equation (i) above gives

$$\bar{h}_L = -\frac{0.994(62.4)(0.08792)(14300)(0.9980)}{4(20)} \ln\left(\frac{240 - 140}{240 - 70}\right) = 517 \text{ BTU/(hr·ft}^2\text{·°R)} \qquad \square$$

6.2.3 THE THERMAL ENTRY LENGTH

The previous two formulations lead to the conclusion that the local Nusselt number based on a local convection coefficient is a significant parameter. We can thus perform a dimensional analysis for the heat-transfer problem and obtain data that we believe will predict the Nusselt number. The prediction will be of value in both the fully developed and entrance regions. Beginning with

$$h_z = h_z(V, \rho, \mu, D, g_c, c_p, k_f, z) \qquad (6.29)$$

and performing a dimensional analysis gives

$$\frac{h_z D}{k_f} = h_z\left(\frac{\rho V D}{\mu g_c}, \frac{v}{\alpha}, \frac{z}{D}\right)$$

or

$$Nu_D = Nu_D\left(Re_D, Pr, \frac{z}{D}\right) \qquad (6.30)$$

Solutions to the energy equation for the constant-temperature and constant-wall-flux problems have been obtained. Equation 6.30 indicates how the variables of the solution can be correlated. The results are provided graphically in Figure 6.8, where the Nusselt number is graphed vs. the parameter $z/D \, Re_D \, Pr$. The Graetz number Gz is introduced as being equal to $D \, Re_D \, Pr/z$.

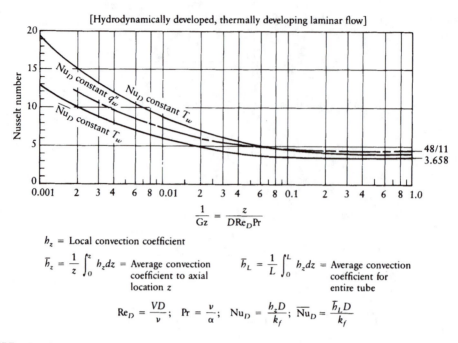

$$\frac{1}{Gz} = \frac{z}{DRe_D Pr}$$

h_z = Local convection coefficient

$\bar{h}_z = \dfrac{1}{z}\displaystyle\int_0^z h_z\,dz$ = Average convection coefficient to axial location z

$\bar{h}_L = \dfrac{1}{L}\displaystyle\int_0^L h_z\,dz$ = Average convection coefficient for entire tube

$$Re_D = \frac{VD}{\nu}; \quad Pr = \frac{\nu}{\alpha}; \quad Nu_D = \frac{h_z D}{k_f}; \quad \overline{Nu}_D = \frac{\bar{h}_L D}{k_f}$$

FIGURE 6.8 Nusselt-number variation with dimensionless distance for heat transfer to or from laminar flow of a Newtonian fluid in a circular duct.

Note that three curves are shown in Figure 6.8. One is the Nusselt-number variation for the case of constant wall temperature discussed earlier. Another is the Nusselt-number variation for the case of constant wall flux, also discussed earlier. The third line is a second case of the Nusselt-number variation for constant wall temperature, in which the Nusselt number is based on the average convection coefficient for the tube:

$$\overline{Nu}_D = \frac{\bar{h}_L D}{k_f} \tag{6.31}$$

As the inverse Graetz number approaches infinity, the average Nusselt number $\overline{Nu}_D$ approaches 3.658. Also, as the inverse Graetz number approaches zero, all Nusselt numbers approach infinity. Thus far we have assumed a thermally developed temperature profile when $1/Gz = 0.05$ or greater, and as seen in the figure, at this point all Nusselt numbers are within 5% of their fully developed values. The graph of Figure 6.8 summarizes results obtained from the solution of the differential equation of energy.

Example 6.3

Milk [$k \approx 0.6$ W/(m·K), $c_p = 3.85$ kJ/(kg·K), $\rho = 1\,030$ kg/m³, $\mu = 2.12 \times 10^{-3}$ N·s/m²]* is pasteurized in a continuous process where it is to be heated from 20° to 71.7°C and maintained at this temperature for 15 s. A hydrodynamically developed flow of milk in a 1/2 standard stainless steel tube moves at an average velocity of 0.1 m/s. An electric resistance heater is placed around the tube and supplies a constant heat flux. Determine the heat flux required to heat the milk for a 6-m heating length. Calculate the entrance length required for the flow to become fully developed hydrodynamically and thermally. Determine the variation of wall temperature with length up to the

---

* Geonkoplis, Christie J., *Transport Processes and Unit Operations*, Allyn & Bacon, Inc., Boston, 1978, pp. 628–629.

point where the flow becomes fully developed. Take the dimensions of 1/2 standard stainless steel tubing to be the same as those of 1/2 standard type K seamless copper water tubing.

Solution

The electric resistance heater supplies a constant wall flux, with the objective being to heat milk from 20° to 71.7°C. Beyond that point other steps can be taken to maintain the milk at the required temperature.

Assumptions

1. Milk is incompressible and has constant properties.
2. Steady-state conditions exist.
3. Stainless steel tubing has the same dimensions as copper water tubing.
4. The stainless tubing wall is thin, and temperature within does not vary in the radial direction.

From Appendix Table F.2 for 1/2 standard type K tubing

$$OD = 1.588 \text{ cm} \qquad ID = 1.340 \text{ cm}$$
$$A = 1.410 \text{ cm}^2$$

We first determine if laminar conditions exist by calculating the Reynolds number:

$$Re_D = \frac{\rho VD}{\mu g_c} = \frac{1\,030(0.1)(0.013\,40)}{2.12\times10^{-3}} = 651$$

Thus, the flow is laminar. The entry length required for hydrodynamically developed conditions to exist is given by Equation 6.1:

$$z_e = 0.05D\,Re_D = 0.05(0.013\,40)(651) = 0.436 \text{ m}$$
$$z_e = 43.6 \text{ cm}$$

The heat balance performed for the tube yields the following:

$$q_w'' PL = \dot{m}c_p\Delta T$$

$$q_w'' \pi DL = \rho\frac{\pi D^2 V}{4}c_p(T_{bo} - T_{bi})$$

Simplifying gives for the heat required

$$q_w'' = \frac{\rho VD c_p}{4L}(T_{bo} - T_{bi})$$

Substituting, we get

$$q_w'' = \frac{(1\,030)(0.1)(0.013\,40)(3\,850)(71.7 - 20)}{4(6)}$$

$$q_w'' = 11\,447 \text{ W/m}^2 \quad (= \text{a constant})$$

The power required is also constant and equal to

$$q = q_w'' A = q_w'' PL = q_w'' \pi DL$$

$$q = 11\ 447\pi(0.013\ 40)(6) = 2\ 891.3 \text{ W}$$

Thus, a 3000-W heater would suffice.

The thermal entry length required is given by

$$z_t = 0.05D \text{ Re}_D \text{ Pr}$$

The Prandtl number is calculated as

$$\text{Pr} = \frac{\nu}{\alpha} = \frac{\mu g_c}{\rho} \frac{\rho c_p}{k_f} = \frac{c_p \mu g_c}{k_f}$$

Substituting and solving,

$$\text{Pr} = \frac{3\ 850(2.12 \times 10^{-3})}{0.6} = 13.6$$

The length required for the flow to be thermally developed is

$$z_t = 0.05(0.013\ 40)(651)(13.6)$$

$$z_t = 5.9 \text{ m}$$

The wall-temperature variation can be determined by using Figure 6.8. For the constant-wall-flux line, we read the data provided in the first two columns of Table 6.1. Entries in the third column are of axial length z found with

$$z = \left(\frac{1}{\text{Gz}}\right)(D \text{ Re}_D \text{ Pr}) = \frac{1}{\text{Gz}}(0.013\ 40)(651)(13.6)$$

$$z = \left(\frac{1}{\text{Gz}}\right)(118.6)$$

TABLE 6.1
Summary of Calculations to Find the Wall Temperature of the Tube Described in Example 6.3

$\frac{1}{\text{Gz}} = \frac{z}{D \text{ Re}_D \text{ Pr}}$	$\text{Nu}_D = \frac{h_z D}{k_f}$	$z = 118.6\frac{1}{\text{Gz}}$ (m)	$h_z = \frac{k_f \text{Nu}_D}{D}$ W/(m²·K)	T_{bz} (°C)	T_{wz} (°C)
0.002	12.0	0.237	537	22.0	43.3
0.004	10.0	0.475	448	24.1	49.7
0.01	7.5	1.19	336	30.3	64.4
0.04	5.2	4.75	233	60.9	110.0
0.05	4.5	5.93	201	71.1	128.1

The local coefficient h_z is shown in the fourth column. By definition of the local coefficient, we can write

$$q_w'' = h_z(T_w - T_b)\big|_z = h_z(T_{wz} - T_{bz}) \tag{i}$$

So before finding the wall temperature T_w at any z location (T_{wz}), we must first determine how the bulk-fluid temperature varies with z (T_{bz}). This is given by Equation 6.9b:

$$T_{bz} - T_{bi} = \frac{2q_w'' \alpha z}{V k_f R}$$

Rearranging and substituting gives for the bulk-fluid temperature

$$T_{bz} = 20 + \frac{2(11\ 447)(0.6)z}{1\ 030(3\ 850)(0.1)(0.6)(0.012\ 40/2)}$$

or

$$T_{bz} = 20 + 8.617z$$

The bulk-fluid-temperature variation is tabulated in column 5 of Table 6.1. Combining with Equation (i) above to find the wall-temperature variation gives

$$T_{wz} = T_{bz} + \frac{q_w''}{h_z} = T_{bz} + \frac{11\ 447}{h_z}$$

with results shown in column 6 of Table 6.1. The data of Table 6.1 are displayed graphically in Figure 6.9. Remember that, for any z, the product of h_z and ($T_{wz} - T_{bz}$) is constant and equal to q_w'' .   ❑

Example 6.4

An indoor ice rink consists of a series of tubes that convey a refrigerant; the tubes are submerged under a volume of water. The water cools from an ambient temperature of 65° to 32°F, where it solidifies, then cools to 15°F. The heat transferred during solidification is critical, however; to properly size the cooling unit required, it is necessary to prepare an analysis for the time when the water changes from liquid to solid.

   Consider one submerged tube. It is made of 3/8 standard type K copper water tubing and conveys Freon-12 refrigerant. The tube is 5 ft in length. Freon-12 enters the tube at –40°F and leaves at –4°F. The volume of water that the tube is responsible for cooling is 5 ft long × 2 in wide × 3 in deep, and 144 BTU/(lbm of water) must be removed to solidify the water. Graph the variation of wall and bulk-fluid temperatures with length. How long will it take for the water volume to solidify? What is the mass flow rate of Freon-12 required?

Solution
While the Freon-12 flows through the tube, it becomes heated. The water on the outside tube surface is a saturated, compressed liquid. Any further heat removal causes the water to solidify, and this process is isothermal. The Freon-12 is a liquid but, as it absorbs heat, it vaporizes.

Assumptions

1. Steady-state conditions exist within the tube.
2. When the water solidification process occurs, the entire tube is at 32°F.

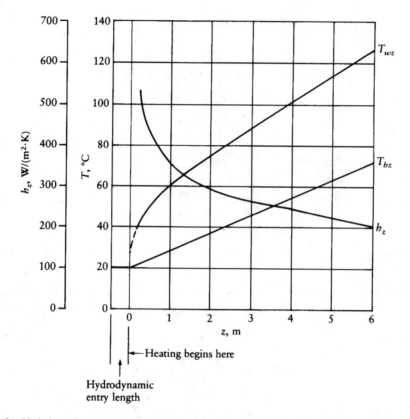

FIGURE 6.9 Variation of temperature and local convection coefficient with axial distance for the constant-wall-flux tube of Example 6.3.

3. The flow is hydrodynamically developed within the copper tube.
4. The Freon-12 remains a saturated liquid within the tube.
5. The displaced volume of the submerged copper tube is comparatively negligible.

The average bulk temperature of the Freon-12 is $[-40 + (-4)]/2 = -22°F$ and, at this temperature, as we find from Appendix Table C.3,

$$\rho = 1.489(62.4) \text{ lbm/ft}^3 \qquad k_f = 0.04 \text{ BTU/(hr·ft·°R)}$$
$$c_p = 0.2139 \text{ BTU/(lbm·°R)} \qquad \alpha = 2.04 \times 10^{-3} \text{ ft}^2/\text{hr}$$
$$\nu = 0.272 \times 10^{-5} \text{ ft}^2/\text{s} \qquad Pr = 4.8$$

From Appendix Table F.2, for 3/8 standard type K copper tubing,

$$OD = 0.5 \text{ in} \qquad\qquad ID = 0.03350 \text{ ft}$$
$$A = 0.0008814 \text{ ft}^2$$

Laminar conditions are presumed to exist, which remains to be verified. We can apply Equation 6.24a, which relates temperature to the convection coefficient for the constant wall temperature problem:

$$\frac{T_w - T_{bo}}{T_w - T_{bi}} = \exp\left(-\frac{2\alpha L \bar{h}_L}{V k_f R}\right)$$

The unknown parameters are the average velocity and the average convection coefficient. We can work with the above equation and Figure 6.8 to determine them. Rearranging and substituting into the above equation gives

$$-\frac{2\alpha L \bar{h}_L}{V k_f R} = \ln\left[\frac{(T_w - T_{bo})}{(T_w - T_{bi})}\right]$$

$$-\frac{2(2.04 \times 10^{-3})(5)\bar{h}_L}{V(0.04)(0.0335/2)} = \ln\left[\frac{32 - (-4)}{32 - (-40)}\right]$$

Solving,

$$V = \frac{\bar{h}_L}{2.277 \times 10^{-2}}$$

where the units of velocity must be in ft/hr. To combine this result with Figure 6.8, we first calculate the inverse Graetz number:

$$\frac{1}{Gz} = \frac{z}{D \, Re_D \, Pr} = \frac{z}{D(VD/\nu)(\nu/\alpha)} = \frac{z\alpha}{VD^2}$$

Substituting,

$$\frac{1}{Gz} = \frac{5(2.04 \times 10^{-3})}{V(0.0335)^2} = \frac{9.09}{V} \tag{ii}$$

We now assume a velocity, calculate $(1/Gz)$, determine $\overline{Nu}_D \; (= \bar{h}_L D / k_f)$ from Figure 6.8, calculate $\bar{h}_L$, and use Equation (i) to calculate a new velocity. The process is repeated until convergence is achieved. The results are summarized in the following table:

Assumed V (ft/hr)	$\frac{1}{Gz}$ (Eq. ii)	$\overline{Nu}_D$ (Fig. 8.8)	$\bar{h}_L = \overline{Nu}_D k_f D$ [BTU/(hr·ft²·°R)]	V (Eq. i) (ft/hr)
100	0.0909	4.7	5.61	246
246	0.037	5.8	6.92	304
304	0.03	6.2	7.40	325
325	0.028	6.3	7.52	330
330	0.0275	6.4	7.64	336
336	0.0271	~6.4	7.64	336 close enough

Convergence is slowly achieved. Thus,

$$V = 336 \text{ ft/hr} = 0.0933 \text{ ft/s}$$

and

$$\bar{h}_L = 7.64 \text{ BTU/(hr·ft}^2\text{·°R)}$$

We now check the laminar-flow assumption by calculating the Reynolds number:

$$Re_D = \frac{VD}{v} = \frac{0.0933(0.0335)}{02.72 \times 10^{-5}} = 1149$$

The flow is laminar. The mass flow rate of Freon-12 is

$$\dot{m} = \rho AV = 1.489(62.4)(0.0008814)(0.0933)$$

or

$$\dot{m} = 7.64 \times 10^{-3} \text{ lbm/s} = 27.5 \text{ lbm/hr}$$

Equation 6.25 gives the heat transferred in terms of the average convection coefficient as

$$q = \bar{h}_L A_s \frac{[(T_w - T_{bo}) - (T_w - T_{bi})]}{\ln[(T_w - T_{bo})/(T_w - T_{bi})]}$$

where $A_s = \pi(ID)L$. Substituting,

$$q = 7.64\pi(0.0335)(5)\frac{32 - (-4) - [32 - (-40)]}{\ln(36/72)}$$

Solving,

$$q = 208.8 \text{ BTU/hr}$$

This is the heat-removal rate, that is, the heat gained by the Freon-12 during the time when the water is solidifying. As a check,

$$q = \dot{m} c_p (T_{bo} - T_{bi}) = 27.5(0.2139)[-4 - (-40)]$$

$$q = 211.8 \text{ BTU/hr}$$

From Appendix Table C.11 for water at 32°F

$$\rho = 1.002(62.4) \text{ lbm/ft}^3 = 62.5 \text{ lbm/ft}^3$$

The mass of water in the volume prescribed (neglecting the volume occupied by the submerged tube) is

$$m = 62.5(5)(2/12)(3/12) = 13.0 \text{ lbm}$$

With 144 BTU/(lbm of water) the required amount of heat to be removed, we calculate the time as

$$t = 144\frac{\text{BTU}}{\text{lbm}} \times 13.0 \text{ lbm} \times \frac{\text{hr}}{208.8 \text{ BTU}}$$

or

$$t = 9.0 \text{ hr}$$

This is the time required for the water at 32°F to change to ice at 32°F.

To obtain the variation of wall and bulk temperatures with axial distance, we use Equation 6.24 and Figure 6.8. Equation 6.24b relates the bulk temperature to the convection coefficient:

$$\frac{T_w - T_{bz}}{T_w - T_{bi}} = \exp\left(-\frac{2\alpha z \bar{h}_z}{V k_f R}\right) \tag{6.24b}$$

where $\bar{h}_z$ is the average convection coefficient to any axial location z and where the corresponding bulk temperature is T_{bz}. It is important to remember that the units for diffusivity are ft²/hr, so again velocity must be in ft/hr. Substituting, we get

$$\frac{32 - T_{bz}}{32 - (-40)} = \exp\left[-\frac{2(2.04 \times 10^{-3}) z \bar{h}_z}{336(0.04)(0.0335/2)}\right]$$

Rearranging and solving for bulk temperature gives

$$32 - T_{bz} = 72 \exp(-1.182 \times 10^{-2} z \bar{h}_z)$$

$$T_{bz} = 32 - 72 \exp(-0.01812 z \bar{h}_z) \tag{iii}$$

From Figure 6.8, we read the data that appear in the first two columns of Table 6.2. The last entry for $1/\text{Gz} = 0.271$ corresponds to $z = 5$ ft. Calculations for length in the third column are found from the first column by using

$$z = \frac{1}{\text{Gz}} D \, \text{Re}_D \, \text{Pr} = (1/\text{Gz})(0.0335)(1149)(4.8)$$

or

$$z = \frac{1}{\text{Gz}} 185$$

The average convection coefficient applicable from 0 to z is tabulated in column 4 of the table and determined from the Nusselt number:

TABLE 6.2
Summary of Data for Example 6.4

$\frac{1}{\text{Gz}} = \frac{z}{D \, \text{Re}_D \, \text{Pr}}$	$\text{Nu}_{D|z} = \frac{h_z D}{k_f}$	$z = 185 \frac{1}{\text{Gz}}$ (ft)	$h_z = 1.19 \frac{\overline{\text{Nu}_{D|z}}}{D}$ (BTU/hr·ft²·°R)	T_{bz} (°F)
0.001	19.3	0.185	23.0	−34.7
0.004	12.1	0.74	14.4	−27.4
0.01	8.9	1.85	10.6	−18.5
0.015	7.7	2.77	9.16	−13.5
0.02	7.1	3.70	8.45	− 8.9
0.0271	6.4	5.0	7.62	− 4.1

$$\bar{h}_z = \overline{\mathrm{Nu}}_{D|z}(k_f/D) = \overline{\mathrm{Nu}}_{D|z}(0.04/0.0335)$$

$$\bar{h}_z = 1.19\overline{\mathrm{Nu}}_{D|z}$$

The bulk temperature at location z, T_{bz}, is found with Equation (iii) above and listed in column 5 of Table 6.2.

The hydrodynamic entry length is calculated with Equation 6.1:

$$z_t = 0.05D\ \mathrm{Re}_D = 0.05(0.0335)(1149)$$

$$z_t = 1.92\ \mathrm{ft}$$

The results are displayed graphically in Figure 6.10.   ❏

6.2.4 THE COMBINED-ENTRY-LENGTH PROBLEM FOR LAMINAR FLOW IN A CIRCULAR DUCT

The preceding paragraphs present results obtained from exact solutions to equations for constant-wall-flux and constant-wall-temperature problems. In those cases, it is assumed that steady laminar flow exists in a circular duct and that the flow is hydrodynamically developed before heating begins. This last assumption is somewhat limiting. So, in this discussion, we examine results for the case where the flow is still laminar but not developed hydrodynamically or thermally. This is referred to as the *combined-entry-length* problem. For this case, the heat transfer in the entrance region is more sensitive to the Prandtl number. Therefore, the solutions presented are correlated as Nusselt number vs. the inverse Graetz number, in keeping with our previous developments, with Prandtl number as an independent variable. Figure 6.11 gives the results for the combined-entry-length

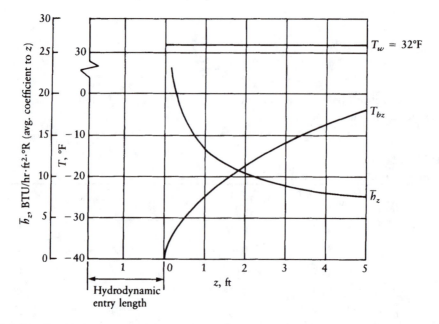

FIGURE 6.10 Graphical summary of the solution to the constant-wall-temperature tube of Example 6.4.

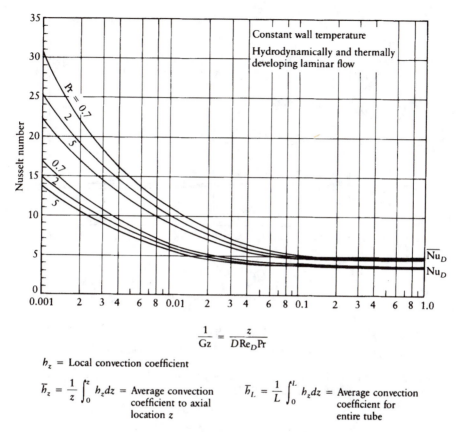

$$\frac{1}{Gz} = \frac{z}{D\,Re_D\,Pr}$$

h_z = Local convection coefficient

$\bar{h}_z = \dfrac{1}{z}\displaystyle\int_0^z h_z\,dz$ = Average convection coefficient to axial location z

$\bar{h}_L = \dfrac{1}{L}\displaystyle\int_0^L h_z\,dz$ = Average convection coefficient for entire tube

$$Re_D = \frac{VD}{\nu}; \quad Pr = \frac{\nu}{\alpha}; \quad Nu_D = \frac{h_z D}{k_f}; \quad \overline{Nu}_D = \frac{\bar{h}_L D}{k_f}$$

Note: All Nu_D and $\overline{Nu}_D$ lines approach 3.658 as $(1/Gz)$ approaches ∞.

FIGURE 6.11 Nusselt-number variation with dimensionless length for laminar flow of a Newtonian fluid in a circular duct of constant wall temperature, with developing velocity and temperature profiles.

problem in which wall temperature is constant. Figure 6.12 corresponds to the constant-heat-flux problem.

Figure 6.11 is a graph of Nusselt number vs. the inverse Graetz number for laminar flow of a Newtonian fluid in a circular duct. The wall temperature is constant, and both velocity and temperature profiles are developing. Curves for Prandtl numbers of 0.7, 2, and 5 are presented. As length approaches infinity, the Nusselt number approaches 3.66 for all cases. Three curves are for the Nusselt number based on a local convection coefficient ($Nu_D = h_z D/k_f$), and the others are for the Nusselt number based on an average coefficient ($\overline{Nu}_D = \bar{h}_L D/k_f$), where

$$\bar{h}_L = \frac{1}{L}\int_0^L h_z\,dz \tag{6.32}$$

Figure 6.12 is a graph of the local Nusselt number ($h_z D/k_f$) vs. the inverse Graetz number for the laminar flow of a Newtonian fluid in a circular duct. The wall flux is constant, and both velocity and temperature profiles are developing. Curves for Prandtl numbers of 0.01, 0.7, and 10 are provided. As the inverse Graetz number approaches infinity, the curves approach 48/11.

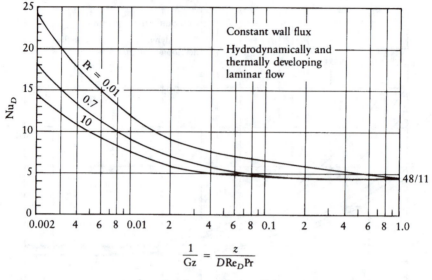

$$\frac{1}{Gz} = \frac{z}{D Re_D Pr}$$

h_z = Local convection coefficient

$$Re_D = \frac{VD}{\nu}; \quad Pr = \frac{\nu}{\alpha}; \quad Nu_D = \frac{h_z D}{k_f}$$

Note: All Nu_D lines approach 48/11 as $(1/Gz)$ approaches ∞.

FIGURE 6.12 Nusselt-number variation with dimensionless length for laminar flow of a Newtonian fluid in a circular duct having a constant wall flux, with developing velocity and temperature profiles.

Example 6.5

A relatively new design for a water heater consists of 3 m of 1 standard type K water tubing that is surrounded with a heating element. The element provides a constant heat flux at the tube wall. Water is heated from a temperature of 20° to 80°C. The volume flow rate of 20°C water to be heated is 80 oz in 120 s. Determine the power required to do this and the wall temperature at the outlet.

Solution

As the water is heated, its properties change. The volume flow of water at 20°C will therefore be different from the volume flow at 80°C.

Assumptions

1. The flow is steady.
2. The water is incompressible.
3. The liquid properties are constant within the tube and are evaluated at the mean temperature of (80 + 20)/2 = 50°C.

From Appendix Table F.2 for 1 standard type K copper water tubing,

OD = 2.858 cm   ID = 2.528 cm
                        A = 5.019 cm²

The volume flow rate of water (at 20°C) is

$$Q = \frac{80 \text{ oz}}{120 \text{ s}} \times 2.957 \times 10^{-5} \frac{\text{m}^3}{\text{oz}} = 1.97 \times 10^{-5} \text{ m}^3/\text{s}$$

where the conversion factor is obtained from Appendix Table A.2. From Appendix Table C.11, for water at 20°C,

$$\rho = 1.000(1\ 000) \text{ kg/m}^3$$

The mass flow rate is constant through the tube and is found to be

$$\dot{m} = (\rho Q)|_{20°C} = 1\ 000(1.97 \times 10^{-5}) = 0.019\ 7 \text{ kg/s}$$

From Appendix Table C.11 for water at 50°C we interpolate to find

$\rho = 0.990(1\ 000) \text{ kg/m}^3 \qquad k_f = 0.640 \text{ W/(m·K)}$
$c_p = 4\ 181 \text{ J(kg·K)} \qquad \alpha = 1.533 \times 10^{-7} \text{ m}^2/\text{s}$
$v = 0.586 \times 10^{-6} \text{ m}^2/\text{s} \qquad \text{Pr} = 3.68$

The power required is calculated as

$$q_w'' = \frac{q}{A_s} = \frac{\dot{m} c_p}{A_s}(T_{bo} - T_{bi})$$

where $A_s = \pi(ID)L$. By substitution,

$$q_w'' = \frac{0.019\ 7(4\ 181)(80 - 20)}{\pi(0.025\ 28)(3)}$$

or

$$q_w'' = 2.074 \times 10^4 \text{ W/m}^2$$

Also,

$$q = q_w'' A_s = 4\ 942 \text{ W}$$

So a 5-kW heater should be sufficient.

To find the wall temperature, we use the local convection coefficient:

$$q_w'' = h_z(T_w - T_b)|_{\text{any } z}$$

We have made no assumptions regarding the developed condition of the flow. We can use Figure 6.12 to obtain the information needed on Nusselt number and to find h_z provided the flow is laminar. The average velocity based on 50°C conditions is

$$V = \frac{\dot{m}}{\rho A} = \frac{0.019\ 7}{990(5.019 \times 10^{-4})} = 3.97 \times 10^{-2} \text{ m/s}$$

The Reynolds number is

$$\text{Re}_D = \frac{VD}{v} = \frac{0.039\ 7(0.025\ 28)}{0.568 \times 10^{-6}} = 1\ 765$$

Thus, the flow is laminar.

The inverse Graetz number at tube end, based on 50°C conditions, is

$$\frac{1}{\text{Gz}} = \frac{z}{D\ \text{Re}_D\ \text{Pr}} = \frac{3}{0.025\ 28(1\ 765)(3.68)} = 0.018\ 3$$

From Figure 6.12, in which the location of Pr = 3.68 is estimated, we find

$$\text{Nu}_D = \frac{h_z D}{k_f} = 6.9$$

The local convection coefficient is

$$h_z = \frac{6.9(0.640)}{0.252\ 8} = 174.7\ \text{W/(m}^2\cdot\text{K)}$$

The outlet wall temperature is now calculated as

$$T_{wo} = \frac{q_w''}{h_z} + T_{bo} = \frac{2.074 \times 10^4}{174.7} + 80$$

$$T_{wo} = 199°\text{C}$$

❑

It is often impractical to have graphs from which to obtain the Nusselt number for the types of problems discussed here. Alternatively, it is desirable to have an equation, especially if a computer or calculator is available. For the combined problem of developing velocity and temperature profiles, the Seider-Tate Equation is a useful correlation:

$$\overline{\text{Nu}}_D = \frac{\bar{h}_L D}{k_f} = 1.86\left(\frac{D\ \text{Re}_D\ \text{Pr}}{L}\right)^{1/3}\left(\frac{\mu}{\mu_w}\right)^{0.14} \tag{6.33}$$

(constant wall temperature; hydrodynamically and thermally developing laminar flow)

in which μ_w is the viscosity of the fluid evaluated at the wall temperature, and the remaining fluid properties are all evaluated at the average of the bulk temperatures $(T_{bo} + T_{bi})/2$. The Seider-Tate Equation applies to steady laminar flow of a Newtonian fluid in a tube having constant wall temperature under the following conditions:

$$\left.\begin{array}{c} T_w = \text{a constant} \\ 0.48 < \text{Pr} < 16700 \\ 0.004\ 4 < (\mu/\mu_w) < 9.75 \end{array}\right\} \tag{6.34}$$

In addition to having heat-transfer correlations as in the above discussion, it is important also to be able to calculate a pressure drop within a circular duct under laminar conditions. From Equations 6.3 and 6.4, we have, for the wall shear stress

$$\tau_w = \left(-\frac{dp}{dz}\right)\frac{R}{2} \tag{6.3}$$

and for the average velocity

$$V = \left(-\frac{dp}{dz}\right)\frac{R^2}{8\mu}$$

We now introduce a friction factor that is defined in terms of the wall shear stress and average velocity as

$$f = \frac{4\tau_w g_c}{\frac{1}{2}\rho V^2} \text{ *} \tag{6.35}$$

Substituting Equation 6.3 gives

$$f = \frac{4(-dp/dz)(R/2)g_c}{\frac{1}{2}\rho V^2}$$

or

$$dp = -\frac{\rho V^2}{2g_c}\frac{f\,dz}{D} \tag{6.36}$$

where $D = 2R$ is the pipe diameter. Integrating from point 1 (where $p = p_1$ and $z = 0$) to point 2 downstream (where $p = p_2$ and $z = L$) gives

$$\int_{p1}^{p2} dp = -\frac{\rho V^2}{2g_c}\frac{f}{D}\int_0^L dz$$

$$p_1 - p_2 = +\frac{\rho V^2}{2g_c}\frac{fL}{D} \tag{6.37}$$

The pressure drop due to friction is thus a fraction of the kinetic energy of the flow. To determine the friction factor, solve Equation 6.4 for dp/dz and combine with Equation 6.36; from 6.4,

$$\frac{dp}{dz} = -\frac{8\mu V}{R^2}$$

---

* The friction factor defined in Equation 6.35 is known as the Darcy-Weisbach friction factor. The Fanning friction factor, f', is used in some texts and is defined as

$$f' = \frac{\tau_w g_c}{\frac{1}{2}\rho V2}$$

The Darcy-Weisbach equation is preferred here because the 4 in Equation 6.35 becomes included in f, which yields a simpler formulation for the equations that follow.

Combining with Equation 6.36,

$$-\frac{dp}{dz} = \frac{\rho V^2}{2g_c}\frac{f}{D} = \frac{8\mu V}{R^2}$$

Rearranging, with $2R = D$, we obtain

$$\frac{\rho V^2}{2g_c}\frac{f}{D} = \frac{8\mu V}{D^2/4}$$

or

$$f = \frac{64\mu g_c}{\rho VD} = \frac{64}{\mathrm{Re}_D} \qquad (6.38)$$

This result applies to steady laminar flow of a Newtonian fluid in a circular duct.

6.3 HEAT TRANSFER TO AND FROM TURBULENT FLOW IN A CIRCULAR CONDUIT

The turbulent-flow heat-transfer problem we consider is similar to that discussed in Section 6.2. Fluid enters a tube and travels downstream. The velocity profile changes and requires anywhere up to 50 diameters before hydrodynamically developed conditions exist. The entrance length for turbulent flow is shorter than that for laminar flow because of turbulent mixing effects. It is not possible to obtain a closed-form solution for the velocity profile in turbulent flow.

6.3.1 CONSTANT HEAT FLUX AT THE WALL AND CONSTANT WALL TEMPERATURE

The heat-transfer problem considered is one in which the wall-heat flux $q_w'' = q/A_s = h_z(T_w - T_b)$ is a constant, or alternatively, when the wall temperature T_w is a constant. Once hydrodynamically developed conditions exist, another 1 to 30 diameters are required for the flow to become thermally developed. The thermal entry length required depends on the Reynolds number and on the Prandtl number. The equations that follow deal almost exclusively with fully developed conditions in turbulent flow.

Because of the complex nature of turbulent flow, it is not possible to derive equations for temperature profile. Instead, we rely on dimensional analysis and experimental data to obtain a relationship between Nusselt number and the pertinent variables. A dimensional analysis for duct-flow problems was performed earlier. The results indicate that the Nusselt number depends on the Reynolds and Prandtl numbers. Combining with experimental results, the following equation (again due to Seider and Tate) is found to apply:

$$\mathrm{Nu}_D = 0.027\,\mathrm{Re}_D^{4/5}\,\mathrm{Pr}^{1/3}\!\left(\frac{\mu}{\mu_w}\right)^{0.14} \qquad (6.39)$$

(constant wall flux or constant wall temperature; hydrodynamically and thermally developed turbulent flow)

$$\left.\begin{array}{ll} 0.7 \leq \mathrm{Pr} \leq 16700 & \mathrm{Pr} = \dfrac{v}{\alpha} \\[2mm] \mathrm{Re}_D \geq 10\ 000 & \mathrm{Re}_D = \dfrac{VD}{v} \\[2mm] L/D \geq 60 & \mathrm{Nu}_D = \dfrac{h_z D}{k_f} \end{array}\right\}$$ (6.40)

The viscosity μ_w is to be evaluated at the wall temperature and all other properties at the average of the bulk temperatures $(T_{bo} + T_{bi})/2$. An alternative expression that is useful when the wall temperature is unknown is the Dittus-Boelter Equation:

$$\mathrm{Nu}_D = 0.023\ \mathrm{Re}_D^{4/5}\ \mathrm{Pr}^n$$ (6.41)

(constant wall flux or constant wall temperature; hydrodynamically and thermally developed turbulent flow)

$$\left.\begin{array}{ll} n = 0.4 \text{ if } T_w > T_b & \text{(heating)} \\ n = 0.3 \text{ if } T_w < T_b & \text{(cooling)} \\ 0.7 \leq \mathrm{Pr} \leq 160 & \mathrm{Pr} = v/\alpha \\ \mathrm{Re}_D \geq 10\ 000 & \mathrm{Re}_D = VD/v \\[2mm] L/D \geq 60 & \mathrm{Nu}_D = \dfrac{h_z D}{k_f} \end{array}\right\}$$ (6.42)

Both correlations apply for the constant-wall-flux or constant-wall-temperature case.

Since the derivation of Equation 6.39, more data have been obtained by various researchers, and other correlations have been derived. These can be found in the literature, and a few are provided in the exercises of this chapter.

Example 6.6

Air-conditioning ductwork is located in the attic space of a house. Air is moved with a fan through an evaporator. After leaving the evaporator, the air is ducted to the ceilings of various rooms of the house. It is proposed that the ducts be insulated, but it is first desired to determine the heat loss through an uninsulated duct so that an economic analysis can be performed.

The longest duct is to be used. When the system is set up and running at steady state, measurements indicate that the average velocity of the air at the duct exit is 10 ft/s. At the inlet the air temperature is 45°F, while at the exit the air has warmed to 72°F. The attic air surrounding the duct is at 105°F, and the convection coefficient between the outside duct wall and the attic air is 1 BTU/(hr·ft²·°R). The duct is made from sheet metal that is 0.020 in thick, rolled into a circular cylinder that has a 7-in diameter and 40-ft length. Determine the heat gained by the air within the duct and the wall temperature at the duct exit.

Solution

The air within the duct gains energy from air outside the duct.

Assumptions

1. Steady-state conditions exist.
2. Air properties are constant and are to be evaluated at $(72 + 45)/2 = 58.5°F$.

3. The convection coefficient outside the duct is a constant that applies over the entire surface.
4. The wall thickness of the duct is small enough that it presents a negligible resistance to radial heat flow.
5. The air behaves as an ideal gas.
6. One-dimensional heat flow exists in the radial direction.

Figure 6.13 illustrates the situation. Also shown is a one-dimensional resistance circuit for the radial heat loss at the end.

From Appendix Table D.1 for air at an average bulk temperature of 58.5°F = 518.5°R we obtain the following interpolated values:

$$\rho = 0.077 \text{ lbm/ft}^3 \qquad\qquad k_f = 0.0146 \text{ BTU/(hr·ft·°R)}$$
$$c_p = 0.240 \text{ BTU/(lbm·°R)} \qquad \alpha = 0.776 \text{ ft}^2/\text{hr}$$
$$v = 15.28 \times 10^{-5} \text{ ft}^2/\text{s} \qquad \text{Pr} = 0.711$$

At the outlet, where $T_{bo} = 72°F = 532°R$, we read

$$\rho = 0.0748 \text{ lbm/ft}^3$$

The mass flow, which is a constant in the duct, can be evaluated at the outlet, where the velocity is known. The duct cross-sectional area is

$$A = \frac{\pi}{4}\left(\frac{7}{12}\right)^2 = 0.267 \text{ ft}^2$$

The mass flow then is

$$\dot{m} = \rho A V\big|_o = 0.0748(0.267)(10) = 0.2 \text{ lbm/s}$$

The average velocity evaluated by using the average bulk temperature is

$$V = \frac{\dot{m}}{\rho A}\bigg|_b = \frac{0.2}{0.077(0.267)} = 9.72 \text{ ft/s}$$

The Reynolds number is

$$\text{Re}_D = \frac{VD}{v} = \frac{9.72\left(\frac{7}{12}\right)}{15.28 \times 10^{-5}} = 3.712 \times 10^4$$

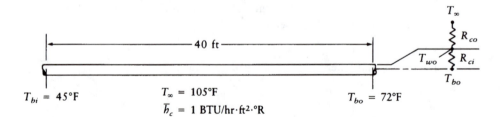

$T_{bi} = 45°F$ $T_\infty = 105°F$ $T_{bo} = 72°F$

$\bar{h}_c = 1 \text{ BTU/hr·ft}^2\text{·°R}$

FIGURE 6.13 Air flow in a duct and accompanying thermal circuit for radial heat loss.

The flow is thus turbulent. The heat gain by the air is calculated to be

$$q = \dot{m}c_p(T_{bo} - T_{bi}) = 0.2(0.240)(72 - 45)$$
$$q = 1.296 \text{ BTU}$$

For the heat flux at the exit, we have, from the thermal circuit,

$$q_w''(L) = \frac{T_\infty - T_{bo}}{R_{co} + R_{ci}} = \frac{T_\infty - T_{bo}}{\dfrac{1}{\bar{h}_c} + \dfrac{1}{h_z}}$$

where $q_w''(L)$ is the heat flux evaluated at $L = 40$ ft and h_z is the local coefficient inside the duct evaluated also at L. The difference in the inside area and outside area is neglected, because the wall thickness is small. Note that neither the heat flux nor the wall temperature is a constant for this problem. The local coefficient at the duct end is found from the Nusselt number, which in turn is calculated with Equation 6.41, because the wall temperature is unknown. The L/D ratio is $40/(7/12) = 68.6$, which is greater than 60, as required in the conditions of Equation 6.42. Therefore,

$$\text{Nu}_D = \frac{h_z D}{k_f} = 0.023 \, \text{Re}_D^{4/5} \, \text{Pr}^{0.4}$$

After substitution, we get

$$h_z = 0.023(3.712 \times 10^4)^{4/5}(0.711)^{0.4}\frac{0.0146}{(7/12)} = 2.27 \text{ BTU/(hr·ft}^2\text{·°R)}$$

The wall flux at L becomes

$$q_w'' = \frac{105 - 72}{\dfrac{1}{1} + \dfrac{1}{2.27}}$$

or

$$q_w'' = 22.9 \text{ BTU/(hr·ft}^2\text{·°R)}$$

From the thermal circuit we can also write

$$q_w'' = \frac{T_{wo} - T_{bo}}{R_{ci}}$$

or

$$22.9 = \frac{T_{wo} - 72}{\dfrac{1}{2.27}}$$

Solving, we find, for the wall temperature at exit

$$T_{wo} = 82.1°F \qquad \square$$

The pressure drop that exists in a fluid flowing under turbulent conditions through a duct is calculated with Equation 6.37:

$$p_1 - p_2 = + \frac{\rho V^2}{2g_c} \frac{fL}{D}$$

(6.37)

Just as for laminar flow, the pressure drop is a function of the kinetic energy of the flow. The difference is in how the friction factor is determined. Because a closed-form solution for velocity is not available for turbulent flow, an equation for friction factor is not derivable. Instead, we rely on dimensional analysis and extensive data on pressure drop. The friction factor is a function of several variables:

$$f = f(\rho, V, D, \mu, \varepsilon, g_c)$$

(6.43a)

where ε is the average roughness of the pipe wall. Performing a dimensional analysis yields

$$f = f\left(\frac{\rho VD}{\mu g_c}, \frac{\varepsilon}{D}\right) = f\left(\mathrm{Re}_D, \frac{\varepsilon}{D}\right)$$

(6.43b)

Figure 6.14 is a graph of friction factor vs. Reynolds number with ε/D as an independent variable. The graph is known as a Moody diagram. Values of the average roughness ε for various materials are provided in Table 6.3.

TABLE 6.3
Values of ε for Various Materials

Pipe material	ε in ft	ε in cm
Riveted steel	0.003–0.03	0.09–0.9
Concrete	0.001–0.01	0.03–0.3
Wood stave	0.0006–0.003	0.018–0.09
Cast iron	0.0085	0.025
Galvanized surface	0.0005	0.015
Asphalted cast iron	0.0004	0.012
Commercial steel or wrought iron	0.00015	0.0046
Drawn tubing	0.000005	0.00015

Several equations have been developed that are curve fits of the Moody diagram. Among those that have been written are the following:

The Hagen-Poiseuille Equation is:

$$f = \frac{64}{\mathrm{Re}} \quad \text{(laminar flow, circular pipe)}$$

(6.44)

The following equations are all valid for $\mathrm{Re} \geq 2100$.

The Chen Equation is:

$$f = \left[-2.0\log\left\{ \frac{\varepsilon}{3.7065D} - \frac{5.0452}{\mathrm{Re}}\log\left(\frac{1}{2.8257}\left[\frac{\varepsilon}{D}\right]^{1.1098} + \frac{5.8506}{\mathrm{Re}^{0.8981}}\right)\right\}\right]^{-2}$$

(6.45)

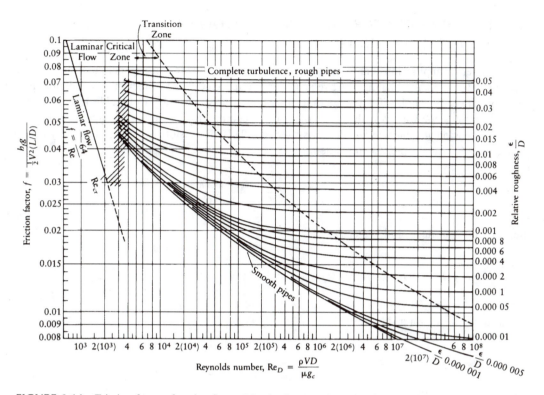

FIGURE 6.14 Friction factors for pipe flow—Moody diagram. *Source:* adapted from Moody, L., "Friction Factors in Pipe Flow," *Trans. ASME,* v. 66, November 1944, p. 672).

The Churchill Equation is:

$$f = 8\left\{\left[\frac{8}{Re}\right]^{12} + \frac{1}{(B+C)^{1.5}}\right\}^{\frac{1}{12}} \tag{6.46}$$

where

$$B = \left[2.457\ln\frac{1}{\left(\frac{7}{Re}\right)^{0.9} + \left(\frac{0.27\varepsilon}{D}\right)}\right]^{16} \quad \text{and} \quad C = \left(\frac{37\,530}{Re}\right)^{16} \tag{6.47}$$

The Haaland Equation is:

$$f = \left\{-0.782\ln\left[\frac{6.9}{Re} + \left(\frac{\varepsilon}{3.7D}\right)^{1.11}\right]\right\}^{-2} \tag{6.48}$$

The Swamee-Jain Equation is:

$$f = \frac{0.250}{\left\{\log\left[\frac{\varepsilon}{3.7D} + \frac{5.74}{Re^{0.9}}\right]\right\}^2} \tag{6.49}$$

6.4 HEAT-TRANSFER CORRELATIONS FOR FLOW IN NONCIRCULAR DUCTS

In a number of applications energy is transferred to a fluid flowing in a noncircular duct. Some heating and air-conditioning ducts are rectangular in cross section. The outer duct in a double-pipe heat exchanger is annular in cross section. In any case it is necessary to be able to analyze the systems in the same way as with the circular duct. Accordingly, it is to our advantage to define a single dimension to characterize the cross section. This dimension is known as the hydraulic diameter D_h (from Chapter 5) and is defined as

$$D_h = \frac{4A}{P} = 4 \times \frac{\text{cross-sectional area}}{\text{perimeter}}$$

So, for a circular duct of diameter D,

$$D_h = 4\left(\frac{\pi D^2/4}{\pi D}\right) = D$$

For a square duct s on a side,

$$D_h = \frac{4s^2}{4s} = s$$

When dealing with noncircular ducts, the hydraulic diameter is used in the expressions for calculating parameters such as the Reynolds number and the Nusselt number.

6.4.1 THE CONCENTRIC ANNULAR DUCT

Figure 6.15 gives the cross section of a concentric annulus showing definitions appropriate to the geometry. Note that each tube has a finite wall thickness. The annular flow area is bounded by the outer diameter of the inner tube (D_{io}) and the inner diameter of the outer tube (D_{oi}). The hydraulic diameter for the cross section is found as

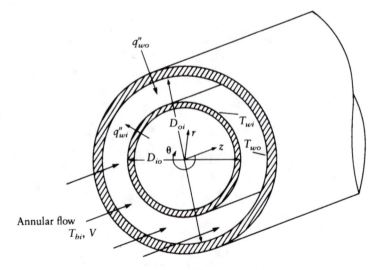

FIGURE 6.15 Concentric annular duct showing pertinent geometric parameters.

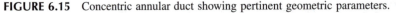

$$D_h = \frac{4A}{P} = \frac{4[\pi(D_{oi}^2 - D_{io}^2)/4]}{\pi D_{oi} + \pi D_{io}} = D_{oi} - D_{io} \tag{6.50}$$

Also shown in Figure 6.15 is the fluid at a bulk temperature of T_b flowing at an average velocity of V. The annular fluid can receive heat through the outer wall as determined from

$$q_{wo}'' = h_o(T_{wo} - T_b) \tag{6.51}$$

where h_o is the local convection coefficient applied at the outer wall. The annular fluid can receive heat also from the inner tube:

$$q_{wi}'' = h_i(T_{wi} - T_b) \tag{6.52}$$

Heat transferred to the fluid is positive, as seen in these equations. Separate (and not necessarily equal) convection coefficients are associated with the heat transferred to the fluid. The Nusselt number at each wall is, by definition,

$$\text{Nu}_D = \frac{h_o D_h}{k_f} \tag{6.53}$$

and

$$\text{Nu}_D = \frac{h_i D_h}{k_f} \tag{6.54}$$

The equations of continuity, momentum, and energy have been solved for the case where fully developed laminar flow exists in the annulus. The heat-transfer problems considered are for one surface insulated and the other at *constant temperature*. The corresponding Nusselt numbers have been determined and are graphed in Figure 6.16. For a given configuration, the Nusselt numbers for the constant-wall-temperature problem can be found from Figure 6.16, and the convection coefficients can be determined with Equations 6.53 and/or 6.54. The heat received by the annular fluid is then found with Equations 6.51 and/or 6.52.

For the case of constant wall flux at both surfaces, the descriptive equations have been solved for fully developed (hydrodynamically and thermally) laminar flow. The heat flux is given by Equations 6.51 and 6.52. The Nusselt numbers in terms of the local convection coefficients are given in Equations 6.53 and 6.54. The solutions for the Nusselt numbers are

$$\text{Nu}_o = \frac{\text{Nu}_{oo}}{1 - (q_{wi}''/q_{wo}'')\theta_i^*} \tag{6.55}$$

$$\text{Nu}_i = \frac{\text{Nu}_{ii}}{1 - (q_{wo}''/q_{wi}'')\theta_o^*} \tag{6.56}$$

The quantities θ_i^* and θ_o^* are called *influence coefficients*. Nu_{ii} is the inner-tube Nusselt number when only the inner tube is heated. On the other hand, Nu_{oo} is the outer-tube Nusselt number when only the outer tube is heated. The individual Nusselt numbers and influence coefficients are graphed in Figure 6.17 as a function of diameter ratio for fully developed flow.

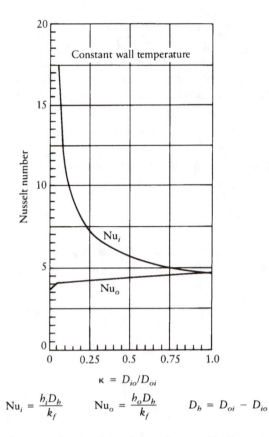

$$\mathrm{Nu}_i = \frac{h_i D_h}{k_f} \qquad \mathrm{Nu}_o = \frac{h_o D_h}{k_f} \qquad D_h = D_{oi} - D_{io}$$

FIGURE 6.16 Inner- and outer-wall Nusselt-number variation with diameter ratio for fully developed (hydrodynamically and thermally) laminar flow of a Newtonian fluid in a concentric annular area. The solutions are for one surface insulated and the other at constant temperature.

The entrance length required for the flow to become fully developed hydrodynamically in an annulus is

$$z_e = 0.05 D_h \mathrm{Re}_D \tag{6.1}$$

where the Reynolds number is based on the hydraulic diameter

$$\mathrm{Re}_D = \frac{V D_h}{\nu}$$

and

$$D_h = D_{oi} - D_{io}$$

The entrance length required for the flow to become fully developed thermally is given by

$$z_t = 0.05 D_h \mathrm{Re}_D \mathrm{Pr} \tag{6.57}$$

Solutions are available also for the combined-entry-length problem. The solutions for turbulent flow are quite complex because the influence coefficients are a function of Reynolds and Prandtl numbers. Such solutions are available in texts on convection heat transfer.[*] For a method that

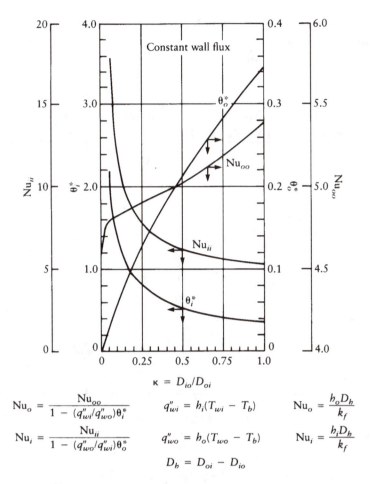

$$Nu_o = \frac{Nu_{oo}}{1 - (q''_{wi}/q''_{wo})\theta_i^*} \qquad q''_{wi} = h_i(T_{wi} - T_b) \qquad Nu_o = \frac{h_o D_h}{k_f}$$

$$Nu_i = \frac{Nu_{ii}}{1 - (q''_{wo}/q''_{wi})\theta_o^*} \qquad q''_{wo} = h_o(T_{wo} - T_b) \qquad Nu_i = \frac{h_i D_h}{k_f}$$

$$D_h = D_{oi} - D_{io}$$

FIGURE 6.17 Solution for constant heat rate applied to fully developed (hydrodynamically and thermally) laminar flow of a Newtonian fluid in a concentric circular annulus.

involves a good engineering approximation to calculating heat-transfer rates to or from the turbulent flow of a fluid in an annulus, see Section 9.2 on double-pipe heat exchangers.

The friction factor for laminar flow in an annulus can be derived in the same manner as was done for a circular duct. The result is

$$\frac{1}{f} = \frac{Re_D}{64}\left[\frac{1 + \kappa^2}{1 - \kappa} + \frac{1 + \kappa}{\ln(\kappa)}\right] \tag{6.58}$$

where $\kappa = D_{io}/D_{oi}$. Laminar conditions exist for Reynolds numbers (based on hydraulic diameter) of less than 2 100. For turbulent flow the Moody diagram can be used to find the friction factor with small error if $\kappa \le 0.75$.

6.4.2 RECTANGULAR CROSS SECTIONS

Solutions for various rectangular sections have been derived. Again the problems of constant wall flux and of constant wall temperature are considered. The Nusselt number for fully developed

---

* See, for example, Kays, W.M., and H.C. Perkins, Chapter 7 of *Handbook of Heat Transfer Fundamentals,* ed. W.M. Rohsenow, J.P. Hartnett, and L.N. Ganic, McGraw-Hill, NY, 1985.

(hydrodynamically and thermally) flow of a Newtonian fluid through rectangular ducts for both problems is provided in Table 6.4. The Nusselt number is expressed in terms of a local convection coefficient that is constant under fully developed conditions.

TABLE 6.4
Nusselt Number for Fully Developed (Hydrodynamically and Thermally) Laminar Flow of a Newtonian Fluid in Tubes of Various Rectangular Cross Sections

Cross-sectional shape	$\dfrac{w}{h} = n$	$D_h = \dfrac{4A}{P} = \dfrac{2nh}{n+1}$	Nu_D Constant axial heat rate	Nu_D Constant wall temperature
Square	1.0	h	3.61	2.98
	1.43	$1.18h$	3.73	3.08
	2.0	$1.33h$	4.12	3.39
h	3.0	$1.50h$	4.79	3.96
w	4.0	$1.60h$	5.33	4.44
	8.0	$1.77h$	6.60	5.95
	∞	$2h$	8.235	7.54
Two dimensional				
Insulated surface	∞	$2h$	5.385	4.86

The flow is considered laminar if the Reynolds number based on hydraulic diameter is less than 2 100; that is,

$$Re_D = \frac{VD_h}{\nu} < 2\,100 \quad \text{(laminar flow)}$$

and

$$D_h = \frac{4A}{P}$$

The equations written earlier for entrance length are applicable here provided hydraulic diameter is substituted into the expression for Reynolds number:

$$z_e = 0.05 D_h \, Re_D \text{ (hydrodynamic entrance length)} \tag{6.1}$$

$$z_t = 0.05 D_h \, Re_D \, Pr \text{ (thermal entrance length)} \tag{6.8}$$

For turbulent flow in rectangular ducts, the Seider-Tate and Dittus-Boelter Equations give reasonably accurate results if the Prandtl number is greater than 0.7.

The Seider-Tate Equation is:

$$Nu_D = 0.027 \, Re_D^{4/5} \, Pr^{1/3} (\mu/\mu_w)^{0.14} \tag{6.39}$$

(constant wall flux or constant wall temperature; hydrodynamically and thermally developed turbulent flow)

where

$$
\left.\begin{array}{ll}
0.7 \leq \mathrm{Pr} \leq 16\ 700 & \mathrm{Pr} = \dfrac{v}{\alpha} \\[2mm]
\mathrm{Re}_D \geq 10\ 000 & \mathrm{Re}_D = \dfrac{VD_h}{v} \\[2mm]
L/D_h \geq 60 & \mathrm{Nu}_D = \dfrac{h_z D_h}{k_f} \\[2mm]
\mu_w @ T_w &
\end{array}\right\} \qquad (6.40)
$$

The Dittus-Boelter Equation is:

$$
\mathrm{Nu}_D = 0.023 \mathrm{Re}_D^{4/5} \mathrm{Pr}^n \qquad (6.41)
$$

(constant wall flux or constant wall temperature; hydrodynamically and thermally developed turbulent flow)

where

$$
\left.\begin{array}{ll}
n = 0.4 \text{ if } T_w > T_b & \text{(heating)} \\[1mm]
n = 0.3 \text{ if } T_w < T_b & \text{(cooling)} \\[1mm]
0.7 \leq \mathrm{Pr} \leq 160 & \mathrm{Pr} = v/\alpha \\[1mm]
\mathrm{Re}_D \geq 10\ 000 & \mathrm{Re}_D = VD_h/v \\[1mm]
L/D_h \geq 60 & \mathrm{Nu}_D = \dfrac{h_z D_h}{k_f}
\end{array}\right\} \qquad (6.42)
$$

It is important to realize that at any axial location, the local convection coefficient varies around the periphery and approaches zero at the corners. Therefore the convection coefficients obtained from Table 6.4 or from the convection correlations are averages that apply about the perimeter.

For laminar flow in a rectangular duct the friction factor varies with Reynolds number, as shown in Figure 6.18. As shown, when the height- to-width ratio of the duct is known, the $f \mathrm{Re}_D$ product can be read from the graph. For turbulent flow, the Moody diagram can be used if hydraulic diameter is substituted into the appropriate expressions.

6.5 SUMMARY

In this chapter we have examined flow in ducts of various cross sections. The circular-duct problem was presented first and two heat-transfer-problem types were discussed: constant wall flux and constant wall temperature. These were presented in detail for laminar conditions. Convection equations were stated for turbulent flow, and more correlations are given in the exercises that follow.

Flow in annular and rectangular cross sections was then discussed. Solutions for laminar and turbulent fully developed conditions were stated.

Pressure drop for laminar and turbulent flows was also brought out but discussed only briefly because it is assumed that the reader has such information available from a study of fluid mechanics. A summary of correlations for fluid flow and heat transfer in ducts of various cross sections is provided in Table 6.5.

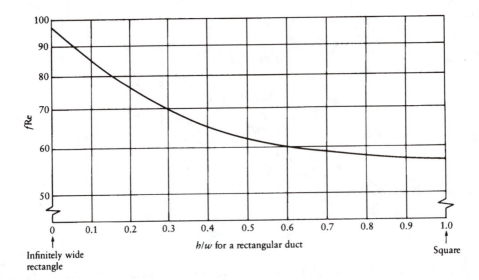

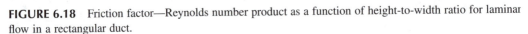

FIGURE 6.18   Friction factor—Reynolds number product as a function of height-to-width ratio for laminar flow in a rectangular duct.

6.6   PROBLEMS

6.6.1   ENTRANCE LENGTH

1. A liquid flows in a circular duct at a mass flow rate of 0.01 kg/s. Determine the entrance length required for the flow to become fully developed hydrodynamically if the fluid is
   (a) Water
   (b) Engine oil
   (c) Ethylene glycol
   Take the fluid temperature in all cases to be 40°C and the duct to be 2 standard type M copper tubing.

2. A gas flows in a 3 nominal schedule 40 pipe at a mass flow rate of 10 lbm/hr. Determine the entrance length required for the flow to become fully developed hydrodynamically if the gas is
   (a) Air
   (b) Hydrogen
   (c) Carbon dioxide
   Take the gas temperature in all cases to be 540°R.

3. Glycerin at an average temperature of 40°C flows in a heated 4 standard type M copper tube at a rate of 20 kg/s. Calculate the entrance length required for the flow to become fully developed
   (a) Hydrodynamically
   (b) Thermally

4. Liquid Freon-12 flows in a capillary tube located in the circuit of an air-conditioning system. The capillary tube has an inside diameter of 0.0625 in and is 1 ft long.
   (a) If the flow is fully developed hydrodynamically at tube end, determine the mass flow rate, assuming the average temperature of the Freon-12 is –4°F.
   (b) If the tube is being cooled and the average temperature of the Freon-12 is –4°F, what would the corresponding thermal entrance length be at the above-determined mass flow rate?

TABLE 6.5
Summary of Equations of Chapter 6

Cross section	D_h	f	Laminar flow $Re_D < 2100$		f	Turbulent flow $Re_D > 2100$	
			Nu_D Constant wall flux	Nu_D Constant wall temperature		Nu_D Constant wall flux	Nu_D Constant wall temperature
	D	$\dfrac{64}{Re_D}$	4.364	3.658	Fig. 6.14	Eq. 6.39	Eq. 6.39
			Hydrodynamically and thermally developed		Eq. 6.48	Eq. 6.41	Eq. 6.41
					Table 6.3	Prob. 28	Prob. 30
			Fig. 6.8	Fig. 6.8		Hydrodynamically and thermally developed	
			Hydrodynamically developed; thermally developing				
			Fig. 6.12	Fig. 6.11 Eq. 6.33			
			Hydrodynamically and thermally developing				
$\kappa = \dfrac{D_{io}}{D_{oi}}$	$D_{oi} - D_{io}$	Eq. 6.58	Fig. 6.17	Fig. 6.17	Fig. 6.14	See Section 9.2 on double-pipe heat exchangers	
			Hydrodynamically and thermally developed		Eq. 6.48		
					Table 6.3		
					$\kappa \le 0.75$		
	$\dfrac{2wh}{h+w}$	Fig. 6.18	Table 6.4	Table 6.4	Fig. 6.14	Eq. 6.39	Eq. 6.39
			Hydrodynamically and thermally developed		Eq. 6.48	Eq. 6.41	Eq. 6.41
					Table 6.3		

Liquid metals: see Prob. 27; Internally finned tubes: see Prob. 31; Curved tubes: see Prob. 32.

Constant wall flux: $q_w'' = h_z(T_w - T_b)$ $T_{bo} - T_{bi} = \dfrac{2q_w'' \alpha L}{V k_f R}$

Constant wall temperature: $q = \bar{h}_L A_s \dfrac{[(T_w - T_{bo}) - (T_w - T_{bi})]}{\ln[(T_w - T_{bo})/(T_w - T_{bi})]}$ $\dfrac{T_w - T_{bo}}{T_w - T_{bi}} = \exp\left(\dfrac{-2\alpha L \bar{h}_L}{V k_f R}\right)$

Hydrodynamic entrance length: $z_e = 0.05D\,Re_D$ Thermal entrance length: $z_t = 0.05D\,Re_D\,Pr$
(laminar flow)   (laminar flow)

$Re_D = \dfrac{V D_h}{v}$ $Pr = \dfrac{v}{\alpha}$ $Nu_D = \dfrac{h_z D_h}{k_f}$ $\overline{Nu}_D = \dfrac{\bar{h}_L D_h}{k_f}$

5. An automobile is tested indoors. Exhaust gases are removed via a 4-in-ID duct through a wall to the outside. The duct is 18 ft long, and the mass flow of gas is 0.005 lbm/s. The exhaust gases have the same properties as carbon dioxide. At the tube inlet the gases are at an average temperature of 440°F.
   (a) Determine the entrance length required for the flow to become fully developed hydrodynamically.
   (b) Is the flow fully developed thermally at tube end?
6. Liquid at an average temperature of 20°C flows in a 1 nominal schedule 40 pipe at a rate of 50×10^{-4} kg/s. Determine the hydrodynamic entrance length if the liquid is
   (a) Water
   (b) Engine oil
7. Determine the thermal entrance length for the liquids and conditions in Problem 6.

6.6.2   CONSTANT WALL FLUX

8. Water is conveyed at 2 cm³/s in a 1 standard type M copper tube that is 2 m long. Water enters the tube at 20°C and at tube end reaches 60°C. If heat is added at a constant rate and the flow is fully developed hydrodynamically and thermally at tube inlet, find the rate of heat addition. Calculate the maximum wall temperature.

9. A 15-ft-long, 1-1\2 nominal schedule 40 pipe conveys engine oil. The oil is cooled in the pipe from a temperature of 150° to 130°F. The heat removal rate is constant and equal to 25 W/(m² of pipe inside surface area). For hydrodynamically and thermally developed conditions throughout the tube, determine the mass flow rate of oil and the wall temperature at tube outlet.

10. Oxygen flows in a tube where it is to be heated from an inlet temperature of 250 K to 450 K. The tube is 3 standard type K copper tubing, and heating is to be effected by a constant wall flux of 1 500 W/(m² of inside tube area). For a flow of oxygen that can be varied from 0 to 9 kg/hr, determine a suitable tube length; graph mass flow rate vs. required tube length. Determine also the tube-wall-outlet temperature and graph its variation with mass flow rate. The oxygen is fully developed hydrodynamically and thermally.

11. Example 6.1 deals with ethylene glycol being heated by a constant wall flux. Fluid properties are evaluated at the inlet temperature, but for better results, the average bulk temperature should be used. With the results obtained in the example, obtain a better estimate of the properties and rework the example.

12. Methyl chloride is used as an anesthetic and as a refrigerant. Liquid methyl chloride is heated in a continuous process in which its temperature is raised from –30° to 0°C. A hydrodynamically developed flow of methyl chloride travels at an average velocity of 3 cm/s through a 3/4 standard type K copper tube. An electric resistance heater wound around the tube provides a constant wall flux of 2 250 W/m². Determine the tube length required to effect the required temperature rise. Graph the variation of wall temperature with length over the heated portion of the tube.

13. Sulfur dioxide is used in the manufacture of sulfuric acid. Liquid sulfur dioxide is to be heated in a copper tube that is surrounded by a heater that supplies a constant wall flux of 450 BTU/(hr·ft²). The average velocity of the fluid is 0.2 ft/s. The liquid is hydrodynamically developed by the time it reaches the heated section, which is 16 ft long. The inlet temperature of the fluid is –40°F, and it is heated to 20°F. Select a suitable tube size. Calculate the wall temperature at tube outlet.

6.6.3   CONSTANT WALL TEMPERATURE

14. Example 6.2 deals with a double-pipe heat exchanger in which steam condenses on the outside surface of a tube conveying water. Keeping all other parameters the same, what effect does an increase in tube diameter have? Repeat the calculations in the example for 1-1/2 standard type M, 2 standard type M, 2-1/2 standard by M, and 3 standard type M. Construct a graph of nominal diameter vs. $\bar{h}_L$.

15. A 1-m-long double-pipe heat exchanger is used to transfer heat from a condensing refrigerant to a hydrogen stream. Hydrogen travels at 1.2×10^{-4} kg/s through a 1-1/2 standard type M copper tube. The hydrogen inlet temperature is 250 K, and the wall temperature is kept constant at –10°C by the condensing refrigerant. Calculate the outlet temperature of the hydrogen and the average convection coefficient.

16. Ammonia or ammonia compounds are used extensively to manufacture fertilizers, in baking powders, in smelling salts, as a soldering flux, and in producing textiles. Consider a double-pipe heat exchanger in which a refrigerant stream is condensing at 75°F in the annulus and heating ammonia in the inner tube. Ammonia enters the exchanger at a

temperature of 14°F and is to be heated to a temperature of 50°F at the outlet. The copper tube is 1-1/2 standard type M, 15 ft long. What is the corresponding mass flow rate of ammonia? Assume the flow to be fully developed hydrodynamically.

17. Suppose the milk of Example 6.3 is to be heated under the same conditions, but instead of using a constant wall flux, a constant wall temperature is to be maintained. Determine the required constant wall temperature.

18. Oxygen is used in an oxyacetylene cutting torch. The oxygen is mixed with the acetylene, and the mixture is ignited. The flame heats metal to the melting temperature, and then a high-velocity, pure-oxygen jet aimed at the molten material removes it. Oxygen for the cutting process is supplied from a tank.

    Consider an oxygen-bottling operation where oxygen is separated from atmospheric air by cooling an air mixture to a very low temperature and removing the condensed oxygen. The oxygen is then allowed to vaporize and is piped to tanks used in the cutting operation. Oxygen enters a 1 standard type K copper tube at an inlet temperature of 150 K. Over a 3-m length of tube, the oxygen is to be heated to room temperature (300 K) so an accurate mass flow rate can be measured and the oxygen can be bottled. Heating is accomplished by passing the oxygen through the inner tube of a double-pipe heat exchanger. The fluid on the outside (i.e., in the annulus) can be a condensing refrigerant. Selection of a refrigerant, however, depends on the wall temperature required. Prepare a graph of wall temperature vs. mass flow rate of oxygen for laminar conditions.

19. Example 6.4 deals with the freezing of water, which is effected by passing Freon-12 through a submerged tube. Suppose that a different refrigerant is used, however. Repeat the first part of the example, using the same conditions except substituting eutectic calcium chloride solution for the Freon-12. Which refrigerant freezes the water faster? What are the significant parameters that affect the results? Calculate also the pressure drop experienced by the Freon-12 and the calcium chloride.

20. An air compressor takes in atmospheric air and compresses it to a pressure of 150 psig. The air temperature increases to 135°F, which is too warm to use for most industrial applications. The warmed compressed air is channeled to a double-pipe heat exchanger (called an *intercooler*), where water circulates through the annulus. The water-flow rate is very high, and the wall of the tube carrying the air is kept roughly at a constant temperature of 65°F. The intercooler is to cool the air to 75°F. How long an intercooler is necessary if the mass flow of air is 5 lbm/hr and the tube is 5/8 standard type K copper? What is the air pressure at the end of the intercooler?

21. Oxygen is to be heated in a heat exchanger where the tube-wall temperature is constant. The mass flow of oxygen is 0.001 kg/s, and the tube is 1-1/2 nominal schedule 40 pipe, 3 m long. Calculate the oxygen outlet temperature for both of the following cases:
    (a) Oxygen inlet temperature = 300 K
    (b) Oxygen inlet temperature = 400 K
    For both cases, take the tube wall temperature to be equal to the oxygen-inlet temperature plus 50 K. Use properties evaluated at the inlet temperature.

22. Nitrogen is heated as it flows in a 2 standard type M copper tube at an average velocity of 0.2 m/s. Nitrogen inlet temperature is 250 K, and the outlet temperature is 400 K. The wall temperature is constant and equal to 500 K. The tube length is 6 m. Use the Seider-Tate correlation (Equation 6.33) to calculate the convection coefficient. Compare the result to that obtained from any of the graphs in the chapter that apply.

23. Water enters a 1 nominal schedule 40 galvanized pipe at a bulk fluid temperature of 60°F. The tube length is 20 ft.
    (a) Use Equation 6.33 for laminar flow ($Re_D \leq 2100$) to construct a graph of mass flow rate vs. outlet temperature.

(b) Use Equation 6.39 for turbulent flow ($10\ 000 \leq \mathrm{Re}_D \leq 20\ 000$) to construct a graph of mass flow rate vs. outlet temperature.

In both cases, evaluate properties at the inlet temperature and take the wall temperature to be constant at 200°F.

(c) Graph also mass flow rate vs. pressure drop over the same ranges specified in (a) and (b).

6.6.4 EMPIRICAL CORRELATIONS

24. In Example 6.6, the Dittus-Boelter Equation is used to find the wall flux at tube end. Use the Seider-Tate Equation (6.33) to obtain a refined estimate of the wall flux.

25. Freon-12 flows at a rate of 1.2 kg/mm through a 1/4 standard type K drawn-copper tube that is 1.4 m long. The Freon inlet temperature is –35°C, and the tube-wall temperature is maintained at 15°C. Estimate
    (a) The local convection coefficient at tube end
    (b) The Freon temperature at exit
    (c) The log-mean temperature difference
    (d) The heated added to the fluid
    (e) The average convection coefficient
    (f) The pressure drop due to friction

    Take the properties of Freon-12 at the average of the inlet temperature and the wall temperature.

26. Engine oil flowing at 0.2 lbm/s is cooled by passing it through finned tubes over which air moves. The inside tube through which the oil flows is 1/2 standard type M stainless steel (assume dimensions the same as copper tubing), 3.5 ft long. The oil inlet temperature is 180°F, and the air keeps the wall temperature at nearly a constant of approximately 68°F. Estimate the oil outlet temperature and the pressure drop experienced by the oil due to friction.

27. Liquid metals have very low Prandtl numbers (e.g., from Appendix C, the Prandtl number of mercury is 0.024 9 at 20°C, while that of water is 7.02 at 20°C). An equation that can be used to predict the Nusselt number for thermally developed turbulent flow of a liquid metal is

$$\mathrm{Nu}_D = 6.3 + 0.016\ 7\ \mathrm{Re}_D^{0.85}\ \mathrm{Pr}^{0.93}$$

    valid if $\mathrm{Pr} < 0.03$.

    Consider fully developed (hydrodynamically and thermally) flow of mercury in a 2-m-long drawn-metal tube that has an inside diameter of 1 cm and an outside diameter of 1.4 cm. The tube is made of glass and kept cool by submerging it in a cold water bath of temperature 15°C. The convection coefficient between the water and the outside tube wall is 800 W/(m²·K). The mercury is cooled from a temperature of 30° to 20°C. Determine the wall temperature at the duct exit. Take the mass flow of mercury to be 220 kg/hr, and determine also the pressure drop due to friction.

28. The Petukhov Equation has been derived to describe fully developed turbulent flow of a Newtonian liquid in a smooth tube:

$$\mathrm{Nu}_D = \frac{(f/8)\,\mathrm{Re}_D\,\mathrm{Pr}}{1.07 + 12.7(f/8)^{1/2}(\mathrm{Pr}^{2/3} - 1)(\mu/\mu_w)^n}$$

where

$n = 0.11$ if $T_w > T_b$

$n = 0.25$ if $T_w < T_b$

$n = 0$ if wall flux is constant

μ_w evaluated at T_w

μ evaluated at $(T_{bo} + T_{bi})/2$

All other properties evaluated at $(T_w + T_b)/2$

$0.5 < \text{Pr} = v/\alpha < 200$ (6% accuracy)

$200 < \text{Pr} < 2\ 000$ (10% accuracy)

$10^4 < \text{Re}_D = VD/v < 5 \times 10^6$

$0 < (\mu/\mu_w) < 40$

Consider a fully developed flow of water in a 3 nominal schedule 40 pipe. The inlet water temperature is 200°F and, at exit, the temperature is 80°F. The tube is 20 ft long, and the mass flow of water is 6 lbm/s. Using the Petukhov Equation, determine the convection coefficient for a wall temperature of 68°F. Compare to that obtained with the Seider-Tate Equation. Take the relative roughness of the wall (ε/D) to be that for a smooth-walled duct. Determine also the pressure drop.

29. The Hausen Equation has been obtained from empirical considerations for fully developed laminar flow in tubes having a constant wall temperature:

$$\overline{\text{Nu}}_D = 3.66 + \frac{0.0668(D/L)\ \text{Re}_D\text{Pr}}{1 + 0.4[(D/L)\text{Re}_D\ \text{Pr}]^{2/3}}$$

in which the heat-transfer coefficient $\bar{h}_L$ $(\overline{\text{Nu}}_D = \bar{h}_L D/k_f)$ is an average value over the entire length of tube, and fluid properties are evaluated at the average bulk temperature.

Consider the flow of glycerine in a 1/2 standard type M copper tube. Glycerine is to be heated from a temperature of 20° to 40°C by keeping the wall temperature at a constant value of 100°C. How long a tube is required to do this? The mass flow of glycerine is 2.2 kg/s.

30. Another relation proposed by Seider and Tate for laminar-flow heat transfer in tubes of constant wall temperature is

$$\overline{\text{Nu}}_D = 1.86(\text{Re}_D\ \text{Pr})^{1/3}(D/L)(\mu/\mu_w)^{0.14}$$

which is valid if $\text{Re}_D\ \text{Pr}(D/L) > 10$. The average heat-transfer coefficient $\bar{h}_L$ $(\overline{\text{Nu}}_D = \bar{h}_L D/k_f)$ is based on the average of the inlet and outlet temperature differences (wall minus bulk for heating). The heat transferred is thus calculated with

$$q = \bar{h}_L A_s\left(T_w - \frac{T_{bi} + T_{bo}}{2}\right)$$

The fluid viscosity μ_w is based on the wall temperature, and the remaining properties are evaluated at the average bulk temperature of the fluid.

Engine oil flows in the inner tube of a double-pipe heat exchanger. An evaporating refrigerant in the annulus absorbs heat from the oil and maintains the wall temperature at 70°F. The oil enters the exchanger at a temperature of 120°F and should be cooled to 90°F. The inner tube is 1 standard type M copper tubing. How long must it be? The mass flow of oil is 3 lbm/s.

31. Figure P6.1 is a photograph of a heat exchanger in which the inner tube contains fins. The spiraled fins are an integral part of the tube wall. The Carnavas empirical relations apply to such configurations; they are

$$\mathrm{Nu}_D = 0.023\,\mathrm{Re}_D^{0.8}\,\mathrm{Pr}^{0.4}\left[\frac{D_i}{D_e}\left(1-\frac{2\delta}{D_i}\right)\right]^{-0.2}\left(\frac{D_iD_h}{D_e^2}\right)^{0.5}\sec^3\theta$$

for heat transfer, and

$$f\,\mathrm{Re}_D^{0.2} = 0.046\frac{D_e}{D_i}\sec^{3/4}\theta$$

for wall friction. The parameters are defined as

θ = helix angle from tube axis
D_i = diameter to base of fins
D_e = tube diameter if fins are melted down and attached to inside wall
δ = fin height
D_h = hydraulic diameter
$\mathrm{Re}_D = VD_h/\nu > 2100$ (turbulent flow)
$\overline{\mathrm{Nu}_D} = \bar{h}_L D_h/k_f$

Consider the tube illustrated in Figure P6.1. The tube itself has the same basic dimensions of 3/4 standard type K tubing. The spiraled internal fins are assumed to be rectangular in cross section (for calculation purposes). They are 1-1/2 times the wall thickness in height and equal to the wall thickness in width. The tube contains 10 equally spaced fins, and the effect of the helix angle of the fins is to be examined. The exchangers, available in stock lengths of 5 ft, are designed to transfer heat from water (inner tube) to a condensing refrigerant. Water travels through the tube at a rate of 0.1 lbm/s, and the inlet temperature is 140°F. The tube wall is at a constant temperature of 65°F.

FIGURE P6.1   *Source:* courtesy of Wolverine Tube, Inc.

(a) Determine the outlet temperature of the water as a function of the helix angle θ $(0° < \theta < 70°)$.

(b) Determine the friction factor and corresponding pressure drop.

(c) Determine the friction factor and corresponding pressure drop if the tube contains no fins and is made of drawn tubing.

(d) Determine the outlet temperature if the tube contains no fins. (Make any reasonable assumptions to obtain the convection coefficient.)

(e) For equal pumping power, are the fins effective at reducing the outlet temperature?

32. Oil in a tank is to be heated by passing steam through a submerged curved tube. The tube is shown schematically in Figure P6.2. The steam flow within the tube can be either laminar or turbulent. Correlations for both types of flow have been derived. These are

$$\text{Dean number} = \text{De} = \frac{VD}{\nu} \frac{(D/D_c)^{1/2}}{2} = \frac{\text{Re}_D}{2}\left(\frac{D}{D_c}\right)^{1/2}, \quad \frac{f}{f_{\text{straight}}} = \left\{1 - \left[1 - \left(\frac{5.8}{\text{De}}\right)^{0.45}\right]^{2.22}\right\}^{-1}$$

valid if

$$5.8 < \text{De} < 1000$$

$$\text{Nu}_D = \frac{h_z D}{k_f} = 0.13 \, \text{Pr}^{1/3}(f \, \text{Re}_D^2/8)^{1/3}$$

where D is the pipe diameter, D_c is the diameter of the pipe curvature, and h_z is the local convection coefficient.

On the other hand, for turbulent flow the Ito empirical relations apply:

Turbulent flow:

$$\frac{f}{f_{\text{straight}}} = \left[\text{Re}_D\left(\frac{D}{D_c}\right)^2\right]^{0.05}$$

$$\text{Re}_D = \left(\frac{D}{D_c}\right)^2 > 6$$

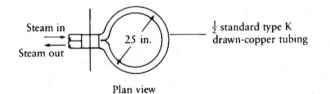

Steam in

Steam out

25 in.

$\frac{1}{2}$ standard type K drawn-copper tubing

Plan view

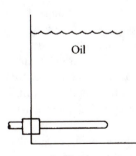

Oil

Profile view

FIGURE P6.2

where, for turbulent flow,

$$Re_D = 2 \times 10^4 \left(\frac{D}{D_c}\right)^{0.32}$$

The corresponding Seban-McLaughlin Equation for heat transfer is

$$Nu_D = \frac{h_z D}{k_f} = \frac{f \, Re_D \, Pr^{0.4}}{8}$$

Consider superheated steam entering the curved tube of the oil container (Figure P6.2) at a temperature of 350°F. Evaluate the properties at the inlet temperature and determine the convection coefficient. If the heat lost by the steam is 25 BTU/hr, calculate the steam and wall temperatures at outlet, using the convection coefficient determined earlier. The steam flow is 1 lbm/hr.

6.6.5 NONCIRCULAR CROSS SECTIONS

33. A double-pipe heat exchanger is set up to take data on a constant-wall-temperature cooling problem. Water moves through a 3/8-in-ID tube, and Freon condenses on the outside surface of the tube (1/2-in OD). The following data are obtained:

| Test no. | Water, °F | | Q (gpm) | Average Freon-12 bulk temperature |
	T_{in}	T_{out}		
1	110.12	98.15	0.875	42.35
2	103.46	93.92	0.975	47.30
3	114.17	104.18	0.975	48.2
4	119.12	105.80	0.975	51.8
5	108.68	98.69	1.05	39.38

The *L/D* ratio for the water flow is 26.67. Determine the Nusselt number from the data and again by using the Dittus-Boelter Equation. Calculate percentage errors between the two Nusselt numbers for each test. (Note that the Freon data is the *fluid* temperature, which may be taken to be equal to the wall temperature. The actual wall temperature may be about 1° to 13°F higher than the Freon-12 temperature, however.) Evaluate fluid properties for all tests at 104°F.

34. Repeat Problem 33, but use the Seider-Tate Equation instead.

35. Repeat Problem 33, but use the Petukhov Equation (see Problem 28) instead. The tubing used is drawn-copper tubing.

36. A 3-m-long annular flow area consists of a 1/2 standard tube placed within a 1 standard tube, both type M and made of copper. Air flowing through the annulus receives heat from a vaporizing refrigerant in the inner tube that keeps the outside diameter of the inner tube at 10°C. The outer tube is submerged in ice, which begins melting and maintains the inside-wall temperature of the outer tube at 0°C. A hydrodynamically developed flow of air at 125°C enters the annulus at a mass flow rate of 1×10^{-3} kg/s. Determine the expected outlet temperature. Calculate also the friction factor and pressure drop.

37. Repeat Problem 36, using the same information except that the inside tube is insulated rather than maintained at constant temperature.

38. Repeat Problem 36, using the same information except that the outside tube wall is insulated rather than maintained at constant temperature.

39. Construct a graph of Nusselt number vs. *w/h* ratio for the constant-heat-flux data of Table 6.4.

40. Construct a graph of Nusselt number vs. *w/h* ratio for the constant-wall-temperature data of Table 6.4.

41. Air that is hydrodynamically and thermally developed flows in a rectangular duct of inside dimensions 2 × 1 ft. The inlet air temperature is 70°F, and at outlet the air temperature is 90°F. The mass flow of air is 0.1 lbm/min, and the duct length is 8 ft.
    (a) How much heat in the form of a constant wall flux is required to heat the air?
    (b) Calculate the pressure drop due to friction.
    (c) Repeat the calculations for a circular duct of the same cross-sectional area.

42. Using the data of Problem 41, determine the constant wall temperature that would be required to heat the air.

43. Two rectangular ducts are available for heating air. One duct is 20 x 10 cm, and the other has the same cross-sectional area but dimensions of 40 × 5 cm. Air is to be heated in one of these ducts from a temperature of 60° to 100°F by maintaining the wall flux constant. If the mass flow of air is 6 lbm/hr, and the ducts are 10 ft long, determine the heat flux required in both cases to heat the air. Assume hydrodynamically and thermally developed flow. Calculate also the friction factor for both cases.

6.6.6 DERIVATIONS AND THEORETICAL PROBLEMS

44. Construct a graph of $V_z/V_{z\,max}$ vs. *r/R* for laminar flow in a circular duct. On the same axes, construct a graph of $R\tau_{rz}/\mu V_{z\,max}$ (dimensionless shear stress) vs. *r/R*, and $V/V_{z\,max}$ vs. *r/R*.

45. Equation 6.20 is derived by first assuming fully developed (hydrodynamically) flow of a Newtonian fluid in a circular duct. The velocity profile that applies in terms of average velocity is parabolic:

$$V_z = 2V\left(1 - \frac{r^2}{R^2}\right)$$

Instead of using the above equation, suppose that the profile selected is a constant:

$$V_z \cong V$$

The flow described by this equation is called *slug flow*. Using the slug-flow description, perform an analysis similar to that provided for the constant-wall-flux problem illustrated in Figure 6.3 by following the steps below:
(a) Perform a heat balance for a slice of the tube (Figure 6.4) and derive Equation 6.9b.
(b) Differentiate the equation for thermally developed flow and derive Equation 6.14.
(c) Verify that the applicable differential equation is

$$V_z\frac{\partial T}{\partial z} = \frac{\alpha}{r}\frac{\partial}{\partial r}\left(r\frac{\partial T}{\partial r}\right)$$

(d) Substituting appropriately for V_z and $\partial T/\partial z$, show that the above differential equation becomes

$$\frac{d}{dr}\left(r\frac{dT}{dr}\right) = \frac{2q_w'' r}{k_f R}$$

(e) Integrate the above expression twice to obtain

$$T = \frac{q_w'' r^2}{2 k_f R} + C_1 \ln(r) + C_2$$

(f) Substitute the boundary conditions:
$r = 0; \; T = $ finite
$r = R; \; T = T_w$
and derive the following temperature distribution:

$$T - T_w = \frac{q_w'' R}{2 k_f}\left(\frac{r^2}{R^2} - 1\right)$$

(g) Integrate the above profile over the cross-sectional area and show that the bulk-fluid temperature is

$$T_b - T_w = -\frac{q_w'' R}{4 k_f} = -\frac{q_w'' D}{8 k_f}$$

(h) Substitute for the wall flux in terms of the local convection coefficient; that is,

$$q_w'' = h_z(T_w - T_b)$$

and show that the local Nusselt number becomes

$$\mathrm{Nu}_D = \frac{h_z D}{k_f} = 8$$

46. In Problem 45, the slug-flow assumption is made, in which the instantaneous velocity is a constant: $V_z = V$. Another approximation to the parabolic profile that can be made is that the profile is linear. That is,

$$V_z = 3V\left(1 - \frac{r}{R}\right)$$

Using the linear profile, follow the steps expressed in Problem 45 and show that, for the constant-wall-flux problem,

$$\mathrm{Nu}_D = 1.263$$

47. Starting with Equation 6.17, derive Equation 6.18.
48. Starting with Equation 6.29, derive Equation 6.30.
49. The velocity in an annulus is

$$V_z = \frac{R^2}{4\mu}\left(-\frac{dp}{dz}\right)\left[1 - \frac{r^2}{R^2} - \frac{1 - \kappa^2}{\ln(\kappa)}\ln\left(\frac{r}{R}\right)\right]$$

where $\kappa = D_{io}/D_{oi}$. Follow the same line of reasoning as that used in deriving Equation 6.38 to show that, for an annulus, the friction factor for laminar flow is

$$\frac{1}{f} = \frac{\mathrm{Re}}{64}\left[\frac{1+\kappa^2}{1-\kappa} + \frac{1+\kappa}{\ln(\kappa)}\right]$$

50. The equations for the constant-wall-temperature problem can be combined to yield a surprisingly simple result. The heat-balance equations are

$$q = \dot{m}c_p(T_{bo} - T_{bi})$$

and

$$q = \bar{h}_L A_s \frac{[(T_w - T_{bo}) - (T_w - T_{bi})]}{\ln[(T_w - T_{bo})/(T_w - T_{bi})]}$$

(a) Setting these two expressions equal to each other and solving for mass flow rate, show that for a circular duct

$$\dot{m} = \frac{\bar{h}_L \pi DL}{c_p \ln[(T_w - T_{bo})/(T_w - T_{bi})]}$$

(b) Substitute from the definition of h_L in terms of Nusselt number to obtain

$$\dot{m} = \frac{\overline{\mathrm{Nu}_D} k_f \pi L}{c_p \ln[(T_w - T_{bo})/(T_w - T_{bi})]}$$

(c) From the definition of Reynolds number derive an expression for Reynolds number in terms of the mass flow rate. Use the result with the definition of the Prandtl number to show that the inverse Graetz number becomes

$$\frac{1}{\mathrm{Gz}} = \frac{L}{D\,\mathrm{Re}_D\,\mathrm{Pr}} = \frac{L\rho\pi\alpha}{4\dot{m}}$$

(d) Solve the above equation for mass flow rate and set equal to that in part (b) to show that

$$\frac{T_w - T_{bo}}{T_w - T_{bi}} = \exp[-4(1/\mathrm{Gz})(\overline{\mathrm{Nu}_D})]$$

7 Convection Heat Transfer in Flows Past Immersed Bodies

CONTENTS

7.1   INTRODUCTION

In this chapter, we will examine a number of problems in which a fluid stream moves relative to a body. The fluid and the body-surface temperatures will be different, so a net energy transfer will take place. The objective is to determine the parameters that influence the energy transfer and to calculate the actual quantity of heat transferred.

The problems discussed include flow past a flat plate, cylinders of various cross sections, a sphere, and tube banks. The method of approach here is the same as that in the previous chapter. The pertinent fluid-mechanics details are summarized, and the heat-transfer problem is investigated. Applications are stressed in the examples and in the problems.

7.2   LAMINAR-BOUNDARY-LAYER FLOW OVER A FLAT PLATE

As discussed in Chapter 5, a number of fluid-mechanics problems exist in which the flow field is divided into two regimes. One applies to the fluid far from a solid surface. The other regime applies to the fluid in the vicinity of a bounding surface or wall. The flow regime near a bounding surface is called *boundary-layer flow*. The boundary-layer equations were derived in Chapter 5.

Figure 7.1 illustrates laminar flow over a flat plate. When the boundary-layer equations are applied and solved for this system, the Blasius solution is obtained. The Blasius solution is not derived in this text, but the solution itself is presented. For the flow to be laminar, the Reynolds number based on length x must be less than 5×10^5; that is,

$$\mathrm{Re}_x = \frac{V_\infty x}{\nu} \leq 5 \times 10^5 \quad \text{(laminar flow)} \tag{7.1}$$

Transition takes place over a range of values, but 5×10^5 is assumed arbitrarily here for purposes of calculation.

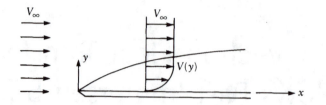

FIGURE 7.1   Laminar flow of a Newtonian fluid over a flat plate.

The boundary-layer equations are given in Chapter 5. For a Newtonian fluid flowing over a flat plate, the two-dimensional equations are

continuity

$$\frac{\partial V_x}{\partial x} + \frac{\partial V_y}{\partial y} = 0 \tag{7.2}$$

x component, Navier-Stokes Equation

$$V_x \frac{\partial V_x}{\partial x} + V_y \frac{\partial V_x}{\partial y} = \nu \frac{\partial^2 V_x}{\partial y^2} \tag{7.3}$$

y component

$$\frac{\partial p}{\partial y} = 0 \tag{7.4}$$

The boundary conditions are

$$\left.\begin{array}{llll}
\text{B.C.1.} & y = 0, & V_x = 0 \\
\text{B.C.2.} & y = 0, & V_y = 0 \\
\text{B.C.3.} & y = \infty, & V_x = V_\infty
\end{array}\right\} \tag{7.5}$$

The Blasius solution (for V_x) of the above equations, subject to the boundary conditions, is given graphically in Figure 7.2 and tabulated in Table 7.1.

The functional solution to flow over a flat plate has no known analytical form. The solution is actually a power series. It is customary to define a boundary-layer thickness at the location where V_x is over 99% of the free-stream velocity V_∞. This occurs at $\eta \approx 5$. Therefore, we can determine an expression for the thickness of the hydrodynamic boundary layer as

$$\eta = \delta \left(\frac{V_\infty}{\nu x}\right)^{1/2} = 5.0 \quad \text{(Table 7.1)}$$

Rearranging,

$$\delta = 5.0 \left(\frac{\nu x^2}{V_\infty x}\right)^{1/2}$$

or

$$\delta = \frac{5.0x}{\sqrt{Rex}} \tag{7.6}$$

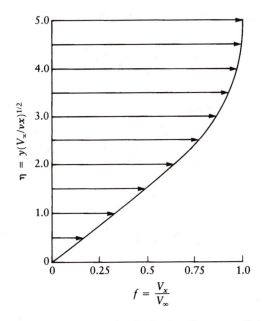

FIGURE 7.2 Blasius dimensionless velocity profile for laminar flow past a flat plate.

TABLE 7.1
Boundary-Layer Velocity Profile for Laminar Flow Past a Flat Plat (Blasius Solution)

$\eta = y\left(\dfrac{V_\infty}{vx}\right)^{1/2}$	$f = V_x/V_\infty$	$\eta = y\left(\dfrac{V_\infty}{vx}\right)^{1/2}$	$f = V_x/V_\infty$
0.0	0.0	2.8	0.811 52
0.2	0.066 41	3.0	0.846 05
0.4	0.132 77	3.2	0.876 09
0.6	0.198 94	3.4	0.901 77
0.8	0.264 71	3.6	0.923 33
1.0	0.329 79	3.8	0.941 12
1.2	0.393 78	4.0	0.955 52
1.4	0.456 27	4.2	0.966 96
1.6	0.516 76	4.4	0.975 87
1.8	0.574 77	4.6	0.982 69
2.0	0.629 77	4.8	0.987 79
2.2	0.681 32	5.0	0.991 55
2.4	0.728 99	∞	1.000 00
2.6	0.772 46		

The velocity profile in the laminar boundary layer at any x location is represented by Figure 7.2 and Table 7.1. This implies that all velocity profiles, regardless of x location, are similar.

As fluid moves past the plate, friction causes a force to exist that tends to move the plate in the flow direction. For the flat-plate problem, this is called the *skin-friction drag force*. To determine an expression for skin-friction drag, we use the Newtonian expression for shear stress at the surface:

$$\tau_w = \mu\left(\frac{\partial V_x}{\partial y}\right)\bigg|_{y=0}$$

The chain rule is used to evaluate the slope at the wall. Thus,

$$\left(\frac{\partial V_x}{\partial y}\right)\bigg|_{y=0} = \left(\frac{\partial V_x}{\partial f}\right)\left(\frac{\partial f}{\partial \eta}\right)\bigg|_{y=0}\left(\frac{\partial \eta}{\partial y}\right) \tag{7.7}$$

From the Blasius solution, we evaluate

$$\left(\frac{\partial f}{\partial \eta}\right)\bigg|_{y=0} = \frac{0.066\ 41 - 0.0}{0.2 - 0.0} = 0.332$$

Also, with $f = V_x/V_\infty$ and

$$n = y\left(\frac{V_x}{vx}\right)^{1/2}$$

we determine

$$\frac{\partial V_x}{\partial f} = V_\infty$$

and

$$\left(\frac{\partial n}{\partial y}\right) = \left(\frac{V_x}{vx}\right)^{1/2}$$

Substituting into Equation 7.7 gives

$$\left(\frac{\partial V_x}{\partial y}\right)\bigg|_{y=0} = V_\infty(0.332)\left(\frac{V_x}{vx}\right)^{1/2}$$

Manipulating the right-hand side, we obtain

$$\left(\frac{\partial V_x}{\partial y}\right)\bigg|_{y=0} = V_\infty(0.332)\left(\frac{V_x}{vx} \cdot \frac{x}{x}\right)^{1/2}$$

or

$$\left(\frac{\partial V_x}{\partial y}\right)\bigg|_{y=0} = \frac{0.332 V_\infty}{x}\left(\frac{V_x x}{v}\right)^{1/2}$$

Finally, we get

$$\left(\frac{\partial V_x}{\partial y}\right)\bigg|_{y=0} = \frac{0.332 V_\infty}{x}\sqrt{Re_x}$$

The wall shear stress then becomes

$$\tau_w = \frac{0.332 V_\infty \mu}{x} \sqrt{Re_x}$$

The total skin-friction drag is obtained by integrating the wall shear stress over the plate area. With b defined as the plate width, we get

$$D_f = \int_0^L \tau_w b \, dx = \int_0^L \frac{0.332 V_\infty \mu b}{x} \left(\frac{V_\infty x}{v}\right)^{1/2} dx$$

Solving,

$$D_f = 0.664 V_\infty \mu b \sqrt{Re_L} \tag{7.8}$$

where

$$Re_L = \frac{V_\infty L}{v}$$

It is customary to express drag data in terms of a drag coefficient C_D, defined as the ratio of drag force per area to kinetic energy of the free stream. Thus,

$$C_D = \frac{D_f}{\rho V_\infty^2 bL/2g_c} = \frac{2 D_f g_c}{\rho V_\infty^2 bL}$$

Substituting Equation 7.8 gives

$$C_D = \frac{1.328}{\sqrt{Re_L}} \tag{7.9a}$$

By appropriately combining the above equations, the drag coefficient can be expressed in terms of the wall shear stress as

$$C_D = \frac{4\tau_w g_c}{\rho V_\infty^2} \left(\frac{x}{L}\right)^{1/2} \tag{7.9b}$$

Example 7.1

Air at 27°C is moved through a rectangular duct by a fan, as shown in Figure 7.3. Flow straighteners are placed somewhere downstream. Each flow straightener is 0.5 m long and 1 m wide (into the page). For an air velocity of 1 m/s past the plates

(a) Determine and graph the boundary-layer growth with length.
(b) Graph the velocity distribution at $x = 0.25$ m.
(c) Determine the skin-friction drag exerted on one plate.

Solution

The boundary-layer equations apply to flow of an incompressible fluid. Air is compressible, but because the velocity here is quite low, the air can be considered incompressible.

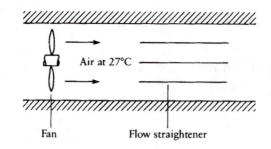

FIGURE 7.3   Air moving through a rectangular duct.

Assumptions

1. The flow is steady and incompressible.
2. Air is a Newtonian fluid having constant properties.

From Appendix Table D.1, for air at 27°C = 300 K,

$$\rho = 1.177 \text{ kg/m}^3 \qquad v = 15.68 \times 10^{-6} \text{ m}^2/\text{s}$$

The Reynolds number at plate end is

$$\text{Re}_L = \frac{V_\infty L}{v} = \frac{1(0.5)}{15.68 \times 10^{-6}} = 3.19 \times 10^4 < 5 \times 10^5$$

The flow is therefore laminar over the entire length, and the Blasius solution applies.

(a) The boundary-layer thickness is given by Equation 7.6:

$$\delta = \frac{5.0x}{\sqrt{\text{Re}_x}} = \frac{5.0x}{(V_\infty x/v)^{1/2}} = 5.0(xv/V_\infty)^{1/2}$$

After substitution, we get

$$\delta = 5.0[x(15.68 \times 10^{-6})/1]^{1/2}$$

$$\delta = 1.98 \times 10^{-2} x^{1/2}$$

This equation is graphed in Figure 7.4.

(b) The velocity distribution is given by the data of Table 7.1. The independent variable for this problem becomes

$$\eta = y\left(\frac{V_\infty}{vx}\right)^{1/2} = y\left[\frac{1}{15.68 \times 10^{-6}(0.25)}\right]^{1/2}$$

$$\eta = 5.05 \times 10^2 y$$

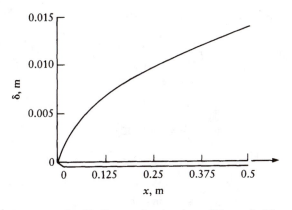

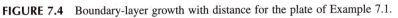

FIGURE 7.4 Boundary-layer growth with distance for the plate of Example 7.1.

The corresponding velocity is

$$V_x = V_\infty f = (1)f = f$$

Using the data from Table 7.1, the information in Table 7.2 is determined.

TABLE 7.2
Results of Calculations for Example 7.1

η	y, m	$f = V_x$, m/s
0	0	0
1	1.98×10^{-3}	0.329 79
2	3.96×10^{-3}	0.629 77
3	5.94×10^{-3}	0.846 05
4	7.92×10^{-3}	0.955 52
5	9.90×10^{-3}	0.991 55

The velocity distribution is given in Figure 7.5.

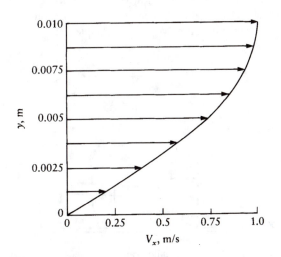

FIGURE 7.5 Velocity distribution for the air of Example 7.1 at $x = 0.25$ m.

(c) The drag exerted on one side of one plate is given by Equation 7.8.

$$D_f\big|_1 = 0.664 V_\infty \mu b \sqrt{Re_L}$$

The absolute viscosity is found as

$$\mu = \frac{\rho v}{g_c} = 1.177(15.68 \times 10^{-6})$$

or

$$\mu = 1.846 \times 10^{-5} \text{ N·s/m}^2$$

The drag force then is

$$D_f\big|_1 = 0.664(1)(1.846 \times 10^{-5})(1)(3.19 \times 10^4)^{1/2}$$

Solving,

$$D_f\big|_1 = 2.19 \times 10^{-3} \text{ N}$$

Including both sides of one plate, we finally get

$$D_f = 4.38 \times 10^{-3} \text{ N} \qquad\qquad \square$$

7.2.1 CONSTANT WALL TEMPERATURE

The corresponding heat-transfer problem that we consider is one where the plate is maintained at a constant temperature T_w, as illustrated in Figure 7.6. As shown, the flow is uniform, having a free-stream velocity of V_∞ and a free-stream temperature of T_∞. We assume that a fictitious *thermal boundary layer* of thickness δ_t develops. Within the thermal boundary layer, temperature varies from T_w at the surface to (within 99% of) T_∞ at δ_t. The descriptive equations to be solved are Equations 7.2 and 7.3, subject to boundary conditions of Equation 7.5. In addition, the energy equation of the boundary layer must be solved (from Chapter 5):

$$V_x\frac{\partial T}{\partial x} + V_y\frac{\partial T}{\partial y} = \alpha\frac{\partial^2 T}{\partial t^2} \tag{7.10}$$

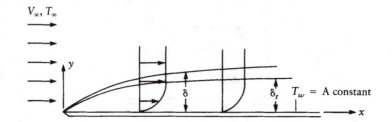

FIGURE 7.6 Laminar flow of a Newtonian fluid over a flat plate where the surface temperature T_w is maintained constant.

The applicable boundary conditions are

$$
\left.
\begin{array}{llll}
\text{B.C.1.} & y = 0, & T = T_w \\
\text{B.C.2.} & y = \infty, & T = T_\infty \\
\text{B.C.3.} & x = 0, & T = T_\infty
\end{array}
\right\}
\tag{7.11}
$$

The solution to the energy equation is more easily found if a change in variables is made. Let

$$
\theta(\eta) = \frac{T - T_w}{T_\infty - T_w}
$$

and

$$
\eta = y \sqrt{\frac{V_\infty}{\nu x}}
$$

We select the above form of η based on experience with the Blasius solution. The temperature profiles, regardless of x location, are expected to be similar. The energy equation itself resembles closely the momentum-boundary-layer Equation 7.3. The boundary conditions also have the same form. In terms of the new variables θ and η, the energy equation becomes, after considerable simplification,

$$
\frac{d^2\theta}{d\eta^2} + \frac{\text{Pr}}{2} f \frac{d\theta}{d\eta} = 0
\tag{7.12}
$$

where $f = V_x/V_\infty$ as in the Blasius solution. By introducing the variables θ and η, the energy equation (a partial differential equation) has been transformed into an ordinary differential equation. The boundary conditions become

$$
\begin{array}{llll}
\text{B.C.1.} & \eta = 0, & \theta = 0 \\
\text{B.C.2.} & \eta = \infty, & \theta = 1
\end{array}
\tag{7.13}
$$

The three boundary conditions, Equations 7.11, thus reduce to only two, which satisfy the original three. Derivation of the solution to the energy equation is beyond the scope of this discussion. The solution itself, however, is presented graphically in Figure 7.7, where $(T - T_\infty)/(T_w - T_\infty)$ is graphed as a function of η. The Prandtl number appears on the graph as an independent variable that ranges from 0.6 to 1 000. Note that liquid metals with Prandtl numbers less than 0.6 are not represented. The graph of Figure 7.7 is for fluids whose hydrodynamic-boundary-layer thickness is greater than the thermal-boundary-layer thickness. Liquid metals exhibit the opposite trend.

With temperature profile known, it is now possible to determine the heat transferred. Fourier's Law is written as

$$
q_w'' = -k_f \frac{\partial T}{\partial t}\bigg|_{y=0}
\tag{7.14}
$$

With 0 defined as $(T - T_w)/(T_\infty - T_w)$, we write

$$
\frac{\partial T}{\partial y}\bigg|_{y=0} = \frac{\partial T}{\partial \theta}\left(\frac{\partial \theta}{\partial \eta}\bigg|_{\eta=0}\right)\frac{\partial \eta}{\partial y} = (T_\infty - T_w)\left(\frac{V_\infty}{\nu x}\right)^{1/2}\left(\frac{\partial \theta}{\partial \eta}\bigg|_{\eta=0}\right)
$$

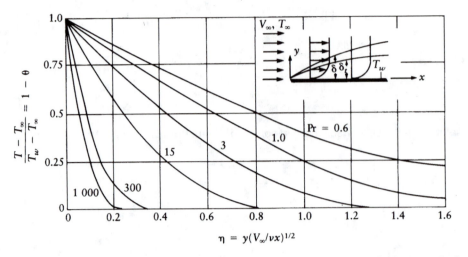

FIGURE 7.7   Dimensionless temperature profiles for laminar flow of a Newtonian fluid over an isothermal flat plate. (The Pr = 1 curve also represents $1 - V_x/V_\infty$ vs. η.)

From the solution represented in Figure 7.7,

$$\frac{d\theta}{d\eta}\bigg|_{\eta=0} = \frac{\partial\theta}{\partial\eta}\bigg|_{\eta=0} \cong 0.332 \; Pr^{1/3}$$

By substitution then,

$$\frac{\partial T}{\partial y}\bigg|_{y=0} = (T_\infty - T_w)(0.332 \; Pr^{1/3})\left(\frac{V_\infty}{\nu x}\right)^{1/2} \tag{7.15}$$

The heat transferred at the wall then is

$$q_w'' = -k_f(T_\infty - T_w)(0.332 \; Pr^{1/3})\left(\frac{V_\infty}{\nu x}\right)^{1/2} \tag{7.16}$$

We can also obtain an equation for the local heat-transfer coefficient h_x. From Newton's Law of Cooling we have

$$q_w'' = h_x(T_w - T_\infty) \tag{7.17}$$

Combining with Equation 7.16 and solving for h_x, we get

$$h_x = k_f(0.332 \; Pr^{1/3})\left(\frac{V_\infty}{\nu x}\right)^{1/2} \tag{7.18a}$$

Customarily expressing the convection coefficient in terms of appropriate dimensionless groups yields

$$\frac{h_x x}{k_f} = 0.332 \; Pr^{1/3}\left(\frac{V_\infty}{\nu x}\right)^{1/2} x = 0.332 \; Pr^{1/3}\left(\frac{V_\infty x}{\nu}\right)^{1/2}$$

or

$$\text{Nu}_x = 0.332 \, \text{Pr}^{1/3} \, \text{Re}_x^{1/2} \tag{7.18b}$$

Equations 7.18 contain the *local* convection coefficient. Integrating the local coefficient over the entire surface area of the plate and dividing by the area gives the average heat-transfer coefficient; that is,

$$\bar{h}_L = \frac{\int_0^L \int_0^b h_x \, dz \, dx}{\int_0^L \int_0^b dz \, dx}$$

where b is the width of the plate. Substituting Equation 7.18a, integrating, and simplifying, we get

$$\bar{h}_L = \frac{b \int_0^L k_f (0.332 \, \text{Pr}^{1/3}) \left(\dfrac{V_\infty}{vx}\right)^{1/2} dx}{bL} = \frac{k_f}{L}(0.332 \, \text{Pr}^{1/3}) \left(\frac{V_\infty}{v}\right)^{1/2} \int_0^L x^{-1/2} dx$$

Integrating leads to

$$\frac{\bar{h}_L L}{k_f} = 0.332 \, \text{Pr}^{1/3} \left(\frac{V_\infty}{v}\right)^{1/2} 2x^{1/2} \Big|_0^L$$

$$\frac{\bar{h}_L L}{k_f} = 0.664 \, \text{Pr}^{1/3} \left(\frac{V_\infty L}{v}\right)^{1/2} \tag{7.19}$$

In terms of dimensionless groups,

$$\overline{\text{Nu}_L} = 0.664 \, \text{Re}_L^{1/2} \, \text{Pr}^{1/3} \tag{7.20}$$

Thus, the average convection coefficient and corresponding Nusselt number are twice those of the local values existing at the end of the plate.

The preceding analysis applies to laminar flow of a Newtonian fluid over an isothermal flat plate in which fluid properties are assumed constant. In the real case, temperature varies, and so do fluid properties within the boundary layer. It is therefore recommended that properties be evaluated at what is called the *film temperature* T_f, defined as the average of the wall and free-stream temperatures:

$$T_f = \frac{T_w + T_\infty}{2} \tag{7.21}$$

In addition, two cases can exist, although only one is presented above. As depicted in Figure 7.6, the thermal boundary layer is shorter in height than the hydrodynamic boundary layer, although the opposite case does exist. Here we consider only the first case, where $\delta_t < \delta$.

One important consequence of the solution to Equation 7.12 is in determining the thickness of the thermal boundary layer. Although not shown here, this thickness is given by:

$$\frac{\delta_t}{\delta} = \text{Pr}^{-1/3} \tag{7.22}$$

This result is shown graphically in Figure 7.8. The graph shows the variation of δ_t/δ with Prandtl number for $0.6 \le \mathrm{Pr} \le 50$. The lowest Prandtl number for gases is about 0.67, with lower values (about 0.01) for liquid metals. Note that if $\mathrm{Pr} = 0.67$, Equation 7.22 gives $\delta_t/\delta = 1.11$, which violates the requirement that $\delta_t/\delta < 1$, but not by an intolerable amount. For $\mathrm{Pr} = 0.01$, $\delta_t/\delta = 4.53$, which is unacceptable. Thus, Equation 7.22 holds for gases and all liquids, except liquid metals.

Example 7.2

Water at 40°F flows at 1.2 ft/s over a metal plate that is maintained at a constant temperature of 100°F. The plate is 2 ft long (in the flow direction) and 18 in wide. Calculate the heat transferred to the water over the first foot and over the entire length of the plate. Graph the temperature profile at $x = 1$ ft.

Solution

The water increases in temperature over the plate length and we expect that more heat is transferred to the water over the first foot than over the last foot.

Assumptions

1. Steady-state conditions exist.
2. Water properties evaluated at the film temperature are constant.

From Appendix Table C.11 for water at $(40 + 100)12 = 70°F \approx 68°F$

$\rho = 62.4 \ \mathrm{lbm/ft^3}$ $k_f = 0.345 \ \mathrm{BTU/(hr \cdot ft \cdot °R)}$
$c_p = 0.9988 \ \mathrm{BTU/(lbm \cdot °R)}$ $\alpha = 5.54 \times 10^{-3} \ \mathrm{ft^2/hr}$
$v = 1.083 \times 10^{-5} \ \mathrm{ft^2/s}$ $\mathrm{Pr} = 7.02$

At $x = 1$ ft the Reynolds number is

$$\mathrm{Re}_x = \frac{V_\infty x}{v} = \frac{1.2(1)}{1.083 \times 10^{-5}} = 1.108 \times 10^5 < 5 \times 10^5$$

The flow is laminar, and Equation 7.19 applies:

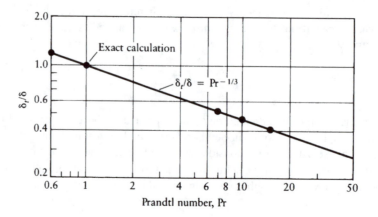

FIGURE 7.8 Influence of Prandtl number on the boundary-layer thickness.

$$\frac{\bar{h}_L L}{k_f} = 0.664 \, \mathrm{Pr}^{1/2} \, \mathrm{Re}_L^{1/2}$$

where $L = x = 1$ ft. Substituting and solving for the average convection coefficient gives

$$\bar{h}_L = \frac{0.345}{1}(0.664)(7.02)^{1/3}(1.108 \times 10^5)^{1/2} = 146 \text{ BTU/(hr} \cdot \text{ft}^2 \cdot {}^\circ\text{R})$$

The heat transferred to the water over the first foot then is

$$q = \bar{h}_L A (T_w - T_\infty) = 146(1)(1.5)(100 - 40) = 1.314 \times 10^4 \text{ BTU/hr}$$

Repeating the calculations for $x = 2$ ft, we write

$$\mathrm{Re}_x = \frac{V_\infty x}{\nu} = \frac{1.2(2)}{1.083 \times 10^{-5}} = 2.216 \times 10^5$$

$$\bar{h}_L = \frac{0.345}{2}(0.664)(7.02)^{1/3}(2.216 \times 10^5)^{1/2} = 103.2 \text{ BTU/(hr} \cdot \text{ft}^2 \cdot {}^\circ\text{R})$$

The heat transferred over the entire plate then is

$$q = 103.2(2)(1.5)(100 - 40) = 1.858 \times 10^4 \text{ BTU/hr}$$

To graph the temperature profile at $x = 1$ ft, it is necessary to obtain values for dimensionless temperature from Figure 7.7. By interpolating values from that graph at Pr = 7, we obtain the values in the first two columns of Table 7.3. At $x = 1$ ft, y is found with

$$\eta = y\sqrt{\frac{V_\infty}{\nu x}} = y\sqrt{\frac{1.2}{1.083 \times 10^{-5}(1)}}$$

or

$$y = 3.00 \times 10^{-3} \, \eta$$

TABLE 7.3
Solution Chart for Example 7.2

$\eta = \sqrt{\dfrac{V_\infty}{\nu x}}$	$\theta' = \dfrac{T - T_\infty}{T_w - T_\infty}$	y, ft	T, °F
0	1	0	100
0.2	0.75	6.01×10^{-4}	85
0.4	0.51	1.2×10^{-3}	70.6
0.6	0.31	1.8×10^{-3}	58.6
0.8	0.17	2.4×10^{-3}	50.2
1.0	0.08	3.0×10^{-3}	44.8
1.2	0.01	3.6×10^{-3}	40.6

Calculated values of y are provided in the third column of Table 7.2. Corresponding temperatures are found with

$$\frac{T - T_\infty}{T_w - T_\infty} = \theta'$$

$$T = \theta'(100 - 40) + 40$$

or

$$T = 60\theta' + 40$$

Calculations made with the above equation are given in the fourth column of Table 7.3. A graph of temperature vs. y appears in Figure 7.9. ❏

7.2.2 CONSTANT WALL FLUX

In a number of practical problems the wall flux is essentially a constant rather than the wall temperature. For flow over a flat plate, the continuity, momentum, and energy equations have been solved for the constant-wall-flux problem. The method of solution (which is not presented here) leads to the following result:

$$\mathrm{Nu}_x = \frac{h_x x}{k_f} = 0.453\ \mathrm{Re}_x^{1/2}\ \mathrm{Pr}^{1/3} \tag{7.23}$$

The average temperature difference between the plate and the free stream is given by

$$(T_w - T_\infty)\big|_{\mathrm{avg}} = \frac{1}{bL}\int_0^L \int_0^b (T_w - T_\infty) dz\ dx \tag{7.24}$$

From Newton's Law of Cooling

$$q_w'' = h_x(T_w - T_\infty)$$

or

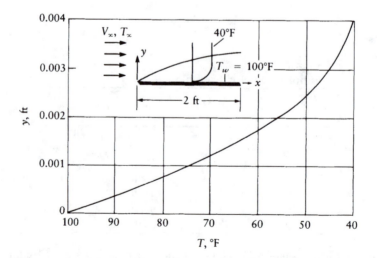

FIGURE 7.9 Temperature variation (at $x = 1$ ft) within the boundary layer for the water of Example 7.2.

$$(T_w - T_\infty) = \frac{q_w'' \, x k_f}{h_x \, x k_f} = \frac{x q_w''}{Nu_x k_f}$$

Substituting into Equation 7.24 gives

$$(T_w - T_\infty)\big|_{avg} = \frac{1}{L}\int_0^L \frac{x q_w''}{Nu_x k_f} dx$$

If Equation 7.23 is next substituted for the local Nusselt number and the above equation is integrated, the result is

$$(T_w - T_\infty)\big|_{avg} = \frac{q_w'' L}{k_f(0.679\ 5)\ Re_L^{1/2}\ Pr^{1/3}} \tag{7.25}$$

The preceding equations for the constant-wall-flux case apply to laminar flow of a Newtonian fluid, in which properties are evaluated at the film temperature $T_f = (T_w + T_\infty)/2$.

The boundary-layer equations are actually approximations to the continuity, momentum, and energy equations derived in Chapter 5. The analysis discussed thus far is known as the exact solution (to the approximate equations).

Example 7.3

Air at 300 K moves at 5 m/s past a warmed isothermal flat plate. The plate is 1 m long, 1/2 m wide and maintained at 400 K. Determine the average convection coefficient and the total drag force exerted on the plate. Calculate the local convection coefficient and the hydrodynamic- and thermal-boundary-layer thicknesses at plate end.

Solution

Air is warmed by the plate, and the heat flux will vary if surface temperature is constant. Air temperature will increase with increasing distance along the plate.

Assumptions

1. Air is a Newtonian fluid having constant properties.
2. Air properties are evaluated at the film temperature.
3. The system is at steady state.

From Appendix Table D.1 for air at $(300 + 400)/2 = 350$ K

$\rho = 0.998$ kg/m^3	$k_f = 0.030\ 03$ W/(m·K)
$c_p = 1\ 009$ J/(kg·K)	$\alpha = 0.298\ 3 \times 10^{-4}$ m^2/s
$\nu = 20.76 \times 10^{-6}$ m^2/s	$Pr = 0.697$

The Reynolds number at plate end is

$$Re_L = \frac{V_\infty L}{\nu} = \frac{5(1)}{20.76 \times 10^{-6}} = 2.41 \times 10^5 < 5 \times 10^5$$

So the flow is laminar at plate end. The average convection coefficient is calculated with Equation 7.20:

$$\overline{\text{Nu}}_L = \frac{\bar{h}_L L}{k_f} = 0.664 \, \text{Re}_L^{1/2} \, \text{Pr}^{1/3}$$

Rearranging and substituting,

$$\bar{h}_L = \frac{0.030\ 03}{1}(0.664)(2.41 \times 10^5)^{1/2}(0.697)^{1/3}$$

Solving,

$$\bar{h}_L = 8.68 \ \text{W/(m}^2 \cdot \text{K)} \quad \text{(average over entire plate surface)}$$

The drag force is given in Equation 7.8:

$$D_f = 0.664 V_\infty \mu b \sqrt{\text{Re}_L} = 0.664(5)(0.998)(20.76 \times 10^{-6})(0.5)(2.41 \times 10^5)^{1/2}$$

Solving,

$$D_f = 1.69 \times 10^{-2} \ \text{N}$$

At plate end, the local Nusselt number is used to find the local convection coefficient:

$$\text{Nu}_x = \frac{h_x x}{k_f} = 0.332 \, \text{Re}_x^{1/2} \, \text{Pr}^{1/3}$$

which is just one-half of the Nusselt number used in finding $\bar{h}_L$. Therefore, at plate end,

$$h_x = \frac{1}{2}\bar{h}_L = 4.34 \ \text{W/(m}^2 \cdot \text{K)} \quad \text{(local, at plate end)}$$

The boundary-layer thickness at plate end is calculated with Equation 7.6:

$$\delta = \frac{5.0L}{\sqrt{\text{Re}_L}} = \frac{5(1)}{(2.41 \times 10^5)^{1/2}} = 1.02 \times 10^{-2} \ \text{m} = 1.02 \ \text{cm}$$

The thermal-boundary-layer thickness can be calculated with Equation

$$\frac{\delta_t}{\delta} = \frac{1}{(\text{Pr})^{1/3}} = \frac{1}{(0.697)^{1/3}} = 1.12$$

$$\delta_t = 1.15 \ \text{cm}$$

❑

Example 7.4

Figure 7.10a shows flow over a flat wall, where the wall itself consists of 10 strip heaters (Figure 7.10b) placed side by side. Each strip heater is 1-1/2 in wide and has an effective heating length (into the page) of 10-1/8 in. Each heater provides 10 W of energy to the fluid. The fluid itself is oxygen traveling at 20 ft/s with a free-stream temperature of 300°R. According to the manufacturer,

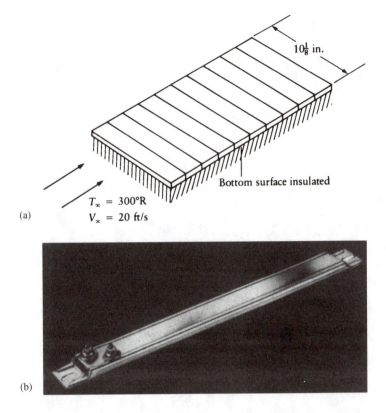

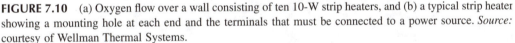

FIGURE 7.10 (a) Oxygen flow over a wall consisting of ten 10-W strip heaters, and (b) a typical strip heater showing a mounting hole at each end and the terminals that must be connected to a power source. *Source: courtesy of Wellman Thermal Systems.*

the heater-surface temperature should not exceed 800°R. Determine the maximum heater-surface temperature for the conditions given.

Solution

As the oxygen passes over the plate, heat is transferred from the heaters to the oxygen. The greatest temperature difference will therefore exist at the front of the plate. The maximum surface temperature correspondingly will exist at plate end.

Assumptions

1. The system is at steady state.
2. Oxygen properties are constant.
3. All heat generated by the heaters is transferred to the oxygen through the upper surface; the insulation makes the heat loss through the lower surface negligible.

Without knowing the film temperature at which we evaluate properties, we therefore assume that fluid properties can be taken at (300°R + 800°R)/2 = 550°R ≈ 540°R as a first estimate. From Appendix Table D.6 for oxygen at 540°R,

$\rho = 0.0812$ lbm/ft³ $\quad k_f = 0.01546$ BTU/(hr·ft·°R)

$c_p = 0.2918$ BTU/(lbm·°R) $\quad \alpha = 0.8862$ ft²/hr

$\nu = 17.07 \times 10^{-5}$ ft²/s $\quad$ Pr = 0.709

The wall flux for any number of heaters is found to be

$$q_w'' = \frac{n(10.0)}{n(1.5)(10.125)} = 0.658 \ \text{W/in}^2 \left(\frac{1}{0.2928} \frac{\text{BTU}}{\text{W}} \frac{144 \ \text{in}^2}{\text{ft}^2} \right) = 324 \ \text{BTU/(hr·ft}^2)$$

Anywhere along the plate we can write

$$q_w'' = h_x(T_w - T_\infty)$$

At plate end, $x = L$ and

$$\frac{q_w'' L}{k_f} = \frac{h_x L}{k_f}(T_w - T_\infty) = \text{Nu}_L(T_w - T_\infty)\big|_L$$

Rearranging and substituting from Equation 7.23 for the Nusselt number gives

$$T_w\big|_L = T_\infty + \frac{q_w'' L}{k_f} \frac{1}{\text{Nu}_L} = T_\infty + \frac{q_w'' L}{k_f} \frac{1}{0.453 \ \text{Re}_L^{1/2} \ \text{Pr}^{1/3}}$$

The Reynolds number at plate end is

$$\text{Re}_L = \frac{V_\infty L}{v} = \frac{20(10)\left(\frac{1.5}{12}\right)}{17.07 \times 10^{-5}} = 1.46 \times 10^5$$

So the flow is laminar, and we are justified in using Equation 7.23. Substituting, we find for the wall temperature at plate end

$$T_w\big|_L = 300 + \frac{324(10)\left(\frac{1.5}{12}\right)}{0.01546} \frac{1}{0.453(1.46 \times 10^5)^{1/2}(0.709)^{1/3}} = 470°\text{R}$$

This result is dependent on the properties of oxygen being evaluated at 540°R. A refined estimate can be obtained by evaluating properties at $(300 + 470)12 = 385°\text{R}$. This is reserved as an exercise.❑

7.2.3 A General Relationship

For laminar flow of a Newtonian fluid over a flat plate, the Churchill-Ozoe Equation applies over the entire range of Prandtl numbers (liquid metals included):

Constant wall temperature

$$\text{Nu}_x = \frac{h_x x}{k_f} = \frac{0.338 \ 7 \ \text{Re}_x^{1/2} \ \text{Pr}^{1/3}}{[1 + (0.046 \ 8/\text{Pr})^{2/3}]^{1/4}} \qquad \text{for} \qquad \text{Pe}_x = \text{Re}_x \text{Pr} = \frac{V_\infty x}{\alpha} > 100 \qquad (7.26)$$

Constant wall flux

$$\mathrm{Nu}_x = \frac{h_x x}{k_f} = \frac{0.463\ 7\ \mathrm{Re}_x^{1/2}\ \mathrm{Pr}^{1/3}}{[1 + (0.020\ 52/\mathrm{Pr})^{2/3}]^{1/4}} \qquad \text{for} \qquad \mathrm{Pe}_x = \mathrm{Re}_x \mathrm{Pr} = \frac{V_\infty x}{\alpha} < 100 \qquad (7.27)$$

Properties in these equations are to be evaluated at the film temperature (= average of wall and free-stream temperatures).

7.2.4 REYNOLDS ANALOGY

The equations for momentum and energy, when applied to flow over a flat plate, are similar in form. For example, consider Equations 7.3 and 7.10:

$$V_x \frac{\partial V_x}{\partial x} + V_y \frac{\partial V_x}{\partial y} = v \frac{\partial^2 V_x}{\partial y^2} \qquad (7.3)$$

$$V_x \frac{\partial T}{\partial x} + V_y \frac{\partial T}{\partial y} = \alpha \frac{\partial^2 T}{\partial y^2} \qquad (7.10)$$

Clearly, if the boundary conditions for each of these equations are the same, then the solutions for V_x and T will resemble one another. (Also, $\mathrm{Pr} = v/\alpha$ must equal 1.)

We can use this result to draw meaningful conclusions regarding the solution form for both equations. Earlier in this chapter, we derived an equation for the shear stress exerted on a flat plate in a uniform flow of fluid as

$$\tau_w = \frac{0.332 V_\infty \mu}{x} (\mathrm{Re}_x)^{1/2}$$

We can define a local drag coefficient as

$$C_d = \frac{\tau_w g_c}{\frac{1}{2}\rho V_\infty^2} = \frac{\text{local shear stress}}{\text{kinetic energy of the flow}}$$

Substituting for shear stress gives

$$C_d = 0.664 \frac{V_\infty \mu g_c}{\rho V_\infty^2 x} (\mathrm{Re}_x)^{1/2}$$

Noting that $\rho V_\infty x / \mu g_c = \mathrm{Re}_x$, the preceding equation simplifies to

$$C_d = 0.664\ \mathrm{Re}_x^{-1/2}$$

or

$$\frac{C_d}{2} = 0.332\ \mathrm{Re}_x^{-1/2} \qquad (7.28)$$

Equation 7.18b is the exact solution for Nusselt number:

$$\mathrm{Nu}_x = 0.332\ \mathrm{Re}_x^{1/2}\ \mathrm{Pr}^{1/3} \qquad (7.18b)$$

Dividing by $\mathrm{Re}_x\ \mathrm{Pr}$ gives

$$\frac{\mathrm{Nu}_x}{\mathrm{Re}_x\ \mathrm{Pr}} = 0.332\ \mathrm{Re}_x^{1/2}\ \mathrm{Pr}^{-2/3} \qquad (7.29)$$

The ratio on the left-hand side is defined as the Stanton number ($= h_x/\rho c_p V_\infty$). Rearranging slightly,

$$\text{St Pr}^{2/3} = 0.332 \, \text{Re}_x^{-1/2}$$

Comparing to Equation 7.28, we see that the equations are equal. Thus,

$$\text{St Pr}^{2/3} \cong \frac{C_d}{2} \tag{7.30a}$$

If this equation is integrated over the plate area, then the local skin-friction coefficient becomes the total drag coefficient C_D:

$$\overline{\text{St}} \, \text{Pr}^{2/3} = \frac{C_D}{2} \tag{7.30b}$$

where

$$\overline{\text{St}} = \frac{\bar{h}_L}{\rho c_p V_\infty}$$

$$\text{Pr} = \frac{v}{\alpha}$$

$$C_D = \frac{2 D_f g_c}{\rho V_\infty^2 b L}$$

Note that Equations 7.30, called the *Reynolds-Colburn Analogy*, relate the drag force D_f to the convection coefficient. Thus, frictional drag measurements could be made and the convection coefficient could be predicted without any heat transfer taking place. The Reynolds-Colburn Analogy applies to laminar and turbulent flow over a flat plate. It can also be modified to apply to turbulent flow in a tube but does not work accurately for laminar tube flow.

Example 7.5

Determine how well the Reynolds-Colburn Analogy fits the data of Example 7.4.

Solution

We are interested in finding out how well Equation 7.30 correlates the parameters of the example.

$$\text{St Pr}^{2/3} \cong \frac{C_d}{2} \tag{7.30a}$$

Evaluating the left-hand side,

$$\text{St Pr}^{2/3} = \frac{h_x}{\rho c_p V_\infty} \left(\frac{v}{\alpha}\right)^{2/3} \tag{i}$$

The calculations of Example 7.4 apply only to the end of the plate. The convection coefficient there is

$$h_x = \frac{q_w''}{(T_w - T_\infty)} = \frac{324}{(470 - 300)} = 1.91 \ \text{BTU/(hr·ft}^2\text{·°R)}$$

Substituting into Equation (i) above gives

$$\text{St Pr}^{2/3} = \frac{1.91/3600}{0.0812(0.2918)(20)}(0.709)^{2/3} = 8.90 \times 10^{-4}$$

The right-hand side of Equation 7.30 is evaluated with

$$\frac{C_d}{2} = 0.332 \ \text{Re}_x^{-1/2} = 0.332(1.46 \times 10^5)^{-1/2} = 8.69 \times 10^{-4}$$

The error is less than 3% owing to the approximations inherent in the solution method and to properties being evaluated at a temperature somewhat higher than the film temperature. In spite of the above sources of error, the agreement is quite good. ❏

7.3 TURBULENT FLOW OVER A FLAT PLATE

Because of the complex nature of turbulent flow, it is not possible to solve the boundary-layer equations to obtain an exact solution for flow over a flat plate. However, other solution methods (not discussed here) have been used successfully. Indeed, solutions have been obtained for a number of cases of interest. For example, the following profile gives a reasonable solution for velocity:

$$\frac{V_x}{V_\infty} = \left(\frac{y}{\delta}\right)^{1/7} \tag{7.31}$$

Equation 7.59 is called the *seventh-root profile*. At the wall, this profile predicts an infinite gradient (dV_x/dy). This is not possible, so the wall shear stress is approximated by the Blasius empirical formula:

$$\tau_w = 0.022 \ 5 \ \frac{\rho V_\infty^2}{g_c}\left(\frac{\nu}{V_\infty \delta}\right)^{1/4} \tag{7.32}$$

which is valid over a Reynolds-number range of 5×10^5 to about 5×10^7. Using these equations and one of the alternative solution methods leads to the following solution for the hydrodynamic-boundary-layer thickness:

$$\delta = \frac{0.368x}{\text{Re}_x^{1/5}} \tag{7.33}$$

The local skin-friction drag coefficient is determined to be

$$C_d = \frac{0.057 \ 3}{\text{Re}_L^{1/5}} \tag{7.34}$$

and the total drag coefficient is

$$C_D = \frac{0.074}{\text{Re}_L^{1/5}} \tag{7.35}$$

Compared to the laminar results, we see that the laminar boundary layer grows with $x^{1/2}$ while the turbulent boundary layer varies with $x^{4/5}$. Also, for turbulent flow random fluctuations due to turbulence have a significant effect. One important result is that the thermal-boundary-layer thickness is about the same as the momentum thickness; that is, $\delta_t \cong \delta$.

The Reynolds-Colburn Analogy applied to turbulent flow over a flat plate leads to the following for the local Nusselt number (flow over a flat plate of constant wall temperature T_w):

$$\text{Nu}_x = \text{St Re}_x \text{Pr} = 0.028\ 7\ \text{Re}_x^{4/5}\ \text{Pr}^{1/3} \tag{7.36}$$

where

$$\text{Nu}_x = \frac{h_x x}{k_f}$$

$$\text{St} = \frac{h_x}{\rho c_p V_\infty}$$

$$\text{Re}_x = \frac{V_\infty x}{\nu}$$

$$\text{Pr} = \frac{\nu}{\alpha}$$

Equation 7.36 is valid over

$$0.6 < \text{Pr} < 60 \quad \text{and} \quad 5 \times 10^5 < \text{Re}_x < 5 \times 10^7$$

Properties are to be evaluated at the film temperature $(T_w + T_\infty)/2$.

7.3.1 Laminar and Turbulent Flow over a Flat Plate

The preceding paragraphs presented equations useful for describing turbulent flow over a flat plate of constant temperature. In a number of real applications flow is turbulent over a large portion of the plate, so the laminar contribution is negligible. For the case where neither entirely laminar nor primarily turbulent conditions can be assumed, it is necessary to account for both effects.

Suppose, for example, that we wish to develop an equation for total drag exerted on a plate over which there is laminar flow at the leading edge and turbulent flow downstream—neither effect dominates. An approximate method for calculating drag is to write

Total skin-friction drag = Laminar drag $(0 \le x \le x_{cr})$

$+$ Turbulent drag $(0 \le x \le L)$

$-$ Turbulent drag $(0 \le x \le x_{cr})$

In writing this expression, it is assumed that transition occurs at a single critical point, denoted as x_{cr} We now substitute Equation 7.9 for laminar drag and Equation 7.35 for turbulent drag to obtain

$$C_D \frac{\rho V_\infty^2 bL}{2g_c} = \frac{1.328}{(\text{Re}_{cr})^{1/2}} \frac{\rho V_\infty^2}{2g_c} bL \frac{x_{cr}}{L} + \frac{0.074}{(\text{Re}_L)^{1/5}} \frac{\rho V_\infty^2}{2g_c} bL - \frac{0.074}{(\text{Re}_L)^{1/5}} \frac{\rho V_\infty^2}{2g_c} bL \frac{x_{cr}}{L}$$

Rearranging and simplifying gives

$$C_D = \frac{1.328(\mathrm{Re}_{cr})^{1/2}}{\mathrm{Re}_{cr}}\frac{x_{cr}}{L} + \frac{0.074}{(\mathrm{Re}_L)^{1/5}} - \frac{0.074(\mathrm{Re}_{cr})^{4/5}}{\mathrm{Re}_{cr}}\frac{x_{cr}}{L}$$

It is evident that

$$\frac{\mathrm{Re}_{cr}}{\mathrm{Re}_L} = \frac{V_\infty x_{cr}}{v}\frac{v}{V_\infty L} = \frac{x_{cr}}{L}$$

Substituting, and letting $\mathrm{Re}_{cr} = 5 \times 10^5$, we obtain after simplification

$$C_D = \frac{0.074}{(\mathrm{Re}_L)^{1/5}} - \frac{1\,743}{\mathrm{Re}_L} \tag{7.37}$$

This equation is valid up to a Reynolds number of about 5×10^7, because the first term on the right-hand side is valid only to that point.

An expression that fits experimental data over a wider range is the Prandtl-Schlichting Equation:

$$C_D = \frac{0.455}{[\log(\mathrm{Re}_L)]^{2.58}} - \frac{1\,700}{\mathrm{Re}_L} \tag{7.38}$$

valid up to a Reynolds number of 10^9. Combining the following equations leads to the graph of Figure 7.11:

$$C_D = \frac{1.328}{(\mathrm{Re}_L)^{1/2}} \quad \text{(laminar)} \quad 0 < \mathrm{Re}_L < 5 \times 10^5 \tag{7.9}$$

$$C_D = \frac{0.074}{\mathrm{Re}_L^{1/5}} \quad \text{(turbulent)} \quad 2 \times 10^5 < \mathrm{Re}_L < 10^7 \tag{7.35}$$

$$C_D = \frac{0.455}{[\log(\mathrm{Re}_L)]^{2.58}} - \frac{1\,700}{\mathrm{Re}_L} \quad \text{(turbulent)} \quad 1 \times 10^7 < \mathrm{Re}_L < 1 \times 10^9 \tag{7.38}$$

Figure 7.11 is of drag coefficient vs. Reynolds number.

Example 7.6

A flat-bottom barge is towed on a water surface at 5 knots. The barge is 12 ft wide and 20 ft long. Estimate the drag due to skin friction on the bottom surface, assuming it is truly flat.

Solution

We can use Figure 7.11 to find the drag coefficient.

Assumptions

1. The system is at steady state.
2. The water temperature is 68°F.

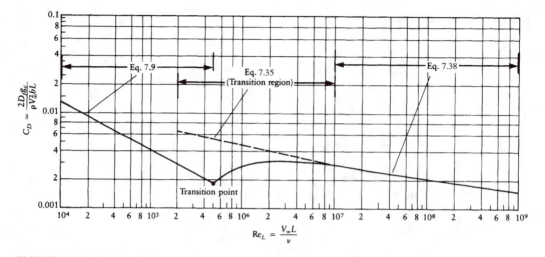

FIGURE 7.11   Drag coefficient as a function of Reynolds number at plate end for flow over a flat plate.

From Appendix Table C.11 for water at 68°F we read

$$\rho = 1.000(62.4) \ \text{lbm/ft}^3$$
$$\nu = 1.083 \times 10^{-5} \ \text{ft}^2\text{/s}$$

Using conversion factors from Appendix Table A.1, the barge velocity is

$$V_\infty = 5 \ \text{knots} \times \frac{5.144 \times 10^{-1}}{3.048 \times 10^{-1}} = 8.44 \ \text{ft/s}$$

The Reynolds number at plate end is

$$\text{Re}_L = \frac{V_\infty L}{\nu} = \frac{8.44(20)}{1.083 \times 10^{-5}} = 1.56 \times 10^7$$

From Figure 7.11 we get

$$C_D = \frac{2D_f g_c}{\rho V_\infty^2 bL} = 0.003$$

The drag force is

$$D_f = \frac{0.003(62.4)(8.44)^2(12)(20)}{2(32.2)}$$

or

$$D_f = 50 \ \text{lbf}$$ ❏

Paralleling the fluid-mechanics problem, we wish to develop an equation for the average convection heat-transfer coefficient for the entire plate. The heat-transfer problem considered is uniform flow of a Newtonian fluid over a flat plate maintained at constant temperature. The local convection coefficient is

$$h_x = h_{x \text{ (laminar)}} \qquad (0 \leq x \leq x_{cr})$$
$$+ h_{x \text{ (turbulent)}} \qquad (x_{cr} \leq x \leq L)$$

Integrating the above equation over the plate area and simplifying gives

$$\bar{h}_L = \frac{1}{L}\int_0^{x_{cr}} h_{x \text{ (lam)}} dx + \frac{1}{L}\int_{x_{cr}}^{L} h_{x \text{ (turb)}} dx$$

where again it is assumed that transition occurs at a single point. We now substitute for the local convection coefficients from the appropriate Nusselt number equations, that is, Equation 7.18b for laminar flow and 7.36 for turbulent flow. We get

$$\bar{h}_L = \frac{1}{L}\int_0^{x_{cr}} 0.332\frac{k_f}{x}(\text{Re}_x^{1/2} \text{ Pr}^{1/3})dx + \frac{1}{L}\int_{x_{cr}}^{L} 0.028\,7\frac{k_f}{x}(\text{Re}_x^{4/5} \text{ Pr}^{1/3})dx$$

Simplifying and integrating gives

$$\bar{h}_L = \frac{0.332k_f \text{ Pr}^{1/3}}{L}\left(\frac{V_\infty}{v}\right)^{1/2}\int_0^{x_{cr}} \frac{dx}{x^{1/2}} + \frac{0.028\,7k_f \text{ Pr}^{1/3}}{L}\left(\frac{V_\infty}{v}\right)^{4/5}\int_{x_{cr}}^{L} \frac{dx}{x^{1/5}}$$

or

$$\bar{h}_L = \frac{0.332k_f \text{ Pr}^{1/3}}{L}\left(\frac{V_\infty}{v}\right)^{1/2}(2x^{1/2})\Big|_0^{x_{cr}} + \frac{0.028\,7k_f \text{ Pr}^{1/3}}{L}\left(\frac{V_\infty}{v}\right)^{4/5}\left(\frac{5x^{4/5}}{4}\right)\Big|_{x_{cr}}^{L}$$

Rearranging and combining terms, we get

$$\frac{\bar{h}_L L}{k} = 0.664\,\text{Re}_{x_{cr}}^{1/2} \text{ Pr}^{1/3} + 0.035\,9[(\text{Re}_L)^{4/5} - (\text{Re}_{x_{cr}})^{4/5}]\,\text{Pr}^{1/3}$$

Letting the critical Reynolds number = 5×10^5, we obtain after simplification

$$\overline{\text{Nu}}_L = 0.035\,9\,\text{Re}_L^{4/5}\,\text{Pr}^{1/3} - 830\,\text{Pr}^{1/3} \qquad (7.39)$$

which is valid over

$$0.6 < \text{Pr} < 60$$
$$5\times10^5 < \text{Re}_L < 10^8$$

Fluid properties evaluated at $(T_\infty + T_w)/2$

Combining the following equations leads to the graph of Figure 7.12:

$$\overline{\text{Nu}}_L = 0.664\,\text{Re}_L^{1/2}\,\text{Pr}^{1/3} \qquad 0 \leq \text{Re}_L \leq 5 \times 10^5 \qquad (7.20)$$

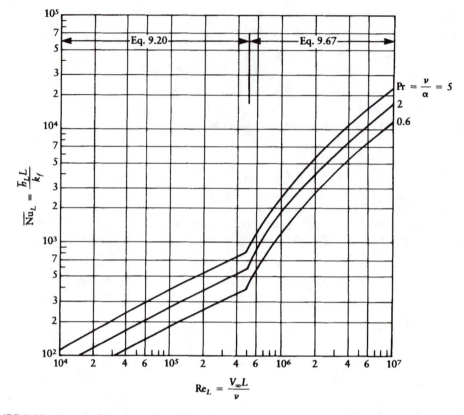

FIGURE 7.12 Average Nusselt number for laminar and turbulent flow of a Newtonian fluid over a flat plate maintained at constant temperature.

$$\overline{Nu}_L = (0.035\ 9\ Re_L^{4/5} - 830)\ Pr^{1/3} \qquad 5 \times 10^5 \le Re_L \le 10^8 \qquad (7.39)$$

The results depicted in Figure 7.12 require that fluid properties be evaluated at the film temperature. An alternative expression, the Zhukauskas-Whitaker Equation, accounts for the temperature dependence of the fluid viscosity:

$$\overline{Nu}_L = 0.036(Re_L^{4/5} - 9\ 200)\ Pr^{0.43}(\mu/\mu_w)^{1/4} \qquad (7.40)$$

where μ_w is evaluated at the wall temperature T_w (which is constant), and all other properties are evaluated at the free-stream temperature T_∞. The equation is valid for

$$\left.\begin{array}{l} 0.7 < Pr < 380 \\ 10^5 < Re_L < 5.5 \times 10^6 \\ 0.26 < (\mu/\mu_w) < 3.5 \end{array}\right\} \qquad (7.41)$$

It is prudent to remember that convection correlations provide only an estimate of what is actually occurring, and that it is fortunate if an accuracy of better than 25% is encountered.

Example 7.7

Carbon dioxide gas at 400 K moves past a flat plate that consists of six strip heaters, as shown in Figure 7.13. The carbon dioxide velocity is 60 m/s. It is desired to maintain the strip-heater-surface

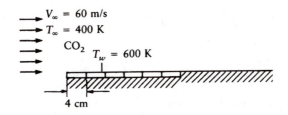

$V_\infty = 60$ m/s
$T_\infty = 400$ K
CO_2 $T_w = 600$ K

4 cm

FIGURE 7.13 Flow past an isothermal flat surface made up of strip heaters.

temperature at 600 K. Determine the wattage requirement for each heater to do this. Each heater is 40 mm wide and has an effective heating length (into the page) of 160 mm.

Solution

Heat will be transferred from each heater to the carbon dioxide, and the carbon dioxide film temperature increases with increasing distance along the plate. Thus, we expect that the heater ratings will not all be equal.

Assumptions

1. Steady-state conditions exist.
2. Carbon dioxide properties are constant.
3. The bottom surface is insulated, and all heat is transferred via convection to the carbon dioxide.

From Appendix Table D.2, for carbon dioxide at a film temperature of $(400 + 600)/2 = 500$ K, we read

$$\rho = 1.073\ 2\ \text{kg/m}^3 \qquad k_f = 0.033\ 52\ \text{W/(m·K)}$$
$$c_p = 1\ 013\ \text{J/(kg·K)} \qquad \alpha = 0.308\ 4 \times 10^{-4}\ \text{m}^2/\text{s}$$
$$\nu = 21.67 \times 10^{-6}\ \text{m}^2/\text{s} \qquad \text{Pr} = 0.702$$

To determine which correlation applies, first find where transition occurs. At transition

$$\text{Re}_{x_{cr}} = 5 \times 10^5 = \frac{V_\infty x_{cr}}{\nu}$$

$$x_{cr} = \frac{5 \times 10^5 (21.67 \times 10^{-6})}{60}$$

or

$$x_{cr} = 0.181\ \text{m} = 18.1\ \text{cm}$$

Therefore, transition occurs on the fifth heater. Equation 7.20 applies to the plates over which laminar flow exists, and Equation 7.67 applies to the turbulent flows. For the first plate, we have

$$\overline{\text{Nu}}_1 = 0.664\ \text{Re}_x^{1/2}\ \text{Pr}^{1/3} \tag{9.20}$$

or

$$\bar{h}_1 = 0.664 \frac{k_f}{x} \left(\frac{V_\infty x_1}{v} \right)^{1/2} \text{Pr}^{1/3}$$

Substituting and simplifying,

$$\bar{h}_1 = 0.664(0.033\ 52) \left(\frac{60}{21.67 \times 10^{-6}} \right)^{1/2} (0.702)^{1/3} \frac{1}{x_1^{1/2}}$$

which becomes

$$\bar{h}_1 = \frac{32.9}{x_1^{1/2}} \tag{i}$$

This equation applies over the entire first plate. With $x_1 = 0.04$ m,

$$\bar{h}_1 = 165 \text{ W/(m}^2\text{K)}$$

The heat transferred by the first heater then is

$$q_1 = h_1(x_1 b)(T_w - T_\infty) = 165(0.04)(0.16)(600 - 400)$$

The first heater must supply

$$q_1 = 211 \text{ W}$$

We can now apply Equation (i) above to determine how much heat the first *two* heaters together will supply. We can use Equation (i) because the physical properties are constant and the average convection coefficient varies only with length. Thus,

$$h_2 = 32.9/(0.08)^{1/2} = 116 \text{ W/(m}^2\cdot\text{K)}$$

Therefore,

$$q_1 + q_2 = 116(0.08)(0.16)(200) = 298 \text{ W}$$

So the second heater is to supply $298 - 211$, or

$$q_2 = 87 \text{ W}$$

Proceeding for the first three heaters,

$$\bar{h}_3 = \frac{32.9}{(0.12)^{1/2}} = 95 \text{ W/(m}^2\cdot\text{K)}$$

$$q_1 + q_2 + q_3 = 95(0.12)(0.16)(200) = 365$$

$$q_3 = 365 - 298$$

or

$$q_3 = 67 \text{ W}$$

For the first four heaters,

$$\bar{h}_4 = \frac{32.9}{(0.16)^{1/2}} = 82.3 \text{ W/(m}^2 \cdot \text{K)}$$

$$q_1 + q_2 + q_3 + q_4 = 82.3(0.16)(0.16)(200) = 421$$

and

$$q_4 = 56 \text{ W}$$

At the end of the fifth heater, the Reynolds number is

$$\text{Re}_5 = \frac{V_\infty x_5}{\nu} = \frac{60[5(0.04)]}{21.67 \times 10^{-6}} = 5.54 \times 10^5$$

which is turbulent. Equation 7.39 applies:

$$\overline{\text{Nu}}_L = \frac{\bar{h}_L x}{k_f} = [0.035\,9(\text{Re}_L)^{4/5} - 830]\,\text{Pr}^{1/3} \qquad (7.39)$$

The average convection coefficient for the first five heaters then is

$$\bar{h}_5 = \frac{0.033\,52}{5(0.04)}[0.035\,9(5.54 \times 10^5)^{4/5} - 830](0.702)^{1/3}$$

$$\bar{h}_5 = 86.7 \text{ W/(m}^2 \cdot \text{K)}$$

The heat transferred from the first five heaters then is

$$q_1 + q_2 + q_3 + q_4 + q_5 = 86.7(0.2)(0.16)(200) = 555$$

which gives

$$q_5 = 134 \text{ W}$$

At the end of the sixth heater,

$$\text{Re}_L = \frac{60[6(0.04)]}{21.67 \times 10^{-6}} = 6.65 \times 10^5$$

The convection coefficient becomes

$$\bar{h}_6 = \frac{0.033\,52}{6(0.04)}[0.035\,9(6.65 \times 10^5)^{4/5} - 830](0.702)^{1/3}$$

$$\bar{h}_6 = 99.7 \text{ W/(m}^2 \cdot \text{K)}$$

The total heat to be supplied by all six heaters is

$$q_1 + q_2 + q_3 + q_4 + q_5 + q_6 = 99.7[6(0.04)](0.16)(200) = 766$$

We now obtain

$$q_6 = 211 \text{ W}$$

An interesting trend is exhibited. When laminar flow exists, the first heater has the largest power requirement, and the power required decreases with increasing distance until transition occurs. Where turbulent flow exists, the last heater has the greatest power requirement. The average convection coefficient follows the same trends.   ❏

7.4 FLOW PAST VARIOUS TWO-DIMENSIONAL BODIES

In the previous sections, flow past a flat surface is discussed, and it is mentioned that the drag force exerted on the surface is due to skin friction. In this section, we examine flow past a number of geometries that contain a curved surface. In such problems, there is skin-friction drag and another kind or type of drag in which we are interested, because the factors that influence drag also influence the heat-transfer rate (Reynolds-Colburn Analogy).

Consider flow past a curved boundary such as that illustrated in Figure 7.14. The surface is stationary, and the free-stream velocity is V_∞. Point A is the nose of the body, where the velocity normal to the surface is zero. Point A is thus referred to as a *stagnation point*, and the pressure measured there is called the *stagnation pressure*. The hydrodynamic boundary layer begins growing from point A. Points B and C have illustrated the associated velocity profiles, which develop much in the same way as in flow over a flat plate. At point D, however, the slope of the velocity profile at the surface is $dV_x/dy = 0$. At points E and F the flow in the vicinity of the surface is no longer traveling in the main flow direction. Thus, at D and beyond, the flow is said to have *separated* from the surface. Separation occurs because the flow cannot negotiate the abrupt change in surface profile; consequently, an *adverse pressure gradient* forms. It is known that, if an external pressure difference from the front of the body to its back is imposed, fluid flows in the direction of decreasing

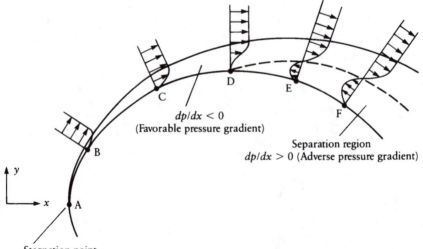

FIGURE 7.14 Uniform flow past a curved surface.

pressure. When separation occurs, the direction of decreasing pressure has reversed. The pressure along the surface within the separation region nearly equals the pressure at the separation (point D). This effect is discussed in more detail for various cases in the paragraphs that follow.

The first body discussed is a very long, circular cylinder immersed in a uniform flow of velocity V_∞. The velocity profile and pressure distribution existing about the cylinder are important. From observations of the flow pattern around a cylinder, streamlines can be sketched, as shown in Figure 7.15. Point A is the stagnation point, and point B is the point of separation. Beyond point B, behind the cylinder, is the separation region or wake. The flow has separated from the cylinder surface, because it cannot negotiate the turn past B. Due to the complex nature of the flow, it is not possible mathematically to derive an equation for velocity at every point on the cylinder. To determine pressure, experimental means are necessary.

One way to determine the pressure distribution is to use a hollow cylinder drilled every $10°$ (see Figure 7.16). The cylinder is then placed in a uniform flow (such as that obtained in a wind tunnel). Each tap is connected to a manometer. The arrangement shown in Figure 7.16 gives the pressure distribution on the cylinder surface. Only half the cylinder is used, because the distribution is symmetric. Typical results of pressure vs. angle are provided in Figure 7.17. The graph shows that at the stagnation point (A) pressure is greater than the reference pressure (atmospheric) because, at A, the velocity is zero. The pressure decreases with θ until, at point 7 or so, the minimum pressure corresponding to maximum velocity is reached. Beyond this, at point 12 or thereabouts, separation occurs. In the separation region, the pressure on the surface is constant and very nearly equals that at point 12. A polar graph of pressure vs. angle is provided in Figure 7.18.

It is apparent from Figure 7.18 that pressure over the front half of the cylinder is different from that over the rear half. This pressure difference, which acts over the projected frontal area of the cylinder, results in a net force acting on the cylinder in the flow direction. This force is called *pressure drag* or *form drag*. Form drag plus skin-friction drag equals the total drag exerted.

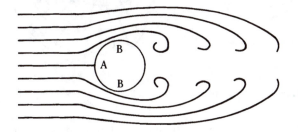

FIGURE 7.15 Streamlines of flow around a circular cylinder.

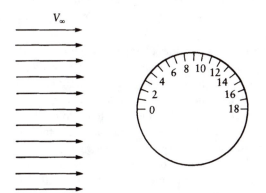

FIGURE 7.16 Experimental arrangement to determine the distribution of pressure on the surface of a cylinder immersed in a uniform flow.

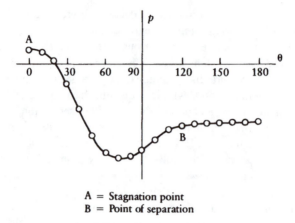

A = Stagnation point
B = Point of separation

FIGURE 7.17 Experimental results of pressure vs. angle for flow past a circular cylinder immersed in a uniform flow.

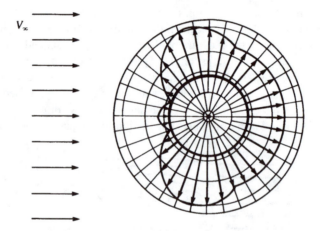

FIGURE 7.18 A polar graph of pressure distribution on the surface of a cylinder immersed in a uniform flow.

As an illustration of the difference between form drag and skin-friction drag, consider two cases of flow past a flat plate, as shown in Figure 7.19. When the plate is aligned parallel to the flow and there is no separation, the total drag is due entirely to skin friction. When the plate is aligned normal to the flow, separation occurs, and the total drag is due primarily to the pressure difference from front to back.

The location of the point of separation affects the magnitude of the drag force, which in turn is influenced by whether the boundary layer is laminar or turbulent. Typical velocity profiles are shown in Figure 7.20. The laminar profile near the wall is more positively sloped than the turbulent profile. There is considerably more mixing in the turbulent case, which tends to distribute the kinetic energy of the flow more evenly. Therefore, the turbulent profile can offer more resistance to an adverse pressure gradient, and we expect that separation occurs farther downstream for a turbulent boundary layer than for a laminar boundary layer. To see how this affects flow past a cylinder, refer to Figure 7.21. In the laminar case, the boundary layer remains laminar up to separation. In the turbulent case, the flow experiences transition from laminar to turbulent within the boundary layer before separation occurs. Separation is thus moved further downstream for the turbulent case, and form drag is reduced. A widely used method of inducing transition is by roughening the surface of the cylinder. A familiar example for a sphere is the dimpled surface of a golf ball.

(a) Plate aligned with flow direction

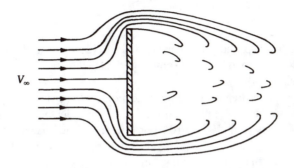

(b) Plate normal to flow direction

FIGURE 7.19   Two cases of flow past a flat plate.

Laminar   **Turbulent**

FIGURE 7.20   Laminar and turbulent boundary-layer velocity profiles.

Separation in laminar flow   Separation in turbulent flow

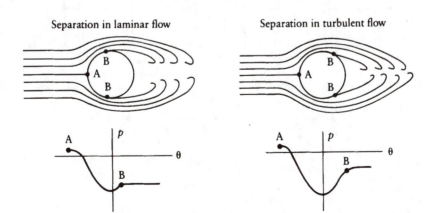

FIGURE 7.21   A comparison of laminar- and turbulent-flow separation.

It is customary practice to express data on drag as drag coefficient vs. Reynolds number, where both are defined as

$$C_D = \frac{2D_f g_c}{\rho V_\infty^2 A} \tag{7.42}$$

$$\mathrm{Re}_D = \frac{\rho V_\infty D}{\mu g_c} = \frac{V_\infty D}{\nu} \tag{7.43}$$

where D_f is the drag force, ρ and μ are density and viscosity of the fluid, V_∞ is the free-stream velocity, A is the projected frontal area or some other characteristic area of the body, and D is the characteristic dimension of the object.

Figure 7.22 is a graph of drag coefficient vs. Reynolds number for flow past a circular cylinder having a large length-to-diameter ratio. The drag coefficient decreases steadily from 60 to 1 as Reynolds number increases from 10^{-1} to 10^3. For a Reynolds number in the range 10^3 to 10^4, C_D remains approximately constant. For a Reynolds number greater than 10^4, a roughened cylinder exhibits a different curve than does a smooth cylinder. The graph shown can be obtained from experimental results measured in a wind tunnel. Drag coefficient for other two-dimensional shapes is provided in Figure 7.23.

Example 7.8

A telephone pole has 12-in diameter and is 30 ft tall. Attached to the pole at its top is a square piece of wood 2 in × 2 in × 4 ft long, as shown in Figure 7.24. A uniform wind velocity of 20 mph flows past the structure. Estimate the force exerted on the pole.

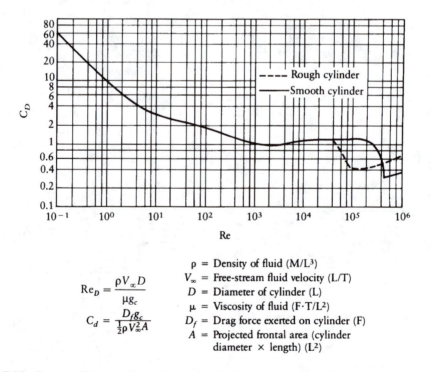

FIGURE 7.22 Drag coefficient vs. Reynolds number for flow past circular cylinders. *Source:* adapted from *Boundary Layer Theory,* by H. Schlichting, 7th ed., McGraw-Hill, New York, 1979, with permission.

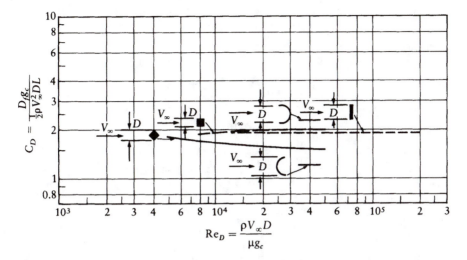

FIGURE 7.23 Drag coefficient of various two-dimensional bodies. *Source:* data taken from several sources; see references.

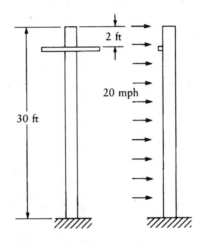

FIGURE 7.24 The telephone pole of Example 7.8.

Solution

The data of Figures 7.22 and 7.23 can be used to determine a drag coefficient, from which a drag force can be calculated. The data of those figures apply to very long two-dimensional cross sections. The geometry of this problem includes an interference between the cylinders where they cross. Moreover, at the ground, the air velocity is zero.

Assumptions

1. Interference where the objects cross has a negligible effect.
2. Air has constant properties.
3. Lacking more specific information, air properties are evaluated at 14.7 psia and 80°F.
4. The velocity of 20 mph is uniform and applies at all locations; the effect of zero velocity at the ground is negligible.

From Appendix Table D.1, at these conditions, we read for air

$$\rho = 0.0735 \text{ lbm/ft}^3 \qquad \nu = 16.88 \times 10^{-5} \text{ ft}^2/\text{s}$$

The flow velocity is

$$V = 20\frac{\text{miles}}{\text{hr}} \times \frac{5280}{3600} = 29.3 \text{ ft/s}$$

The Reynolds number for flow past the pole itself is

$$Re_D = \frac{V_\infty D}{\nu} = \frac{29.3(12/12)}{16.88 \times 10^{-5}} = 1.74 \times 10^5$$

Figure 7.22 shows, for a smooth cylinder,

$$C_D = \frac{2D_f g_c}{\rho V_\infty^2 A} = 1.1$$

The projected frontal area of the pole is that of a rectangle:

$$A = DL = \frac{12}{12}(30) = 30 \text{ ft}^2$$

The drag force exerted on only the pole is

$$D_f = 1.1\left(\frac{1}{2}\right)(0.0735)(29.3)^2(30)/32.2$$

or

$$D_f = 32.3 \text{ lbf} \qquad \text{(pole only)}$$

For the square piece of wood, the Reynolds number is

$$Re_D = \frac{V_\infty D}{\nu} = \frac{29.3(2/12)}{16.88 \times 10^{-5}} = 2.9 \times 10^4$$

Figure 7.23 shows

$$C_D = 2$$

With a projected frontal area of $(2/12)(4) = 0.666 \text{ ft}^2$, the drag force is determined to be

$$D_f = 2\left(\frac{1}{2}\right)(0.0735)(29.3)^2(0.666)/32.2$$

or

$$D_f = 1.31 \text{ lbf} \qquad \text{(crosspiece only)}$$

The total force exerted on the pole and crosspiece then is

$$D_f = 33.6 \text{ lbf} \qquad \text{(total)}$$ ❏

The corresponding heat-transfer problem that we consider is that of uniform flow past an isothermal circular cylinder. Because of the complexities of such a problem, we cannot derive a closed-form solution for the velocity or the temperature of the fluid. Consequently, we resort to experimental results. Our concern will be with determining a convection coefficient from which the heat transferred can be calculated.

Figure 7.25 shows results of measurements taken for flow past a nearly isothermal circular cylinder. On the horizontal axis is θ, measured in degrees from the stagnation point. The vertical axis is the local Nusselt number $Nu_\theta = h_\theta D/k_f$. Several curves are graphed for various Reynolds numbers. Data points obtained directly from experiment are also shown. The two curves for Reynolds numbers of 101 300 and 70 800 are cases where the boundary layer remains laminar up to the point of separation, which occurs roughly at $\theta \approx 80°$. Because D and k_f are essentially constant, these curves actually show how the convection coefficient varies. The other curves, at Reynolds numbers equal to or greater than 140 000, have two minima. The first occurs in the range of 80 to 90° for θ, which represents the transition from a laminar to a turbulent boundary layer. The second minimum occurs at the separation point, $\theta \approx 140°$. For all curves, it is seen that the local convection coefficient h_θ increases moderately in the wake.

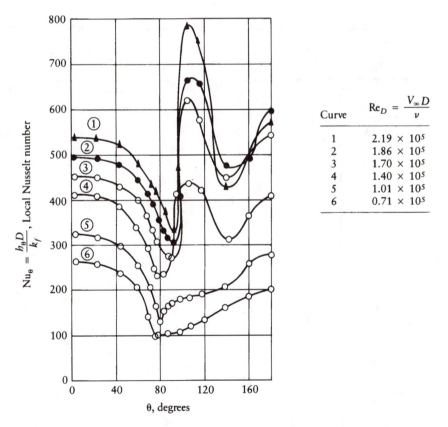

Curve	$Re_D = \dfrac{V_\infty D}{\nu}$
1	2.19×10^5
2	1.86×10^5
3	1.70×10^5
4	1.40×10^5
5	1.01×10^5
6	0.71×10^5

FIGURE 7.25 Local Nusselt number vs. angle for air flow past a nearly isothermal circular cylinder. *Source:* from "Investigation of Variation of Point Unit-Heat-Transfer Coefficient Around a Cylinder Normal to an Air Stream," *Trans. ASME.*, v. 71, pp. 375–381, adapted with permission.

While the curves of Figure 7.25 are useful in observing the behavior of the local Nusselt number, we are better suited to make engineering calculations based on an overall convection coefficient rather than the local coefficient. The overall or average coefficient $\bar{h}_c$ relates the heat transferred to the maximum temperature difference:

$$q_w = \bar{h}_c A_s (T_w - T_\infty)$$

where $A_s = \pi D L$ is the surface area of the cylinder.

A number of empirical correlations have been developed to describe a variety of conditions. The Hilpert Equation relates the Nusselt number to the Reynolds and Prandtl numbers via two constants, C and m:

$$\overline{Nu}_D = \frac{\bar{h}_c D}{k_f} = C\, Re_D^m\, Pr^{1/3} \tag{7.44}$$

where values for C and m are given in Table 7.3. The Hilpert Equation can be applied to cylinders having other than circular cross-sectional shapes. It is necessary only to modify the values of the constants. Table 7.4 provides values of C and m for other cross sections. All properties are to be evaluated at the film temperature, $T_f = (T_w + T_\infty)/2$.

TABLE 7.4
Values of the Constants C and m for Use with Equation 7.44 for Flow Over Cylinders of Various Cross Sections

Cross section	$Re_D = (V_\alpha D)/v$	C	m
	0.4–4	0.989	0.330
	4–40	0.911	0.385
	40–4×10^3	0.683	0.466
	4×10^3–4×10^4	0.193	0.618
	4×10^4–4×10^5	0.027	0.805
	5×10^3–10^5	0.246	0.588
	5×10^3–10^5	0.102	0.675
	4×10^3–1.5×10^4	0.228	0.731

Data from *Heat Transfer*, by M. Jakob, v. 1, John Wiley & Sons, New York, 1949, used with permission.

Results for flow past a cylinder are based on a constant wall temperature. It is, however, impossible to maintain a constant wall temperature in the cases described. An average wall temperature is therefore used. Earlier a distinction was made between constant-wall-temperature and constant-surface-flux problems. In flow past a cylinder, the equations apply to either case, and the distinction is not as significant.

Other equations have been developed for heat transfer to or from flow past a circular cylinder. These are provided in the exercises. You are cautioned to interpret calculations made with such equations as being reasonable estimates of the true result. An accuracy of better than 25% is considered fortunate.

Example 7.9

A hot-wire anemometer is a device that can be used to measure the velocity of a gas. Consider a platinum wire 0.013 cm in diameter and 1 cm long. It is to be placed in a 300 K air stream to determine the velocity. The surface temperature of the wire is maintained at 500 K by means of an external control circuit (such as a Wheatstone Bridge). If the electrical resistivity of platinum is 17 $\mu\Omega$·cm, determine the required current that must pass through the wire if the air velocity is 1 m/s.

Solution

The wire is heated by the current passing through. The warmed wire is cooled by the air, and the cooling rate is proportional to the air velocity. The surface temperature of the wire is probably not constant.

Assumptions

1. Steady-state conditions exist.
2. Air is incompressible and has constant properties.
3. Radiation effects are negligible.
4. The wire-surface temperature is uniform.

The film temperature is $(300 + 500)/2 = 400$ K, and at this temperature Appendix Table D.1 shows the following properties for air:

$\rho = 0.883$ kg/m³ $\qquad k_f = 0.033\ 65$ W/(m·K)

$c_p = 1\ 014$ J/(kg·K) $\qquad \alpha = 0.376 \times 10^{-4}$ m²/s

$\nu = 25.90 \times 10^{-6}$ m²/s $\qquad$ Pr = 0.689

We use Equation 7.44 to find the convection coefficient, from which we calculate the heat transferred. The heat-transfer rate is then related to the required current.

The Reynolds number of flow past the wire, or cylinder, is

$$\mathrm{Re}_D = \frac{V_\infty D}{\nu} = \frac{1(0.000\ 13)}{25.9 \times 10^{-6}} = 5.019$$

From Table 7.3, at this Reynolds number, we read

$$C = 0.911 \qquad m = 0.385$$

Equation 7.22 becomes

$$\overline{Nu}_D = \frac{\overline{h}_c D}{k_f} = 0.911 \, Re_D^{0.385} \, Pr^{1/3}$$

Substituting,

$$\frac{\overline{h}_c(0.000\ 13)}{0.033\ 65} = 0.911(5.019)^{0.385}(0.689)^{1/3} = 1.50$$

Solving,

$$\overline{h}_c = 388 \ W/(m^2 \cdot K)$$

The heat transferred to the air from the wire is calculated as

$$q_w = \overline{h}_c A_s(T_w - T_\infty) = 388(\pi)(0.000\ 13)(0.01)(500 - 300)$$

or

$$q_w = 0.317 \ W$$

This value is the rate at which heat must be generated electrically for equilibrium conditions to exist. The electrical resistance of the wire is

$$R_e = 17 \times 10^{-6}\left(\frac{L}{\pi D^2}\right) = (17 \times 10^{-6})\left[\frac{1}{\pi(0.013)^2}\right] = 0.032 \ \Omega$$

From the definition of power (= [current]²[resistance]), we get

$$q_w = i^2 R_e$$

Rearranging and substituting,

$$i^2 = \frac{q_w}{R_e} = \frac{0.317}{0.032}$$

Solving,

$$i = 3.1 \ A \qquad\qquad \square$$

7.5   FLOW PAST A BANK OF TUBES

A tube bank is a number of parallel equal-diameter tubes that all convey a fluid inside. A second fluid flows past the tube bank, or bundle, external to the tubes. A temperature difference between the fluids causes a net energy transfer. Tube banks are found in conventional boilers used in power plants, and in heat exchangers.

Tubes can be arranged either in an aligned or in a staggered (with respect to the external-fluid-velocity direction) arrangement, as shown in Figure 7.26. Also shown are the geometric variables of importance used in analyzing the external flow. The aligned configuration is characterized by

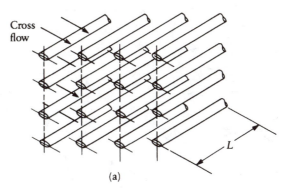

(a)

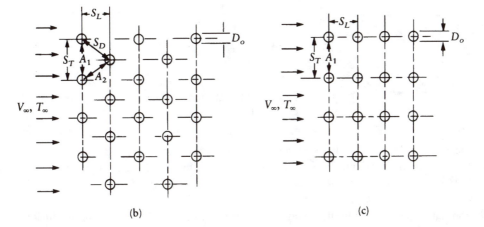

(b)   (c)

FIGURE 7.26   (a) Flow past a tube bank, (b) staggered tube-bank arrangement, and (c) aligned tube-bank arrangement.

the *transverse pitch* S_T, the *longitudinal pitch* S_L, and the outside diameter of the tubes D_o. Both pitches are measured from tube center to center. The flow area is labeled A_1 in the figure, where $A_1 = (S_T - D_o)L$, and where L is the tube length into the page. The staggered configuration is characterized also by a transverse pitch S_T, a longitudinal pitch S_L, and a diagonal pitch S_D, also measured from tube center to center. There are two flow areas of importance for the staggered configuration: A_1 and A_2. The area A_1 is $(S_T - D_o)L$, while $A_2 = (S_D - D_o)L$. In terms of the transverse and longitudinal pitches,

$$S_D = \left[S_L^2 + \left(\frac{S_T}{2} \right)^2 \right]^{1/2}$$

$$A_2 = \left\{ \left[S_L^2 + \left(\frac{S_T}{2} \right)^2 \right]^{1/2} - D_o \right\} L$$

The patterns of flow past the tubes in a bank are illustrated in Figure 7.27.

For in-line tubes, it is seen that, beyond the first row, each tube is in the wake of the tube upstream. For the staggered configuration, the wake behind each tube is reduced in size due to a tube being located where the wake would exist. Clearly, the flow pattern in either case is quite complex, and derivation of an equation for velocity or temperature distribution is not possible. We are seeking an expression useful for calculating the pressure drop experienced by the fluid flowing

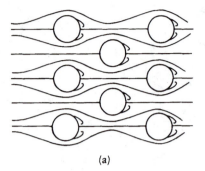

(a)

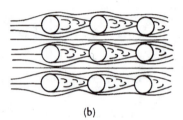

(b)

FIGURE 7.27 Patterns of flow past (a) staggered and (b) aligned tube bundles.

past the tube bank. We are also seeking an equation for calculating the convection coefficient between the external fluid and the outside tube area, the final objective being to calculate the heat transferred. Again, the distinction between a constant-wall-temperature vs. a constant-heat-flux problem is not an important consideration.

From an inspection of the flow patterns illustrated in Figure 7.27, several conclusions might be drawn. Intuitively, it may be inferred that a greater pressure drop exists for the in-line configuration and that the staggered configuration transfers heat much more effectively. Both conclusions may be drawn, based solely on the flow pattern, that is, the size or presence of a wake. However, this is not the case. Both configurations have pressure drops that are roughly in the same range. For the heat-transfer problem, the external convection coefficient is influenced by the degree of turbulence and by the position of the tube in the bank. For the first row the convection coefficient is about the same as that for a single cylinder. Larger coefficients exist for tubes within the inner rows. Little change occurs beyond the 4th or 5th rows.

In general, we are interested in overall effects for pressure drop and heat transfer. For the aligned tube bank we have for the pressure drop

$$\Delta p = N f_1 f_2 \left(\frac{\rho V_{\max}^2}{2 g_c} \right) \qquad \text{(aligned or staggered)} \qquad (7.45)$$

where N is the number or rows of tubes, f_1 and f_2 are factors found by experiment and provided in Figure 7.28, and $V_{\max}$ is the maximum velocity experienced by the fluid as it passes through the bank. Note that f_1 is a friction factor that is a function of the Reynolds number

$$\mathrm{Re}_D = \frac{V_{\max} D_o}{\nu} \qquad (7.46)$$

and f_2 is a geometry factor that depends on the pitch-spacing ratio, defined as $(S_T - D_o)/(S_L - D_o)$. The graph is so constructed that when $S_D = S_T$, f_1 is read directly, and $f_2 = 1$. The maximum velocity

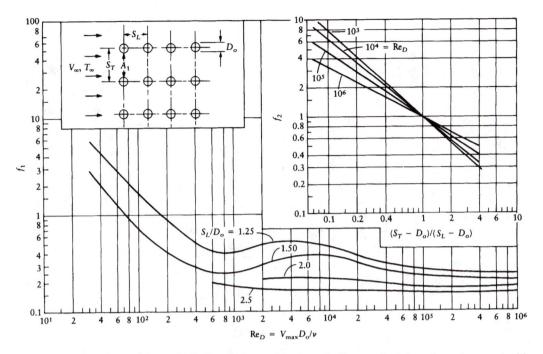

FIGURE 7.28 Friction factors for finding pressure drop in an in-line or aligned configuration, used with Equation 7.45. *Source:* from "Heat Transfer from Tubes in Cross Flow," in *Advances in Heat Transfer,* ed. by J.P. Hartnett and T.F. Irvine, Jr., v. 8, Academic Press, New York, 1972, with permission.

V_{max} will occur at the cross section denoted as A_1 [$= (S_T - D_o)L$] in Figure 7.29. The Grimison Equation can be used to find the average convection coefficient for the in-line configuration:

$$\overline{Nu}_D = \frac{\bar{h}_c D_o}{k_f} = 1.13 C_1 C_2 \, Re_D^m \, Pr^{1/3} \qquad \text{(aligned or staggered)} \qquad (7.47)$$

where C_1, C_2, and m are constants listed in Table 7.5. The constants C, and m are functions of the tube spacings, while C_2 is a function of the number of rows.

TABLE 7.5
Constants for Determining the Convection Coefficient for an In-Line or Aligned Configuration, Used with Equation 7.47.

S_L/D_o	$S_T/D_o = 1.25$		1.50		2.00		3.00	
	C_1	m	C_1	m	C_1	m	C_1	m
1.25	0.348	0.592	0.275	0.608	0.100	0.704	0.063 3	0.752
1.50	0.367	0.586	0.250	0.620	0.101	0.702	0.067 8	0.744
2.00	0.418	0.570	0.299	0.602	0.229	0.632	0.198	0.648
3.00	0.290	0.601	0.357	0.584	0.374	0.581	0.286	0.608

N (rows) =	1	2	3	4	5	6	7	8	9
$C_2 =$	0.64	0.80	0.87	0.90	0.92	0.94	0.96	0.98	0.99

Source: from "Correlation and Utilization of New Data on Flow Resistance and Heat Transfer for Cross Flow of Gases over Tube Banks," by E.D. Grimison, *Trans. ASME*, v. 59, pp. 583–594, 1937. Data on the constant C_2 taken from Stanford University Tech. Rep. No. 15, by W.M. Kays and R.K. Lo, 1952.

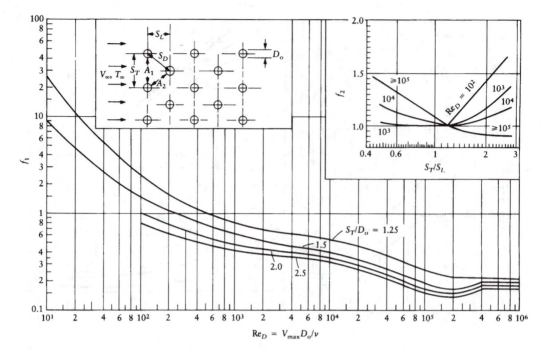

FIGURE 7.29 Friction factors for finding pressure drop in a staggered configuration, used with Equation 7.45. *Source:* from "Heat Transfer from Tubes in Cross Flow," in *Advances in Heat Transfer*, ed. by J.P. Hartnett and T.F. Irvine, Jr., v. 8, Academic Press, New York, 1972, with permission.

For the staggered tube bank, the same equations can be applied. The appropriate modifications are incorporated in the values of the constants. Figure 7.29 shows the factors f_1 and f_2, while Table 7.6 gives the values of the constants for use with Equation 7.47. In all cases, fluid properties are evaluated at the free-stream or inlet temperature.

TABLE 7.6
Constants for Determining the Convection Coefficient for a Staggered Configuration, Used with Equation 7.47.

S_l/D_o	$S_T/D_o = 1.25$ C_1	m	1.50 C_1	m	2.00 C_1	m	3.00 C_1	m
0.600	—	—	—	—	—	—	0.213	0.636
0.900	—	—	—	—	0.446	0.571	0.401	0.581
1.000	—	—	0.497	0.558	—	—	—	—
1.125	—	—	—	—	0.478	0.565	0.518	0.560
1.250	0.518	0.556	0.505	0.554	0.519	0.556	0.522	0.562
1.500	0.451	0.568	0.460	0.562	0.482	0.568	0.488	0.568
2.000	0.404	0.572	0.416	0.568	0.452	0.556	0.449	0.570
3.000	0.310	0.592	0.356	0.580	0.440	0.562	0.428	0.574

N (rows) =	1	2	3	4	5	6	7	8	9
C_2 =	0.68	0.75	0.83	0.89	0.92	0.95	0.97	0.98	0.99

Source: from "Correlation and Utilization of New Data on Flow Resistance and Heat Transfer for Cross Flow of Gases over Tube Banks," by E.D. Grimison, *Trans. ASME*, v. 59, pp. 583–594, 1937. Data on the constant C_2 taken from Stanford University Tech. Rep. No. 15, by W.M. Kays and R.K. Lo, 1952.

Example 7.10

A staggered tube consists of 7 rows of 3/4 standard type K copper tubes. The bank is used to recover heat from water that flows to a nearby stream from an industrial plant. The bank is located in an air duct, as illustrated in Figure 7.30. Air passing over the bank has a velocity of 12 ft/s and a temperature of 70°F. The pitch spacings of the bank are $S_L = 1.5$ in and $S_T = 1.3$ in. Water flowing within the tubes elevates the tube outside-surface temperature to 200°F. Calculate the pressure drop for the air passing over the tubes and the heat transferred to the air. The tube lengths are all 2 ft.

Solution

The air receives heat from the tubes, and the air temperature increases as air passes through.

Assumptions

1. The system is at steady state.
2. Air properties are constant.
3. Radiation effects are negligible.

From Appendix Table D.1, for air at $70 + 460 = 530°R \approx 540°R$, we read

$\rho = 0.0735$ lbm/ft³ $k_f = 0.01516$ BTU/(hr·ft·°R)
$c_p = 0.240$ BTU/(lbm·°R) $\alpha = 0.859$ ft²/hr
$v = 16.88 \times 10^{-5}$ ft²/s   Pr = 0.708

From Appendix Table F.2, for 3/4 standard type K copper tubing,

OD = 0.875 in   ID = 0.06208 ft
$A = 0.003027$ ft²

It is necessary to evaluate V_{max} for the bank. Referring to Figure 7.31, we perform a mass flow balance for the flow entering the bank between two adjacent tubes. The inlet flow area is $(S_T)(L)$,

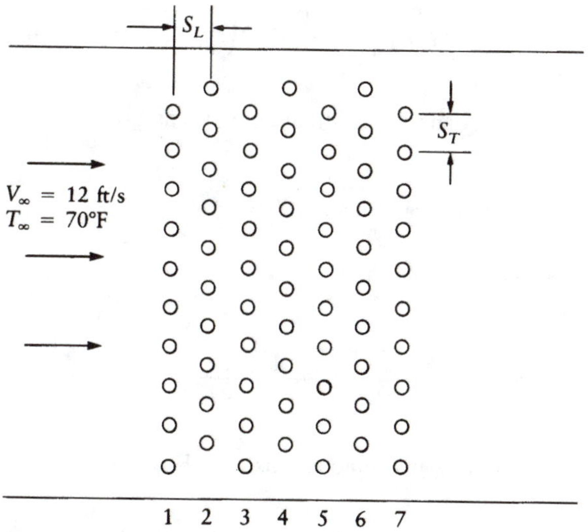

FIGURE 7.30 A 7-row, 70-staggered-tube bank located in an air duct.

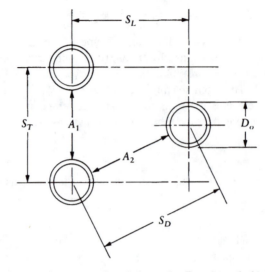

FIGURE 7.31 Sketch for conducting a mass flow balance for flow through the bank.

where L is the tube length into the page. The velocity at this area is $V_\infty = 12$ ft/s. The area labeled A_1 has dimensions of $(S_T - D_o)L$. We therefore write for constant air density

$$V_\infty(S_TL) = V_1A_1 = V_1(S_T - D_o)L$$

Canceling L and solving for V_1 gives

$$V_1 = \frac{S_TV_\infty}{S_T - D_o} = \frac{(1.3/12)(12)}{1.3/12 - 0.875/12} = 36.7 \text{ ft/s}$$

Now the flow that enters between two tubes and travels through A_1 is halved; that is, only half the flow at A_1 passes through the area labeled A_2 in Figure 7.31. The diagonal pitch is calculated as

$$S_D = \left[S_L^2 + \left(\frac{S_T}{2} \right)^2 \right]^{1/2} = \left[1.5^2 + \left(\frac{1.3}{2} \right)^2 \right]^{1/2} = 1.64 \text{ in}$$

A mass flow balance between A_2 and the upstream conditions gives

$$\frac{1}{2}V_\infty(S_TL) = V_2(S_D - D_o)L$$

Solving for V_2 gives

$$V_2 = \frac{S_TV_\infty}{2(S_D - D_o)} = \frac{1.3(12)}{2(1.64 - 0.875)} = 10.2 \text{ ft/s}$$

Because $V_1 > V_2$, we use V_1 as V_{max} in the equations. The Reynolds number is

$$Re_D = \frac{V_{max}D_o}{\nu} = \frac{36.7(0.875/12)}{16.88 \times 10^{-5}} = 1.59 \times 10^4$$

With $S_T/D_o = 1.3/0.875 = 1.49$, Figure 7.29 shows, at the above Reynolds number,

$$f_1 = 0.35$$

For $S_T/S_L = 1.3/1.5 = 0.87$ we also read

$$f_2 = 1.05$$

The pressure drop is calculated with Equation 7.45:

$$\Delta p = Nf_1f_2\left(\frac{\rho V_{max}^2}{2g_c}\right)$$
$$\Delta p = 7(0.35)\,(1.05)\,(0.0735)\,(36.7)^2/2(32.2)$$
$$\Delta p = 3.95 \text{ lbf/ft}^2 = 0.0275 \text{ psi}$$

The convection coefficient is found with Equation 7.47, but we must first determine the constants C_1, C_2, and m. Referring to Table 7.7, we obtain the following values for $S_T/D_0 = 1.49$ and $S_L/D_o = 1.7$:

$$C_1 \approx 0.438$$
$$m \approx 0.565$$
$$C_2 = 0.97$$

Equation 7.47 gives

$$\frac{\bar{h}_c D_o}{k_f} = 1.13 C_1 C_2\, \text{Re}_D^m\, \text{Pr}^{1/3}$$

$$\bar{h}_c = \frac{0.01516}{0.875/12}1.13(0.438)(0.97)(1.59 \times 10^4)^{0.565}(0.708)^{1/3}$$

or

$$\bar{h}_c = 21.0 \text{ BTU/(hr·ft}^2\cdot{}^\circ\text{R)}$$

The heat transferred then becomes

$$q = \bar{h}_c A_s(T_w - T_\infty)$$

where A_s is the outside surface area of 70 tubes, evaluated as

$$A_s = 70\pi D_o L = 70\pi(0.875/12)(2) = 32.1 \text{ ft}^2$$

Thus.

$$q = 21.0(32.1)(200 - 70)$$

or

$$q = 8.76 \times 10^4 \text{ BTU/hr} \qquad \square$$

7.6 FLOW PAST A SPHERE

Flow past a sphere is similar to flow past a cylinder. The flow experiences boundary-layer growth and separation, which give rise to a drag force. Like the cylinder problem, drag on a sphere is represented as a drag coefficient vs. Reynolds number, where

$$C_D = \frac{2D_f g_c}{\rho V_\infty^2 A}$$

$$\mathrm{Re}_D = \frac{V_\infty D}{v}$$

$$(7.48)$$

In these equations, V_∞ is the velocity of the fluid past the sphere (or the sphere velocity through a stagnant fluid), A is the projected frontal area of the sphere ($A = \pi D^2/4$), and D is the sphere diameter. Data on C_D vs. Re_D are provided in Figure 7.32.

The corresponding heat-transfer problem is one where heat is transferred to or from the sphere for the constant-heat-flux and constant-surface-temperature conditions. The Whitaker Equation relates the average Nusselt number to the pertinent flow variables:

$$\overline{\mathrm{Nu}}_D = \frac{\bar{h}_c D}{k_f} = 2 + (0.4\,\mathrm{Re}_D^{1/2} + 0.06\,\mathrm{Re}_D^{2/3})\,\mathrm{Pr}^{0.4}\left(\frac{\mu}{\mu_w}\right)^{1/4}$$

$$(7.49)$$

valid over

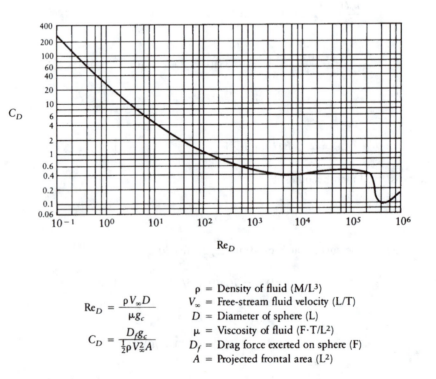

$$\mathrm{Re}_D = \frac{\rho V_\infty D}{\mu g_c}$$

$$C_D = \frac{D_f g_c}{\frac{1}{2}\rho V_\infty^2 A}$$

ρ = Density of fluid (M/L³)
V_∞ = Free-stream fluid velocity (L/T)
D = Diameter of sphere (L)
μ = Viscosity of fluid (F·T/L²)
D_f = Drag force exerted on sphere (F)
A = Projected frontal area (L²)

FIGURE 7.32 Drag coefficient vs. Reynolds number for flow past a sphere. *Source:* adapted from *Boundary Layer Theory,* by H. Schlichting, 7th ed., McGraw-Hill, New York, 1979, with permission.

$$0.71 < Pr < 380$$
$$3.5 < Re_D < 7.6 \times 10^4$$
$$1.0 < (\mu/\mu_w) < 3.2$$

The equation is accurate to within 30%. The viscosity μ_w is evaluated at the surface temperature, and all other properties are evaluated at the free-stream temperature.

7.7 SUMMARY

We have continued the study of convection by examining several forced-convection problems. Flow past a flat plate was discussed and the exact solution was presented. Flows past a cylinder, a bank of tubes, and a sphere were investigated. In each problem, the pertinent fluid mechanics of the system were first discussed, then the corresponding heat-transfer problem was formulated. Table 7.7 contains a summary of equations for the problems presented.

7.8 PROBLEMS

7.8.1 Flow Past a Flat Plate

1. A 5 m/s flow of 20°C air moves past a flat plate. At what point does the flow become turbulent?
2. Helium at 200°F flows over a flat plate that is 10 ft long in the flow direction. What is the required velocity for the flow to become turbulent at the end of the plate?
3. Glycerin at 10°C flows over a flat plate. The glycerin velocity is 10 m/s. How long a plate is required for the flow to become turbulent at plate end?
4. Water at 70°F flows through a rectangular channel that contains flow straighteners. Each straightener is 2 ft long in the flow direction, and the water is 1 ft deep. For a water velocity of 1.2 ft/s,
   (a) Determine and graph the boundary-layer growth with length.
   (b) Graph the velocity distribution at plate end.
   (c) Determine the drag exerted on one plate.
5. Suppose the plates in Example 7.1 are at 50°C. What is the heat-transfer rate to the air from one plate?
6. A truck travels at 25 mph through air at 70°F. The oil pan of the engine extends below the front end of the cab. The bottom of the oil pan is essentially a flat plate that is 12 in wide and 20 in long. Oil in the pan is warmed by the engine and, in turn, the heated oil maintains the outside surface of the pan at 130°F. Calculate the heat-transfer rate to the air from the oil pan.
7. The fins on an air-cooled motorcycle engine are in essence flat plates 4 cm wide and 14 cm long. For a motorcycle speed of 60 km/hr, determine the heat-transfer rate to the air from one fin. The air temperature is 5°C, and the surface temperature of the fins is constant at 130°C.
8. Hydrogen at 150 K flows past one side of a metal plate that is 1 m long and maintained at a constant temperature of 300 K. The plate is 0.7 m wide. Determine the heat-transfer rate to the hydrogen over the first 30 cm of the plate. Graph the temperature profile at plate end. Take the free-stream velocity to be 1.5 m/s.
9. Ethylene glycol at 70°F flows over a flat plate in which a constant heat flux of 100 BTU/ft² has been set up. The fluid velocity is 2 ft/s, and the plate is 4 ft long (in the flow direction) and 1 ft wide.
   (a) Graph the variation of the local convection coefficient with distance.
   (b) Graph the variation of the wall temperature with distance.

TABLE 7.7
Summary of Equations of Chapter 7

Flow past a flat plate

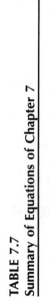

	Laminar flow					
	Constant T_w				Constant q_w	
C_D	δ	Nu_x	$\overline{Nu}_L$	δ_t	Nu_x	
Eq. 7.9a	Eq. 7.6	Eq. 7.18b	Eq. 7.20	Eq. 7.22	Eq. 7.23	
Eq. 7.30	Eq. 7.30	Eq. 7.26 Eq. 7.30	Eq. 7.30		Eq. 7.27	

Turbulent flow

	Constant T_w		
C_D	$\delta = \delta_t$	Nu_x	$\overline{Nu}_L$
Eq. 7.30	Eq. 7.33	Eq. 7.30	Eq. 7.30
Eq. 7.35		Eq. 7.36	

Combined flow
Laminar and turbulent

C_D	$\overline{Nu}_L$
Eq. 7.37	Eq. 7.39
Eq. 7.38	Eq. 7.12
Eq. 7.11	Eq. 7.40

Flow past cylinders

	C_D	$\overline{Nu}_D$
(circle, D)	Fig. 7.22	Eq. 7.44 / Table 7.4 / Problems 24, 25, 26
(diamond, D)	Fig. 7.23	Eq. 7.44 / Table 7.4
(square, D)	Fig. 7.23	Eq. 7.44 / Table 7.4
(plate, D)	Fig. 7.23	Eq. 7.44 / Table 7.4

Flow past tube banks

	Δp	$\overline{Nu}_D$
Staggered	Eq. 7.45 / Fig. 7.29	Eq. 7.47 / Table 7.6
Aligned	Eq. 7.45 / Fig. 7.28	Eq. 7.47 / Table 7.5

Spheres C_D from Fig. 7.32; $\overline{Nu}_D$ from Eq. 7.49.

(c) Determine the average wall-minus-free-stream temperature difference. Evaluate properties at the free-stream temperature.

10. Oxygen at 450°R travels past a heated, constant-temperature flat plate. The plate is 3 ft long, 1 ft wide, and maintained at 650°R. Determine the average convection coefficient, the total drag exerted, the local convection coefficient at plate end, and the hydrodynamic- and thermal-boundary-layer thicknesses at plate end. Make the calculations for velocities of 5, 10, and 15 ft/s.

11. Ammonia at 0°C flows past a flat plate. The plate is maintained at –20°C, is 2 m long and 0.5 m wide. The drag exerted on the plate is measured as 0.002 1 N. Determine the average convection coefficient and the local convection coefficient at plate end.

12. Helium at 33 K flows past a warmed isothermal flat plate at a velocity of 0.8 m/s. The drag exerted is measured to be 2.6×10^{-4} N. Estimate the plate temperature and the local (at plate end) and average convection coefficients. The plate is 20 cm wide and 2 m long.

13. Suppose the air velocity in Example 7.3 is changed. Repeat the calculations if the velocity is halved, and again if the velocity is doubled.

14. Obtain a refined estimate of the wall temperature in Example 7.4 by evaluating fluid properties at the new temperature.

15. Liquid carbon dioxide at –20°C flows over a flat plate maintained at 0°C. The carbon dioxide velocity is 0.01 m/s, and the plate is 3 m long by 1 m wide. Calculate the convection coefficient at plate end, using Equation 7.18b, and compare to the result obtained from the Churchill-Ozoe Equation 7.53.

16. Carbon dioxide gas flows over a flat wall that consists of five strip heaters. Each heater is 1-1/2 in wide and has an effective heating length of 4 in. The carbon dioxide travels at 15 ft/s and has a free-stream temperature of 350 K. The heater-surface temperature is 450 K. Calculate the allowable power rating of each heater. Heat is transferred to the carbon dioxide from only the upper surface of the heaters. How do the ratings of the heaters differ if heating occurs on the upper and lower surfaces?

17. (a) At what velocity does the barge of Example 7.6 exhibit minimum drag?
    (b) What is the minimum drag?
    (c) Repeat the calculations of the example for a velocity of 10 knots and for 2.5 knots.

18. Engine oil at 32°F and traveling at 8 ft/s is heated by passing it over a flat plate that consists of 10 strip heaters placed side by side. All heat from the strip heaters is transferred through their upper surface to the oil; the bottom of the plate is insulated. Each strip heater is 1-1/2 in wide and has an effective heating length of 4 in. Determine the wattage requirement necessary for each individual heater to maintain the surface temperature at 180°F.

19. Steam passes over a flat plate that consists of 18 strip heaters placed side by side. The steam temperature is 450 K, and its velocity is 30 m/s. It is desired to maintain the surface temperature of the heaters at 650 K. Determine the smallest wattage requirement of any of the heaters to do this. The plate bottom is insulated so that all heat is transferred through the upper surface to the steam. Each heater is 6 cm wide and has an effective heating length of 12 cm.

7.8.2 REYNOLDS-COLBURN ANALOGY

20. One side of a flat plate is placed in a flow of engine oil. The liquid velocity is 2 ft/s, and the liquid temperature is 68°F. The plate is 4 ft long and 6 in wide. Determine the total drag force exerted on the plate. Calculate the average convection coefficient, using
    (a) The exact solution for an isothermal plate.
    (b) The Reynolds-Colburn Analogy.
    Evaluate all properties at 68°F.

21. Liquid glycerin flows at 2 m/s past one side of a flat plate of dimensions 1 m long by 0.5 m wide. The glycerin temperature is 10°C. Determine the value of the local convection coefficient at plate end by using the Churchill-Ozoe Equation 7.53 for an isothermal plate and by again using the Reynolds-Colburn Analogy. Evaluate liquid properties at 10°C.

22. Repeat Problem 21 for liquid mercury at 20°C having a velocity of 5 cm/s.

7.8.3 FLOW PAST TWO-DIMENSIONAL BODIES

23. Suppose the diameter of the telephone pole in Example 7.8 is changed. Does the drag force exerted on only the pole double if the diameter is doubled? Does it matter if the pole surface is roughened or smooth? Is the drag force cut in half if the diameter is halved?

24. The Whitaker Equation for flow past a cylinder is

$$\overline{Nu}_D = \frac{\bar{h}_c D}{k_f} = (0.4\, Re_D^{1/2} + 0.06\, Re_D^{2/3})\, Pr^{0.4}(\mu/\mu_w)^{1/4}$$

with
$$0.67 < Pr = v/\alpha < 300$$
$$10 < Re_D = V_\infty D/v < 10^5$$
$$0.25 < (\mu/\mu_w) < 5.2$$

where μ_w is the fluid viscosity evaluated at the wall temperature T_w, and all other properties are evaluated at the free-stream fluid temperature T_∞.

Fluid flows inside a 1/2 nominal schedule 40 pipe that is 6 in long. The pipe is immersed in an air flow having a velocity of 30 ft/s with a free-stream temperature of 300 K. The internal fluid is very hot and maintains the outside-surface temperature of the pipe at roughly a constant 450 K. Estimate the value of the heat transferred to the air, using the Hilpert Equation 7.44 and again using the Whitaker Equation. Calculate the drag exerted on the pipe.

25. The Zhukauskas Equation for flow past a circular cylinder is

$$\overline{Nu}_D = \frac{\bar{h}_c D}{k_f} = C\, Re_D^m\, Pr^n (Pr/Pr_w)^{1/4}$$

with
$$0.7 < Pr = v/\alpha < 500$$
$$Pr \le 10, \text{ then } n = 0.37$$
$$Pr > 10, \text{ then } n = 0.36$$
$$1 < Re_D = V_\infty D/v < 10^6$$

where Pr_w is evaluated at the surface temperature T_w, and all other properties are evaluated at the free-stream temperature T_∞. Values of C and m are listed as

Re_D	C	m
1–40	0.75	0.4
40–10^3	0.51	0.5
10^3–2×10^5	0.26	0.6
2×10^5–10^6	0.076	0.7

Liquid ethylene glycol in a flow circuit is heated by passing it through a duct that contains a cylindrical immersion heater placed perpendicular to the flow direction. The ethylene glycol is at a temperature of 68°F, and it passes an immersion heater that has a 1/2-in diameter, is 8 in long, and heats uniformly. The average surface temperature of the heater is 140°F. Estimate the heat transferred to the ethylene glycol by using the Hilpert Equation 7.44 and the Zhukauskas Equation. Calculate the drag exerted on the heater. The liquid velocity is 3 ft/s.

26. The Churchill-Bernstein Equation for flow past a circular cylinder is

$$\overline{Nu}_D = \frac{\bar{h}_c D}{k_f} = 0.3 + \frac{0.62\,Re_D^{1/2}\,Pr^{1/3}}{[1 + (0.4/Pr)^{2/3}]^{1/4}}\left[1 + \left(\frac{Re_D}{28\,200}\right)^{5/8}\right]^{4/5}$$

$$Re_D\,Pr = \frac{V_\infty D}{\nu}\frac{\nu}{\alpha} = \frac{V_\infty D}{\alpha} < 0.2$$

where all properties are evaluated at the film temperature

$$T_f = \frac{(T_w + T_\infty)}{2}$$

   Ammonia flows past a heated cylindrical immersion heater that is 1 cm in diameter and 10 cm long. The cylinder surface temperature is 0°C, and the ammonia free-stream temperature is −20°C. Estimate the heat transferred to the ammonia by using the Hilpert Equation and again by using the Churchill-Bernstein Equation. Calculate also the drag force exerted. The ammonia velocity is 1 m/s.

27. A 4-in-diameter circular cylinder is immersed in a flow of oxygen. The oxygen velocity is 15 ft/s, and the oxygen free-stream temperature is 360°R.
   (a) For a cylinder surface temperature of 540°R, calculate the drag exerted and the average convection coefficient.
   (b) For a square cylinder oriented as shown in Figure P7.1, determine the dimension D required for the square cylinder to have the same drag as the circular cylinder.
   (c) Determine D if the square cylinder has the same average convection coefficient as the circular cylinder. Use the Hilpert Equation and assume equal free-stream velocities and equal oxygen and wall temperatures between the two cases.

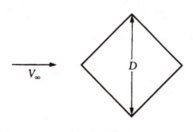

FIGURE P7.1

28. Repeat Problem 27 for a square cylinder oriented as shown in Figure P7.2.

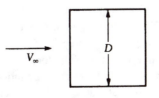

FIGURE P7.2

29. Repeat Problem 27 for the flat plate shown in Table 7.4. Show that a solution exists for equal drag force, but that for equal convection the charted values (Table 7.5) do not extend to the range of Re_D encountered.

30. Example 7.9 deals with a hot-wire anemometer and, using the Hilpert Equation, indicates that 3.1 A are required for the case described. Determine the required current if the following equations are used instead:
    (a) Whitaker Equation (Problem 24).
    (b) Zhukauskas Equation (Problem 25).
    (c) Churchill-Bernstein Equation (Problem 26).

7.8.4   FLOW PAST A TUBE BANK

31. A staggered tube bank consists of six rows of 3/4 standard type K copper tubes 3 ft long used to recover heat. Oxygen to be used for combustion purposes passes over the bank of tubes. The oxygen free-stream velocity is 15 ft/s, and the oxygen free-stream temperature is 68°F. The pitch spacings of the bank are $S_L = 2$ in and $S_T = 2.2$ in. The average outside-surface temperature of the tubes is 100°F. Determine the pressure drop for the oxygen passing over the tubes and the heat transferred to the oxygen. There are 18 tubes.

32. Engine oil used in a hydraulic power system is to be heated by a series of cylindrical immersion heaters. The oil passes through a chamber past the immersion heaters. The heaters themselves are arranged as an in-line tube bundle with $S_L = 4$ cm and $S_T = 6$ cm. Each heater is 2 cm in diameter, 15 cm long, and has an average surface temperature of 100°C. The oil free-stream velocity and temperature are 0.2 m/s and 20°C, respectively. There are 4 rows of heaters and 12 heaters in all. Calculate the pressure drop experienced by the oil in passing through the bank and the total heat transferred to the oil.

33. The tube bank of Example 7.10 is arranged in a staggered configuration. However, before a sizable investment in tubing and labor is involved, an in-line configuration is to be analyzed. How many tubes are required to transfer the same amount of heat, given $S_L = 1.5$ in, $S_T = 1.3$ in, and all other conditions are the same except the tube configuration is in-line rather than staggered? Calculate also the pressure drop for the in-line configuration. The number of tubes per row is limited to 10.

7.8.5   DERIVATIONS

34. Beginning with Equation 7.12, substitute

$$\theta(\eta) = \frac{T - T_w}{T_\infty - T_w}$$

$$\eta = y\left(\frac{V_\infty}{vx}\right)^{1/2}$$

and show that Equation 7.12 reduces to Equation 7.10.

35. Show that Equations 7.13 are equivalent to Equations 7.11.

36. Begin with Equation 7.30b and derive an explicit equation for the average convection coefficient $\bar{h}_c$ in terms of the drag force D_f. Substitute for all dimensionless groups.

37. Figure 7.25 illustrates the behavior of the local Nusselt number with respect to angular coordinate for the flow of air past a circular cylinder. The purpose of this problem is to relate any of the curves in that figure to the Hilpert Equation for the average Nusselt number.

(a) Select any of the curves of Figure 7.25 and by any method so desired (from counting squares to a numerical scheme) determine the area under the curve.

(b) Note that

$$\bar{h}_c = \int_0^{2\pi} h_\theta d\theta$$

or, for constant air properties,

$$\overline{Nu}_D = \int_0^{2\pi} Nu_\theta d\theta$$

Thus, the average Nusselt number can be determined.

(c) Compare to $\overline{Nu}_D$ as calculated from Equation 7.44.

8 Natural-Convection Systems

CONTENTS

8.1 INTRODUCTION

The subject of natural or free convection was discussed in Chapter 5. Exact solutions for several natural-convection problems were derived. In this chapter, we continue the discussion and examine many natural convection problems in detail.

Forced-convection problems involve the use of an external motive device that moves fluid past an object that has a temperature different from that of the fluid. Natural-convection problems contain no such external fluid-moving device. Instead, the fluid in a natural-convection system moves due to a density gradient that is acted upon by a body force. Density gradients are due to a temperature gradient, and the body force has been attributed to gravity.

It is important to note that body forces other than gravity can act to move the fluid in a natural-convection problem. For example, centripetal acceleration is a body force that exists in rotating machinery. The Coriolis force is a body force that is significant in oceanic and atmospheric motions. Although a number of variations can exist, in this text we work primarily with natural-convection problems where the body force is gravitational and the density gradient is due to a temperature gradient.

Compared to forced-convection problems, natural-convection systems are characterized by smaller velocities, small convection coefficients, and correspondingly smaller heat-transfer rates. They are no less significant, however, because of the numerous applications that exist. For instance, a covered pot containing coffee at 150°F loses heat primarily by natural convection. Based on the natural-convection heat loss, a suitable heater must be selected that will maintain the coffee at the proper temperature. A steam pipe suspended from a ceiling loses heat by radiation and by natural convection. The magnitude of the convection heat loss must be calculated to determine insulation requirements. Foods placed in a conventional oven are heated by a natural-convection process. It is therefore important to be able properly to model natural-convection systems.[*]

The geometries we consider in this chapter include natural-convection heat transfer from a heated flat plate, a cylinder, a sphere, and other miscellaneous systems. For a number of forced-convection problems the external forced velocity is comparatively small. In such problems, natural-convection heat transfer may be of the same magnitude as the forced-convection component. The chapter concludes with a brief discussion of these mixed-convection systems.

As in previous chapters, it may sometimes seem that a barrage of equations is thrown at you. It is therefore wise to remember that the objective in each system is to be able to accurately determine a heat-transfer rate.

8.2 NATURAL CONVECTION ON A VERTICAL SURFACE—LAMINAR FLOW

Natural-convection heat transfer from (or to) a heated (or cooled) vertical plate is a classic problem. There can be an induced laminar flow that may also become turbulent. There is an exact solution and a boundary-layer solution. Moreover, the problem can be solved for a constant-wall-flux or a constant-wall-temperature boundary. We will examine some of these cases and follow up with some empirical correlations that have been developed.

Consider the vertical wall illustrated in Figure 8.1a. Shown is a wall of constant temperature T_w in contact with a cooler fluid at temperature T_∞. The x and y axes are also shown. Our interest is in determining a velocity profile, a temperature profile, and the heat transferred to the fluid. Expected velocity and temperature profiles are sketched in the figure, with gravity acting downward in the negative x direction. Figure 8.1b shows the corresponding cooled-plate problem with gravity acting in the positive x direction. The following formulation applies to either case, but the discussion centers about the steady-state flow depicted in Figure 8.1a.

We know that fluid flows in the system under the action of gravity. Density changes are due to a temperature gradient. Our method of approach is to begin with the continuity, momentum, and energy equations stated in Chapter 5. The only equation that will reflect a density gradient will be the momentum equation. There can at most be three velocities:

$$\tilde{V} = \tilde{e}_x V_x + \tilde{e}_y V_y + \tilde{e}_z V_z$$

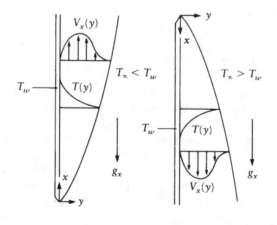

(a) Heated plate   (b) Cooled plate

FIGURE 8.1   Heat transfer from (or to) a heated (or cooled) vertical flat plate.

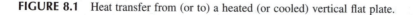

* You should be aware, however, that many cases involving natural-convection heat transfer may also involve radiation heat transfer of a comparable magnitude. While a complete analysis would require consideration of both modes in such problems, we focus our attention in this chapter (for purposes of instruction) on the natural-convection problem.

For the two-dimensional system of Figure 8.1a, $V_z = 0$. With V_x and V_y functions of only x and y, the continuity equation 5.11 reduces to

$$\frac{\partial V_x}{\partial x} + \frac{\partial V_y}{\partial y} = 0 \tag{8.1}$$

The momentum equation 5.28 becomes

$$V_x\frac{\partial V_x}{\partial x} + V_y\frac{\partial V_x}{\partial y} = -\frac{g_c dp}{p\, dx} + v\left(\frac{\partial^2 V_x}{\partial x^2} + \frac{\partial^2 V_x}{\partial y^2}\right) - g_x \tag{8.2}$$

$$V_x\frac{\partial V_y}{\partial x} + V_y\frac{\partial V_y}{\partial y} = v\left(\frac{\partial^2 V_y}{\partial x^2} + \frac{\partial^2 V_y}{\partial y^2}\right) \tag{8.3}$$

Temperature in the fluid varies with at most four variables:

$$T = T(x,y,z,t)$$

and under the assumptions made, we conclude that $T = T(x, y)$ only. The energy equation becomes

$$V_x\frac{\partial T}{\partial x} + V_y\frac{\partial T}{\partial y} = \alpha\left(\frac{\partial^2 T}{\partial x^2} + \frac{\partial^2 T}{\partial y^2}\right) \tag{8.4}$$

The boundary conditions are

$$\left.\begin{array}{rl} y = 0 & V_x = 0 \\ \text{any } x & V_y = 0 \\ & T = T_w \\ y = \infty & V_x = 0 \\ \text{any } x & V_y = 0 \\ & \frac{\partial V_x}{\partial y} = 0 \\ & T = T_\infty \end{array}\right\} \tag{8.5}$$

The differential equations written above cannot be solved, because they contain nonlinear terms. So we attempt a boundary-layer formulation to see if we can reduce the equations to a simpler form. Performing an order-of-magnitude analysis (as described in Chapter 5), the above equations reduce to:

$$\frac{\partial V_x}{\partial x} + \frac{\partial V_y}{\partial y} = 0 \tag{8.6}$$

$$V_x\frac{\partial V_x}{\partial x} + V_y\frac{\partial V_x}{\partial y} = -\frac{g_c dp}{p\, dx} - g_x + v\frac{\partial^2 V_x}{\partial y^2} \tag{8.7a}$$

$$V_x\frac{\partial T}{\partial x} + V_y\frac{\partial T}{\partial y} = \alpha\frac{\partial^2 T}{\partial y^2} \qquad (8.8)$$

As the equations now stand, there appears to be no accounting for a density gradient. We note that the pressure gradient acting along the x direction exists solely due to the weight of the fluid. We therefore write the hydrostatic equation:

$$\frac{dp}{dx} = -\frac{p_\infty g_x}{g_c} \qquad (8.9)$$

The variable that relates density to temperature is the volumetric thermal-expansion coefficient:

$$\beta = -\frac{1}{\rho}\left(\frac{\partial p}{\partial T}\right)_p \qquad (8.10a)$$

Written in approximate form, we have

$$\beta \cong -\frac{1}{\rho}\frac{\rho_\infty - \rho}{T_\infty - T} \qquad (8.10b)$$

Referring to Equation 8.7a, we combine the pressure and gravity terms and substitute Equations 8.9 and 8.10b to obtain

$$-\frac{g_c dp}{\rho\, dx} - g_x = -\frac{1}{\rho}(-\rho_\infty g_x) - g_x = g_x\left(\frac{\rho_\infty}{\rho} - 1\right)$$

$$= \frac{g_x}{\rho}(\rho_\infty - \rho) = \frac{g_x}{\rho}[-\rho\beta(T_\infty - T)]$$

or

$$-\frac{g_c dp}{\rho\, dx} - g_x = +\beta g_x(T - T_\infty) \qquad (8.11)$$

Equation 8.11 combined with 8.7a gives

$$V_x\frac{\partial V_x}{\partial x} + V_y\frac{\partial V_x}{\partial y} = \beta g_x(T - T_\infty) + v\frac{\partial^2 V_x}{\partial y^2} \qquad (8.7b)$$

The differential equations to be solved then are 8.6, 8.7b, and 8.8, subject to

$$\text{B.C. 1} \quad y = 0, V_x = 0, V_y = 0, T = T_W$$

$$\text{B.C. 2} \quad y = \infty, V_x = 0, \partial V_x/\partial y = 0, T = T_\infty$$

Even in the simplified boundary-layer form derived above, the equations have not been solved in closed form. Instead, numerical results have been obtained, known as the Ostrach solution, and are provided in Figure 8.2. Figure 8.2a is a dimensionless velocity profile $xV_x/2v(Gr_x)^{1/2}$ vs. a dimensionless distance $\eta = y(Gr_x/4)^{1/4}/x$ expressed in terms of the Grashof number:

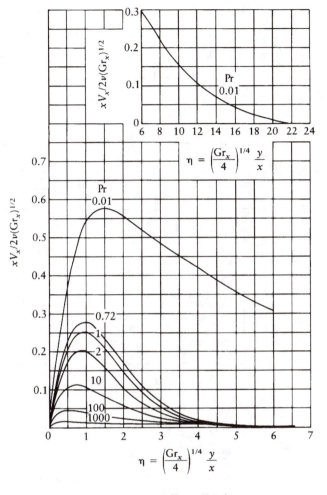

(a) Velocity profile $\mathrm{Gr}_x = \dfrac{g_x \beta (T_w - T_\infty) x^3}{\nu^2}$

FIGURE 8.2a Ostrach solution for laminar convection on a heated vertical flat plate. *Source:* from *NACA Report 1111,* by S. Ostrach, 1957. *Note:* for a detailed discussion and graphs that are scaled differently, see p. 127 of *Convection Heat Transfer,* by Adrian Bejan, John Wiley & Sons, New York, 1984.

$$Gr_x = \frac{g_x \beta (T_w - T_\infty) x^3}{\nu^2} \tag{8.12}$$

Figure 8.2b is a dimensionless temperature $\theta = (T - T_\infty)/(T_w - T_\infty)$ vs. the dimensionless distance. As shown in both graphs, the resulting profile is a function of the Prandtl number. The Ostrach solution applies to both configurations of Figure 8.1, whether $T_w > T_\infty$ or $T_w < T_\infty$. For the cooled plate, the profiles are inverted. In using the charts, fluid properties (except β) are evaluated at the film temperature $T_f = (T_w + T_\infty)/2$. For liquids, β is evaluated at the film temperature, and, for gases, β is evaluated at T_∞.

Figure 8.2 can be used to determine boundary-layer thicknesses. It is apparent that, for a fluid having a Prandtl number of 0.72, details of the numerical results indicate that the effects of the plate are no longer felt at $\eta = 5.0$. Therefore, $\eta = 5.0$ can be used to evaluate a thermal-boundary-layer thickness it δ_t. Thus,

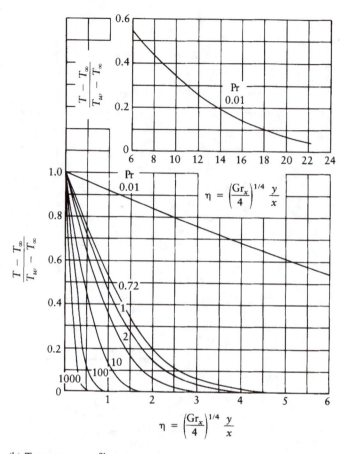

(b) Temperature profile

FIGURE 8.2b   Ostrach solution for laminar convection on a heated vertical flat plate (continued).

$$\eta = 5.0 = \frac{\delta_t}{x}\left(\frac{Gr}{4}\right)^{1/4}$$

$$\delta_t = 5.0x(4/Gr_x)^{1/4} \quad (\text{for Pr} = 0.72)$$

(8.13a)

On the other hand, it could be argued that, at $\eta = 4.0$, temperature and velocity profiles are within 99% of the free-stream values, so

$$\delta_t = 4.0x(4/Gr_x)^{1/4} \quad (\text{Pr} = 0.72)$$

(8.13b)

In keeping with our objective of determining the heat-transfer rate, our next step is to use the Ostrach solution to find the convection coefficient or, as has been the custom, the Nusselt number. Newton's Law of Cooling is

$$q'' = h_x(T_w - T_\infty)$$

(8.14)

where h_x is the local convection coefficient. The heat-transfer rate is given also by Fourier's Law:

$$q''_w = -k_f \frac{\partial T}{\partial y}\bigg|_{y=0} = -k_f \frac{dT}{dy}\bigg|_{y=0} \qquad (8.15)$$

In terms of $\theta(\eta) = [T(y) - T_\infty]/(T_w - T_\infty)$, we find

$$\frac{dT}{dy}\bigg|_{y=0} = \left(\frac{dT}{d\theta}\frac{d\theta}{d\eta}\frac{d\eta}{dy}\right)\bigg|_{y=0} = (T_w - T_\infty)\frac{1}{x}\left(\frac{Gr_x}{4}\right)^{1/4}\frac{d\theta}{d\eta}\bigg|_{\eta=0}$$

Substituting into Equation 8.15 and setting the result equal to 8.14 gives

$$-k_f(T_w - T_\infty)\frac{1}{x}\left(\frac{Gr_x}{4}\right)^{1/4}\frac{d\theta}{d\eta}\bigg|_{\eta=0} = h_x(T_w - T_\infty) \qquad (8.14)$$

Cancelling appropriately and rearranging, we obtain

$$Nu_x = \frac{h_x x}{k_f} = -\left(\frac{Gr_x}{4}\right)^{1/4}\frac{d\theta}{d\eta}\bigg|_{\eta=0} \qquad (8.16)$$

Now $(d\theta/d\eta)|_{\eta=0}$ has been evaluated from the Ostrach solution. The numerical results for $[-(d\theta/d\eta)|_{\eta=0}]$ are given in Table 8.1.

TABLE 8.1
Values of the Slope of Temperature
Profile vs. Prandtl Number for the
Ostrach Solution, for Use with
Equation 8.16

Pr	$\left(-\dfrac{d\theta}{d\eta}\right)\bigg|_{\eta=0}$
0.01	0.081
0.72	0.505
1	0.567
2	0.716
10	1.169
100	2.191
1 000	3.966

The above results can be used to derive an expression for the average convection coefficient for the whole plate. For a plate width (into the page) of b we write the following definition:

$$\bar{h}_L = \frac{1}{Lb}\int_0^L\int_0^b h_x\,dz\,dx$$

For the two-dimensional case of Figure 8.1, h_x does not vary with z (into the page). Substituting from Equation 8.16 yields

$$\bar{h}_L = \frac{1}{L}\int_0^L h_x\,dx = \frac{1}{L}\int_0^L \frac{k_f}{x}\left(\frac{Gr_x}{4}\right)^{1/4}\left(-\frac{d\theta}{d\eta}\right)\bigg|_{\eta=0}dx$$

The local Grashof number is defined as

$$\mathrm{Gr}_x = \frac{g_x \beta (T_w - T_\infty) x^3}{v^2}$$

Substituting and simplifying, we get

$$\bar{h}_L = \frac{1}{L} \int_0^L \frac{k_f}{x} \left[\frac{g_x \beta (T_w - T_\infty) x^3}{4v^2} \right]^{1/4} \left(-\frac{d\theta}{d\eta} \right)\bigg|_{\eta=0} dx$$

$$= \frac{k_f}{L} \left[\frac{g_x \beta (T_w - T_\infty)}{4v^2} \right]^{1/4} \left(-\frac{d\theta}{d\eta} \right)\bigg|_{\eta=0} \int_0^L \frac{dx}{x^{1/4}}$$

Rearranging and integrating, we obtain

$$\frac{\bar{h}_L L}{k_f} = \left[\frac{g_x \beta (T_w - T_\infty)}{4v^2} \right]^{1/4} \left(-\frac{d\theta}{d\eta} \right)\bigg|_{\eta=0} \left(\frac{4 x^{3/4}}{3} \right)\bigg|_0^L$$

or

$$\frac{\bar{h}_L L}{k_f} = \frac{4}{3} \left[\frac{g_x \beta (T_w - T_\infty) L^3}{4v^2} \right]^{1/4} \left(-\frac{d\theta}{d\eta} \right)\bigg|_{\eta=0}$$

In dimensionless terms,

$$\overline{\mathrm{Nu}}_L = \frac{4}{3} \left(\frac{(\mathrm{Gr}_L)^{1/4}}{4} \right) \left(-\frac{d\theta}{d\eta} \right)\bigg|_{\eta=0} \tag{8.17}$$

Combining with Equation 8.16, in which we set $x = L$, we find that the average Nusselt number based on plate length equals 4/3 times the Nusselt number at plate end:

$$\overline{\mathrm{Nu}}_L = \frac{4}{3} Nu_L \tag{8.18}$$

It should be noted that the above solution is for a constant wall temperature. Results for a constant wall flux q_w'' differ from the above by about 5%. Therefore the constant-wall-temperature result can be used with little error for the constant-wall-flux problem.

Natural-convection effects obviously depend on the value of the volumetric thermal-expansion coefficient β. The value of β for liquids and nonideal gases must be obtained from data tables (see Appendix tables C for liquids and D for gases). For ideal gases, we use the ideal gas law to relate pressure and temperature. Thus,

$$\rho = \frac{p}{RT}$$

$$\frac{\partial \rho}{\partial T} = -\frac{p}{RT^2} = -\frac{\rho}{T}$$

Substituting into the definition of β gives

$$\beta = -\frac{1}{\rho}\left(\frac{\partial p}{\partial T}\right)_p = -\frac{1}{\rho}\left(-\frac{\rho}{T}\right)$$

or

$$\beta = \frac{1}{T} \quad \text{(ideal gases)} \tag{8.19}$$

The value of temperature to use depends on the correlation in question. In all cases, absolute temperature must be used.

Example 8.1

In the course of designing a furnace, it is necessary to determine the thickness of wall insulation required. As a first step, the heat transferred to one oven wall must be calculated. The wall itself is 1 ft tall by 2 ft wide. It is determined that the wall temperature is roughly a constant and equal to 120°F when the inside air temperature is 400°F. Determine the heat transferred to the wall.

Solution

Air near the wall becomes cooler and descends under the action of gravity. For no air movement, conduction is the primary mode of heat transfer.

Assumptions

1. Air properties are constant and evaluated at the film temperature of $(400 + 120)/2 = 260°F = 720°R$.
2. The system is at steady state.
3. Only natural-convection heat transfer takes place.
4. Air behaves as an ideal gas.

From Appendix Table D.1 for air at 720°R

$\rho = 0.0551$ lbm/ft^3 $k_f = 0.01944$ BTU/(hr·ft·°R)
$c_p = 0.242$ BTU/(lbm·°R) $\alpha = 1.457$ ft^2/hr
$\nu = 27.88 \times 10^{-5}$ ft^2/s   Pr $= 0.689$

Also, because the fluid is a gas (assumed ideal),

$$\beta = \frac{1}{T_\infty} = \frac{1}{120 + 460} = 0.00116/°R$$

The heat transferred is calculated with

$$q_w = \bar{h}_L A(T_w - T_\infty)$$

and the convection coefficient is found with Equation 8.18. We first determine the Grashof number:

$$Gr_L = \frac{g_x \beta (T_w - T_\infty) L^3}{\nu^2} = \frac{32.2(0.00116)(400 - 120)(1)^3}{(27.88 \times 10^{-5})^2}$$

$$Gr_L = 1.34 \times 10^8$$

The convection coefficient is found with

$$\frac{\bar{h}_L L}{k_f} = \frac{4}{3}\left(\frac{Gr_L}{4}\right)^{1/4}\left(-\frac{d\theta}{d\eta}\right)\Bigg|_{\eta=0}$$

Substituting gives

$$\bar{h}_L = \frac{0.01944}{1}\frac{4}{3}\left(\frac{1.34 \times 10^8}{4}\right)^{1/4}(0.505)$$

where the slope of the temperature profile has been evaluated from Table 8.1 at $Pr = 0.72$. The convection coefficient is

$$\bar{h}_L = 1.00 \text{ BTU/(hr·ft}^2\text{·°R)}$$

This is a small value compared to the forced-convection results of the previous two chapters. The heat transferred is

$$q_w = 1.00(1)(2)(400 - 120)$$

$$q_w = 558 \text{ BTU/hr}$$ ❑

Although the equations of the exact solution can be used to calculate the Nusselt number for the vertical-plate problem, direct reliance of an equation on numerical results is sometimes inconvenient. Consequently, empirical correlations have been developed, based on actual data. The Churchill-Chu Equation for natural convection for a vertical plate of constant wall temperature or having a constant wall flux is

$$\overline{Nu}_L = \frac{\bar{h}_L L}{k_f} = 0.68 + \frac{0.670 Ra_L^{1/4}}{\left[1 + \left(\frac{0.492}{Pr}\right)^{9/16}\right]^{4/9}} \qquad (8.20)$$

where

$$Ra_L = Gr_L Pr = \frac{g_x \beta(T_w - T_\infty)L^3}{\nu\alpha} \le 10^9$$

$$0 < Pr = \frac{\nu}{\alpha} < \infty$$

$$\beta \text{ at } T_f = \frac{T_w + T_\infty}{2} \quad \text{for liquids}$$

$$\beta \text{ at } T_\infty \text{ for gases}$$
$$\text{Other properties at } T_f$$

Example 8.2

A single-pane glass window of a heated dwelling is shown schematically in Figure 8.3. Outside the dwelling, the ambient temperature is 0°C, while the inside temperature is 22°C. Determine the

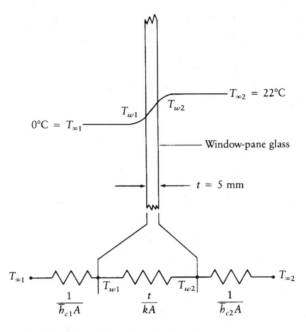

FIGURE 8.3   One-dimensional heat flow through a window glass.

heat lost through the glass per unit area. The glass thickness is 5 mm, and the window height is 60 cm.

Solution

Outside the dwelling we have cold air in contact with a warmed vertical surface. Inside we have warm air in contact with a cooler vertical surface. Natural convection is important on both sides of the glass. Figure 8.3 shows also the expected temperature profile and the thermal circuit.

Assumptions

1. One-dimensional heat flow exists.
2. Air properties are evaluated at the ambient temperatures, because surface temperatures are unknown.
3. Steady-state conditions exist.
4. End and edge effects are negligible.

For the one-dimensional flow of Figure 8.3, we write

$$q = \frac{\text{Overall temperature difference}}{\text{Sum of thermal resistances}}$$

$$= \frac{T_{\infty 2} - T_{\infty 1}}{\dfrac{1}{\bar{h}_{c1} A} + \dfrac{t}{kA} + \dfrac{1}{\bar{h}_{c2A}}}$$

or

$$\frac{q}{A} = q'' = \frac{T_{\infty 2} - T_{\infty 1}}{\dfrac{1}{\bar{h}_{c1}} + \dfrac{t}{k} + \dfrac{1}{\bar{h}_{c2}}} \tag{i}$$

The convection coefficients can be calculated with the Churchill-Chu Equation if $Ra_L < 10^9$. From Appendix Table D.1, for air at $22 + 273 = 295$ K $\cong 300$ K,

$\rho = 1.177$ kg/m³ $k_f = 0.026\ 24$ W/(m·K)
$c_p = 1\ 005$ J/(kg·K) $\alpha = 0.221\ 60 \times 10^{-4}$ m²/s
$v = 15.68 \times 10^{-6}$ m²/s $Pr = 0.708$

Also,

$\beta = 1/295 = 0.003\ 68$/K

For air at $0 + 273 \cong 275$ K, we interpolate to get

$\rho = 1.295$ kg/m³ $k_f = 0.024\ 26$ W/(m·K)
$c_p = 1\ 005.5$ J/(kg·K) $\alpha = 0.176\ 61 \times 10^{-4}$ m²/s
$v = 12.59 \times 10^{-6}$ m²/s $Pr = 0.713$

Also,

$\beta = 1/273 = 0.003\ 66$/K

From Appendix Table B.3, for glass, $k = 0.81$ W/(m·K). The Rayleigh number for use in the Churchill-Chu Equation is

$$Ra_L = \frac{g_x \beta (T_w - T_\infty) L^3}{v \alpha}$$

The wall temperatures are unknown, so we will have to formulate an iterative procedure to solve the problem. Substituting, for the outside wall we find

$$Ra_{L_1} = \frac{9.81(0.003\ 66)(T_{w1} - 0)(0.60)^3}{12.59 \times 10^{-6}(0.176\ 61 \times 10^{-4})} = 3.49 \times 10^7 (T_{w1} - 0)$$

For the inside surface,

$$Ra_{L_2} = \frac{9.81(0.003\ 38)(T_{w2} - 22)(0.60)^3}{15.68 \times 10^{-6}(0.221\ 60 \times 10^{-4})}$$

$$= -2.06 \times 10^7 (T_{w2} - 22)$$

where the negative sign denotes that the surface is cooler than the ambient temperature (i.e., gravity is in the positive x direction). We now assume wall temperatures T_{w1} and T_{w2}, then calculate the Rayleigh and Nusselt numbers. The convection coefficients and the heat-transfer rate (Eq. i) are then determined. From these numbers, we calculate refined estimates of the wall temperatures. The process is repeated until convergence is achieved. Assume $T_{w2} = 18°C$ and $T_{w1} = 4°C$. Then,

$$Ra_{L_1} = 1.396 \times 10^{-8}$$

$$Ra_{L_2} = 8.24 \times 10^7$$

From Equation 8.20,

$$\overline{\mathrm{Nu}}_L = \frac{\overline{h}_L L}{k_f} = 0.68 + \frac{0.67\mathrm{Ra}_L^{1/4}}{\left[1 + \left(\frac{0.492}{\mathrm{Pr}}\right)^{9/16}\right]^{4/9}}$$

we find

$$\overline{h}_{c1} = \frac{0.024\,26}{0.6}\left\{0.68 + \frac{0.67(1.396 \times 10^8)^{1/4}}{\left[1 + \left(\frac{0.492}{0.713}\right)^{9/16}\right]^{4/9}}\right\}$$

$$\overline{h}_{c1} = 2.29\ \mathrm{W/(m^2K)}$$

Also,

$$\overline{h}_{c2} = \frac{0.026\,24}{0.6}\left\{0.68 + \frac{0.67(8.24 \times 10^7)^{1/4}}{\left[1 + \left(\frac{0.492}{0.708}\right)^{9/16}\right]^{4/9}}\right\}$$

$$\overline{h}_{c2} = 2.17\ \mathrm{W/(m^2K)}$$

The heat flux from Equation (i) is

$$q'' = \frac{22 - 0}{1/2.29 + 0.005/0.81 + 1/2.17} = 24.3\ \mathrm{W/m^2}$$

For each surface,

$$q'' = \frac{T_{\infty 2} - T_{w2}}{1/\overline{h}_{c2}}$$

$$q'' = \frac{T_{w1} - T_{\infty 1}}{1/\overline{h}_{c1}}$$

Solving for wall temperature,

$$T_{w2} = 22 - (24.3)(1/2.17) = 10.8\,°\mathrm{C}$$

and

$$T_{w1} = \frac{24.3}{2.29} + 0 = 10.6\,°\mathrm{C}$$

Using these temperatures as second estimates and repeating the calculations, we get

$$\mathrm{Ra}_{L1} = 3.70 \times 10^8,\ \overline{h}_c = 2.92\ \mathrm{W/(m^2K)}$$

$$\mathrm{Ra}_{L2} = 2.31 \times 10^8,\ \overline{h}_c = 2.80\ \mathrm{W/(m^2K)}$$

The heat flux is

$$q'' = \frac{22 - 0}{1/2.92 + 0.005/0.81 + 1/2.80} = 31.2 \text{ W/m}^2$$

The wall temperatures are

$$T_{w2} = 22 - 31.2/2.80 = 10.8°C$$

$$T_{w1} = 10.7°C$$

These values are close enough to the trial values. The method converges rapidly. The heat loss then is

$$q'' = 31.2 \text{ W/m}^2 \qquad\qquad\qquad ❑$$

8.3 NATURAL CONVECTION ON A VERTICAL SURFACE—TRANSITION AND TURBULENCE

Natural-convection boundary layers in contact with a vertical surface can experience transition and exhibit turbulent flow. Natural convection itself is caused by a thermal instability. Turbulence is a series of hydrodynamic instabilities. For a sufficiently long vertical plate, the flow may contain disturbances. When these disturbances are no longer damped out by viscous effects and instead are amplified, the boundary layer has become turbulent.

Based on experiments performed, it is found that transition for natural convection on a vertical plate can be correlated with the Rayleigh number (= Gr Pr). The Rayleigh number at transition is 10^9:

$$\text{Ra}_x = \text{Gr}_x\text{Pr} = \frac{g_x\beta(T_w - T_\infty)x^3}{\nu\alpha} \cong 10^9 \quad \text{(transition)} \qquad (8.21)$$

Therefore, the results of the preceding section apply only for $\text{Ra}_L \leq 10^9$.

While the Churchill-Chu equation 8.20 applies to laminar conditions, a second Churchill-Chu equation can be used that applies over all ranges of Rayleigh numbers for natural convection on vertical plates:

$$\overline{\text{Nu}}_L = \frac{\bar{h}_L L}{k_f} = \left\{ 0.825 + \frac{0.387\text{Ra}_L^{1/6}}{\left[1 + \left(\frac{0.492}{\text{Pr}}\right)^{9/16}\right]^{8/27}} \right\}^2 \qquad (8.22)$$

where

$$\text{Ra}_L = \text{Gr}_L\text{Pr} = \frac{g_x\beta(T_w - T_\infty)L^3}{\nu\alpha} > 0$$

$$0 < \text{Pr} = \frac{\nu}{\alpha} < \infty$$

$$\beta \text{ at} T_f = (T_w + T_\infty)/2 \quad \text{(for liquids)}$$

$$\beta = 1/T_\infty \quad \text{(for gases)}$$

$$\text{Other properties at } T_f$$

The above equation applies to constant-wall-temperature and to constant-wall-flux problems Equation 8.22 is less accurate than the laminar expression (8.20) over the range $0 < \text{Ra}_L \leq 10^9$.

Example 8.3

The cooling unit for a proposed design of a refrigerator must be sized. One step in determining the cooling capacity is finding the heat loss through the walls. Consider one vertical side of the exterior of a refrigerator. It is 5 ft 6 in tall and 2 ft 4 in wide. For an ambient temperature of 90°F and a surface temperature of 70°F, determine the heat loss through the side.

Solution

The outside air is in contact with a cooler surface, so heat flows to the wall from the ambient air. Near the edges the heat flow is affected by the air-flow pattern.

Assumptions

1. Edge effects are negligible.
2. The system is at steady state.
3. One-dimensional heat flow through the walls exists.
4. Air properties are constant.

The method of solution here is much the same as in Example 8.2, but now we must first check for turbulent flow. From Appendix Table D.1, for air at $(80 + 460)/2 = 540°R$, we read

$\rho = 0.0735 \text{ lbm/ft}^3$ $\qquad k_f = 0.01516 \text{ BTU/(hr·ft·°R)}$
$c_p = 0.240 \text{ BTU/(lbm·°R)}$ $\qquad \alpha = 0.859 \text{ ft}^2/\text{hr}$
$v = 16.88 \times 10^{-5} \text{ ft}^2/\text{s}$ $\qquad \text{Pr} = 0.708$

Also, because the fluid is a gas,

$$\beta = 1/(90 + 460) = 0.00182/°R$$

The Rayleigh number is

$$\text{Ra}_L = \frac{g_x \beta (T_w - T_\infty) L^3}{v\alpha} = \frac{32.2(0.00182)(90 - 70)(5.5)^3}{(16.88 \times 10^{-5})(0.859/3600)}$$

$$\text{Ra}_L = 4.84 \times 10^9$$

The convection coefficient is found with Equation 8.22:

$$\bar{h}_c = \frac{k_f}{L} \left\{ 0.825 + \frac{0.387 \text{Ra}_L^{1/6}}{\left[1 + \left(\dfrac{0.492}{\text{Pr}}\right)^{9/16}\right]^{8/27}} \right\}^2$$

$$= \frac{0.01516}{5.5} \left\{ 0.825 + \frac{0.387(4.84\times10^9)^{1/6}}{\left[1 + \left(\dfrac{0.492}{0.708}\right)^{9/16}\right]^{8/27}} \right\}^2$$

or

$$\bar{h}_c = 0.553 \text{ BTU/hr·ft}^2 \cdot °\text{R})$$

The heat gain then is

$$q = \bar{h}_c A(T_w - T_\infty) = 0.553(5.5)(2 + 4/12)(70 - 90)$$

$$q = 142 \text{ BTU/hr}$$ ❏

8.4 NATURAL CONVECTION ON AN INCLINED FLAT PLATE

Figure 8.4 shows a flat plate inclined at some angle θ between the vertical direction and the plate itself. For laminar flow, the vertical-plate correlation can be used if the component of gravity parallel to the heated surface is used; that is,

$$\overline{Nu}_L = \frac{\bar{h}_L L}{k_f} = 0.68 + \frac{0.670 Ra_L^{1/4}}{\left[1 + \left(\frac{0.492}{Pr}\right)^{9/16}\right]^{4/9}} \tag{8.20}$$

where

$$Ra_L = Gr_L Pr = \frac{g_x(\cos\theta)\beta(T_w - T_\infty)L^3}{\nu\alpha}$$

$$0 < Pr = \frac{\nu}{\alpha} < \infty$$

$$0° < 0 < 60°$$

$$\beta \text{ at } T_f = (T_w + T_\infty)/2 \text{ (for liquids)}$$

$$\beta = 1/T_\infty \text{ (for gases)}$$

Other properties at T_f

Laminar-turbulent transition occurs at a critical Rayleigh number roughly given by the Vliet Equation:

$$Ra_L = 3 \times 10^5 \exp[0.136\ 8\cos(90 - \theta)] \tag{8.23}$$

Heated plate

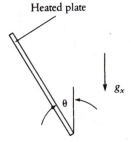

g_x

θ

FIGURE 8.4   Nomenclature for natural convection on an inclined flat plate.

Transition actually occurs over a range of Rayleigh numbers beginning at about 1/10 of the above and ending at about 10 times the value.

For turbulent natural convection on an inclined plate where θ varies from $0°$ to $60°$ (with the vertical), the vertical-plate correlation (Equation 8.22) can be used without modification. The component of gravity parallel to the plate is not significant. The correlations apply to the constant-wall-temperature and to the constant-wall-flux problem.

Example 8.4

The roof of a dwelling receives energy from the sun during the day. The roof temperature becomes too hot to touch. For a roof-surface temperature of $65°C$ and an ambient air temperature of $20°C$, determine the variation of average convection coefficient with distance. The roof makes an angle of $45°$ with the vertical and has a length of 5 m.

Solution

A typical roof surface is shingled, not smooth, and the surface itself affects the flow of air. Winds also affect the heat transfer. Moreover, radiation from the roof surface may be significant.

Assumptions

1. The roof-surface characteristics do not affect the air flow.
2. Steady-state conditions exist.
3. The roof loses heat only to the air and only by natural convection.
4. Edge effects are negligible.
5. Air properties are constant and evaluated at $(65 + 20)/2 = 42.5°C \cong 315$ K.

From Appendix Table D.1, for air at $T_f = 315$ K,

$\rho = 1.123$ kg/m^3 $k_f = 0.027\ 38$ W/(m·K)
$c_p = 1\ 006.7$ J/(kg·K) $\alpha = 0.244\ 6 \times 10^{-4}$ m^2/s
$v = 17.204 \times 10^{-6}$ m^2/s   Pr $= 0.703$

For air,

$\beta = 1/(20 + 273) = 0.003\ 41/$K

We first calculate the value of x along the surface, where transition occurs, as given by Equation 8.23:

$$\mathrm{Ra}_L = \frac{g_x(\cos\theta)\beta(T_w - T_\infty)x^3}{v\alpha} = 3 \times 10^5 \exp[0.136\ 8 \cos(90 - \theta)]$$

Substituting,

$$\frac{9.81(\cos 45°)(0.003\ 41)(65 - 20)x^3}{(17.204 \times 10^{-6})(0.244\ 6 \times 10^{-4})} = 3 \times 10^5 \exp(0.136\ 8 \cos 45°)$$

Solving for x, we get

$$x^3 = \frac{3.305 \times 10^5}{2.53 \times 10^9}$$

$$x = 0.051 \text{ m}$$

Thus, turbulent flow will exist over most of the roof. Equation 8.20, for laminar conditions, is used up to $x = 0.051$ m. Results of calculations are summarized in Table 8.2.

TABLE 8.2
Length vs. Convection–Coefficient Data for the
Inclined–Roof Problem of Example 8.4

x, m	Ra_L	$\bar{h}_c, \mathrm{W}/(\mathrm{m}^2 \cdot \mathrm{K})$
0.02	2.024×10^4	9.32
0.04	1.62×10^4	4.43
0.051	3.36×10^5	7.00
0.051	4.75×10^5	7.29
0.1	3.58×10^6	6.39
1.0	3.58×10^9	4.99
3.0	9.67×10^{10}	4.73
5.0	4.48×10^{11}	4.66

Equation 8.22 is used for $0.051 \text{ m} \le x \le 10 \text{ m}$ with results shown in Table 8.2. We apply the following equations:

Laminar flow

$$\mathrm{Ra}_L = \frac{g_x(\cos\theta)\beta(T_w - T_\infty)x^3}{\nu\alpha} = 2.53 \times 10^9 x^3$$

$$\bar{h}_c = \frac{k_f}{x}\left\{0.68 + \frac{0.670\mathrm{Ra}_L^{1/4}}{\left[1 + \left(\frac{0.492}{\mathrm{Pr}}\right)^{9/16}\right]^{4/9}}\right\} = \frac{0.027\,38}{x}(0.68 + 0.514\mathrm{Ra}_L^{1/4})$$

Turbulent flow

$$\mathrm{Ra}_L = \frac{g_x\beta(T_w - T_\infty)x^3}{\nu\alpha} = 3.58 \times 10^9 x^3$$

$$\bar{h}_c = \frac{k_f}{x}\left\{0.825 + \frac{0.387\mathrm{Ra}_L^{1/6}}{\left[1 + \left(\frac{0.492}{\mathrm{Pr}}\right)^{9/16}\right]^{8/27}}\right\}^2 = \frac{0.027\,38}{x}(0.825 + 0.324\mathrm{Ra}_L^{1/6})^2$$

Note that the convection coefficient is the average over the surface up to the corresponding value of x. ❏

8.5 NATURAL CONVECTION ON A HORIZONTAL FLAT SURFACE

The natural-convection problems considered thus far had a principal body dimension aligned primarily with the direction of action of the gravity force. The resultant flow patterns were parallel to the surface. For the case of a horizontal flat surface, the principal body dimension is perpendicular to the gravity vector and hence to the direction of action of the buoyant forces. Figure 8.5a is of a heated plate whose warmed surface faces downward. Heated fluid in the vicinity of the surface becomes less dense, flows laterally to the edge, then flows upward. Figure 8.5b is of a cooled plate whose cooler surface faces upward. Cooled fluid in the vicinity of the surface becomes more dense, makes its way to the edge, then moves downward. The flow pattern for Figure 8.5a appears to be inverted from that of 8.5b.

Figure 8.5c is of a heated surface facing upward. In this case, the tendency of the warmed, less-dense fluid to rise is acted against by cooler and denser fluid above it. This situation is unstable, and therefore other factors (such as a slight plate inclination) can strongly influence the flow pattern. Analogous comments can be made about Figure 8.5d, which is of a cooled plate facing downward.

Correlations have been formulated for these cases and are as follows:

Heated plate facing downward and cooled plate facing upward:

$$\overline{Nu}_{L_c} = 0.27 Ra_{L_c}^{1/4} \qquad 3 \times 10^5 < Ra_{L_c} < 3 \times 10^{10} \tag{8.24a}$$

Heated plate facing upward and cooled plate facing downward:

$$\overline{Nu}_{L_c} = 0.54 Ra_{L_c}^{1/4} \qquad 2.6 \times 10^4 < Ra_{L_c} < 10^7 \tag{8.24b}$$

$$\overline{Nu}_{L_c} = 0.15 Ra_{L_c}^{1/3} \qquad 10^7 < Ra_{L_c} < 3 \times 10^{10} \tag{8.24c}$$

where

$$L_c = \text{Characteristic length}$$

$$= \frac{\text{Plate surface area (warmed or cooled)}}{\text{Plate perimeter or circumference}}$$

Heated surface

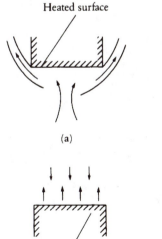

(a)   (b)

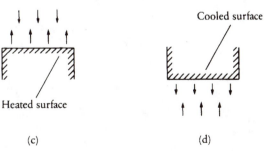

(c)   (d)

FIGURE 8.5   Natural-convection flow patterns about a horizontal flat surface.

$$\overline{Nu}_{Lc} = \frac{\bar{h}_L L_c}{k_f}$$

$$Ra_{L_c} = \frac{g_x \beta (T_w - T_\infty) L_c^3}{\nu \alpha}$$

$$\beta \text{ at} T_f = (T_w + T_\infty)/2 \quad \text{(for liquids)}$$

$$\beta = 1/T_\infty \quad \text{(for gases)}$$

Other properties at T_f

Example 10.5

A laboratory oven has outside dimensions of 2 ft wide × 2 ft tall × 2 ft long, as shown in Figure 8.6. Under certain operating conditions, the outside-surface temperature of the oven reaches 100°F. The air temperature around the oven is 60°F. Does the oven lose more heat through its upper surface or one of its vertical sides?

Solution

Heat is transferred by convection from the sides and the top to the ambient air. Flow patterns are affected at the edges of each surface.

Assumptions

1. The system is at steady state.
2. Air properties are constant and evaluated at the film temperature
   $T_f = (100 + 60)/2 = 80°F = 540°R$.
3. Edge effects are negligible.
4. Heat is transferred from the surface to the air by convection, which is the primary mode of heat transfer.

From Appendix Table D.1, for air at 540°R,

$\rho = 0.0735$ lbm/ft^3 $k_f = 0.01516$ BTU/(hr·ft·°R)
$c_p = 0.240$ BTU/(lbm·°R) $\alpha = 0.859$ ft^2/hr
$\nu = 16.88 \times 10^{-5}$ ft^2/s   Pr = 0.708

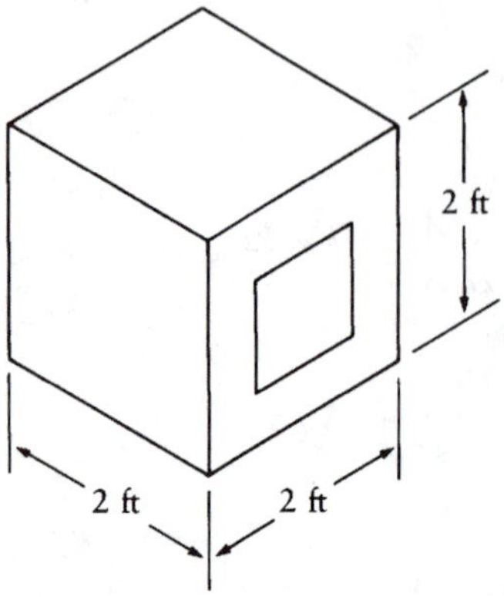

2 ft

2 ft   2 ft

FIGURE 8.6 The laboratory oven of Example 8.5.

Also,

$$\beta = 1/(60 + 460) = 0.00192/°R$$

We now apply the appropriate convection correlation and calculate the convection coefficients. For one vertical side, we use Equation 8.20. The Rayleigh number is

$$Ra_L = \frac{g_x\beta(T_w - T_\infty)L^3}{\nu\alpha} = \frac{32.2(0.00192)(100 - 60)(2)^3}{(16.88\times10^{-5})(0.859/3600)}$$

$$Ra_L = 4.92 \times 10^8$$

The flow is laminar, because $Ra_L < 10^9$.

The Nusselt number is calculated with

$$\overline{Nu}_L = \frac{\bar{h}_L L}{k_f} = 0.68 + \frac{0.670Ra_L^{1/4}}{\left[1 + \left(\dfrac{0.492}{Pr}\right)^{9/16}\right]^{4/9}}$$

Rearranging and substituting gives

$$\bar{h}_L = \frac{0.01516}{2}\left\{0.68 + \frac{0.67(4.92\times10^8)^{1/4}}{\left[1 + \left(\dfrac{0.492}{0.708}\right)^{9/16}\right]^{4/9}}\right\}$$

$$\bar{h}_L = 0.586 \text{ BTU/(hr} \cdot \text{ft}^2 \cdot °R)$$

The heat transferred from one side then is

$$q_{1\ side} = \bar{h}_L A_s(T_w - T_\infty) = 0.586(2)(2)(100 - 60)$$

$$q_{1\ side} = 93.7 \text{ BTU/hr}$$

For the top, we have a heated surface facing upward. The characteristic length is

$$L_c = \frac{A}{P} = \frac{(2)(2)}{2 + 2 + 2 + 2} = \frac{4}{8} = 0.5$$

The Rayleigh number based on the characteristic length is

$$Ra_{L_c} = \frac{g_x\beta(T_w - T_\infty)L_c^3}{\nu\alpha} = \frac{32.2(0.00192)(100 - 60)(0.5)^3}{(16.88 \times 10^{-5})(0.859/3600)}$$

$$Ra_{L_c} = 7.67 \times 10^6$$

Thus, because of the value of the Rayleigh number, Equation 8.24b applies:

$$\overline{Nu}_{L_c} = \frac{\bar{h}_L L_c}{k_f} = 0.54\ Ra_{L_c}^{1/4}$$

So

$$\bar{h}_L = \frac{0.01516}{0.5}(0.54)(7.67 \times 10^6)^{1/4}$$

or

$$\bar{h}_L = 0.862 \text{ BTU}/(\text{hr} \cdot \text{ft}^2 \cdot {}^\circ\text{R})$$

The corresponding heat transferred is

$$q_{top} = 0.862(2)(2)(100 - 60)$$

$$q_{top} = 138 \text{ BTU/hr}$$

The top therefore transfers heat at a higher rate than does one side.   ❑

8.6   NATURAL CONVECTION ON CYLINDERS

Cylinder configurations can be the same as those for plates—vertical, horizontal, or inclined. Furthermore, the cylinder surface can be at a constant temperature or have a constant wall flux. In this section, we examine a few representative cases.

8.6.1   VERTICAL CYLINDERS

The equations of continuity, momentum, and energy can be written for this problem. These equations can be reduced to a boundary-layer formulation via an order-of-magnitude analysis. Here, however, we will merely state equations that have been derived, based on experimental work. For vertical cylinders with a constant heat flux,

$$\overline{\text{Nu}}_D = \frac{\bar{h}_c D}{k_f} = 0.6\left(\text{Ra}_D\frac{D}{L}\right)^{0.25} \qquad \text{Ra}_D\frac{D}{L} \geq 10^4 \qquad\qquad (8.25\text{a})$$

$$\overline{\text{Nu}}_D = 1.37\left(\text{Ra}_D\frac{D}{L}\right)^{0.16} \qquad 0.05 \leq \text{Ra}_D\frac{D}{L} \leq 10^4 \qquad\qquad (8.25\text{b})$$

$$\overline{\text{Nu}}_D = 0.93\left(\text{Ra}_D\frac{D}{L}\right)^{0.05} \qquad \text{Ra}_D\frac{D}{L} \leq 0.05 \qquad\qquad (8.25\text{c})$$

where

$$\text{Ra}_D = \frac{g_x \beta(T_w - T_\infty)D^3}{\nu\alpha}$$

$$D = \text{cylinder diameter}$$

$$L = \text{cylinder length}$$

$$\beta \text{ at } T_f = (T_w + T_\infty)/2 \quad \text{(for liquids)}$$

$$\beta = 1/T_\infty \quad \text{(for gases)}$$

Other fluid properties at T_f

The Rayleigh number actually should be calculated by using an average temperature difference; however, the results differ from the constant-surface-temperature problem only by about 5%. So the above relationships apply also for the constant-wall-temperature case.

8.6.2 HORIZONTAL CYLINDERS

Natural convection from a horizontal cylinder has been analyzed. The resultant expression (another Churchill-Chu Equation) based on experimental results is

$$\overline{Nu}_D = \frac{\bar{h}_c D}{k_f} = \left\{ 0.60 + \frac{0.387 Ra_D^{1/6}}{\left[1 + \left(\frac{0.559}{Pr} \right)^{9/16} \right]^{8/27}} \right\}^2 \tag{8.26}$$

where

$$10^{-5} < Ra_D = \frac{g_x \beta (T_w - T_\infty) D^3}{\nu \alpha} < 10^{12}$$

$$0 < Pr = \nu/\alpha < \infty$$

$$\beta \text{ at } T_f = (T_w + T_\infty)/2 \quad \text{(for liquids)}$$

$$\beta = 1/T_\infty \quad \text{(for gases)}$$

Other fluid properties at T_f

Example 8.6

An insulated steam pipe is routed horizontally through an unheated room, then vertically to the ceiling. The pipe is insulated, and the outside diameter of the insulation is 15 cm. The horizontal and vertical lengths of pipe are both 4 m long. The outside-surface temperature of the insulation is 50°C, and the room temperature is 5°C. How much heat is lost from the insulation by convection?

Solution

The steam loses energy as it travels through the pipe, and it is likely that the surface temperature of the insulation does not remain constant. The insulation probably loses a portion of energy by radiation to the surroundings as well as by convection. The insulation surface may not be perfectly cylindrical.

Assumptions

1. The entire surface of insulation is at 50°C.
2. The air has constant properties evaluated at (50 + 5)/2 = 27.5°C or $\cong$ 300 K.
3. Heat is lost only by convection from the insulation surface.
4. The insulation surface is a cylinder.
5. The system is at steady state.

From Appendix Table D.1, for air at 300 K,

$\rho = 1.177 \text{ kg/m}^3$ $k_f = 0.026\ 24 \text{ W/(m·K)}$
$c_p = 1\ 005.7 \text{ J/(kg·K)}$ $\alpha = 0.221\ 60 \times 10^{-4} \text{ m}^2/\text{s}$
$\nu = 15.68 \times 10^{-6} \text{ m}^2/\text{s}$ $Pr = 0.708$

For air,

$$\beta = 1/(5 + 273)= 0.003\ 6/K$$

The Rayleigh number for the vertical or the horizontal runs of pipe is calculated as

$$\mathrm{Ra}_D = \frac{g_x\beta(T_w - T_\infty)D^3}{v\alpha} = \frac{9.81(0.003\ 6)(50 - 5)(0.15)^3}{(15.68 \times 10^{-6})(0.221\ 60 \times 10^{-4})}$$

or

$$\mathrm{Ra}_D = 1.54 \times 10^7$$

For the horizontal pipe, we use Equation 8.26:

$$\overline{\mathrm{Nu}}_D = \frac{\bar{h}_c D}{k_f} = \left\{ 0.60 + \frac{0.387\ \mathrm{Ra}_D^{1/6}}{\left[1 + \left(\frac{0.559}{\mathrm{Pr}}\right)^{9/16}\right]^{8/27}} \right\}^2$$

Rearranging and substituting,

$$\bar{h}_c = \frac{0.026\ 24}{0.15} = \left\{ 0.6 + \frac{0.387(1.54\times10^7)^{1/6}}{\left[1 + \left(\frac{0.559}{0.708}\right)^{9/16}\right]^{8/27}} \right\}^2$$

Solving,

$$\bar{h}_c = 5.62\ W/(m^2 \cdot K)$$

The heat transferred from the horizontal length of 4 m is

$$q_{\mathrm{hor}} = \bar{h}_c A_s (T_w - T_\infty) = 5.62(\pi)(0.15)(4)(50 - 5)$$

$$q_{\mathrm{hor}} = 476\ W$$

For the vertical pipe, we use Equation 8.25a:

$$\overline{\mathrm{Nu}}_D = \frac{\bar{h}_c D}{k_f} = 0.6\left(\mathrm{Ra}_D\frac{D}{L}\right)^{1/4}$$

Rearranging and substituting,

$$\bar{h}_c = \frac{0.026\ 24}{0.15}(0.6)\left(1.54 \times 10^7 \frac{0.15}{4}\right)^{1/4}$$

or

$$\bar{h}_c = 2.89 \text{ W/(m}^2 \cdot \text{K)}$$

The heat loss from the vertical length of 4 m is

$$q_{\text{ver}} = 2.89(\pi)(0.15)(4)(50-5) = 245 \text{ W}$$

The total heat lost by the pipe is $q_{\text{ver}} + q_{\text{ver}}$, or

$$q = 720 \text{ W} \qquad \qquad \square$$

8.6.3 INCLINED CYLINDERS

Inclined cylinders are not discussed here. The interested reader is referred to the literature* for further information.

8.7 NATURAL CONVECTION AROUND SPHERES AND BLOCKS

Natural convection from a sphere to a fluid can be formulated by beginning with the equations of motion. Here, however, we rely on correlations that are based on experimental data. The Raithby-Hollands Equation for natural convection around spheres is

$$\overline{\text{Nu}}_D = \frac{\bar{h}_c D}{k_f} = 2 + 0.56\left(\frac{\text{Pr}}{0.846 + \text{Pr}}\right)^{1/4} \text{Ra}_D^{1/4} \qquad (8.27)$$

where

$$1 < \text{Ra}_D = \frac{g_x \beta (T_w - T_\infty)D^3}{\nu\alpha} < 10^5$$

$$0 < \text{Pr} = \nu/\alpha < \infty$$

$$\beta \text{ at} T_f = (T_w + T_\infty)/2 \quad \text{(for liquids)}$$

$$\beta = 1/T_\infty \quad \text{(for gases)}$$

$$\text{Other properties at } T_f$$

Rectangular blocks are not well suited to write continuity, momentum, and energy equations for. Thus such systems lend themselves almost exclusively to experimental studies. The King Equation has been developed for natural convection around rectangular blocks:

$$\overline{\text{Nu}}_L = \frac{\bar{h}_c L}{k_f} = 0.6\text{Ra}_L^{1/4} \qquad (8.28)$$

where

$$10^4 < \text{Ra}_L = \frac{g_x \beta (T_w - T_\infty)L^3}{\nu\alpha} < 10^9$$

---

* Morgan, V.T., "The Overall Convective Heat Transfer from Smooth Circular Cylinders," *Adv. Heat Transf.,* v. 11, pp. 199–264, 1975.

$$0 < \text{Pr} = v/\alpha < \infty$$

$$\beta \text{ at} T_f = (T_w + T_\infty)/2 \quad \text{(for liquids)}$$

$$\beta = 1/T_\infty \quad \text{(for gases)}$$

Other properties at T_f

L = characteristic length found with $\dfrac{1}{L} = \dfrac{1}{L_h} + \dfrac{1}{L_v}$

L_h = horizontal dimensions

L_v = vertical dimensions

Equation 8.28 is a general expression that can be applied to a variety of miscellaneous shapes. Spheres, blocks, horizontal cylinders, and vertical plates can all be modeled, but not to a high degree of accuracy. For a sphere, L becomes the radius. For a horizontal cylinder, L becomes the diameter. For a vertical cylinder, L_v could be height and L_h the diameter.

Example 8.7

An edible snack known as a "snowball" consists of shaved ice in a cup, over which flavored syrup is poured. Ice is delivered to a "snowball stand" in the form of a cube that is 1 ft wide × 1.2 ft tall × 1 ft long and is set down on four stainless steel pads positioned at each corner. Determine the convection coefficient about the ice cube. Take the surface temperature of ice to be 0°F and the ambient air temperature to be 70°F.

Solution

The ice block is probably near a surface (such as a table top) that will affect the natural-convection air-flow pattern. The ice will begin to melt, which also affects the flow pattern. We have separate equations for the vertical and horizontal surfaces, and also a single equation for the entire block.

Assumptions

1. Natural-convection air flows are not affected by nearby surfaces.
2. Phase change has a negligible effect during the time when the calculated convection coefficient applies.
3. Air properties are constant and evaluated at a film temperature of $(0 + 70)/2 = 35°F = 495°R$.
4. The system is at steady state.

From Appendix Table D.1 for air at 495°R

$\rho = 0.0809$ lbm/ft³ $k_f = 0.01402$ BTU/(hr·ft·°R)
$c_p = 0.240$ BTU/(lbm·°R) $\alpha = 0.685$ ft²/hr
$v = 13.54 \times 10^{-5}$ ft²/s   Pr $= 0.712$

Also,

$$\beta = 1/(70 + 460) = 0.00189/°R$$

We will use the King Equation 8.28. The characteristic length is found as

$$\frac{1}{L} = \frac{1}{1 \text{ ft}} + \frac{1}{1.2 \text{ ft}}$$

$$L = 0.545 \text{ ft}$$

The Rayleigh number is

$$\mathrm{Ra}_L = \frac{g_x \beta (T_w - T_\infty) L^3}{\nu \alpha} = \frac{32.3(0.00189)(70-0)(0.545)^3}{(13.54 \times 10^{-5})(0.685/3600)}$$

$$\mathrm{Ra}_L = 2.68 \times 10^7$$

The Nusselt number is given by

$$\overline{\mathrm{Nu}_L} = \frac{\bar{h}_c L}{k_f} = 0.60 \, \mathrm{Ra}_L^{1/4}$$

The convection coefficient then is

$$\bar{h}_c = \frac{0.01402}{0.545}(0.60)(2.68 \times 10^7)^{1/4}$$

$$\bar{h}_c = 1.11 \text{ BTU/(hr·ft}^2\text{·°R)} \qquad \square$$

8.8 NATURAL CONVECTION ABOUT AN ARRAY OF FINS

Chapter 2 contains an analysis of fins. A number of fin types are discussed there. One topic of interest is in optimizing an array of fins. An optimum fin thickness and spacing exists such that maximum heat is transferred.

Fins are attached to a surface to increase the surface area, providing a means for transferring more heat away. The addition of a fin, however, also introduces a conduction resistance to heat transfer. The interaction between the convection and conduction heat-transfer modes is therefore an important consideration in fin design.

In this section, we examine flow patterns about, and heat transfer from, an array of rectangular fins such as that shown in Figure 8.7. In agreement with the notation of Chapter 2, the fins are the straight rectangular type, of length L and width b. The fin thickness is $2\delta'$, which is not to be confused with the hydrodynamic-boundary-layer thickness δ. The wall temperature is T_w.

Figure 8.8 shows the flow pattern that exists about two adjacent fins. The fins are separated by a distance S. Also shown are the boundary layers that form along each surface. If the fins are too close together, the fluid cannot move freely enough to make the fins effective. If the fins are too far apart, then efficiency suffers because more fins could have been used to transfer away more heat.

For forced-convection heat transfer, the boundary-layer thickness δ might vary from 0 to 0.25 cm (0.1 in). For natural convection, the boundary-layer thickness might vary to 1.2 cm (0.5 in). If no interference between adjacent boundary layers is permitted, then the fin spacing S should be just twice as large as the maximum thickness of the boundary layer ($S = 2\delta_{max}$). Tests on such arrays have shown that mutual interference is only slight. The Bar-Cohen Equations, based on experimental work with fins, have been developed to predict optimum fin spacing S, optimum fin thickness $2\delta'$, and optimum heat-rejection rate for the array of Figure 8.7. For natural convection, the Bar-Cohen Equations are

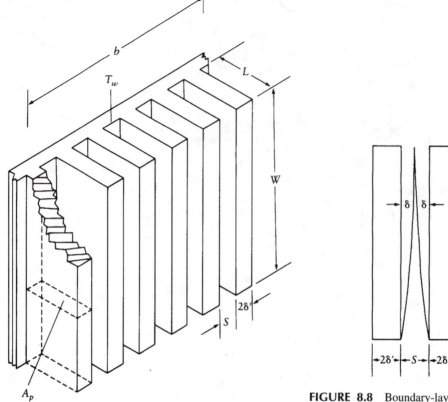

FIGURE 8.8 Boundary-layer growth pattern for natural convection about two adjacent fins.

FIGURE 8.7 An array of fins having a rectangular profile.

$$\zeta^4 = \frac{W v^2}{g_x \beta (T_w - T_\infty) \mathrm{Pr}}$$

$$L = 1.54 \left(\frac{k}{k_f} \right)^{1/2} \zeta$$

$$S = 2\delta' = 2.89\zeta$$

$$\frac{q}{b W (T_w - T_\infty)} = \frac{1.3 (k k_f)^{1/2}}{6\zeta}$$

$$(8.29)$$

where

k = thermal conductivity of fin material

k_f = thermal conductivity of fluid

β at $T_f = [(T_w + T_\infty)/2]$ (for liquids)

$\beta = 1/T_\infty$ (for gases)

Other fluid properties at T_f

Example 8.8

A tank containing warm oil is to be cooled by natural convection. It is proposed to add aluminum fins to the outside rear of the tank, which is a vertical surface 24 cm tall and 46 cm wide. The surface temperature is expected to reach 100°C, and the ambient temperature is 35°C. Investigate the possibility of placing fins on the surface in a natural-convection cooling arrangement by determining the maximum amount of heat that fins can transfer.

Solution

The fins will help transfer heat away from the tank. Air flow near the rear of the tank should not be obstructed. We will attempt to use the configuration of Figure 8.7 and fill the entire available area with a vertically oriented fin array.

Assumptions

1. Air properties are constant and evaluated at $(100 + 35)/2 = 67.5°C$, which is roughly equal to 350 K.
2. Aluminum properties are constant.
3. The system is at steady state.
4. Heat is transferred from the fins only by convection.

$\rho = 0.998$ kg/m³ $k_f = 0.030\ 03$ W/(m·K)
$c_p = 1\ 009.0$ J/(kg·K) $\alpha = 0.298\ 3 \times 10^{-4}$ m²/s
$\nu = 20.76 \times 10^{-6}$ m²/s $Pr = 0.697$

Also,

$\beta = 1/(35 + 273) = 0.003\ 25/K$

From Appendix Table B.1, for aluminum

$\rho = 2\ 702$ kg/m³ $k = 236$ W/(m·K)
$c_p = 896$ J/(kg·K) $\alpha = 97.5 \times 10^{-6}$ m²/s

Referring to Figure 8.7 for this problem, we make $b = 46$ cm and $W = 24$ cm. If an unworkable result emerges, we can reduce the dimensions of the fin array accordingly or, alternatively, conclude that natural convection does not provide the cooling required for *maximum* heat transfer.

Applying the Bar-Cohen Equations (8.29), we find

$$\zeta^4 = \frac{W\nu^2}{g_x\beta(T_w - T_\infty)Pr} = \frac{0.24(20.76\times10^{-6})^2}{9.81(0.003\ 25)(100 - 35)(0.697)} = 7.17\times 10^{-11}\ \text{m}^4$$

or

$$\zeta = 2.91 \times 10^{-3}\ \text{m}$$

The fin length is

$$L = 1.54\left(\frac{k}{k_f}\right)^{1/2}\zeta = 1.54\left(\frac{236}{0.030\ 03}\right)^{1/2}(2.91 \times 10^{-3})$$

or

$$L = 0.397 \text{ m}$$

The fin spacing and thickness are

$$S = 2\delta' = 2.89\zeta = 0.008\ 41 \text{ m}$$

The heat-transfer rate is

$$q = bW(T_w - T_\infty)(1.3)(kk_f)^{1/2}/6\zeta$$

$$= 0.46(0.24)(100 - 35)(1.3)[236(0.030\ 03)]^{1/2}/6(2.91 \times 10^{-3})$$

or

$$q = 1\ 420 \text{ W}$$

The fins can be made of sheet aluminum, and the exact thickness will probably be dictated by what is available commercially. The fin length of 39.7 cm (for *maximum* heat transfer for the situation) is probably too long. Thus, space requirements might prohibit making the fins so long. Alternatives are as follows:

1. Shorten the fins and use other parts of the cabinet to transfer heat away.
2. Use a different fin material so that k becomes smaller and so does length.
3. Use a forced-convection arrangement so that the convection coefficient increases and the need for such a large area decreases.

With a width of 0.46 m, there can be at most $0.46/2(0.008\ 41) \cong 27$ fins.   ❏

8.9   COMBINED FORCED- AND NATURAL-CONVECTION SYSTEMS

A combined forced- and natural-convection system is one in which the forced-convection heat-transfer effects are small enough that natural convection plays an important role. Thus, the term *forced convection* implies that natural-convection effects are negligible.

When there is no externally forced flow, natural convection is the only mode of heat transfer occurring that induces fluid movement. If a weak, externally forced flow is present, it plays an important role in moving fluid near a heated surface, although natural convection might still be dominant. If the externally forced flow is quite large, forced-convection effects are dominant. If both effects are present and no detailed information is provided on the convection coefficient for the combined problem, current engineering practice is to evaluate the coefficients for forced and for natural convection separately, then use the larger value.

As evident from the above discussion, a more quantitative way is needed for determining if both natural- and forced-convection effects are important. Detailed studies show that the dimensionless group Gr/Re^2 can be used as an indicator. In general,

$$\frac{Gr}{Re^2} \gg 1 \rightarrow \text{natural convection dominates}$$

$$\frac{Gr}{Re^2} \approx 1 \rightarrow \text{combined problem}$$

$$\frac{\text{Gr}}{\text{Re}^2} \ll 1 \rightarrow \text{forced convection dominates}$$

where

$$\frac{\text{Gr}}{\text{Re}^2} = \frac{g_x \beta (T_w - T_\infty) L^3}{v^2} \frac{v^2}{V^2 L^2} = \frac{g_x \beta (T_w - T_\infty) L}{V^2}$$

Another factor to consider is whether the flow is what is called *aiding* or *opposing*. In the combined problem, if the buoyancy-induced motion is in the same direction as the forced motion, then aiding flow exists. If the buoyant motion acts against the forced motion, then opposing flow exists. For aiding flow past a vertical plate, a rule of thumb applicable for any value of the Prandtl number is

$$\text{Nu}_{\text{combined}} \approx \left(\text{Nu}_{\text{natural}}^3 + \text{Nu}_{\text{forced}}^3 \right)^{1/3}$$

Thus, the Nusselt number for the combined problem would depend on the Nusselt number for the natural-convection problem (such as Equation 8.20) and for the forced-convection problem (Equation 7.18b or 7.20).

8.10 SUMMARY

In this chapter, we have considered several natural-convection problems. Vertical, inclined, and horizontal plates are discussed, as are vertical and horizontal cylinders. Some practical design results are presented for various fin arrays. Finally, the combined problem is mentioned.

Again, although it may seem as if a barrage of equations is hurled at the reader, it is important to remember the objective in such problems: determining the heat-transfer rates. Expressing empirical correlations in terms of dimensionless groups is merely a more convenient and elegant way of trying to achieve the desired result. Additional pertinent correlations are provided in the exercises, and all correlations are tabulated in Table 8.3.

8.11 PROBLEMS

8.11.1 NATURAL CONVECTION-VERTICAL PLANE SURFACES

1. Water is stored in an outdoor storage tank. One vertical wall of the tank is warmed by the sun to a temperature of 60°C. The water temperature is 20°C. For laminar flow, construct a graph of water-velocity and temperature profiles that are induced in the vicinity of the wall at a location where $x = 0.1$. Use the available value of the coefficient of thermal expansion.
2. Calculate the local convection coefficient for the data of Problem 1.
3. Steam at 1100°R is in contact with a wall whose temperature is 880°R. For laminar flow construct a graph of steam-velocity and temperature distributions induced in the vicinity of the wall at a height of 2 ft. Assume that, at these temperatures, steam behaves as a perfect gas.
4. Calculate the convection coefficient for the steam of Problem 3.
5. Cooling of special parts is to be conducted in an inert atmosphere, but which cooling fluid to use must first be decided. The choice is nitrogen or helium, and calculations are to be made for both fluids in contact with a vertical heated wall that is 1 m tall. The

TABLE 8.3
Summary of Equations and Correlations for Chapter 8

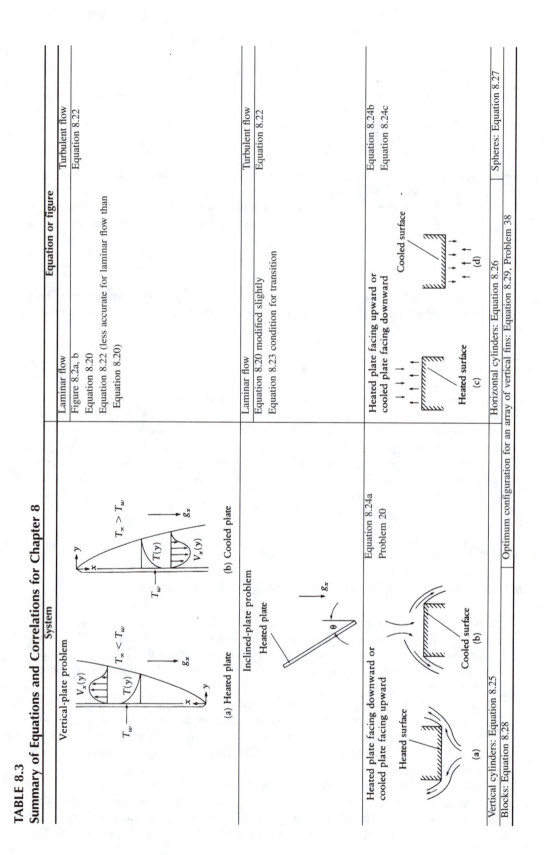

vertical wall is maintained at 900 K, and the fluid has a temperature of 700 K. Which fluid results in the higher overall convection coefficient?

6. Estimate the thickness of the thermal and the hydrodynamic boundary layers in Example 8.1.

7. Gaseous oxygen at 540°R is in contact with a 6-in-tall vertical plate maintained at 900°R. Calculate the natural-convection coefficient by using Equation 8.17 and again by using Equation 8.20.

8. Figure P8.1 is of a 1/2-in-thick vertical wall divider made of plywood. On one side is air at 170°F, and on the other is air at 80°F. Determine the heat transferred through the wall if it is 2 ft tall and 1 ft 6 in wide.

9. Figure P8.2 is of a vertical wall that is part of a dwelling. The outside common brick is 10 cm thick, and the inside panel is 1.3 cm-thick plasterboard. The brick and plasterboard are separated by 9.5 cm of glass-fiber insulation. On the brick side is air at 2°C, while on the plasterboard side is air at 27°C. The wall is 2.5 m tall. How much heat is transferred through wall per unit width (into the page)?

10. A vertical wall is maintained at a temperature of 40°C. A cooler fluid is in contact with the wall. Determine the vertical length required for the induced flow to become turbulent if the fluid is
    (a) Ammonia
    (b) Ethylene glycol
    (c) Mercury
    In all cases, take the fluid temperature to be 0°C, and use the available value of the coefficient of thermal expansion.

11. A vertical wall is maintained at a temperature of 100°F. A cooler fluid is in contact with the wall. Determine the vertical length required for the induced flow to become turbulent if the fluid is
    (a) Liquid carbon dioxide
    (b) Gaseous carbon dioxide
    In both cases, take the fluid temperature to be 36°F.

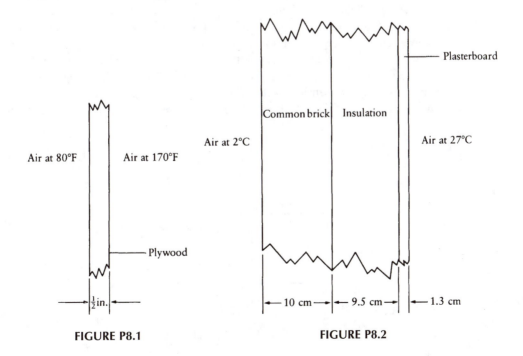

FIGURE P8.1   FIGURE P8.2

12. One vertical wall of an aquarium is 0.5 m tall and made of plexiglass that is 2.5 cm thick. On one side is water (about 0.5 m deep) at a temperature of 20°C, while on the other is air at 27°C. Estimate the plexiglass surface temperatures.

13. A convection heater consists of two stamped sheet metal sides that are soldered together at the edges such that oil can circulate between the plates. Essentially, each side can be considered a flat plate, where hot oil is on one side and air is on the other (see Figure P8.3). The metal is 65 cm tall. Use the King Equation 8.28 to estimate the heat transferred from the oil to the air per unit depth into the page. Neglect end effects and the resistance offered by the metal.

8.11.2   NATURAL CONVECTION—INCLINED SURFACES

14. Referring to Example 8.4, it is found that, for a roof angle of 45°, the average convection coefficient over the laminar portion of the roof is 7.00 W/(m²·K). Determine the average convection coefficient over the roof area where laminar-flow conditions exist for roof angles of 30° and 60°. Compare to the example.

15. Figure P8.4 is of a conventional coffee pot made of 3-mm-thick glass (assume window-pane-glass properties). It is filled to a height of 9.5 cm with brewed coffee (assume properties the same as water). The coffee pot is covered to retard evaporative losses (assume negligible). The pot sits on a flat heater that is to maintain the coffee at a temperature of 80°C. Ambient air temperature is 27°C. Estimate the size of the heater required, provided that the surface area through which heat is transferred is the glass in contact with the coffee, and that curvature effects are negligible. List all assumptions.

8.11.3   NATURAL CONVECTION—HORIZONTAL PLANE SURFACES

16. A huge pot is used to boil water. On the underside of the cover is steam at 225°F. Air on the top side is at 80°F. Neglecting effects of condensation on the underside, determine the heat transferred through the top if it is made of stainless steel and is essentially a flat disk having a diameter of 12 in.

17. The roof of a "hatchback" car is 2 m wide and 4 m long. Treating it as a flat plate that has received enough energy (from the sun) to elevate its surface temperature to 55°C, calculate the convection heat-transfer coefficient. The ambient air temperature is 20°C.

18. An air-conditioned closet contains air at a temperature of 65°F. The ceiling is made of 1/2-in plasterboard of dimensions 4 ft × 8 ft. On the other side of the ceiling (which is uninsulated) is air at 95°F. Calculate the heat transferred through the ceiling per unit

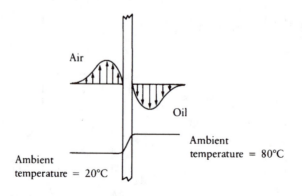

FIGURE P8.3

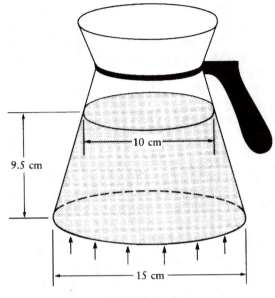

9.5 cm

10 cm

15 cm

FIGURE P8.4

area. Evaluate all fluid properties (except the coefficient of thermal expansion) at 80°F [= (65 + 95)/2)].

19. Repeat the calculations of Problem 18, except include a 3-1/2-in-thick glass-wool-insulation layer.

20. Equations different from those given in the chapter have been developed for heated surfaces facing downward. For a square of side 2L,

$$\overline{Nu}_L = \frac{\overline{h}_c L}{k_f} = 0.816 Ra_L^{1/5}$$

For a circle of radius R,

$$\overline{Nu}_R = \frac{\overline{h}_c R}{k_f} = 0.818 Ra_R^{1/5} Pr^{0.034}$$

For an infinite strip of width 2W,

$$\overline{Nu}_w = \frac{\overline{h}_c W}{k_f} = 0.5 Ra_w^{1/5}$$

where

$$Ra_n = \frac{g_x \beta (T_w - T_\infty) n^3}{\nu \alpha} \quad (n = L \text{ or } R \text{ or } W \text{ as appropriate})$$

$$\beta \text{ at} T_f = (T_w + T_\infty)/2 \quad \text{(for liquids)}$$

$$\beta = 1/T_\infty \quad \text{(for gases)}$$

Other properties at T_f

A flat plate of dimensions 1×1 m is used as a floating cover for a tank of glycerin. The glycerin is at a temperature of 10°C. The tank is in the sun, and the cover becomes heated until it reaches a temperature of 30°C. Determine the heat transferred to the glycerin, using Equation 8.24 as appropriate and the applicable equation above. Calculate percent differences.

21. A 1-m-diameter circular disk is used as a cover for an oil tank (assume engine oil). The oil is at 0°C, and the cover is at 40°C. Determine the heat transferred to the oil, using Equation 8.24 and the applicable equation from Problem 20.

8.11.4 NATURAL CONVECTION—CYLINDERS

22. A furnace door is to have a handle attached that is essentially a vertical cylinder 1 cm in diameter and 11 cm long (see Figure P8.5). It is known that 0.2 W of heat is lost through the handle. Determine the surface temperature of the handle. The ambient air temperature is 27°C.

23. A towel-heating rack can be found in the bathrooms of many English hotels. The rack consists of a tube through which hot water is circulated. The tube becomes warm and heats towels that are hung on it. Figure P8.6 is a sketch of a rack. An important concern is the temperature of the outside tube surface when no towels are hung on the rack. Serious burns can result if the surface is too hot. Consider the rack of Figure P8.6. Hot water enters the tube at 155°F and experiences a decrease of $\Delta T = 30°F$ after going through the tube. Determine the outside-tube-wall temperature. Evaluate water properties at 140°F and let the ambient air temperature be 80°F. Take the volume flow rate of water to be 2 ft³/min.

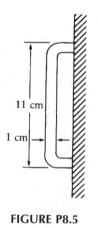

FIGURE P8.5

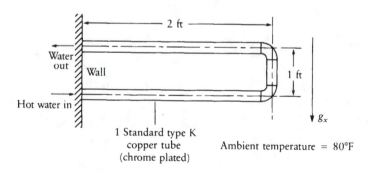

FIGURE P8.6

24. Repeat Problem 23 for the rack of Figure P8.7. Keep all other parameters the same.

25. The heating element in the bottom of a teapot is sketched in Figure P8.8. When first turned on, the surface temperature of the element is 65°C, and the surrounding water temperature is 15°C. Determine the wattage rating of the heater.

26. Example 8.6 shows that, for two cylinders of the same length, diameter, and temperature, the horizontal pipe transfers away more heat than the vertical one. In this problem, we investigate the effect of diameter. Repeat the calculation of convection coefficient for the horizontal pipe of the example for diameters of 13, 17, and 19 cm.

27. Repeat Problem 26 for the vertical pipe of Example 8.6.

28. A cigarette is 100 mm long and 1.1 cm in diameter. After 1 cm of the cigarette has been consumed by an avid smoker, the average surface temperature of the paper becomes 40°C. For an ambient air temperature of 20°C, determine the convection coefficient between the surface and the air. Use the King Equation 8.28 and compare to the horizontal-cylinder-equation results.

8.11.5 NATURAL CONVECTION—MISCELLANEOUS GEOMETRIES AND PROBLEMS

29. A refrigerator is sketched as a rectangular box in Figure P8.9. It is desired to obtain a first estimate of the size of the cooling unit required for the design. Assuming the outside-surface temperature is everywhere constant and equal to 70°F, and assuming the ambient

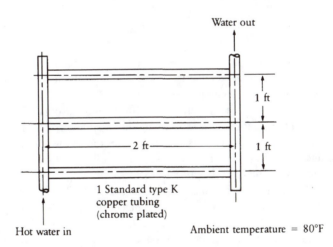

FIGURE P8.7

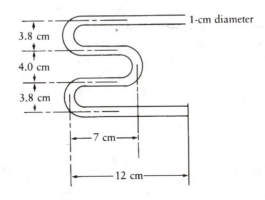

FIGURE P8.8

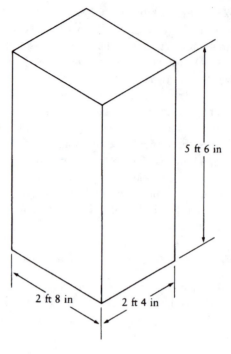

5 ft 6 in

2 ft 8 in   2 ft 4 in

FIGURE P8.9

air temperature is 90°F, determine the heat transferred via natural convection from each surface individually. Then add these values to find the total heat transferred. The surfaces include top, bottom, and four sides.

30. A cube $1 \times 1 \times 1$ m is at a temperature of 100°C and surrounded by air at 50°C.
    (a) Use the King Equation 8.28 to estimate the overall convection coefficient.
    (b) Use the equations for flat plates to estimate the convection coefficient for each surface. Average the surface coefficients together to obtain an overall value for the cube Compare the results to those of part (a).

31. Ceramic spheres of 1-3/4 in diameter and marketed as Pac-man figures are oven fired during manufacture. When the figures are placed in the oven, the sphere-surface temperature is 65°F, and the surrounding air temperature is 450°F. While the figures are in the oven, the sphere-surface temperature eventually reaches 300°F, after which the spheres are removed. During heating, does the convection coefficient change considerably? Determine the convection coefficient for surface temperatures of 65°, 100°, 200°, and 300°F. Tabulate $\bar{h}_c$ vs. temperature difference. Evaluate properties at 450°F.

32. What is the average fluid velocity required in Example 8.1 before it is necessary to consider it a combined-convection problem?

33. What is the average fluid velocity required in Example 8.3 before it is necessary to consider it a combined-convection problem?

34. What is the average fluid velocity required in Example 8.6 before it is necessary to consider it a combined-convection problem?

8.11.6 NATURAL CONVECTION—FINS

35. The fin length calculated in Example 8.8 is far too long. Repeat the calculations for 1C steel fins and determine if a more workable configuration results.

36. A vertical array of rectangular fins is used to transfer heat to oil (assume engine oil) in a tank. The fins are purchased "off the shelf," so there is little opportunity to design them

for optimum performance. There is, however, control available for maintaining the fin-surface temperature at any desired level between temperatures of 40° and 100°C. The fins are 7 mm in thickness and spaced 7 mm apart. There are 20 vertical fins, and each is 10 cm long × 15 cm tall. The fins are made of aluminum. The oil temperature is 20°C. Determine the fin-surface temperature for optimum or near optimum performance. Although the fins are made of aluminum, is there a "better" material to use?

37. Cast-iron fins are to be made in a vertical configuration and used to transfer heat from flat vertical sides of rectangular transformers. The fins are to occupy a space that is 3 ft tall and 24 in wide. For a fin-surface temperature of 140°F and an ambient air temperature of 20°F, what is the maximum amount of heat the fin array will transfer away? What are the optimum fin dimensions?

38. A different theoretical development of the vertical-plate fin-array problem results in an equation for the overall Nusselt number, known as the Elenbaas Equation:

$$\overline{Nu}_s = \frac{\overline{h}_c S}{k_f} = \frac{\xi}{24}[1 - \exp(-35/\xi)]^{3/4}$$

where

$$\xi = Ra_s(S/W)$$

$$Ra_s = \frac{g_x \beta(T_w - T_\infty)S^3}{\nu\alpha}$$

$$\beta \text{ at } T_f = (T_w + T_\infty)/2 \quad \text{(for liquids)}$$

$$\beta = 1/T_\infty \quad \text{(for gases)}$$

Other properties at T_f

(a) Graph the above equation as $\overline{Nu}_s$ vs. ξ. Use log-log paper and let ξ vary from 0.1 to 10^4.

(b) Differentiate the above equation and show that the maximum heat transferred occurs at $\xi = 45.9$.

(c) Substitute this result back into the Elenbaas Equation and show that

$$\overline{Nu}_s\big|_{opt} = 1.19$$

(d) Rework Example 8.8, using the Elenbaas results. Keep the spacing and fin dimensions the same as in the example. The heat transferred is given by

$$q = \overline{h}_c A_s(T_w - T_\infty)$$

and

$$A_s = \text{Number of fins} \cdot \text{Surface area of 1 fin}$$

8.11.7 DERIVATIONS

39. Begin with the continuity, momentum, and energy equations of Chapter 5 and derive Equations 8.1 through 8.8. List all assumptions.

40. Starting with Equation 8.9, verify the derivation of Equation 8.7b.

41. Figure P8.10 is an illustration of a vertical cylinder immersed in an infinite fluid (as discussed in Section 8.6). The cylinder is at a different temperature than the fluid, so natural convection will take place. Also shown in the figure are expected velocity and temperature profiles for $T_w > T_\infty$. Note the r, θ, z axes.

(a) Give reasons to indicate that radial and axial velocities are nonzero.

(b) Assuming that V_r and V_z are both functions of r and z only, show that the continuity equation (from Chapter 5) becomes

$$\frac{1}{r}\frac{\partial}{\partial r}(rV_r) + \frac{\partial}{\partial z}(V_z) = 0$$

(c) Show that the momentum equations reduce to

$$V_r\frac{\partial V_r}{\partial r} + V_z\frac{\partial V_r}{\partial z} = -\frac{g_c}{\rho}\frac{\partial p}{\partial r} + v\left(\frac{\partial}{\partial r}\left[\frac{1}{r}\frac{\partial}{\partial r}(rV_r)\right] + \frac{\partial^2 V_r}{\partial x^2}\right) - \frac{g_c}{\rho r}\frac{\partial p}{\partial \theta} = 0$$

and

$$V_r\frac{\partial V_z}{\partial r} + V_z\frac{\partial V_z}{\partial z} = -\frac{g_c}{\rho}\frac{\partial p}{\partial z} + v\left[\frac{1}{r}\frac{\partial}{\partial r}\left(r\frac{\partial V_z}{\partial r}\right) + \frac{\partial^2 V_z}{\partial x^2}\right] + g_z$$

(d) Assuming temperature varies with r and z, show that the energy equation reduces to

$$\rho c_p\left(V_r\frac{\partial T}{\partial r} + V_z\frac{\partial T}{\partial z}\right) = k_f\left[\frac{1}{r}\frac{\partial}{\partial r}\left(r\frac{\partial T}{\partial r}\right) + \frac{\partial^2 T}{\partial z^2}\right]$$

Neglect viscous dissipation.

(e) Write the boundary conditions for the system.

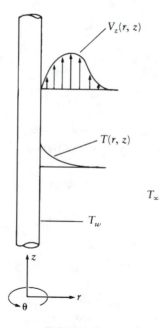

FIGURE P8.10

42. Consider a vertical plate in contact with a fluid at a different temperature (i.e., the vertical plate problem). We seek to use an alternative (to what is in this chapter) scheme to determine a velocity profile.

We have seen in previous chapters that the boundary-layer method is a powerful tool. It has the advantage that an explicit form is obtained for the velocity and temperature profiles. The disadvantage is that the result is only an approximation. We now apply the method to the laminar-flow solution of natural convection on a vertical plate.

We select assumed profiles for velocity and they are to satisfy the boundary conditions. Substituting the derived profiles into the boundary-layer equations then leads to a solution for the boundary-layer thickness.

The velocity profile can be subjected to five boundary conditions:

B.C.1. $y = 0$, $V_x = 0$
                                     (nonslip condition)

B.C.2. $y = \delta$, $V_x = 0$
                                     (stagnant fluid at the edge of the boundary layer)

B.C.3. $y = \delta$, $\dfrac{\partial V_x}{\partial y} = 0$
                                     (absence of a shear stress at the edge of the boundary layer)

B.C.4. $y = 0$, $g_x \beta(T_w - T_\infty) + v\dfrac{\partial^2 V_x}{\partial y^2} = 0$
                                     (momentum equation evaluated at the wall by using B.C.1)

B.C.5. $y = \delta$, $\dfrac{\partial^2 V_x}{\partial y^2} = 0$
                                     (momentum equation evaluated at the edge of the boundary layer by using B.C.2)

We assume a velocity distribution of the form

$$\frac{V_x}{V_r} = C_o + C_1(y/\delta) + C_2(y/\delta)^2 + C_3(y/\delta)^3 + C_4(y/\delta)^4$$

where δ is the thickness of the boundary layer, V_r is a reference velocity (or a parameter with dimensions of velocity), and the Cs are constants to be determined from application of the boundary conditions. Note that the form of B.C.4 is somewhat different from that of the other four. Furthermore, if B.C.4 is not included in the boundary conditions that are selected, then the constants (Cs) all become zero. Thus, B.C.4 is significant with regard to the natural-convection problem.

(a) Show that, by selecting boundary conditions 1, 2, and 4 above, the velocity profile becomes

$$\frac{V_x}{V_r} = \frac{g_x \beta(T_w - T_\infty)\delta^2}{2V_r v}\left(\frac{y}{\delta} - \frac{y^2}{\delta^2}\right)$$

or

$$\frac{2vV_x}{g_x\beta(T_w - T_\infty)\delta^2} = \frac{y}{\delta}\left(1 - \frac{y}{\delta}\right) \quad \text{(B.C. 1, 2, 4)}$$

It is seen that V_r, the reference velocity, does not appear explicitly, and the parameter $g_x\beta(T_w - T_\infty)\delta^2/2v$ has dimensions of velocity.

(b) Show that, by using boundary conditions 1, 2, 3, and 4, the velocity profile becomes

$$\frac{4vV_x}{g_x\beta(T_w - T_\infty)\delta^2} = \frac{y}{\delta}\left(1 - \frac{y}{\delta}\right)^2 \quad \text{(B.C. 1, 2, 3, 4)}$$

(c) If all five boundary conditions are used, show that the velocity profile becomes

$$\frac{6vV_x}{g_x\beta(T_w - T_\infty)\delta^2} = \frac{y}{\delta}\left(1 - \frac{y}{\delta}\right)^3 \quad \text{(B.C. 1, 2, 3, 4, 5)}$$

(d) For purposes of comparison, graph the above profiles on the same set of axes as $vV_x/[g_x\beta(T_w - T_\infty)\delta^2]$ vs. y/δ. Include also the numerical results of the Ostrach solution for the case of $Pr = 1$. (Note that the second-order profile is parabolic and is not a good model. The third- and fourth-order approximations are better.)

43. In this problem, we repeat Problem 42 to determine a temperature profile using the boundary layer method. The boundary layer thickness here is denoted as δ_t, and we assume that $\delta_t = \delta$. That is, we assume that the hydrodynamic and thermal boundary layer thickness are equal. We assume a profile given by

$$\frac{T - T_\infty}{T_w - T_\infty} = B_0 + B_1(y/\delta) + B_2(y/\delta)^2 + B_3(y/\delta)^3 + B_4(y/\delta)^4$$

where the Bs are constants yet to be determined. The profile can be subjected to the following conditions:

B.C.1. $y = 0$, $T = T_w$
                                   (wall temperature)

B.C.2. $y = \delta$, $T = T_\infty$
                                   (quiescent fluid temperature)

B.C.3. $y = \delta$, $\dfrac{\partial T}{\partial y} = 0$

                                   (temperature profile no longer varies with y at the edge of the boundary layer)

B.C.4. $y = 0$, $\dfrac{\partial^2 T}{\partial y^2} = 0$

                                   (obtained from the energy equation evaluated at the wall, using B.C.1 from the velocity-profile conditions)

B.C.5. $y = \delta,$ $\dfrac{\partial^2 T}{\partial y^2} = 0$

(obtained from the energy equation evaluated at the edge of the boundary layer, using B.C.2 from the velocity-profile conditions)

A number of possible temperature profiles can be derived, depending on the boundary conditions selected. Derived profiles are summarized in the following table:

Summary of Possible Temperature Profiles

Boundary conditions used	Profile
1, 2	$\dfrac{T - T_\infty}{T_w - T_\infty} = (1 - y/\delta)$
1, 2, 3	$\dfrac{T - T_\infty}{T_w - T_\infty} = (1 - y/\delta)^2$
1, 2, 3, 4	$\dfrac{T - T_\infty}{T_w - T_\infty} = 1 - \dfrac{3}{2}\dfrac{y}{\delta} + \dfrac{1}{2}\dfrac{y^3}{\delta^3}$
1, 2, 3, 4, 5	$\dfrac{T - T_\infty}{T_w - T_\infty} = 1 - \dfrac{2y}{\delta} + \dfrac{2y^3}{\delta^3} - \dfrac{y^4}{\delta^4}$
1, 2, 3, 5	$\dfrac{T - T_\infty}{T_w - T_\infty} = (1 - y/\delta)^3$

(a) Derive the profile(s) assigned by your instructor.
(b) Graph each assigned profile with the Ostrach solution for purposes of comparison. Proceeding further is beyond the scope of this discussion. However, the interested reader is referred to the literature.[*]

8.12 PROJECT PROBLEMS

1. Figure P8.11 is a sketch of a device used to obtain natural-convection cooling data for a vertically inclined flat aluminum plate. The plate is suspended by two rigid wire rods

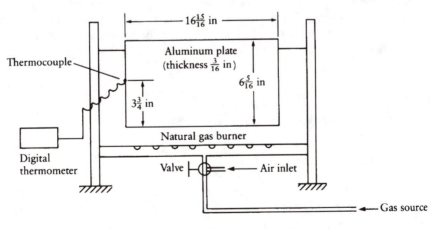

FIGURE P8.11

---

[*] See, for example, *Heat Transfer*, by A.J. Chapman, 4th ed., Macmillan Publishing Co., 1984, pp. 311–314.

attached to the stands on either side. Below the plate is a tube drilled with a number of holes and connected to a natural-gas line. Gas is fed to the tube and ignited at the holes. Combustion of the gas heats the plate to a predetermined temperature, as indicated by the thermocouple. The gas is then shut off, and the plate cools in air. A temperature reading is taken every few seconds. The data are then reduced and used to calculate Nusselt numbers.

Temperature readings were actually taken at 5-s intervals. The following table is a summary of the readings spaced at 60-s intervals:

T_w, °F	t, min	T_w, °F	t, min
367.6	0	217.4	8
340.6	1	205.0	9
318.0	2	194.8	10
295.8	3	184.8	11
278.4	4	174.6	12
260.4	5	167.0	13
243.2	6	159.8	14
230.4	7	152.8	15

Note: T_∞ = 70.2°F

Considering the plate as a system, the rate of heat loss by the plate is equal to its change of internal energy:

$$q = \frac{du}{dT}$$

$$q = mc\frac{dT_w}{dt} \tag{i}$$

where m = mass of plate, c = specific heat, and dT_w/dt = time rate of change of plate-surface temperature. Thus the heat loss at any instant is proportional to the slope of the temperature-time curve. The plate loses heat by convection to the surrounding air and by radiation to the walls of the room. The convective heat loss is

$$q_c = \bar{h}_c A_w (T_w - T_\infty) \tag{ii}$$

The radiative heat loss is given by

$$q_r = \sigma \left(\frac{1}{\frac{1}{\varepsilon_w} + \frac{1}{\varepsilon_s} - 1} \right) A_w (T_w^4 - T_s^4) \tag{iii}$$

where σ is the Stefan-Boltzmann constant (= 5.67×10^{-8} W/(m^2·K^4) = 0.1714×10^{-8} BTU/(hr·ft^2·°R^4)), T_s is the temperature of the surrounding walls of the room (equal to T_∞), ε_w is the emissivity of the aluminum plate (assume rough surface), and ε_s is the emissivity of the walls (assume rough brick). Once q_r is evaluated for any data point, a radiation coefficient is determined from

$$q_r = \bar{h}_r A_w (T_w - T_s) = \bar{h}_r A_w (T_w - T_\infty) \tag{iv}$$

Equations (ii) and (iv) added together equal Equation (i). Thus

$$\rho \Psi_s c \frac{dT_w}{dt} = (\bar{h}_c + \bar{h}_r) A_w (T_w - T_\infty)$$

Solving for the convection coefficient, we get

$$\bar{h}_c = \frac{\rho \Psi_s c}{A_w} \frac{dT_w}{dt} \frac{1}{T_w - T_\infty} - \bar{h}_r \tag{v}$$

The Nusselt number for the vertical-plate problem then is

$$\overline{Nu}_L = \frac{\bar{h}_c L}{k_f} \tag{vi}$$

where L is the plate length in the vertical direction. At this point it is assumed that the plate temperature is uniform (Biot number $\ll 1$). However, the Biot number is defined as

$$Bi = \frac{\bar{h}_c \Psi_s}{k A_w}$$

and $\bar{h}_c$ is yet to be determined. Therefore, the Biot number must be calculated at the end of the problem.

Data Reduction
(a) Construct a graph of temperature vs. time and determine in any convenient way dT_w/dt for each temperature given in the data table. *(Suggestions:* It can be done graphically or by curve fitting (least squares method) of the equation $T = C_0 + C_1 \log t + C_2[\log t]^2$. Once the constants are known, it is an easy matter to find dT_w/dt for any t.)
(b) Determine the radiative heat loss q_r from Equation (iii) for each data point. Remember to use absolute-temperature values in that equation, and remember that the plate has six sides.
(c) Calculate the radiative coefficient using Equation (iv).
(d) Calculate the convection coefficient using Equation (v).
(e) Calculate the Nusselt number for each data point.
(f) Calculate the Rayleigh number for each data point.
(g) Use the appropriate Churchill-Chu Equation and calculate Nusselt numbers. Compare the results to those obtained in part (e).
(h) Calculate the Biot number for each data point.
2. The experiment described in Project Problem 1 can be repeated with a solid, circular metal rod in place of a flat plate. The following data are for a 1C steel rod with a 1/2-in diameter:

T_w,°F	t, min	T_w,°F	t, min
300.0	0	152.2	10
273.6	1	145.6	11
253.0	2	139.4	12
232.8	3	133.8	13
216.6	4	128.8	14
202.0	5	124.2	15
189.8	6	120.8	16
178.2	7	116.4	17
168.8	8	113.0	18
160.0	9	110.4	19

Note: $T_\infty = 70°F$

Data Reduction

(Equation numbers refer to those in Project Problem 1, which should be read thoroughly before proceeding.)

(a) Construct a graph of temperature vs. time and determine in any convenient way dT_w/dt for each temperature given in the data table.

(b) Determine the radiative heat loss from Equation (iii) for each data point. Remember to use absolute-temperature values in that equation.

(c) Calculate the radiative coefficient, using Equation (iv).

(d) Calculate the convection coefficient, using Equation (v).

(e) Calculate the Nusselt number for each data point.

(f) Calculate the Rayleigh number for each data point.

(g) Use the appropriate correlation of the chapter and calculate Nusselt numbers. Compare graphically or in tabular form to the results obtained in part (e).

(h) Calculate the Biot number for each data point.

3. The experiment described in Project Problem 1 can be repeated with a solid, circular metal rod in place of a flat plate. The following data are for an (assume pure) aluminum rod with a 3/8-in diameter:

T_w,°F	t, min	T_w,°F	t, min
300.0	0	135.4	8
256.4	1	127.4	9
225.6	2	120.4	10
202.4	3	115.0	11
184.0	4	110.0	12
168.4	5	105.8	13
155.6	6	102.2	14
144.8	7	99.2	15

Note: $T_\infty = 70°F$

Data Reduction

(Equation numbers refer to those in Project Problem 1, which should be read thoroughly before proceeding.)

(a) Construct a graph of temperature vs. time and determine in any convenient way dT_w/dt for each temperature given in the data table.

(b) Determine the radiative heat loss from Equation (iii) for each data point. Remember to use absolute-temperature values in that equation.

(c) Calculate the radiative coefficient, using Equation (iv).

(d) Calculate the convection coefficient, using Equation (v).

(e) Calculate the Nusselt number for each data point.

(f) Calculate the Rayleigh number for each data point.

(g) Use the appropriate correlation of the chapter and calculate Nusselt numbers. Compare graphically or in tabular form to the results obtained in part (e).

(h) Calculate the Biot number for each data point.

4. The experiment described in Project Problem 1 can be repeated with a solid, circular metal rod in place of a flat plate. The following data are for an (assume pure) aluminum rod with a 3/4-in diameter:

T_w°F	t, min	T_w°F	t, min
300.0	0	165.6	10
275.0	1	158.6	11
255.2	2	152.4	12
239.0	3	146.8	13
224.4	4	141.8	14
211.8	5	137.0	15
201.0	6	132.6	16
191.0	7	128.6	17
182.0	8	125.2	18
173.4	9	121.8	19

Note: $T_\infty = 70$°F

Data Reduction

(Equation numbers refer to those in Project Problem 1, which should be read thoroughly before proceeding.)

(a) Construct a graph of temperature vs. time and determine in any convenient way dT_w/dt for each temperature given in the data table.

(b) Determine the radiative heat loss from Equation (iii) for each data point. Remember to use absolute-temperature values in that equation.

(c) Calculate the radiative coefficient, using Equation (iv).

(d) Calculate the convection coefficient, using Equation (v).

(e) Calculate the Nusselt number for each data point.

(f) Calculate the Rayleigh number for each data point.

(g) Use the appropriate correlation of the chapter and calculate Nusselt numbers. Compare graphically or in tabular form to the results obtained in part (e).

(h) Calculate the Biot number for each data point.

9 Heat Exchangers

CONTENTS

9.1 INTRODUCTION

Heat exchangers is a broad term used in reference to devices designed for exchanging heat. Most often the heat is transferred from one fluid to another. A fluid that is discharged from a useful process might contain high energy, and it may be desired to recover some of the energy that would ordinarily be discarded. One example of this is in remotely located plants that generate their own electricity via steam turbines. Condensed steam that is still at a relatively high temperature might be routed through a small exchanger where it is used to heat water. A common example of a heat exchanger that dissipates unwanted energy is the radiator of a car.

Heat exchangers can be classified in a number of ways, depending on their construction or on how the fluids move relative to each other through the device. A *double-pipe* heat exchanger consists of two concentric pipes or tubes. One fluid—the warmer one, for example—flows through the inner pipe. Another fluid flows through the annulus. Due to a temperature difference between the fluids, heat is transferred. The fluid streams (in the pipe and in the annulus) could be traveling in the same direction (*parallel flow*) or in opposite directions (*counterflow*). Double-pipe heat exchangers are discussed in the next section.

A *shell-and-tube* exchanger consists of a huge outer cylinder (called the *shell*) within which are contained many tubes. Generally, the shell and tube exchanger can handle fluid-flow rates that are many times as large as those in a double-pipe exchanger. Shell-and-tube exchangers are discussed in Section 9.3.

A *crossflow* heat exchanger is one in which the two fluid paths cross each other, usually at right angles. Common examples of crossflow exchangers include an automobile radiator, an oil cooler, and the coil and the evaporator of an air-conditioning system. It is interesting to note that all these devices have different names, but all perform the same function; moreover, any two of them could be identical. Crossflow exchangers are discussed in Section 9.5.

We will deal only with fluid-to-fluid exchangers. We will not work with condensing or evaporating fluids in this chapter.

9.2 DOUBLE-PIPE HEAT EXCHANGERS

A double-pipe heat exchanger is illustrated in Figure 9.1. As indicated, it consists of one tube within another. One fluid flows through the inner tube. Another fluid flows through the annulus.

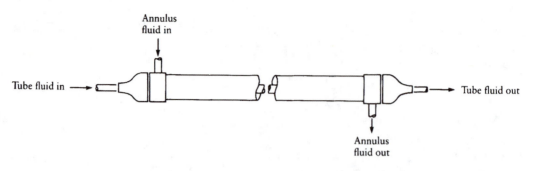

FIGURE 9.1 A double-pipe heat exchanger.

Usually, common copper water tubing is used to construct the exchanger, although pipe and specialty tubing can also be used. Our objective in analyzing a double-pipe heat exchanger will be to predict its performance—the amount of heat transferred, the outlet temperature of one (or both) fluid stream(s), or the required area, depending on what is known. For an existing exchanger, the fluid-flow rates, inlet and outlet temperatures, and physical data (diameters, etc.) are known or can be measured.

Figure 9.2 is a sketch of temperature vs. length for two systems: counterflow and parallel flow. As shown, the fluids flow in opposite directions for counterflow, and in the same direction for parallel flow. (Note that the notation for temperature is upper case for the warmer fluid and lower case for the colder fluid.) In both cases, the annulus fluid is being heated, and the pipe fluid is being cooled. The "1" subscript refers to an inlet temperature, while "2" refers to an outlet temperature.

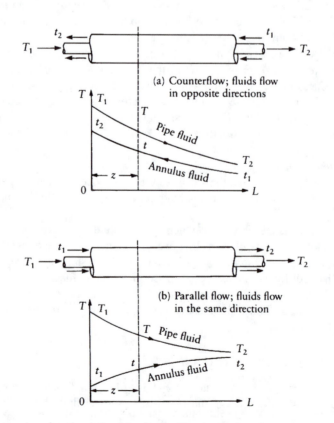

FIGURE 9.2 Temperature–length relationships for double-pipe heat exchangers.

The two fluid-stream temperatures vary with distance z within the exchanger. It is usually assumed that all heat lost by the warmer fluid is transferred to the cooler fluid. The heat being transferred thus encounters a tube-film resistance, a tube-wall resistance, and an annulus-film resistance:

$$\Sigma R = \text{Film resistance for the inside tube-surface area}$$
$$+ \text{Tube-wall conduction resistance}$$
$$+ \text{Film resistance for the outside tube-surface area} \qquad (9.1)$$

Equation 9.1 must contain appropriate areas to evaluate the resistances. The film coefficient that applies to the inside of the inner tube ($\bar{h}_i$) is based on the inside tube-surface area, while the film coefficient $\bar{h}_o$ is based on the outside surface area of the inner tube. It is customary to use the outside surface area of the inner tube, and rewrite the sum of the resistances in terms of an overall heat-transfer coefficient U_o:

$$\frac{1}{U_o} = \frac{1}{\bar{h}_i(A_i/A_o)} + \frac{D_o}{2k}\ln\frac{D_o}{D_i} + \frac{1}{\bar{h}_o} \qquad (9.2)$$

where D_o is the outer diameter (OD) of the inner tube, D_i is the inside diameter of the inner tube, $A_i = \pi D_o L$, and $A_o = \pi D_o L$. The pipe itself is made of metal and thus offers a comparatively negligible resistance to heat transfer except in the case of condensation or vaporization. (For the phase-change fluid, the convection coefficient is very high, and the corresponding resistance has a magnitude that is comparable to that of the tube wall.) Neglecting the tube-wall resistance, Equation 9.2 can be written as

$$\frac{1}{U_o} = \frac{D_o}{h_i D_i} + \frac{1}{h_o} \qquad (9.3a)$$

where the overbar on the convection coefficients has been dropped, although it should be remembered that the overall coefficients based on the tube length appear in the equation. We introduce the coefficient h_{io}, which is defined as being equal to $h_i D_i/D_o$. This new coefficient in effect is the inside-wall-area convection coefficient referred to the outside-wall area. The overall coefficient can now be written as

$$\frac{1}{U_o} = \frac{1}{h_{io}} + \frac{1}{h_o} \qquad (9.3b)$$

The heat transferred for the entire exchanger then is

$$q = U_o A_o \Delta t \qquad (9.4)$$

where Δt is an as-yet-undetermined temperature difference, which should reflect the fact that we have two inlet and two outlet temperatures. The overall heat-transfer coefficient can be calculated if the film coefficients are known, using empirical correlations such as those found in Chapter 6.

At any z location within the exchanger (Figure 9.2), the temperature difference is $T - t$. This difference varies throughout the exchanger. It is desirable to determine the overall heat transfer, however, rather than a local value. Furthermore, it is convenient to use the inlet and outlet temperatures again rather than local values because the inlet and outlet temperatures can be easily measured. Thus, it is convenient to express the driving potential Δt of Equation 9.4 in terms of the inlet and outlet temperatures of both streams rather than a local temperature difference.

Although there will be heat transferred in either parallel or counterflow configurations, the temperature difference is not identical (numerically) for both cases. Beginning with a counterflow arrangement, we will derive the appropriate temperature difference Δt under the following assumptions:

1. The overall heat-transfer coefficient U_o (based on the outside surface area of the inner tube A_o) is a constant over the length of the exchanger.
2. Steady flow exists.
3. Fluid properties are constant.
4. There are no phase changes in the system.
5. There are no heat losses; that is, all heat transferred from the warmer fluid goes to the cooler fluid.

The magnitude of the heat transferred at any z location (Figure 9.2) is

$$dq = U_o(T-t)dA_o \qquad (9.5)$$

The heat lost by the warmer fluid over a differential length is

$$dq = \dot{m}_h c_{ph}dT \qquad (9.6)$$

where $\dot{m}_h$ is the mass flow rate of the warmer fluid, and C_{ph} is its specific heat. For the cooler fluid, the magnitude of the heat gained is

$$dq = \dot{m}_c c_{pc}dt \qquad (9.7)$$

Setting Equations 9.6 and 9.7 equal to one another and integrating from the end that has the cold fluid inlet to any point within the exchanger gives

$$\int_{T_2}^{T} \dot{m}_h c_{ph}dT = \int_{t_1}^{t} \dot{m}_c c_{pc}dt$$

or

$$\dot{m}_h c_{ph}(T - T_2) = \dot{m}_c c_{pc}(t - t_1) \qquad (9.8a)$$

Rearranging and solving for the temperature of the warmer fluid, we get

$$T = T_2 + \frac{\dot{m}_c c_{pc}}{\dot{m}_h c_{ph}}(t - t_1) \qquad (9.8b)$$

Substituting into Equation 9.5, we obtain

$$dq = U_o\left[T_2 + \frac{\dot{m}_c c_{pc}}{\dot{m}_h c_{ph}}(t - t_1) - t\right]dA_o$$

Having eliminated the warmer fluid temperature T as a variable, we next set the above equation equal to Equation 9.7 for the cooler fluid:

$$\dot{m}_c c_{pc} dt = U_o \left[T_2 + \frac{\dot{m}_c c_{pc}}{\dot{m}_h c_{ph}} (t - t_1) - t \right] dA_o$$

The cooler fluid temperature and surface area are the only differential variables. Separating terms and integrating, we get

$$\int_0^{A_o} \frac{U_o dA_o}{\dot{m}_c c_{pc}} = \int_{t_1}^{t_2} \frac{dt}{\left[T_2 + \frac{\dot{m}_c c_{pc}}{\dot{m}_h c_{ph}} (t - t_1) - t \right]}$$

or

$$\frac{U_o A_o}{\dot{m}_c c_{pc}} = \left[\frac{1}{(\dot{m}_c c_{pc} / \dot{m}_h c_{ph}) - 1} \right] \times \ln \left[\frac{T_2 - \frac{\dot{m}_c c_{pc}}{\dot{m}_h c_{ph}} t_1 + \left(\frac{\dot{m}_c c_{pc}}{\dot{m}_h c_{ph}} - 1 \right) t_2}{T_2 - \frac{\dot{m}_c c_{pc}}{\dot{m}_h c_{pc}} t_1 + \left(\frac{\dot{m}_c c_{pc}}{\dot{m}_h c_{ph}} - 1 \right) t_1} \right] \tag{9.9}$$

Equation 9.8b, when integrated to L, becomes

$$T_1 = T_2 + \frac{\dot{m}_c c_{pc}}{\dot{m}_h c_{ph}} (t_2 - t_1) \tag{9.8c}$$

Solving for T_2, substituting into Equation 9.9, and simplifying gives

$$\frac{U_o A_o}{\dot{m}_c c_{pc}} = \frac{1}{(\dot{m}_c c_{pc} / \dot{m}_h c_{ph}) - 1} \ln \frac{T_1 - t_2}{T_2 - t_1} \tag{9.10}$$

From Equation 9.8c

$$\frac{\dot{m}_c c_{pc}}{\dot{m}_h c_{ph}} = \frac{T_1 - T_2}{t_2 - t_1}$$

Substituting into Equation 9.10 gives

$$\frac{U_o A_o}{\dot{m}_c c_{pc}} = \frac{1}{[(T_1 - T_2)/(t_2 - t_1)] - 1} \ln \frac{T_1 - t_2}{T_2 - t_1}$$

$$= \frac{t_2 - t_1}{(T_1 - t_2) - (T_2 - t_1)} \ln \frac{T_1 - t_2}{T_2 - t_1}$$

Rearranging,

$$\dot{m}_c c_{pc} (t_2 - t_1) = U_o A_o \frac{(T_1 - t_2) - (T_2 - t_1)}{\ln[(T_1 - t_2)/(T_2 - t_1)]} \tag{9.11a}$$

The left-hand side is recognized as the total heat gained by the cooler fluid. Therefore,

$$q = U_o A_o \frac{(T_1 - t_2) - (T_2 - t_1)}{\ln[(T_1 - t_2)/(T_2 - t_1)]} \tag{9.11b}$$

Comparing with Equation 9.4, we conclude that the temperature difference or driving potential for heat transfer in a double-pipe heat exchanger is

$$\Delta t = \frac{(T_1 - t_2) - (T_2 - t_1)}{\ln\left[(T_1 - T_2)/(T_2 - t_1)\right]}$$

$$= \text{LMTD} \quad \text{(counterflow)} \tag{9.12a}$$

Equation 9.12 is often called the *log-mean temperature difference*, abbreviated LMTD or ΔT_{lm}. It is sometimes written as

$$\text{LMTD} = \frac{\Delta t_2 - \Delta t_1}{\ln\left(\Delta t_2/\Delta t_1\right)} \tag{9.12b}$$

where Δt_2 is the temperature difference between the two fluids at one end of the exchanger, and Δt_1 is the difference at the other end ($\Delta t_2 = T_1 - t_2$ and $\Delta t_1 = T_2 - t_1$ for counterflow). Thus, for the counterflow double-pipe exchanger,

$$q = U_o A_o \text{LMTD}$$

We can repeat the preceding derivation for parallel flow. The basic equations are the same as those above. The results are

$$q = U_o A_o \frac{(T_1 - t_1) - (T_2 - t_2)}{\ln\left[(T_1 - t_1)/(T_2 - t_2)\right]} = U_0 A_0 \text{LMTD}$$

$$= U_o A_o \frac{\Delta t_2 - \Delta t_1}{\ln \Delta t_2/\Delta t_1} \tag{9.12c}$$

where $\Delta t_1 = T_1 - t_1$, $\Delta t_2 = T_2 - t_2$ for parallel flow (compare to Equation 9.12a for counterflow). The term Δt_1 is the temperature between the two fluids at one end of the exchanger, and Δt_2 is the difference at the other end, although defined somewhat differently than that for counterflow. In the case of parallel flow the result is still a log-mean temperature difference.

Example 9.1

A hot fluid at 100°C enters a double-pipe heat exchanger and is cooled to 75°C. A cooler fluid at 5°C enters the exchanger and is warmed to 50°C. Determine the LMTD for both counterflow and parallel-flow configurations.

Solution

The driving potential for both configurations is independent of fluid properties.

Given

$$T_1 = 100°C \quad t_1 = 5°C$$

$$T_2 = 75°C \quad t_2 = 50°C$$

For counterflow,

$$\text{LMTD} = \frac{(T_1 - t_2) - (T_2 - t_1)}{\ln\left[(T_1 - t_2)/(T_2 - t_1)\right]} = \frac{(100 - 50) - (75 - 5)}{\ln(50/70)}$$

$$\text{LMTD} = 59.4\,°C \qquad\qquad \text{(counterflow)}$$

For parallel flow,

$$\text{LMTD} = \frac{(T_1 - t_1) - (T_2 - t_2)}{\ln\left[(T_1 - t_1)/(T_2 - t_2)\right]} = \frac{(100 - 5) - (75 - 50)}{\ln(95/25)}$$

$$\text{LMTD} = 52.4\,°C \qquad\qquad \text{(parallel flow)}$$

For any exchanger, area is constant, and, because the heat-transfer coefficient is the same for counterflow and parallel-flow configurations, then more heat is transferred by using counterflow. ❏

Example 9.2

A hot fluid at 250°F enters a double-pipe exchanger and is cooled to 150°F. A cooler fluid enters the exchanger at 100°F and is to be warmed to 150°F. Determine the LMTD for both counterflow and parallel-flow configurations.

Solution

Both fluids are to have the same outlet temperature.

Given

$$T_1 = 250\,°F \qquad t_1 = 100\,°F$$
$$T_2 = 150\,°F \qquad t_2 = 150\,°F$$

For counterflow,

$$\text{LMTD} = \frac{(250 - 150) - (150 - 100)}{\ln\,(100/50)} = 72.1\,°F \qquad\qquad \text{(counterflow)}$$

For parallel flow,

$$\text{LMTD} = \frac{(250 - 100) - (150 - 150)}{\ln\left(\dfrac{150}{0}\right)} = 0\,°F \qquad\qquad \text{(parallel flow)}$$

For a finite heat-transfer rate and a finite overall heat-transfer coefficient, if parallel flow is to give equal outlet temperatures, then the area needed must be infinite. This is not feasible economically. ❏

Example 9.2 illustrates (as do the temperature profiles of Figure 9.2) that a counterflow arrangement can be used to raise the outlet temperature of the cooler fluid above the outlet temperature of the warmer fluid. In parallel flow, however, the upper limit of the outlet temperature of the cooler fluid is that of the warmer fluid. Generally, there is a disadvantage to the use of parallel flow over counterflow.

The above equations for double-pipe exchangers can be used to predict outlet temperatures, which are often unknown. In a number of instances the inlet temperature of a warm stream and of a cold stream are known. Outlet temperatures are not. Equation 9.11a relates the heat gained by the cooler fluid to the LMTD for counterflow:

$$\dot{m}_c c_{pc}(t_2 - t_1) = U_o A_o \frac{(T_1 - t_2) - (T_2 - t_1)}{\ln[(T_1 - t_2)/(T_2 - t_1)]} \tag{9.11a}$$

Rearranging,

$$\ln \frac{(T_1 - t_2)}{T_2 - t_1} = \frac{U_o A_o}{\dot{m}_c c_{pc}} \left(\frac{T_1 - T_2}{t_2 - t_1} - 1 \right) \tag{9.13}$$

From the heat balance of Equation 9.8a,

$$\frac{\dot{m}_c c_{pc}}{\dot{m}_h c_{ph}} = \frac{T_1 - T_2}{t_2 - t_1} = R \tag{9.14a}$$

where R is introduced as the ratio of the product of mass flow times specific heat (cooler fluid to warmer fluid). Substituting into Equation 9.13, we get

$$\ln \frac{(T_1 - t_2)}{T_2 - t_1} = \frac{U_o A_o}{\dot{m}_c c_{pc}}(R - 1)$$

Exponentiating both sides yields

$$\frac{T_1 - t_2}{T_2 - t_1} = \exp[U_o A_o(R - 1)/\dot{m}_c c_{pc}] \tag{9.15}$$

Solving Equation 9.14a for t_2 in terms of R gives

$$t_2 = t_1 + \frac{T_1 - T_2}{R} \qquad \text{(counterflow or parallel flow)} \tag{9.14b}$$

Substituting into Equation 9.15 and solving for T_2, we get

$$T_2 = \frac{T_1(R - 1) - Rt_1[1 - \exp(U_o A_o(R - 1)/\dot{m}_c c_{pc})]}{R \exp[U_o A_o(R - 1)/\dot{m}_c c_{pc}] - 1} \qquad \text{(counterflow)} \tag{9.16}$$

The ratio R can be found by knowing the mass flows and specific heats of the fluids (Equation 9.14a). The coefficient U_o is calculated by using the correlations of Chapter 6. Then Equation 9.16 gives the outlet temperature of the warmer fluid, knowing the inlet temperatures. With T_2 determined, we use Equation 9.14b to find the outlet temperature of the cooler fluid.

For parallel flow a development similar to the one above can be generated. The result is

$$T_2 = \frac{\{R + \exp[U_o A_o(R + 1)/\dot{m}_c pc]\}T_1 + Rt_1\{\exp[U_o A_o(R + 1)/\dot{m}_c c_{pc}] - 1\}}{(R + 1)\exp[U_o A_o(R + 1)/\dot{m}_c c_{pc}]}$$

$$\text{(parallel flow)} \tag{9.17}$$

Again, for the cooler fluid Equation 9.14b is used when T_2 is known. We now proceed in the analysis of double-pipe heat exchangers by presenting equations for predicting convection coefficients and ultimately the overall heat-transfer coefficient.

9.2.1 CIRCULAR DUCT

It is instructive to rewrite the expressions from Chapter 6 that apply to flow in conduits. For a circular duct, we will use a modified form of the Seider-Tate Equation for laminar flow and the Dittus-Boelter Equation for turbulent flow:

$$\mathrm{Nu}_D = \frac{h_L D}{k_f} = 1.86\left(\frac{D\,\mathrm{Re}_D\mathrm{Pr}}{L}\right)^{1/3} {}^{*}$$

where

$\mathrm{Re}_D = VD/v < 2\,200$

$0.48 < \mathrm{Pr} = v/\alpha < 16\,700$

μ changes moderately with temperature

Properties at the average fluid temperature [=(inlet + outlet)/2]

$$\mathrm{Nu}_D = \frac{h_L D}{k_f} = 0.023(\mathrm{Re}_D)^{4/5}\,\mathrm{Pr}^n \tag{9.18b}$$

where

$n = 0.4$ (fluid being heated)

$n = 0.3$ (fluid being cooled)

$\mathrm{Re}_D = VD/v \geq 10\,000$

$0.7 \leq \mathrm{Pr} = v/\alpha \leq 160$

$L/D \geq 60$

Properties at the average fluid temperature [=(inlet + outlet)/2]

The above equations can be used, respectively, for laminar and turbulent flow (but not transition flow) to find the convection coefficient that applies to the inside surface area of a duct.

Another important consideration is the pressure drop a fluid experiences while traveling through a double-pipe heat exchanger. The pressure drop is a function of the friction factor. From Chapter 6, pressure drop is calculated with

$$p_1 - p_2 = \frac{fL}{D}\frac{\rho V^2}{2g_c} \tag{9.19}$$

where the friction factor f is found with the Moody diagram (Figure 6.14) or appropriate curve fit equations (6.45 through 6.49).

9.2.2 ANNULAR DUCT

For an annulus, we must modify the formulations somewhat. For frictional effects, we introduced the concept of hydraulic diameter:

$$D_h = \frac{4A}{P}$$

Referring to Figure 9.3 for an annulus, we write

---

* The overbar notation on the convection coefficient is dropped.

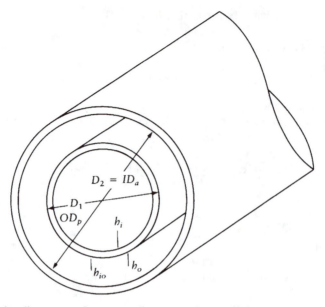

FIGURE 9.3   Annulus diameters and corresponding convection coefficients.

$$D_h = \frac{4\pi(D_2^2 - D_1^2)}{4\pi(D_2 + D_1)} = D_2 - D_1 \quad \text{(friction)} \tag{9.20}$$

Note that D_2 is the ID of the outer tube and D_1 is the OD of the inner tube (different from the notation of Chapter 6). The Reynolds number for use in finding the friction factor in an annulus is calculated by using the hydraulic diameter of Equation 9.20.

Chapter 6 does not contain an equation for calculating convection coefficient for turbulent flow in an annular duct. Such correlations are complex and difficult to use. For heat-transfer purposes, however, we can *estimate* the convection coefficient, using the correlations for a circular duct, if we first find an equivalent diameter (analogous to the friction problem). Defining the *equivalent diameter for heat transfer* as

$$D_e = \frac{4A}{P}$$

we get for heat transfer

$$D_e = \frac{\pi(D_2^2 - D_1^2)}{\pi D_1}$$

$$D_e = \frac{D_2^2 - D_1^2}{D_1} \quad \text{(heat transfer)} \tag{9.21}$$

The flow area is the same as for the hydraulic diameter, but the heat-transfer surface (perimeter P) is that only of the inner tube. The Reynolds number for use in finding the convection coefficient in an annulus is calculated by using the equivalent diameter of Equation 9.21. We thus have two Reynolds numbers (one for frictional effects and the other for heat transfer) associated with flow in an annulus, and they are not equal.

For flow in an annulus, then, the equations we will use are again the (modified) Seider-Tate Equation and the (modified) Dittus-Boelter Equation with equivalent diameter D_e substituted for diameter:

$$\text{Nu}_D = \frac{h_L D_e}{k_f} = 1.86\left(\frac{D_e \text{Re}_D \text{Pr}}{L}\right)^{1/3} \qquad (9.18a)$$

$\text{Re}_D = VD_e/v < 2\ 200$
$0.48 < \text{Pr} = v/\alpha < 16\ 700$
μ changes moderately with temperature
Properties at the average fluid temperature [=(inlet + outlet)/2]

and

$$\text{Nu}_D = \frac{h_L D_e}{k_f} = 0.023(\text{Re}_D)^{4/5}\text{Pr}^n \qquad (9.18b)$$

$n = 0.4$ (fluid being heated)
$n = 0.3$ (fluid being cooled)
$\text{Re}_D = VD_e/v \geq 10\ 000$
$0.7 \leq \text{Pr} = v/\alpha \leq 160$
$L/D \geq 60$
Properties at the average fluid temperature [=(inlet + outlet)/2]

It must be stressed that the above equations provide only an *estimate* of the convection coefficient.

Frictional pressure losses in the annulus must include an entrance and an exit loss as well as wall friction. The pressure drop can be found with

$$p_1 - p_2 \cong \frac{\rho V^2}{2g_c}\left(\frac{fL}{D_h} + 1\right) \qquad (9.22)$$

where f is obtained from the Moody diagram (Figure 6.14) or the curve fit equations. The "1" in the parentheses is included to account for entrance and exit effects, which are treated essentially as minor losses. Note that the diameter appearing in Equation 9.22 is the hydraulic diameter.

The above equations are organized together in the Summary section of this chapter. The equations are placed in a suggested order for performing calculations on a double-pipe exchanger. The following example illustrates the method.

Example 9.3

Steam passes through a turbine into a condenser. Liquid water from the condensed steam is used to heat ethylene glycol. The water is available at 195°F with a mass flow rate of 5000 lbm/hr. The ethylene glycol has a temperature of 85°F and a mass flow rate of 12,000 lbm/hr. It is proposed to use a double-pipe heat exchanger made of 2 × 1-1/4 standard type M copper tubing, with soldered fittings, that is 20 ft long. Determine the outlet temperature of the ethylene glycol for counterflow.

Solution

The water loses heat only to the ethylene glycol, and as both fluids change temperature, their properties change. With outlet temperatures unknown, we can evaluate properties either at the inlet temperatures or at the average of both inlet temperatures. It is prudent to place the fluid with the higher flow rate into the passage having the greater area so that pressure losses are minimized.

Assumptions

1. Steady-state conditions exist.
2. Fluid properties are constant and evaluated at the average of the inlet temperatures $(195 + 85)/2 = 140°F$.

We now follow the suggested outline of calculations appearing in the Summary section.

From Appendix Table C.11, for water at 140°F,

$\rho = 0.985(62.4)$ lbm/ft³ $k_f = 0.376$ BTU/(hr·ft·°R)
$c_p = 0.9994$ BTU/lbm·°R $\alpha = 6.02 \times 10^{-3}$ ft²/hr
$v = 0.514 \times 10^{-5}$ ft²/s $Pr = 3.02$

Also, $\dot{m}_h = 5000$ lbm/hr and $T_1 = 195°F$.

For ethylene glycol at 140°F, Appendix Table C.5 shows

$\rho = 1.087(62.4)$ lbm/ft³ $k_f = 0.150$ BTU/(hr·ft·°R)
$c_p = 0.612$ BTU/lbm·°R $\alpha = 3.61 \times 10^{-3}$ ft²/hr
$v = 5.11 \times 10^{-5}$ ft²/s $Pr = 51$

Also, $\dot{m}_h = 12000$ lbm/hr and $t_1 = 85°F$.

From Appendix Table F.2, for seamless copper water tubing (subscripts: a = annulus, p = inner pipe or tube), we get

2 standard type K $ID_a = 0.1674$ ft
1-1/4 standard type K $ID_p = 0.1076$ ft
                                   $OD_p = 0.375/12 = 0.1146$ ft

I. Flow Areas

$$A_p = \pi(0.1076^2)/4 = 0.00909 \text{ ft}^2$$

$$A_a = \pi(0.1674^2 - 0.1146^2)/4 = 0.0117 \text{ ft}^2$$

Because $A_a > A_p$, route the ethylene glycol (which has the higher flow rate) through the annulus.

II. Annulus Equivalent Diameters

$$D_h = 0.1674 - 0.1146 = 0.0528 \text{ ft} \qquad \text{(friction)}$$

$$D_e = (0.1674^2 - 0.1146^2)/0.1146 = 0.1299 \text{ ft} \qquad \text{(heat transfer)}$$

III. Reynolds Numbers

$$\text{Water } Re_p = \frac{5000}{3600}\frac{1}{0.985(62.4)(0.00909)}\frac{0.1076}{0.514 \times 10^{-5}} = 5.2\times10^4$$

$$\text{E.G. } Re_a = \frac{12,000}{3600}\frac{1}{1.087(62.4)(0.0117)}\frac{0.1299}{5.11 \times 10^{-5}} = 1.07\times10^4$$

IV. Nusselt Numbers

$$\text{Water } Nu_p = 0.023(5.2 \times 10^4)^{4/5}(3.02)^{0.3} = 190$$

$$\text{E.G. } Nu_a = 0.023(1.07 \times 10^4)^{4/5}(51)^{0.4} = 185$$

V. Convection Coefficients

$$\text{Water } h_i = \frac{190(0.376)}{0.1076} = 664$$

$$h_{io} = \frac{664(0.1076)}{0.1146} = 623 \text{ BTU/(hr·ft}^2\text{·°R)}$$

$$\text{E.G. } h_o = \frac{185(0.150)}{0.1299} = 214 \text{ BTU/(hr·ft}^2\text{·°R)}$$

VI. Exchanger Coefficient

$$\frac{1}{U_o} = \frac{1}{623} + \frac{1}{214} \qquad U_o = 159 \text{ BTU/(hr·ft}^2\text{·°R)}$$

Outlet temperatures (subscripts: c = cooler fluid or fluid being heated, h = warmer fluid or fluid being cooled)

$$R = \frac{\dot{m}_c c_{pc}}{\dot{m}_h c_{ph}} = \frac{12000(0.612)}{5000(0.9994)} = 1.47$$

$$A_o = \pi(0.1146)(20) = 7.2 \text{ ft}^2$$

$$T_1 = 195°F \qquad t_1 = 85°F$$

$$T_2 = \frac{195(0.47) - 1.47(85) \times \{1 - \exp[159(7.2)(0.47)/(12,000)(0.612)]\}}{1.47 \exp[159(7.2)(0.47)/(12,000)(0.612)] - 1}$$

$$T_2 = 174°F \quad \text{(water)}$$

$$t_2 = 85 + \frac{195 - 174}{1.47} = 99.3°F \quad \text{(E.G.)}$$

Heat balance (as a check on the calculations)

$$\text{Water} \quad q_h = 5000(0.9994)(195 - 174)$$
$$= 1.05 \times 10^5 \text{ BTU/hr}$$

$$\text{E.G.} \quad q_h = \dot{m}_c c_{pc} \Delta T = 12,000(0.612)(99.3 - 85)$$
$$= 1.05 \times 10^5 \text{ BTU/hr}$$

VII. Summary of Requested Information
Outlet temperature of ethylene glycol for counterflow = 99.3°F

If the outlet temperature of the ethylene glycol is not high enough, it is possible to use bigger tube diameters to increase the surface area; or to increase the mass flow rates, thereby increasing the Reynolds numbers and correspondingly the Nusselt numbers; or to use several of this size exchanger in series. The decision will depend in part on the pressure drop encountered by each fluid stream. Note that we evaluated water properties at 140°F. the calculations should be repeated at this point, with water properties evaluated at (195 + 174)/2 = 184.5°F. Ethylene glycol properties should be evaluated at (85 + 99.3)/2 = 92.2°F. ❑

9.2.3 Fouling Factors

In the preceding example, the outlet temperatures calculated for the two fluid streams will exist when the exchanger is new and the tube surfaces are clean. After a time, substances (such as minerals) dissolved within the fluid streams become deposited on the tube walls. The result is that two more resistances must be added into the calculation of the overall heat-transfer coefficient.

Figure 9.4 shows a cross section of two concentric tubes and the additional resistances. Also shown are the film coefficients: h_i for the inside surface of the inner tube, h_{io} [$= h_i(ID_p/OD_p)$], and h_o. For a clean-wall exchanger the overall heat-transfer coefficient is, from Equation 9.3,

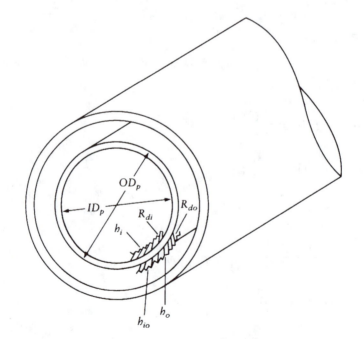

FIGURE 9.4   Sketch used to illustrate the effect of dirt-and-scale build-up on the surface of a tube.

$$\frac{1}{U_o} = \frac{OD_p}{h_i ID_p} + \frac{1}{h_o} = \frac{1}{h_{io}} + \frac{1}{h_o} \tag{9.3}$$

or

$$U_o = \frac{h_{io} h_o}{h_{io} + h_o} \tag{9.23}$$

The coefficient that should be used for design purposes is the one that includes the dirt or fouling factors R_{do} for the inside surface and R_{do} for the outside surface

$$\frac{1}{U} = \frac{1}{U_o} + R_{di} + R_{do} = \frac{1}{U_o} + R_d \tag{9.24}$$

where U is the design coefficient. Note that R_{di} should be referred to ID_p rather than OD_p, but by custom it is not, because the values available for the dirt factors are the best estimates. The dirt factors R_{di} and R_{do} are usually measured for various combinations of fluids and metals. Typical design values for the fouling factors are provided in Table 9.1. The numbers represent a yearly effect. Thus when combined with $1/U_o$, as in Equation 9.24, the resulting U is the coefficient that exists in the exchanger after one year has passed. Clearly, with all the variables involved, the factors provided in Table 9.1 are only estimates (based on measurements). An exchanger designed with fouling factors included will be protected from delivering less than the required heat load for at least a year.

TABLE 9.1
Approximate Fouling Factors for Heat Exchanger Tubes

	R_d	
Fluid	ft²·hr·°R/BTU	(m²·K)/W
Seawater and distilled water	0.0005–0.001	0.000 1–0.000 2
Engine oil	0.001	0.000 2
Alcohol vapors	0.0005	0.000 1
Steam (oil free)	0.0005	0.000 1
Refrigerant vapors	0.002	0.000 4
Refrigerant liquids	0.001	0.000 2
Air	0.002	0.000 4
Diesel engine exhaust	0.01	0.002
Organic vapors	0.0005	0.000 1

Source: from Standards of Tubular Exchanger Manufacturers Association, 1978.

One difficulty that might be encountered is in the use of soldered fittings to construct an exchanger. It is not practical to "unsolder" an exchanger for cleaning of the tube surfaces. (The annulus fluid will foul the inner surface of the outer tube and the outer surface of the inner tube.) Consequently, it is desirable to use threaded compression fittings. The pressure losses may be considered to be the same as for soldered fittings, however.

Example 9.4

A bank of four diesel engines is used with alternators for the generation of electricity. The exhaust from the engines is discharged to the atmosphere. It is proposed to use all or part of the exhaust for heating air, which can then be used for space heating to reduce costs. From measurements of velocity, it is determined that the mass flow rate of exhaust available from the engines is 90 kg/hr. The exhaust-gas temperature is 600 K. The air is available at 20°C and is not of much use unless it can be heated to at least 80°C at a mass flow rate of 100 kg/hr. There are a number of 4 × 3 double-pipe exchangers that are 2 m long and made of type K copper tubing with compression fittings. Determine (a) how many of them are required, (b) the overall coefficient of (all) the exchanger(s), and (c) the pressure drop for each stream. Assume a counterflow arrangement.

Solution

The air is heated by the diesel exhaust with air inlet and outlet temperatures known. The diesel exhaust properties are not known.

Assumptions

1. Air properties are constant and evaluated at the average of the inlet and outlet temperatures.
2. Diesel exhaust has the same properties as carbon dioxide gas.
3. Diesel exhaust properties are constant. For lack of better information (i.e., outlet temperature of the exhaust gas), evaluate the exhaust properties at the average of the inlet temperatures: $[600 + (20 + 273)]/2 = 480 \cong 500$ K.
4. The system is at steady state.

From Appendix Table D.1, for air at 323 K,

$\rho = 1.088$ kg/m^3 $\qquad\qquad k_f = 0.028\ 14$ W/(m·K)
$c_p = 1\ 007$ J/(kg·K) $\qquad\quad \alpha = 0.26 \times 10^{-4}$ m^2/s
$v = 18.2 \times 10^{-6}$ m^2/s $\qquad$ Pr $= 0.703$

Also, $\dot{m}_c = 100$ kg/hr, $t_1 = 20°C = 293$ K, and $t_2 = 80°C = 353$ K.

From Appendix Table D.2, for carbon dioxide gas at 500 K,

$\rho = 1.073\ 2$ kg/m^3 $\qquad\qquad k_f = 0.033\ 52$ W/(m·K)
$c_p = 1\ 013$ J/(kg·K) $\qquad\quad \alpha = 0.308\ 4 \times 10^{-4}$ m^2/s
$v = 21.67 \times 10^{-6}$ m^2/s $\qquad$ Pr $= 0.702$

Also, $\dot{m}_h = 90$ hr and $T_1 = 600$ K.

From Appendix Table F.2, for seamless copper tubing,

4 standard type K $\qquad\qquad ID_a = 9.8$ cm
3 standard type K $\qquad\qquad ID_p = 7.384$ cm
$\qquad\qquad\qquad\qquad\qquad OD_p = 7.938$

Composing the calculations in a systematic order, we write

I. Flow Areas

$$A_p = \pi(0.073\ 84)^2/4 = 4.28 \times 10^{-3} \text{ m}^2$$

$$A_a = \pi(0.098^2 - 0.079\ 38)^2/4 = 2.59 \times 10^{-3} \text{ m}^2$$

Because $A_p > A_a$, direct the air through the pipe (inner tube).

II. Heat Balance

$$\text{Air} \quad q = \frac{100}{3\,600}(1\,007)(80 - 20) = 1.68 \times 10^3 \text{ W}$$

$$\text{CO}_2 \quad T_2 = T_1 - q/\dot{m}_h c_{ph} = 600 - \frac{1.68 \times 10^3(3\,600)}{90(1\,013)} = 534 \text{ K}$$

III. Log-Mean Temperature Difference

$$\text{LMTD} = \frac{(600 - 353) - (534 - 293)}{\ln\,(247/241)} = 244 \text{ K}$$

IV. Annulus Equivalent Diameters

$$D_h = 0.098 - 0.079\,38 = 0.018\,62 \text{ m}$$

$$D_e = (0.098^2 - 0.079\,38^2)/0.079\,38 = 0.0416 \text{ m}$$

V. Reynolds Numbers

$$\text{Air} \quad \text{Re}_p = \frac{100}{3\,600}\frac{1}{1.088(0.004\,28)}\frac{0.073\,84}{18.2 \times 10^{-6}} = 2.42 \times 10^4$$

$$\text{CO}_2 \quad \text{Re}_a = \frac{90}{3\,600}\frac{1}{1.073\,2(0.002\,59)}\frac{0.041\,6}{21.67 \times 10^{-6}} = 1.73 \times 10^4$$

VI. Nusselt Numbers

$$\text{Air} \quad \text{Nu}_p = 0.023(2.42 \times 10^4)^{4/5}(0.703)^{0.4} = 64.2$$

$$\text{CO}_2 \quad \text{Nu}_a = 0.023(1.73 \times 10^4)^{4/5}(0.702)^{0.3} = 50.8$$

VII. Film Coefficients

$$\text{Air} \quad h_i = \frac{64.2(0.028\,14)}{0.073\,84} = 24.5$$

$$h_{io} = 24.5\frac{0.073\,84}{0.079\,38} = 22.8 \text{ W/(m}^2\text{·K)}$$

$$\text{CO}_2 \quad h_o = \frac{50.8(0.033\,52)}{0.041\,6} = 40.9 \text{ W/(m}^2\text{·K)}$$

VIII. Fouling Factors

$$\text{Air} \quad R_{di} = 0.000\ 4\ (\text{m}^2 \cdot \text{K})/\text{W}$$

$$CO_2 \quad R_{do} = (0.002\ )(\text{m}^2 \cdot \text{K})/\text{W}$$

IX. Exchanger Coefficients

$$\frac{1}{U_o} = \frac{1}{22.8} + \frac{1}{40.9} \qquad\qquad U_o = 14.6\ \text{W}/(\text{m}^2 \cdot \text{K})$$

$$\frac{1}{U_o} = \frac{1}{14.6} + 0.000\ 4 + 0.002 \quad U_o = 14.1\ \text{W}/(\text{m}^2 \cdot \text{K})$$

X. Area Required

$$q = UA_o\text{LMTD}; \ \ A_o = \frac{1.68 \times 10^3}{(14.1)(244)} = 0.49\ \text{m}^2$$

One exchanger has a surface area of $A_o = \pi D_o L$. So,

$$L = \frac{0.49}{\pi(0.079\ 38)} = 1.96\ \text{m}$$

Each available exchanger is 2 m in length. Therefore, only one exchanger is required.

XI. Friction Factors

$$\text{Air} \quad \left.\begin{array}{l} \text{Re}_p = 2.42 \times 10^4 \\ \text{smooth-wall tubing} \end{array}\right\} f_p = 0.0245 \qquad\qquad \text{(Figure 6.14)}$$

$$CO_2 \quad \text{Re}_a = \frac{90}{3\ 600}\frac{1}{1.073\ 2(0.002\ 59)}\frac{0.018\ 62}{21.67 \times 10^{-6}}$$

$$\left.\begin{array}{l} = 7.73 \times 10^3 \\ \text{smooth-wall tubing} \end{array}\right\} f_a = 0.033$$

$$\qquad\qquad\qquad\qquad\qquad\qquad\qquad\qquad \text{(Figure 6.14)}$$

XII. Velocities $(= \dot{m}/\rho A)$

$$\text{Air} \quad V_p = \frac{100/3\ 600}{1.088(0.004\ 28)} = 5.97\ \text{m/s}$$

$$CO_2 \quad V_a = \frac{90/3\ 600}{1.073(0.002\ 59)} = 9.0\ \text{m/s}$$

XIII. Pressure Drops $(L = 2$ m actual length)

$$\text{Air } \Delta p_p = \frac{0.024\ 5(2)\,1.088(5.97)^2}{0.073\ 84\qquad 2} = 12.87\text{ Pa}$$

$$CO^2 \ \Delta p_a = \frac{1.073(9)^2}{2}\left[\frac{0.033(2)}{0.018\ 62} + 1\right] = 200\text{ Pa}$$

XIV. Summary of Requested Information

(a) One exchanger required
(b) $U_o = 14.6$, $U = 14.1$ W/(m²·K)
(c) Air pressure drop = 12.87 Pa
      Diesel exhaust pressure drop = 200 Pa

Because the length of each exchanger is longer than that required by the conditions of the problem, more heat than specified will be exchanged. This means that the outlet temperature of the air will be greater than 80°C, and the outlet temperature of the diesel exhaust will be lower than 583 K. The actual outlet temperatures can be calculated with Equations 9.16 and 9.14.   ❏

Double-pipe heat exchangers can be set up in a multitude of ways. The fluids can travel through a bank of exchangers all in series. Alternatively, the annulus fluid can travel through a bank arranged in parallel, while the tube fluid can travel through the same bank in series; or vice versa. Such modifications are used for increased heat recovery or to reduce the pressure drop of one of the streams.*

9.3   SHELL-AND-TUBE HEAT EXCHANGERS

Double-pipe heat exchangers are useful as heat-transfer devices for relatively low flow rates and moderate temperature differences. Where a high flow rate is involved, the number of double-pipe exchangers required becomes prohibitive—both in ground area required and in funds expended. When high heat-transfer rates are required, an alternative apparatus, known as a *shell-and-tube* heat exchanger, can be used.

A shell-and-tube exchanger consists of a large-diameter pipe (on the order of 12 nominal to 24 nominal and larger), inside which is placed a number of tubes (ranging from about 20 to over 1 000 tubes). One fluid is directed through the tubes, and another inside the shell but outside the tubes. An example of a shell-and-tube exchanger is shown in Figure 9.5.

The essential components are

1. The shell, containing inlet and outlet ports, called nozzles
2. Tubes
3. Tube sheets into which the tubes are attached in a fluid-tight manner
4. Channel covers
5. Transverse baffles for directing the flow of the shell fluid
6. Baffle spacers (not shown in Figure 9.5) onto which the baffles are bolted to keep them positioned properly

---

* Further details are available in *Process Heat Transfer,* by D.Q. Kern, McGraw-Hill, 1950, Chapter 6.

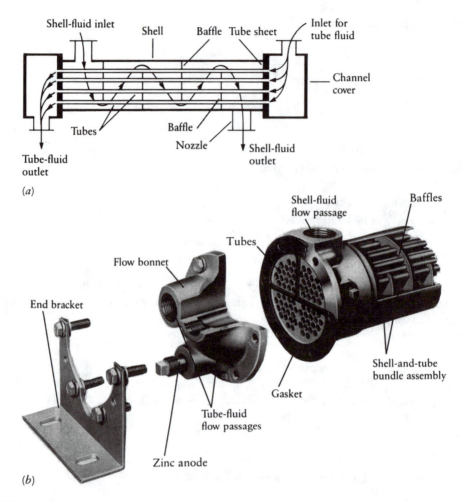

FIGURE 9.5 (a) Simple shell-and-tube heat exchanger and (b) components of a shell-and-tube exchanger. *Source:* courtesy of Young Radiator Co., a Motive Power Industries Company.

9.3.1 Shells

Usually, shells are made from standard steel or wrought-iron pipe (with dimensions given in Appendix Table F.1), but special metals are used for certain fluids when corrosion is a problem. Wall thicknesses required are a function of the operating pressure.

9.3.2 Tubes

Heat-exchanger tubes, also referred to as *condenser tubes,* are available in a variety of metals. Tubes are specified somewhat differently from copper water tubes and the like. A 1-in condenser tube has an outside diameter of 1 inch. Wall thicknesses are standardized in terms of the *Birmingham Wire Gage,* or BWG, of the tube. Table 9.2 gives data on condenser tubes.

How close adjacent tubes can be is a function of how close holes can be located in the tube sheets. When too close, the tube sheet becomes structurally weak. However, standards have been established regarding tube locations. Tube orientation with respect to the external (or shell) fluid is also standardized and is shown in Figure 9.6. Based on the above factors, the number of tubes that can be placed within a shell is therefore limited. Table 9.3 provides what are called *tube counts,*

TABLE 9.2
Physical Dimensions of Condenser Tubes in Terms of BWG

Tube OD in			ID in	
in	(cm)	BWG	in	(cm)
1/2	(1.27)	12	0.282	(0.716)
		14	0.334	(0.848)
		16	0.374	(0.940)
		18	0.402	(1.02)
		20	0.435	(1.09)
3/4	(1.91)	10	0.482	(1.22)
		11	0.510	(1.29)
		12	0.532	(1.35)
		13	0.560	(1.42)
		14	0.584	(1.48)
		15	0.606	(1.54)
		16	0.620	(1.57)
		17	0.634	(1.61)
		18	0.652	(1.66)
1	(2.54)	8	0.670	(1.70)
		9	0.704	(1.79)
		10	0.732	(1.86)
		11	0.760	(1.93)
		12	0.782	(1.99)
		13	0.810	(2.06)
		14	0.834	(2.12)
		15	0.856	(2.17)
		16	0.870	(2.21)
		17	0.884	(2.25)
		18	0.902	(2.29)

Source: reprinted from *Process Heat Transfer,* by D.Q. Kern, McGraw-Hill, 1950, p. 843, with permission.

which are actually the maximum number of tubes of a given orientation that can be used inside a shell. For example, consider that we have 3/4-in-OD tubes laid out on a 1-in square pitch with a shell inside diameter of 25 in. If the tube arrangement is that for two passes (denoted as 2-P), then Table 9.3 shows that the maximum number of tubes is 394. If the tube arrangement is that for six passes (6-P), then the maximum number of tubes is 356.

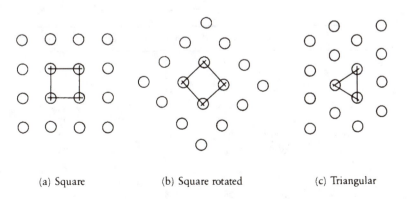

(a) Square   (b) Square rotated   (c) Triangular

FIGURE 9.6 Possible condenser-tube orientations.

TABLE 9.3
Tube Counts for Shell-and-Tube Equipment

Shell ID, in	1-P	2-P	4-P	6-P	8-P
	3/4-in-OD tubes on 1-in square pitch				
8	32	26	20	20	
10	52	52	40	36	
12	81	76	68	68	60
13-1/4	97	90	82	76	70
15-1/4	137	124	116	108	108
17-1/4	177	166	158	150	142
19-1/4	224	220	204	192	188
21-1/4	277	270	246	240	234
23-1/4	341	324	308	302	292
25	413	394	370	356	346
27	481	460	432	420	408
29	553	526	480	468	456
31	657	640	600	580	560
33	749	718	688	676	648
35	845	824	780	766	748
37	934	914	886	866	838
39	1049	1024	982	968	948
	1-in-OD tubes on 1-1/4-in square pitch				
8	21	16	14		
10	32	32	26	24	
12	48	45	40	38	36
13-1/4	61	56	52	48	44
15-1/4	81	76	68	68	64
17-1/4	112	112	96	90	82
19-1/4	138	132	128	122	116
21-1/4	177	166	158	152	148
23-1/4	213	208	192	184	184
25	260	252	238	226	222
27	300	288	278	268	260
29	341	326	300	294	286
31	406	398	380	368	358
33	465	460	432	420	414
35	522	518	488	484	472
37	596	574	562	544	532
39	665	644	624	612	600
	3/4-in-OD tubes on 15/16-in triangular pitch				
8	36	32	26	24	18
10	62	56	47	42	36
12	109	98	86	82	78
13-1/4	127	114	96	90	86
15-1/4	170	160	140	136	128
17-1/4	239	224	194	188	178
19-1/4	301	282	252	244	234
21-1/4	361	342	314	306	290
23-1/4	442	420	386	378	364
25	532	506	468	446	434

TABLE 9.3
Tube Counts for Shell-and-Tube Equipment (continued)

Shell ID, in	1-P	2-P	4-P	6-P	8-P
\multicolumn{6}{c}{3/4-in-OD tubes on 15/16-in triangular pitch (continued)}
27	637	602	550	536	524
29	721	692	640	620	594
31	847	822	766	722	720
33	974	938	878	852	826
35	1102	1068	1004	988	958
37	1240	1200	1144	1104	1072
39	1377	1330	1258	1248	1212
\multicolumn{6}{c}{3/4-in-OD tubes on 1-in triangular pitch}
8	37	30	24	24	
10	61	52	40	36	
12	92	82	76	74	70
13-1/4	109	106	86	82	74
15-1/4	151	138	122	118	110
17-1/4	203	196	178	172	166
19-1/4	262	250	226	216	210
21-1/4	316	302	278	272	260
23-1/4	384	376	352	342	328
25	470	452	422	394	382
27	559	534	488	474	464
29	630	604	556	538	508
31	745	728	678	666	640
33	856	830	774	760	732
35	970	938	882	864	848
37	1074	1044	1012	986	870
39	1206	1176	1128	1100	1078
\multicolumn{6}{c}{1-in-OD tubes on 1-1/4-in triangular pitch}
8	21	16	16	14	
10	32	32	26	24	
12	55	52	48	46	44
13-1/4	68	66	58	54	50
15-1/4	91	86	80	74	72
17-1/4	131	118	106	104	94
19-1/4	163	152	140	136	128
21-1/4	199	188	170	164	160
23-1/4	241	232	212	212	202
25	294	282	256	252	242
27	349	334	302	296	286
29	397	376	338	334	316
31	472	454	430	424	400
33	538	522	486	470	454
35	608	592	562	546	532
37	674	664	632	614	598
39	766	736	700	688	672

Source: reprinted from *Process Heat Transfer,* by D.Q. Kern, McGraw-Hill, 1950, pp. 841–842, with permission.

9.3.3 BAFFLES

Baffles are used inside the shell to direct the shell fluid past the tubes in such a way that heat transfer is enhanced. Most of the shell fluid flows primarily at nearly right angles to the tube axes. Considerable turbulence is caused, and higher heat-transfer coefficients, and hence higher rates, result. The center-to-center distance between adjacent baffles is known as the *baffle spacing* or *baffle pitch*. Baffle spacing influences the velocity of the shell fluid as it moves past the tubes. Baffle spacing usually is made to vary from 1/5 to 1 times the inside diameter of the shell. Baffles are bolted to baffle spacers that are attached to the tube sheets.

Several designs for baffles are employed in shell-and-tube exchangers. Figure 9.7a illustrates a segmental baffle. These can be arranged to direct the shell fluid in "up-and-down" or "side-to-side" flow. Segmental baffles are known also as 25% cut baffles. Figure 9.7b illustrates a disc-and-doughnut style, while Figure 9.7c shows an orifice baffle. We will work exclusively with 25% cut baffles in this text.

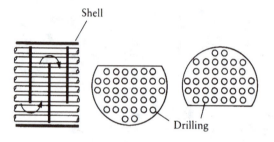

(a) Segmental baffles (25% cut)

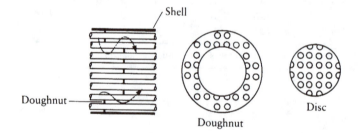

(b) Disc-and-doughnut baffles

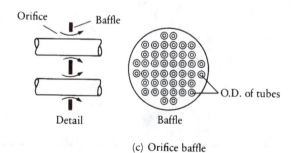

(c) Orifice baffle

FIGURE 9.7 Types of baffles used in shell-and-tube exchangers. *Source:* from *Process Heat Transfer,* by D.Q. Kern, McGraw-Hill, 1950, pp. 130–131, with permission.

9.3.4 Modifications

Figure 9.5 is of a simple shell-and-tube exchanger. As can be discerned from that figure, the tube fluid passes through the exchanger once, and the shell fluid passes through only once. This arrangement seems ideal because a true counterflow system can be set up. This is known as a 1–1 counterflow exchanger. From a practical viewpoint, however, the large number of tubes makes it difficult to obtain a high velocity ($\rightarrow$ high Reynolds number$\rightarrow$ high Nusselt number$\rightarrow$ high film coefficient) through all the tubes in a single pass. This difficulty is overcome by inserting one (or more) partition(s) in the end channel(s) so that the tube fluid passes through the exchanger more than once. Figure 9.8 illustrates a 1–2 exchanger, where the shell fluid passes through once and the tube fluid twice. One of the end channels contains a partition that directs the fluid through half the tubes in one direction and through the other half on the way back. The tube fluid thus enters and exists at one end of the exchanger. Other designs include 1–4, 1–6, and 1–8 exchangers. An odd number of tube passes is seldom used.

Figure 9.8 shows an exchanger with fixed-tube sheets, that is, fixed to the channels and the shell. The number of tube passes is controlled by placing flow dividers or diverters in the end channels. The left-hand and right-hand end channels are manufactured in different ways, and appropriate combinations will yield 1–2, 1–4, and other exchangers. The advantage is that the same tubes and shell can be used for several configurations. One problem with the fixed-tube-sheet design is that, when a heated fluid passes through, the tubes may wish to expand by a different amount from the shell. This problem can be amplified if the shell and tubes are made from different materials. Expansion problems may be critical in shell-and-tube exchangers. Some designs that alleviate this problem are illustrated in Figure 9.9. Note also that the U-bend and the pull-through exchangers can be disassembled for cleaning of the tubes (inside and outside surfaces) and the shell.

We now turn our attention to the task of trying to model a shell- and-tube exchanger. As with double-pipe exchangers, we are seeking to be able to predict outlet temperatures or heat-transfer rates or an overall heat-transfer coefficient, depending on the nature of the problem. We will write equations based on actual data that are useful for predicting film coefficients and pressure drops.

9.3.5 Tube Side

For the tubes in a shell-and-tube exchanger, the equations of Chapter 6 can be applied, as was done for the double-pipe exchanger:

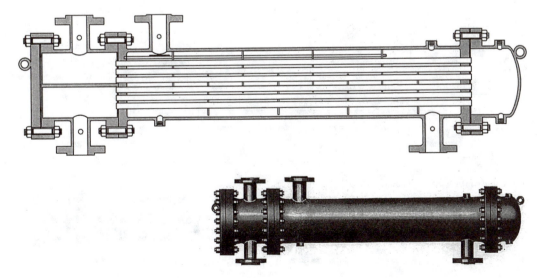

FIGURE 9.8 A 1–2 straight-tube (fixed tube-sheet) heat exchanger. The tube bundle is nonremovable.

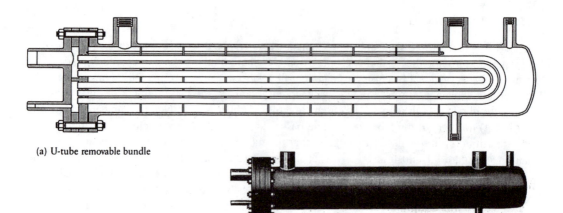

(a) U-tube removable bundle

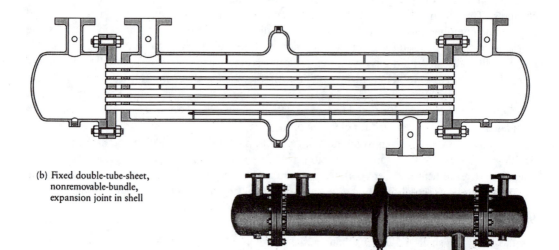

(b) Fixed double-tube-sheet,
    nonremovable-bundle,
    expansion joint in shell

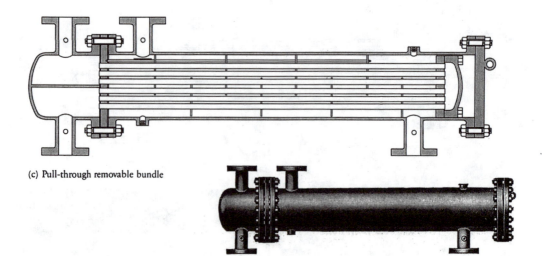

(c) Pull-through removable bundle

FIGURE 9.9   Shell-and-tube heat exchangers.

$$\text{Nu}_D = 0.023 \ \text{Re}_D^{4/5} \ \text{Pr}^n \qquad \text{(circular duct)}$$
$$n = 0.4 \qquad \text{(fluid being heated)}$$
$$n = 0.3 \qquad \text{(fluid being cooled)}$$
$$0.7 \le \text{Pr} = v/\alpha \le 160$$
$$\text{Re}_D = VD/v \ge 10\ 000 \qquad D = ID_t$$
$$\text{Nu}_D = \frac{h_L D}{k_f}$$

Properties at the average fluid temperature [= (inlet + outlet)/2]

The pressure drop experienced by the tube fluid must include the effects of all the tubes. The pressure drop for the tube fluid is given by

$$\Delta p_f = \frac{fLN_p}{ID_t} \frac{\rho V^2}{2g_c}$$

where f is the friction factor obtained from the Moody diagram (smooth curve line for condenser tubes), L is the tube length, N_p is the number of tube passes ID_t is the tube inside diameter, ρ is the fluid density, and V is the average fluid velocity in the tubes. It is necessary to account for the number of passes the tube fluid makes. The tube fluid will experience an additional loss due to sudden expansions and contractions that the tube fluid undergoes during a return. Experiment shows that the return-pressure loss is given by

$$\Delta p_r = 4N_p \frac{\rho V^2}{2g_c}$$

where N_p is the number of tube passes (2 in a 1–2 exchanger), and the quantity $4N_p$ is the return-pressure-loss coefficient (i.e., treated as a minor loss). The total pressure drop experienced by the tube fluid becomes

$$\Delta p_t = \frac{\rho V^2}{2g_c}\left(\frac{fLN_p}{ID_t} + 4N_p\right) \qquad (9.25)$$

9.3.6   SHELL SIDE

The shell-fluid velocity changes continuously through the shell because the flow area is not constant. The shell width and number of tubes varies from top to bottom (or side to side). The direction of flow in the shell is partly along the tube axes (at the baffle ends) and partly at nearly right angles to the tube axes (between the baffles). However, it is desired to use a single characteristic length to represent the shell geometry to retain the format of the equations written for an annulus. Figure 9.10 shows a square- and a triangular-pitch layout. For either pitch layout, we define an equivalent length as

$$D_e = \frac{4 \times \text{Area}}{\text{Wetted perimeter}}$$

For the square pitch the area is a "square" minus the areas of 4 quarter circles (the hatched section in Figure 9.10). The wetted perimeter is the circumference of 4 quarter circles. Thus,

$$D_e = \frac{4[S_T^2 - \pi(OD_t)^2/4]}{\pi(OD_t)} \qquad \text{(square pitch)} \qquad (9.26)$$

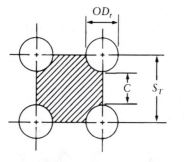

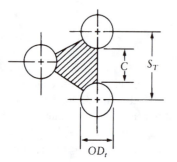

FIGURE 9.10 Square- and triangular-pitch tube layouts.

For the triangular pitch, we write

$$D_e = \frac{4[(S_T/2)(0.86S_T) - \pi(OD_t)^2/8]}{\pi(OD_t)/2} \quad \text{(triangular pitch)} \quad (9.27)$$

It will be necessary to calculate a shell-side-fluid velocity, given a mass flow rate. This can be done if we have an equation for the shell-side characteristic area. The variables that influence the velocity are the shell inside diameter D_s, the clearance C between adjacent tubes (see Figure 9.10), the baffle spacing B, and the pitch S_T. The shell characteristic area is given by

$$A_s = \frac{D_s C B}{S_T} \quad (9.28)$$

The shell-side velocity is then found with*

$$V_s = \frac{\dot{m}}{\rho A_s}$$

Remember that the shell-side velocity is not a constant, and the above formulation yields an estimate that is useful in calculating a film coefficient.

The Nusselt number for the shell side is given by an empirical equation that has been verified on a number of commercial exchangers:

$$Nu_s = \frac{h_o D_e}{k_f} = 0.36 \, Re_s^{0.55} Pr^{1/3} \quad (9.29)$$

---

* In the literature concerning heat exchangers, a term called the *mass velocity G* is widely used: $G = \dot{m}/A = \rho V$. The dimensions of mass velocity are $M/T \cdot L^2$. G is also called the *mass flux*, because its dimensions are mass flow rate (M/T) per unit area ($1/L^2$). We are not using G in this text, but you should be aware of its existence.

where

$$2 \times 10^3 \leq \mathrm{Re}_s = V_s D_e / v \leq 1 \times 10^6$$

$$\mathrm{Pr} = v/\alpha > 0$$

μ changes moderately with temperature

Properties at the average fluid temperature = [(inlet + outlet)/2]

The pressure drop experienced by the shell fluid is a function of the number of times the fluid passes the tube bundle between baffles and the distance traveled during each crossing. There is always an odd number of bundle crossings if the shell nozzles are at opposite sides, and an even number when the nozzles are at the same side. Based on data obtained on actual exchangers, the shell-side pressure drop can be calculated with

$$\Delta p_s = f(N_b + 1) \frac{D_s \rho V_s^2}{D_e \, 2g_c} \tag{9.30}$$

where N_b is the number of baffles, and $(N_b + 1)$ is the number of times the shell fluid crosses the tube bundle. The friction factor f for the shell is given by

$$f = \exp(0.576 - 0.19 \ln \mathrm{Re}_s) \tag{9.31}$$

where

$$400 \leq \mathrm{Re}_s = V_s D_e v \leq 1 \times 10^6$$

The friction factor accounts also for entrance and exit losses.

9.3.7 TRUE TEMPERATURE DIFFERENCE

We now examine how temperatures of the fluids vary within the exchanger and the corresponding effect on the heat-transfer rate. Figure 9.11 shows two 1–2 exchangers and the associated temperature profiles for the fluids. In Figure 9.11a, it is seen that the cold fluid enters the tubes at t_1, travels in "parallel flow" with the shell fluid, then returns in "counterflow" with the shell fluid to the outlet at t_2. In Figure 9.11b, the nozzles are changed from Figure 9.11a, but a parallel-counterflow arrangement is still encountered.

We recall from the double-pipe exchanger discussion that greater temperature differences (i.e., greater LMTD) are obtained for counterflow than for parallel flow. However, the 1–2 shell-and-tube exchanger is a combination of both, so the LMTD for either flow alone cannot be used for the true temperature difference. It is necessary to derive a new relationship for shell-and-tube equipment when combined flows exist. The following assumptions are applicable:

1. The shell-fluid temperature is an average and equal to a constant at any cross-sectional location (any z location in Figure 9.11).
2. An equal amount of heating occurs in each pass.
3. The overall coefficient of heat transfer is constant.
4. The flows are steady.
5. Fluid properties are constant.

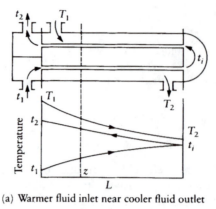

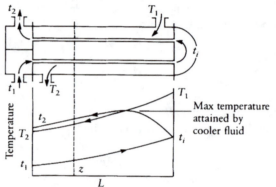

(a) Warmer fluid inlet near cooler fluid outlet

(b) Warmer fluid outlet near cooler fluid inlet

FIGURE 9.11 Temperature variation with distance for both fluids in a shell-and-tube heat exchanger. Two possible flow arrangements are illustrated.

6. No phase changes occur.

7. All heat lost by the warmer fluid is gained by the cooler fluid.

Based on the above assumptions, a true temperature difference for a 1–2 shell-and-tube exchanger can be derived. The derivation is lengthy and not presented here.[*] The results are summarized, however.

We introduce a temperature factor S, defined as

$$S = \frac{t_2 - t_1}{T_1 - t_1} \tag{9.32}$$

and restate R from Equation 9.14a:

$$R = \frac{\dot{m}_c c_{pc}}{\dot{m}_h c_{ph}} = \frac{T_1 - T_2}{t_2 - t_1} \tag{9.14}$$

The technique is to use the LMTD for counterflow (as a "best possible" case), then use R and S to obtain a correction factor F_t (not derived here):

---

[*] For details of the derivation, see *Process Heat Transfer*, by D.Q. Kern, McGraw-Hill, 1950, pp. 139–147.

$$F_t = \frac{\sqrt{(R^2 + 1)} \ln[(1 - S)/(1 - RS)]}{(R - 1)\ln\left[\dfrac{2 - S(R + 1 - \sqrt{R^2 + 1})}{2 - S(R + 1 + \sqrt{R^2 + 1})}\right]} \tag{9.33}$$

The true temperature difference then becomes

$$\Delta t_{1-2exch} = F_t \text{LMTD}_{counterflow}$$

where for counterflow

$$\text{LMTD} = \frac{(T_1 - t_2) - (T_2 - t_1)}{\ln\left[(T_1 - t_2)/(T_2 - t_1)\right]} \quad \text{(counterflow)} \tag{9.12a}$$

The function (or correction factor) F_t is graphed in Figure 9.12. It turns out that the correction factor F_t of Equation 9.33 applies to exchangers that have one shell pass and two or more tube passes. It is not practical to use a 1–2 heat exchanger in which F_t has been determined to be less than 0.75.

It is now possible to write the heat-transfer equation for a 1–2 shell- and-tube exchanger as

$$q = U_o A_o F_t \text{LMTD}_{counterflow} \tag{9.34}$$

where A_o is the external surface area of all the tubes, and U_o is obtained from

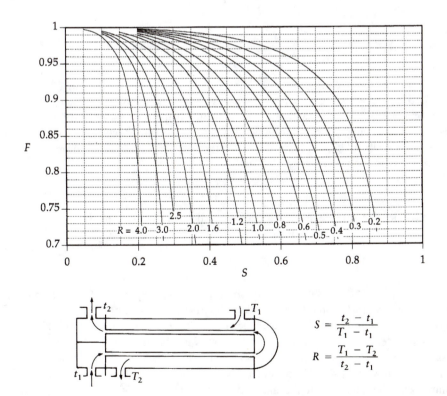

FIGURE 9.12 Correction factor F_t for a shell-and-tube heat exchanger—one shell pass and two or more tube passes.

$$\frac{1}{U_o} = \frac{1}{h_i(ID_t/OD_t)} + \frac{1}{h_o} = \frac{1}{h_{io}} + \frac{1}{h_o} \qquad (9.35)$$

As with double-pipe exchangers, shell-and-tube-exchanger surfaces are subject to fouling. The fouling factors of Table 9.1 apply, and the overall heat-transfer coefficient becomes

$$\frac{1}{U} = \frac{1}{U_o} + R_{di} + R_{do} = \frac{1}{U_o} + R_d \qquad (9.24)$$

In a number of cases involving shell-and-tube equipment, the inlet temperatures of the fluid streams are known, and the outlet temperatures are to be determined. By appropriately combining the defining equation for the correction factor F_t (Equation 9.33) with the definition of the log-mean temperature difference and the overall heat-transfer coefficient, it is possible to derive the following equation for 1–2 shell-and-tube exchangers:

$$\left(\frac{UA_o}{\dot{m}_c c_{pc}}\right)_{actual} = F_t \left(\frac{UA_o}{\dot{m}_c c_{pc}}\right)_{counterflow}$$

$$= \frac{1}{\sqrt{R^2+1}} \ln\left[\frac{2 - S(R + 1 - \sqrt{R^2+1})}{2 - S(R + 1 + \sqrt{R^2+1})}\right] \qquad (9.36)$$

$$R = \frac{\dot{m}_c c_{pc}}{\dot{m}_h c_{ph}} = \frac{T_1 - T_2}{t_2 - t_1}$$

$$S = \frac{t_2 - t_1}{T_1 - t_1}$$

Equation 9.36 is graphed in Figure 9.13 as S vs. $(UA_o/\dot{m}_c c_{pc})_{actual}$ with R as an independent variable. For an existing 1–2 exchanger, A_o and $\dot{m}_c c_{pc}$ are known, R can be calculated, and U can be computed from the film coefficients. Then, using Figure 9.13 gives S, from which t_2 for the cooler fluid can be determined. Finally, the outlet temperature for the warmer fluid is found from a heat balance:

$$T_2 = T_1 - R(t_2 - t_1) \qquad (9.37)$$

A suggested order for performing calculations on shell-and-tube exchangers is provided in the Summary section.

Example 9.5

In an electricity-generating facility, steam leaves a turbine and is piped to a condensing unit. After condensation occurs, it is desired to further cool the (distilled) water by means of a shell-and-tube exchanger. The water enters the heat exchanger at 110°F with a flow rate of 170,000 lbm/hr. The heat will be transferred to raw water from a nearby river. The raw water is available at 65°F, and the mass flow rate is 150,000 lbm/hr. It is proposed to use a heat exchanger that has a 17-1/4-in-ID shell and 3/4-in-OD, 18-BWG tubes that are 16 ft long. The tubes are laid out on a 15/16-in triangular pitch. The tube fluid will make two passes. The shell contains baffles that are spaced 1 ft apart. Determine the outlet temperature of the distilled water and the pressure drop for each stream.

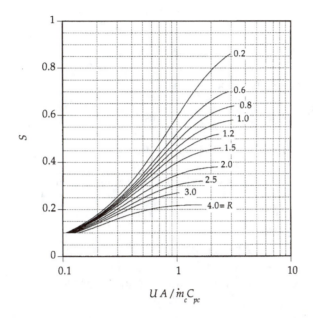

$$U A / \dot{m}_c C_{pc}$$

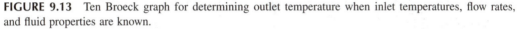

FIGURE 9.13 Ten Broeck graph for determining outlet temperature when inlet temperatures, flow rates, and fluid properties are known.

Solution

The flow rates here are much larger than those encountered with double- pipe exchangers.

Assumptions

1. Fluid properties are constant and evaluated at or near the inlet temperatures.
2. Raw- and distilled-water properties can be obtained from the same water-property tables.
3. The system is at steady state.
4. All heat lost by the distilled water is gained by the raw water.

From Appendix Table C.11, for (distilled) water at 104°F (an estimate), we read

$\rho = 0.994 \ (62.4) \ \text{lbm/ft}^3$ $k_f = 0.363 \ \text{BTU/(hr·ft·°R)}$
$c_p = 0.998 \ \text{BTU/lbm·°R}$ $\alpha = 5.86 \times 10^{-3} \ \text{ft}^2\text{/hr}$
$v = 0.708 \times 10^{-5} \ \text{ft}^2\text{/s}$ $Pr = 4.34$

Also, $T_1 = 110°F$ and $\dot{m}_h = 170{,}000 \ \text{lbm/hr}$.

From Appendix Table C.11 for (raw) water at 68°F (estimated):

$\rho = 62.4 \ \text{lbm/ft}^3$ $k_f = 0.345 \ \text{BTU/lbm·ft·°R}$
$c_p = 0.9988 \ \text{BTU/lbm·°R}$ $\alpha = 5.54 \times 10^{-3} \ \text{ft}^2\text{/hr}$
$v = 1.083 \times 10^{-5} \ \text{ft}^2\text{/s}$ $Pr = 7.02$

Also, $t_1 = 65°F$ and $\dot{m}_c = 150{,}000 \ \text{lbm/hr}$

From Table 9.2 for 3/4-in-OD, 18-BWG tubes,

$OD_t = 3/4\text{-in} = 0.0625 \ \text{ft}$ $ID_t = 0.652 \ \text{in} = 0.0543 \ \text{ft}$

Also,

$N_t = 224$ (Table 9.3) $N_p = 2$ tube passes

Shell dimensions and other miscellaneous data:

$D_s = 17.25$ in $= 1.438$ ft $N_b = 15$ baffles
$B = 1$ ft $S_T = 15/16$ in $= 0.0781$ ft
$C = S_T - OD_t = 15/16 - 3/4 = 0.1875$ in $= 0.0156$ ft

Calculations are performed in steps (t subscript = tubes; s = shell; formulas are in Summary section):

I. Flow Areas

$$A_t = 224\pi(0.0543)^2/4(2) = 0.259 \text{ ft}^2$$

$$A_s = 1.438(0.0156)(1)/0.0781 = 0.287 \text{ ft}^2$$

Place the raw water through the tubes, because it has the lower flow rate.

II. Shell Equivalent Diameter

$$D_e = \frac{3.44(0.0781)^2 - \pi(0.0625)^2}{\pi(0.0625)} = 0.0444 \text{ ft}$$

III. Reynolds Numbers

Raw water $Re_t = \dfrac{150,000}{3600(62.4)(0.259)}\dfrac{0.05423}{1.083 \times 10^{-5}} = 1.29 \times 10^4$

Distilled water $Re_t = \dfrac{170,000}{3600(0.994)(62.4)(0.287)}\dfrac{0.0444}{0.708 \times 10^{-5}} = 1.66 \times 10^4$

IV. Nusselt Numbers

Raw water $Nu_t = 0.023(1.29 \times 10^4)^{4/5}(7.02)^{0.4} = 97.4$

Distilled water $Nu_s = 0.36(1.66 \times 10^4)^{0.55}(4.34)^{1/3} = 123$

V. Convection Coefficients

$$h_i = \frac{97.4(0.345)}{0.0543} = 619;$$

$$h_{io} = \frac{619(0.0543)}{0.0625} = 538 \text{ BTU}(/\text{hr·ft}^2\cdot{}^\circ\text{R});$$
Raw water

Distilled water $h_o = \dfrac{123(0.363)}{0.0444} = 1005 \text{ BTU}(/\text{hr·ft}^2\cdot{}^\circ\text{R});$

VI. Exchanger Coefficient (clean or new)

$$\frac{1}{U_o} = \frac{1}{538} + \frac{1}{1005} \qquad U_o = 350 \ \text{BTU}/(\text{hr·ft}^2\cdot°\text{R})$$

Outlet temperatures

$$R = \frac{150{,}000(0.9988)}{170{,}000(0.998)} = 0.833$$

$$A_o = 224\pi(0.0625)(16) = 703.7 \ \text{ft}^2$$

$$\frac{U_o A_o}{\dot{m}_c c_{pc}} = \frac{350(703.7)}{150{,}000(0.9988)} = 1.64$$

From Figure 9.13,

$$S = 0.58 = \frac{t_2 - t_1}{T_1 - t_1};$$

Raw water $\qquad t_2 = 0.58(110 - 65) + 65; \ t_2 = 91.1°\text{F}$

Distilled water $\qquad T_2 = 110 - 0.883(91.1 - 65); \ T_2 = 86.9°\text{F}$

VII. Friction Factors

Raw water $\qquad \begin{aligned} &\text{Re}_t = 1.29 \times 10^4 \\ &\text{smooth-wall tubing} \end{aligned} \Bigg\} \ f_t = 0.029 \qquad$ (Moody diagram)

$$\text{Re}_s = 166 \times 10^4$$

Distilled water $\qquad f_s = \exp[0.576 - 0.19 \ \ln(1.66 \times 10^4)] = 0.281$

VIII. Velocities

Raw water $\qquad V_t = \dfrac{150{,}000}{3600(62.4)(0.259)} = 2.58 \ \text{ft/s}$

Distilled water $\qquad V_s = \dfrac{170{,}000}{3600(0.994)(62.4)(0.287)} = 2.65 \ \text{ft/s}$

IX. Pressure Drop

Raw water $\quad \Delta p_t = \dfrac{62.4(2.58)^2}{2(32.2)}\left[\dfrac{0.029(16)(2)}{0.0543} + 4(2)\right] = 161.8 \ \text{lbf/ft}^2 = 1.1 \ \text{psi}$

$$\Delta p_s = \frac{(0.994)(62.4)(2.65)^2}{2(32.2)}\frac{1.438}{0.0444}0.281(15 + 1)$$

Distilled water $\qquad = 985 \ \text{lbf/ft}^2 = 6.8 \ \text{psi}$

X. Summary of Requested Information

Outlet temperatures

> Raw water = 91.1°F
> Distilled water = 86.9°F

Pressure drops

> Raw water = 1.1 psi
> Distilled water = 6.8 psi

The pressure drop for the streams is comparatively low, with ~10 psi being an accepted maximum for low pumping costs. If the pressure drop for either fluid is too high, then a decrease in the flow rate of the respective fluid might be an acceptable solution. The outlet temperatures may change along with the decrease in flow rate, however, and less heat will be transferred. Thus there will be a trade-off between trying to recover as much heat as possible versus keeping the pressure drop as low as possible. Such decisions are obviously based on economic factors.   ❏

A number of methods can be used to increase the heat recovered. For example, two 1–2 shell-and-tube exchangers could be connected in series. Two 1–2 exchangers could be piped so that the tube fluid passes through in series while the shell fluid passes through in parallel. The tubes could be arranged so that the tube fluid passes through the same exchanger more than twice, or the shell fluid could pass through more than once (e.g., 1–4, 1–6, 1–8 or 2–4, 2–6, 2–8 types). A number of applications exist; for further information the interested reader is referred to the literature.[*]

9.4   EFFECTIVENESS–NTU METHOD OF ANALYSIS

The effectiveness–NTU method of analysis is an alternative approach to the heat-exchanger analysis just presented. Once the overall coefficient for the exchanger is calculated, the effectiveness–NTU method can be used to predict outlet temperatures for the fluids. To begin, we must define the maximum heat that can be transferred by the exchanger, q_{max} In a counterflow exchanger q_{max} could be achieved only if the exchanger had an infinite length (i.e., infinite surface area). One fluid would experience the maximum possible temperature difference, which is the difference in the fluid inlet temperatures: $\Delta T_{max} = T_1 - t_1$. Consider, for example, a situation where the product of mass flow and specific heat (defined as the capacitance) for the cooler fluid is less than that for the warmer fluid, or

$$\dot{m}_c c_{pc} < \dot{m}_h c_{ph} \tag{9.38}$$

The heat transfer experienced by each fluid is written as

$$\dot{m}_c c_{pc} \Delta T_c = \dot{m}_h c_{ph} \Delta T_h$$

If Equation 9.38 applies, then we conclude that

$$\Delta T_c > \Delta T_h$$

---

[*] *Process Heat Transfer* by D.Q. Kern, McGraw-Hill, 1950, Chapters 7 and 8.

or the temperature change for the cooler fluid is greater than that for the warmer fluid. Hypothetically, because the exchanger length is infinite, then the cooler fluid could be heated to the inlet temperature of the warmer fluid, $t_2 = T_1$. Therefore, we have

$$\dot{m}_c c_{ph} < \dot{m}_c c_{ph} \rightarrow q_{max} = \dot{m}_c c_{pc}(T_1 - t_1)$$

Conversely,

$$\dot{m}_h c_{ph} < \dot{m}_c c_{pc} \rightarrow q_{max} = \dot{m}_h c_{ph}(T_1 - t_1)$$

In general

$$q_{max} = (\dot{m}c_p)_{max}(T_1 - t_1) \tag{9.39}$$

where the capacitance $(\dot{m}c_p)_{min}$ equals that for the cooler to the warmer fluid, whichever is smaller.

Next, we define exchanger effectiveness ε^* as the ratio of actual heat-transfer rate to the maximum possible heat-transfer rate, or

$$\varepsilon = \frac{q}{q_{max}} \tag{9.40}$$

Substituting, we get

$$\varepsilon = \frac{\dot{m}_c c_{pc}(t_2 - t_1)}{(\dot{m}c_p)_{min}(T_1 - t_1)} = \frac{\dot{m}_h c_{ph}(T_1 - T_2)}{(\dot{m}c_p)_{min}(T_1 - t_1)}$$

The above reduces to simpler expressions when the minimum is determined

$$\varepsilon = \frac{t_2 - t_1}{T_1 - t_1} \qquad \text{(cooler fluid has minimum } \dot{m}c_p) \tag{9.41a}$$

$$\varepsilon = \frac{T_1 - T_2}{T_1 - t_1} \qquad \text{(warmer fluid has minimum } \dot{m}c_p) \tag{9.41b}$$

where again upper-case T is for the warmer fluid, lower-case t is for the cooler fluid, and subscripts 1 and 2 are for inlet and outlet, respectively. As can be discerned from its definition, effectiveness ranges from 0 to 1. Also, if effectiveness is known, then the heat-transfer rate can be found with

$$q = \varepsilon(\dot{m}c_p)_{min}(T_1 - t_1) \tag{9.42}$$

Note that those equations above that are involved in the effectiveness definitions are independent of exchanger type. All that remains, then, is to derive an equation for effectiveness for each exchanger of interest.

As will be evident from the development that follows, we will encounter the term $UA/(\dot{m}c_p)_{min}$ This term is defined as the *Number of Transfer Units, NTU*; that is,

$$\text{NTU} = \frac{UA}{(\dot{m}c_p)_{min}} \tag{9.43}$$

NTU is a dimensionless parameter used widely in the analysis of heat exchangers.

---

* Not to be confused with the roughness factor used in pipe-friction calculations, ε/D.

9.4.1 EFFECTIVENESS–NTU EQUATIONS

To determine a specific expression for effectiveness, we now select a counterflow double-pipe heat exchanger, for which we assume that $(\dot{m}c_p)_{min} = \dot{m}_c c_{pc}$. For purposes of simplifying the notation, let $\dot{m}_h c_{ph} = (\dot{m}c_p)_{max}$ for this case. From Equation 9.41,

$$\varepsilon = \frac{t_2 - t_1}{T_1 - t_1} \, {}^* \tag{9.44}$$

The heat-transfer equations are

$$q = \dot{m}_h c_{ph}(T_1 - T_2) = (\dot{m}c_p)_{max}(T_1 - T_2)$$

$$q = \dot{m}_c c_{pc}(t_2 - t_1) = (\dot{m}c_p)_{min}(t_2 - t_1)$$

Setting them equal to each other gives the ratio of the capacitances as

$$\frac{(\dot{m}c_p)_{min}}{(\dot{m}c_p)_{max}} = \frac{T_1 - T_2}{t_2 - t_1} \, {}^\dagger \tag{9.14a}$$

Equation 9.13 for counterflow is

$$\ln\frac{T_1 - t_2}{T_2 - t_1} = \frac{U_o A_o}{\dot{m}_c c_{pc}}\left(\frac{T_1 - T_2}{t_2 - t_1} - 1\right) \tag{9.13}$$

Rewriting 9.13 in terms of NTU and the capacitance ratio gives

$$\ln\frac{T_1 - t_2}{T_2 - t_1} = \text{NTU}\left[\frac{(\dot{m}c_p)_{min}}{(\dot{m}c_p)_{max}} - 1\right]$$

Exponentiating both sides gives

$$\frac{T_1 - t_2}{T_2 - t_1} = \exp\left\{\text{NTU}\left[\frac{(\dot{m}c_p)_{min}}{(\dot{m}c_p)_{max}} - 1\right]\right\} \tag{9.45}$$

Our objective now is to rewrite the left-hand side in terms of the effectiveness. From Equation 9.44,

$$t_2 = t_1 + T_1\varepsilon - t_1\varepsilon \tag{9.46}$$

From Equation 9.14

$$T_2 = T_1 - T_1\varepsilon(\dot{m}c_p)_{min}/(\dot{m}c_p)_{max} + t_1\varepsilon(\dot{m}c_p)_{min}/(\dot{m}c_p)_{max} \tag{9.47}$$

---

* Notice that, when the cooler fluid has the minimum capacitance $\dot{m}c_p$, then $\varepsilon = S$, where S was first defined in Equation 9.36. When the warmer fluid has the minimum capacitance, then $\varepsilon \neq S$. Thus, ε and S are to be considered as two separate entities.

† This ratio is recognized as R, defined first in Equation 9.14.

The outlet temperatures are now specified in terms of the inlet temperatures, the effectiveness, and the fluid properties. Substituting the above outlet-temperature expressions into the left-hand side of Equation 9.45 and simplifying gives

$$\frac{T_1 - t_2}{T_2 - t_1} = \frac{1 - \varepsilon}{1 - \varepsilon (\dot{m}c_p)_{min}/(\dot{m}c_p)_{max}}$$

Combining with the right-hand side of Equation 9.45 and rearranging to solve for effectiveness, we get

$$\varepsilon = \frac{1 - \exp\{NTU[(\dot{m}c_p)_{min}/(\dot{m}c_p)_{max} - 1]\}}{1 - \dfrac{(\dot{m}c_p)_{min}}{(\dot{m}c_p)_{max}} \exp\{NTU[(\dot{m}c_p)_{min}/(\dot{m}c_p)_{max} - 1]\}}$$

(counterflow double-pipe exchanger)

(9.48)

Here, we selected the cooler fluid as having the minimum value of the capacitance $(\dot{m}c_p)$. When the warmer fluid has the minimum product, exactly the same expression results. There is a distinct similarity between Equation 9.48 for effectiveness and Equation 9.16 for the outlet temperature of the warmer fluid in a counterflow exchanger. Equation 9.48 is graphed as ε vs. NTU [$= U_o A_o/(\dot{m}c_p)_{min}$], with $(\dot{m}c_p)_{min}/(\dot{m}c_p)_{max}$ as an independent variable, in Figure 9.14. A corresponding expression is available for parallel flow and is given as an exercise.

Effectiveness–NTU equations have been derived for shell-and-tube exchangers having one or more shell passes and one or more tube passes. For a shell-and-tube exchanger with one shell pass and any integral multiple of two tube passes, we have

$$\varepsilon = 2\left\{1 + \frac{(\dot{m}c_p)_{min}}{(\dot{m}c_p)_{max}} + \left[1 + \frac{(\dot{m}c_p)^2_{min}}{(\dot{m}c_p)^2_{max}}\right]^{1/2} \times \left(\frac{1 + \exp\left\{-NTU\left[1 + \frac{(\dot{m}c_p)^2_{min}}{(\dot{m}c_p)^2_{max}}\right]^{1/2}\right\}}{1 - \exp\left\{-NTU\left[1 + \frac{(\dot{m}c_p)^2_{min}}{(\dot{m}c_p)^2_{max}}\right]^{1/2}\right\}}\right)\right\}^{-1}$$

(9.49)

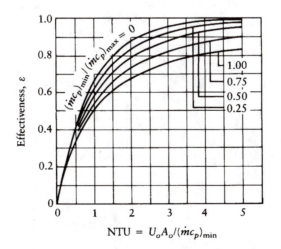

FIGURE 9.14 Effectiveness vs. number of transfer units for a double-pipe, counterflow heat exchanger.

This equation is graphed in Figure 9.15 as ε vs. NTU, with $(\dot{m}c_p)_{min}/(\dot{m}c_p)_{max}$ as an independent variable.

In Figures 9.14 and 9.15, the capacitance ratio $(\dot{m}c_p)_{min}/(\dot{m}c_p)_{max}$ can be as small as zero. The zero case applies to an evaporator or condenser where the heat-capacity rate of the fluid changing phases is considered to be infinite. The usefulness of the effectiveness–NTU approach is illustrated in the following example.

Example 9.6

Use the effectiveness–NTU method to calculate the outlet temperatures of the fluids in Example 9.5.

Solution

Data from Example 9.5 is as follows:

	Distilled water—shell	Raw water—tubes
$\dot{m}$, lbm/hr	170,000	150,000
c_p, BTU/lbm·°R	0.998	0.9988
Inlet temperatures, °F	110	65

Also, $U_o = 350$ BTU/(hr·ft²·°R), and $A_o = 703.7$ ft².

The effectiveness–NTU approach is used when the overall heat transfer coefficient is known. We first determine the capacitance values:

Raw water $\quad (\dot{m}c_p) = 150,000(0.9988) = 149,820$ BTU/(hr·°R)

Distilled water $\quad (\dot{m}c_p) = 170,000(0.998) = 169,660$ BTU/(hr·°R)

The raw water has the minimum capacitance. We next calculate

$$\frac{(\dot{m}c_p)_{min}}{(\dot{m}c_p)_{max}} = \frac{149,820}{169,660} = 0.883$$

$$\frac{U_o A_o}{(\dot{m}c_p)_{min}} = \frac{350(703.7)}{149,820} = 1.64$$

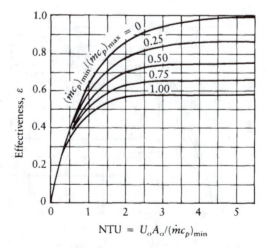

FIGURE 9.15 Effectiveness vs. number of transfer units for a shell-and-tube exchanger having one shell pass and any integral multiple of two tube passes.

With these values, Figure 9.15 gives

$$\varepsilon = 0.58$$

The maximum heat that can be transferred is

$$q_{max} = (\dot{m}c_p)_{min}(T_1 - t_1) = 149,820(110 - 65)$$

or

$$q_{max} = 6.74 \times 10^6 \text{ BTU/hr}$$

The actual heat transferred by the exchanger then is

$$q = \varepsilon q_{max} = 0.58(6.74 \times 10^6) = 3.91 \times 10^6 \text{ BTU/hr}$$

The outlet temperatures are now calculated with the heat-balance equations:

$$q = \dot{m}_c c_{pc}(t_2 - t_1)$$

Substituting,

$$3.91 \times 10^6 = 149,820(t_2 - 65)$$

or

Distilled water $t_2 = 91.1°F$

Also, for the warmer fluid

$$3.91 \times 10^6 = 169,660(110 - T_2)$$

or

Raw water $T_2 = 86.9°F$

These results are identical to those obtained in Example 9.5, wherein the Ten Broeck chart (Figure 9.13) was used. In this example, $\varepsilon = S$ numerically.   ❏

9.5   CROSSFLOW HEAT EXCHANGERS

In crossflow heat exchangers, the directions of the fluid velocities are generally at right angles to each other; the different types that exist are numerous. Crossflow exchangers can be classified according to whether each fluid stream remains *mixed* or *unmixed* as it passes through. Figure 9.16 illustrates the concept of mixed vs. unmixed flows. Figure 9.16a shows a passageway in which the fluid does not encounter internal channels, so fluid particles are free to mix. Figure 9.16b is of a passageway that does contain internal channels that restrict lateral fluid movement; hence the flow is unmixed.

(a) Passageway in which fluid can mix

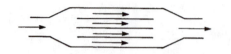

(b) Passageway containing flow dividers to restrict
mixing, called an unmixed passageway

FIGURE 9.16   Illustration of mixed vs. unmixed flows.

Figure 9.17a is a sketch of a crossflow heat exchanger in which one fluid is mixed while the other is unmixed. Figure 9.17b shows an exchanger in which both fluids are unmixed. Such arrangements are more a result of construction method than of planning.

Figure 9.18 shows a finned-tube type of crossflow heat exchanger. A number of very thin (0.005-in) aluminum pieces are stamped with holes. Three copper tubes (slightly larger in diameter than the holes) are pushed through the holes, and fittings are soldered to the tube ends. This type of exchanger is used in air-conditioning units both as an evaporator and as a condenser.

Figure 9.19 shows how a crossflow exchanger, similar to those used as automobile radiators, is constructed. As indicated, a thin copper sheet is bent into a sine-wave cross-sectional shape and placed inside a brass piece that has a rectangular shape. A number of these assemblies are placed side by side, then the faces are dipped into a flux and next into molten solder. The edges of adjacent brass pieces are joined, and the copper pieces are also thus joined to the brass at points of contact. Tanks containing inlet and outlet fittings are next soldered to the ends. The size of the exchanger is controlled by the number of brass-copper assemblies used. The finished piece is known by a variety of names (such as a *radiator* or a *heater core*). Regardless of terminology, its function is to exchange heat.

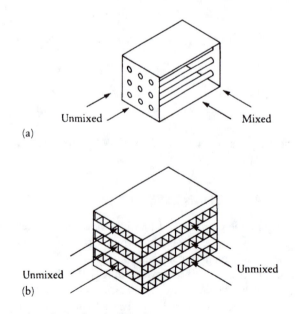

(a)

Unmixed   Mixed

(b)

Unmixed   Unmixed

FIGURE 9.17   Two examples of crossflow heat exchangers.

FIGURE 9.18 Finned-tube crossflow heat exchanger.

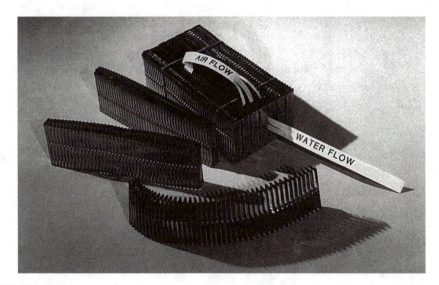

FIGURE 9.19 Construction of a conventional radiator core.

As with other heat exchangers considered thus far, we would want to be able to model the heat-exchange process in a crossflow exchanger. There is available in the literature* an excellent summary of tests on crossflow heat exchangers. A number of conventional types have been tested for heat-transfer and friction characteristics. As an example, consider that we are working with an exchanger made of the brass-copper assemblies shown in Figure 9.19. Such an exchanger has air passing through unmixed. The air-flow passageway is normal to the sine-wave-shaped copper piece. The other fluid is water, and it passes through between adjacent faces of the brass pieces. We could approximate the water passageway as a narrow rectangle (for lack of a better description in which a correlation has been determined). Sine-wave and narrow-rectangle correlations are both available. Using such correlations provides values for the convection coefficient for each fluid stream, from which an overall coefficient U can be calculated. Due to space limitations, it is not possible to reproduce here all the data required for an in-depth analysis necessary to find U. We will therefore assume a knowledge of the numerical value of the overall heat-transfer coefficient for problems of interest.

---

* Compact Heat Exchangers by W.M. Kays and A.L. London, McGraw-Hill, 1964.

Once the overall heat-transfer coefficient is known for a crossflow heat exchanger, the heat transfer can be calculated with

$$q = UAF_t \, \text{LMTD} \tag{9.50}$$

where U is the overall coefficient, A is the area of the exchanger, F_t is a correction factor, and LMTD is the log-mean temperature difference for a double-pipe heat exchanger in counterflow. With crossflow exchangers, it is not uncommon to combine U and A and report only their product. Area is extremely difficult to define because of the construction methods. In a finned-tube exchanger, for example (Figure 9.18), the area outside the tubes should include the surface area of all fins added together minus the area removed by the tubes plus the outside surface area of the tubes. For a radiator (Figure 9.19), the air-side surface area should include (perhaps) the outside surface area of the copper pieces and the outside of the brass rectangles. In both cases, it is far easier to group area with the overall coefficient rather than to calculate it. On the other hand, it is necessary to be able to calculate how large an exchanger would have to be to transfer a required heat load. This is more conveniently identified with a *frontal* area than with an area over which U applies.

Figure 9.20 gives a schematic of a crossflow heat exchanger with the associated temperature profiles for each fluid. As indicated, the warmer fluid enters at temperature T_1 and is cooled to T_2. The cooler fluid enters at t_1 and is warmed to t_2. As evident from the figure, the fluids do not travel through the exchanger in the same manner as they would in a counterflow exchanger. For counterflow the exiting cooler fluid is nearest the entering warmer fluid, where the warmer fluid is at its highest temperature. This, of course, is why counterflow is an efficient arrangement. Experiments have shown that a counterflow arrangement can provide a higher outlet temperature for the cooler fluid than does a crossflow exchanger. So, in analyzing crossflow exchangers, the standard for comparison is the counterflow exchanger. In reference to Equation 9.50, the LMTD that appears there is for counterflow; that is,

$$\text{LMTD} = \frac{(T_1 - t_2) - (T_2 - t_1)}{\ln[(T_1 - t_2)/(T_2 - t_1)]} \quad \text{(counterflow)} \tag{9.12}$$

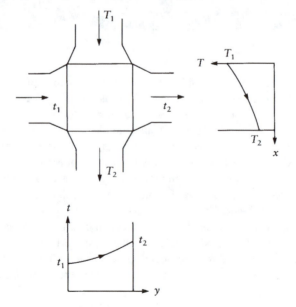

FIGURE 9.20   Temperature profiles for flow in a crossflow heat exchanger.

Equations for the correction factor for crossflow exchangers have been derived,[*] and they involve a double infinite series. Rather than derive the equations here, we present instead correction-factor graphs in Figures 9.21a, 9.22a, and 9.23. In each case, the horizontal axis is $S = (t_2 - t_1)/(T_1 - t_1)$, and the vertical axis is the correction factor F_t, with $R = (T_1 - T_2)/(t_2 - t_1)$ appearing as an independent variable.

In a number of cases, only inlet temperatures are known, and it desired to calculate outlet temperatures. This can be done in several ways, but here we discuss only the effectiveness–NTU method. Equations for effectiveness have been derived for crossflow exchangers, and they are as follows:

$$\varepsilon = 1 - \exp\left[\frac{(\dot{m}c_p)_{min}}{(\dot{m}c_p)_{max}}(\text{NTU})^{0.22} \times \left\{\exp\left[-\frac{(\dot{m}c_p)_{min}}{(\dot{m}c_p)_{max}}(\text{NTU})^{0.78}\right] - 1\right\}\right]$$

(crossflow, both fluids unmixed)   (9.51)

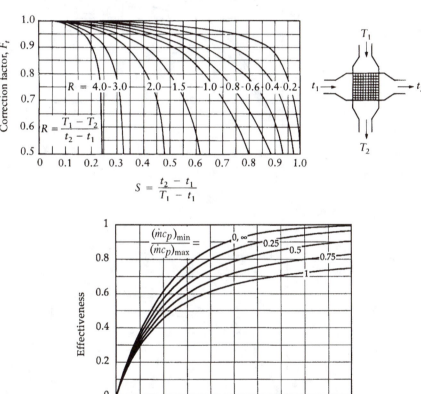

(a)

(b)

FIGURE 9.21 (a) Correction-factor chart for a crossflow heat exchanger with both fluids unmixed. *Source:* reprinted with permission from "Mean Temperature Difference," by R.A. Bowman, A.C. Mueller, and W.M. Nagle, *Trans. ASME*, v. 62, 1940, pp. 283–294. (b) Effectiveness-vs.-NTU graph for a crossflow heat exchanger with both fluids unmixed.

---

[*] *Process Heat Transfer* by D.Q. Kern, McGraw-Hill, 1950, Chapter 16, pp. 546–553.

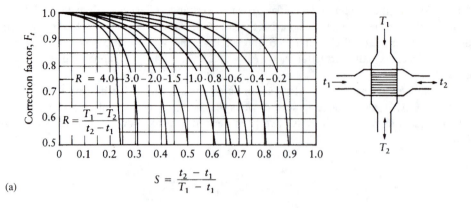

(a)

$$S = \frac{t_2 - t_1}{T_1 - t_1}$$

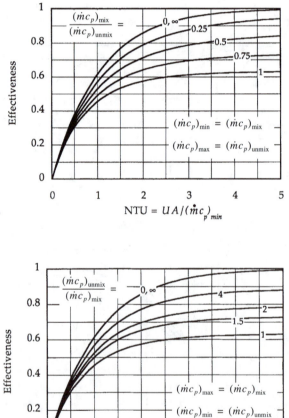

(b)

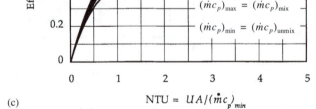

(c)

FIGURE 9.22 (a) Correction-factor chart for a crossflow heat exchanger with one fluid mixed. *Source:* reprinted with permission from "Mean Temperature Difference," by R.A. Bowman, A.C. Mueller, and W.M. Nagle, *Trans. ASME*, v. 62, 1940, pp. 283–294. (b and c) Effectiveness-vs.-NTU graph for a crossflow heat exchanger with one fluid mixed.

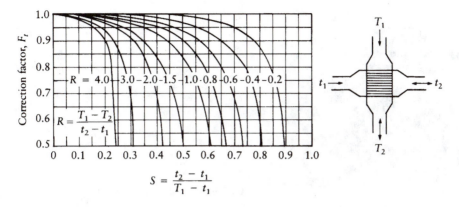

FIGURE 9.23 Correction-factor chart for a crossflow heat exchanger with both fluids mixed. *Source:* reprinted with permission from "Mean Temperature Difference," by R.A. Bowman, A.C. Mueller, and W.M. Nagle, *Trans. ASME*, v. 62, 1940, pp. 283–294.

$$\varepsilon = \left[\frac{(\dot{m}c_p)_{\text{mixed}}}{(\dot{m}c_p)_{\text{unmix}}}\right]\left[1 - \exp\left\{\frac{(\dot{m}c_p)_{\text{unmix}}}{(\dot{m}c_p)_{\text{mixed}}}[1 - \exp(-\text{NTU})]\right\}\right]$$

(crossflow, only one fluid mixed)   (9.52)

or alternatively,

$$\varepsilon = 1 - \exp\left[\frac{(\dot{m}c_p)_{\text{unmix}}}{(\dot{m}c_p)_{\text{mixed}}}\left\{1 - \exp\left[-\frac{(\dot{m}c_p)_{\text{mixed}}}{(\dot{m}c_p)_{\text{unmix}}}(\text{NTU})\right]\right\}\right]$$

(crossflow, only one fluid mixed)   (9.53)

For the case where both fluids are mixed, a closed-form solution is not available. A series solution has been derived and solved numerically. Effectiveness–NTU graphs are given in Figures 9.21b, 9.22b, and 9.22c.

Crossflow exchangers are subject to fouling, as are other exchangers, but it is difficult to apply fouling-factor values to the area in a crossflow exchanger. Consequently, we will not consider fouling effects in crossflow heat exchangers.

The Summary section provides a suggested order of calculations for the analysis of crossflow heat exchangers.

Example 9.7

A crossflow heat exchanger used as an oil cooler and manufactured with a fan is illustrated in Figure 9.24. Performance tests by the manufacturer provided the following data:

	Oil (assume engine oil)	Air
Inlet temperature	190°F	126°F
Outlet temperature	158°F	166°F
Mass flow rate	39.8 lbm/min	67 lbm/min
	Unmixed	Unmixed

FIGURE 9.24   Oil cooler and fan.

Core frontal area on the air side = (9.82 in)(8.00 in)
Core area on the oil side = (3.25 in)(9.82 in)

(a) Determine the *UA* product for the exchanger. (b) Calculate the exit temperatures for the exchanger, assuming that only the inlet temperatures are known, and compare the results to those given above.

Solution

The oil used in testing exchangers is a special oil with carefully measured properties.

Assumptions

1. The oil used is engine oil (for lack of more specific information and properties).
2. Fluid properties are constant and evaluated at the average film temperature [(inlet + outlet)/2].
3. The system is at steady state.

From Appendix Table C.4, for engine oil at $(190 + 158)/2 = 174°F \cong 176°F$, we read

$$\rho = 0.852(62.4) \text{ lbm/ft}^3 \qquad k_f = 0.08 \text{ BTU/(hr·ft·°R)}$$
$$c_p = 0.509 \text{ BTU/(lbm·°R)} \qquad \alpha = 2.98 \times 10^{-3} \text{ ft}^2\text{/hr}$$
$$v = 0.404 \times 10^{-3} \text{ ft}^2\text{/s} \qquad \text{Pr} = 490$$

From Appendix Table D.1, for air at $(126 + 166)/2 = 146°F = 606°R$,

$$\rho = 0.0653 \text{ lbm/ft}^3 \qquad k_f = 0.01677 \text{ BTU/(lbm·ft·°R)}$$
$$c_p = 0.241 \text{ BTU/(lbm·°R)} \qquad \alpha = 1.066 \text{ ft}^2\text{/hr}$$
$$v = 20.89 \times 10^{-5} \text{ ft}^2\text{/s} \qquad \text{Pr} = 0.706$$

Outlining the calculations and following the suggested order in the Summary section, we write the following steps (equations are in Summary section):

I. Heat Balance

$$\text{Air} \qquad q = 67(60)(0.241)(166 - 126) = 3.88 \times 10^4 \text{ BTU/hr}$$

$$\text{Engine oil} \qquad q = 39.8(60)(0.509)(190 - 158) = 3.88 \times 10^4 \text{ BTU/hr}$$

II. Log-Mean Temperature Difference

$$\text{LMTD} = \frac{(190-166)-(158-126)}{\ln[(190-166)/(158-126)]} = 27.8°F$$

III. Frontal Areas for Each Fluid Stream

Air $\quad A_c = 9.82(8.00)/144 = 0.545 \text{ ft}^2$

Engine oil $\quad A_h = 3.25(9.82)/144 = 0.222 \text{ ft}^2$

IV. Correction Factor

$$S = \frac{166-126}{190-126} = 0.625 \quad R = \frac{190-158}{166-126} = 0.8$$

With both fluids unmixed, Figure 9.21a applies. We obtain

$$F_t = 0.87$$

V. Overall Coefficient ($q = UAF_t\text{LMTD}$)

$$UA = \frac{3.88 \times 10^4}{0.87(27.8)} = 1.60 \times 10^3 \text{ BTU/(hr·°R)}$$

VI. Capacitances

Air $\quad \dot{m}c_p = 67(60)(0.241) = 968.8 \text{ BTU/(hr·°R)}$

Engine Oil $\quad \dot{m}c_p = 39.8(60)(0.509) = 1215 \text{ BTU/(hr·°R)}$

$$\frac{(\dot{m}c_p)_{\min}}{(\dot{m}c_p)_{\max}} = \frac{968.8}{1215} = 0.798$$

VII. Number of Transfer Units

$$\text{NTU} = \frac{UA}{(\dot{m}c_p)_{\min}} = \frac{1.6 \times 10^3}{968.8} = 1.65$$

VIII. Effectiveness (Figure 9.21b, both fluids unmixed)

$$\varepsilon = 0.62$$

With the cooler fluid having minimum $\dot{m}c_p$, Equation 9.41a applies:

$$\varepsilon = \frac{t_2 - t_1}{T_1 - t_1}$$

IX. Outlet Temperatures

Air
$$t_2 = (T_1 - t_1)\varepsilon + t_1 = (190 - 126)(0.62) + 126$$

$$t_2 = 165\,°F$$

$$(\dot{m}c_p)_{air}(165 - 126) = (\dot{m}c_p)_{oil}(190 - T_2)$$

Engine oil
$$T_2 = 190 - \frac{968.8}{1215}(39)$$

$$T_2 = 159\,°F$$

X. Summary of Requested Information

(a) $UA = 1.6 \times 10^3$ BTU/hr·°R

(b) Outlet temperatures

	Calculated	Given in problem statement
Air	165°F	166°F
Engine oil	158°F	159°F

❏

9.6   EFFICIENCY OF A HEAT EXCHANGER

It is of interest to establish a standard for comparison in heat exchangers. Specifically, it is important to have a standard of maximum performance against which heat exchangers can be compared. Sometimes the effectiveness is used as a comparator:

$$\varepsilon = \frac{\text{Actual heat removed}}{\text{Maximum heat that might be removed}}$$

Recall, however, that the denominator of the above expression is associated with an exchanger that has an infinite area. This seems somewhat objectionable because an infinite area is neither practical nor realistic, even though using the effectiveness as a measure of exchanger efficiency has merit.

Alternatively, we have shown in this chapter that maximum heat can be transferred in a real exchanger arranged in counterflow. Consequently, there is merit also in using a counterflow exchanger as a basis for comparison. Thus, we can define the efficiency of an exchanger as

$$\eta = \frac{\text{Temperature difference in the exchanger}}{\text{Temperature difference for counterflow}} = F_t$$

which is recognized immediately as the correction factor F_t. The objectionable aspect of this definition is that a counterflow exchanger is assigned an efficiency of 100%. This is not in keeping with the traditional thermodynamic idea of efficiency, in which a value of 100% is related to the concept of entropy. Exchanger efficiency is still an unresolved and somewhat controversial issue.

9.7 SUMMARY

In this chapter, we have examined some of the more common types of heat exchangers. Double-pipe heat exchangers were described and analyzed. It was shown that a counterflow configuration was the most effective of all in exchanging heat. The log-mean temperature difference for a counterflow configuration was subsequently used as a basis for comparison purposes. Shell-and-tube exchangers were described and analyzed. A method of analysis was developed for double-pipe and for shell-and-tube exchangers. A summary of the calculation steps is provided in this section.

Crossflow heat exchangers were also discussed in this chapter. An abbreviated method of analysis was developed, and it is outlined in this section.

The effectiveness–NTU method was presented. Effectiveness–NTU charts for all exchangers discussed were also provided in this chapter. Finally, a brief discussion of exchanger efficiency was included to show the controversial aspects of trying to define this term satisfactorily.

9.7.1 DOUBLE-PIPE HEAT EXCHANGERS SUGGESTED ORDER OF CALCULATIONS

1. Discuss the problem.
2. List assumptions.
3. Obtain fluid properties.
4. Obtain dimensions of the tubes: OD_p, ID_p, ID_a (see below).

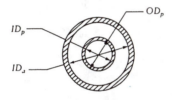

5. Perform the calculations for both the tube and the annulus (subscripts: p = pipe; a = annulus).

I. Flow Areas

$$A_p = \pi (ID_p)^2/4$$

$$A_a = \pi (ID_p^2 - OD_p^2)/4$$

II. Annulus Equivalent Diameters

Friction $\qquad D_h = ID_a - OD_p$

Heat transfer $\qquad D_e = (ID_a^2 - OD_p^2)/OD_p$

III. Reynolds Numbers

$$\mathrm{Re}_p = \frac{V_p ID_p}{v_p} = \frac{\dot{m}_p}{\rho_p A_p} \frac{ID_p}{v_p}$$

$$\mathrm{Re}_a = \frac{V_a D_e}{v_a} = \frac{\dot{m}_a}{\rho_a A_a} \frac{D_e}{v_a}$$

IV. Nusselt Numbers

Laminar flow, Re < 2 200

$$Nu_p = \frac{h_i ID_p}{k_p} = 1.86\left(\frac{ID_p Re_p Pr_p}{L}\right)^{1/3}$$

$$Nu_a = \frac{h_o D_e}{k_p} = 1.86\left(\frac{D_e Re_a Pr_a}{L}\right)^{1/3}$$

$$0.48 < Pr = \nu/\alpha < 16\ 700$$

μ moderately changes with temperature

Turbulent flow, Re > 10000

$$Nu_p = \frac{h_i ID_p}{k_p} = 0.023\ Re_p^{4/5}\ Pr_p^n$$

$$Nu_a = \frac{h_o ID_e}{k_a} = 0.023\ Re_a^{4/5}\ Pr_a^n$$

$$n = 0.4 \text{ if fluid is being heated}$$

$$n = 0.3 \text{ if fluid is being cooled}$$

$$0.7 < Pr = \nu/\alpha < 160$$

$$L/D > 60$$

V. Convection Coefficients

Pipe $\qquad h_i = \dfrac{Nu_p k_p}{ID_p};\ h_{io} = h_i \dfrac{ID_p}{OD_p}$

Annulus $\qquad h_o = \dfrac{Nu_a k_a}{D_e}$

VI. Exchanger Coefficient

$$\frac{1}{U_o} = \frac{1}{h_{io}} + \frac{1}{h_o} \quad \text{(neglecting resistance of inner pipe wall)}$$

Once the above calculations are made, the problem can be completed, depending on what information is sought (i.e., outlet temperatures or a heat balance, etc.)

Outlet temperatures (subscripts: c = cooler fluid or fluid being heated; h = warmer fluid or fluid being cooled)

$$R = \frac{\dot{m}_c c_{pc}}{\dot{m}_h c_{ph}};\ A_o = \pi(OD_p)L$$

$$T_2 = \frac{T_1(R-1) - \mathrm{Rt}_1\{1 - \exp[U_oA_o(R-1)/\dot{m}_cc_{pc}]\}}{R\exp[U_oA_o(R-1)/(\dot{m}_cc_{pc})] - 1} \qquad \text{(counterflow)}$$

$$t_2 = t_1 + \frac{T_1 - T_2}{R}$$

Heat balance

$$\mathrm{LMTD} = \frac{(T_1 - t_2) - (T_2 - t_1)}{\ln[(T_1 - t_2)/(T_2 - t_1)]} \qquad \text{(counterflow)}$$

$$q = \dot{m}_cc_{pc}(t_2 - t_1) = \dot{m}_hc_{ph}(T_1 - T_2)$$

$$q = U_oA_o\mathrm{LMTD} \qquad \text{(clean)}$$

Area required to transfer heat

$$\frac{1}{U} = \frac{1}{U_o} + R_{di} + R_{do} \qquad (R_d \text{ fouling factors from Table 9.1)}$$

$$A_o = \frac{q}{U(\mathrm{LMTD})}$$

$$L = \frac{A_o}{\pi(OD_p)}$$

VII. Friction Factors

$$\left. \begin{array}{l} \mathrm{Re}_p = \dfrac{V_pID_p}{v_p} = \dfrac{\dot{m}_p}{\rho_pA_p}\dfrac{ID_p}{v_p} \\[3em] \dfrac{\varepsilon}{ID_p} \text{ or smooth-wall tubing} \end{array} \right\} \quad \begin{array}{l} f_p \text{ from Figure 6.14 or} \\ \text{curve fit equation} \end{array}$$

$$\left. \begin{array}{l} \mathrm{Re}_a = \dfrac{V_aD_h}{v_a} = \dfrac{\dot{m}_a}{\rho_aA_a}\dfrac{D_h}{v_a} \\[3em] \dfrac{\varepsilon}{D_h} \text{ or smooth-wall tubing} \end{array} \right\} \quad \begin{array}{l} f_a \text{ from Figure 6.18 or} \\ \text{curve fit equation} \end{array}$$

Laminar flow, Re < 2 200

$$f_p = \frac{64}{\mathrm{Re}_p}$$

$$\frac{1}{f_a} = \frac{\mathrm{Re}_a}{64}\left[\frac{1+\kappa^2}{1-\kappa} + \frac{1+\kappa}{\ln\kappa}\right] \quad \text{where } \kappa = \frac{OD_p}{ID_a}$$

VIII. Velocities

$$V_p = \frac{\dot{m}_p}{\rho_p A_p} \qquad V_a = \frac{\dot{m}_a}{\rho_a A_a}$$

IX. Pressure Drops

$$\Delta p_p = \frac{f_p L \rho_p V_p^2}{ID_p\, 2g_c}$$

$$\Delta p_a = \frac{f_a L \rho_a V_a^2}{D_h\, 2g_c}$$

X. Summary of Requested Information

See text for effectiveness–NTU method of analysis.

9.7.2 SHELL-AND-TUBE HEAT EXCHANGERS SUGGESTED ORDER OF CALCULATIONS

1. Discuss the problem.
2. List assumptions.
3. Obtain fluid properties.
4. Obtain tube and shell dimensions:

Tube dimensions

OD_t, ID_t from Table 9.2
N_t = number of tubes or tube count from Table 9.3
N_p = number of passes the tube fluid makes

Shell data

D_s = shell inside diameter
B = baffle spacing; N_b = number of baffles
S_T = tube pitch
$C = S_T - OD_t$ = clearance between adjacent tubes

5. Perform the calculations for the tube and the shell fluids (subscripts: t = tube; s = shell).

I. Flow Areas

$$A_t = N_t \pi (ID_t^2)/4N_p$$

$$A_s = D_s CB/S_T$$

II. Shell Equivalent Diameter

$$D_e = \frac{4[S_T^2 - \pi(OD_t)^2/4]}{\pi(OD_t)} = \frac{4S_T^2 - \pi(OD_t)^2}{\pi(OD_t)} \quad \text{(square pitch)}$$

$$D_e = \frac{4[0.86S_T^2/2 - \pi(OD_t)^2/8]}{\pi(OD_t)/2} = \frac{(3.44S_T^2 - \pi(OD_t)^2)}{\pi(OD_t)} \quad \text{(triangular pitch)}$$

III. Reynolds Numbers

$$\text{Re}_t = \frac{V_t ID_t}{v_t} = \frac{\dot{m}_t}{\rho_t A_t} \frac{ID_t}{v_t}$$

$$\text{Re}_s = \frac{V_s D_e}{v_s} = \frac{\dot{m}_s}{\rho_s A_s} \frac{D_e}{v_s}$$

IV. Nusselt Numbers

Tube side:

Laminar flow, $\text{Re}_t < 2\,200$

$$\text{Nu}_t = \frac{h_i ID_t}{k_t} = 1.86 \left(\frac{ID_t \text{Re}_t \text{Pr}_t}{L} \right)^{1/3}$$

$$0.48 < \text{Pr} = v/\alpha < 16\,700$$

μ moderately changes with temperature

Turbulent flow, $\text{Re}_t > 10\,000$

$$\text{Nu}_t = \frac{h_i ID_t}{k_t} = 0.023\,\text{Re}_t^{4/5} \text{Pr}_t^n$$

$$n = 0.4 \text{ if fluid is being heated}$$

$$n = 0.3(0.4) \text{ if fluid is being cooled}$$

$$0.7 < \text{Pr} = v/\alpha < 160$$

$$L/D > 60$$

Shell side:

$$\mathrm{Nu}_s = \frac{h_o D_e}{k_s} = 0.36\,\mathrm{Re}_s^{0.55}\mathrm{Pr}_s^{1/3}$$

$$2 \times 10^3 < \mathrm{Re}_s = V_s D_e / \nu_s < 1 \times 10^6$$

$$\mathrm{Pr}_s = \nu/\alpha > 0$$

μ changes moderately with temperature

V. Convection Coefficients

$$h_i = \frac{\mathrm{Nu}_t k_t}{ID_t};\ h_{io} = h_i \frac{ID_t}{OD_t}$$

$$h_o = \frac{\mathrm{Nu}_s k_s}{D_e}$$

VI. Exchanger Coefficient

$$\frac{1}{U_o} = \frac{1}{h_{io}} + \frac{1}{h_o}$$

Once the above calculations are made, the problem can be completed, depending on what information is sought (i.e., outlet temperatures or a heat balance, etc.).

Outlet temperatures (subscripts: c = cooler fluid or fluid being heated; h = warmer fluid or fluid being cooled)

$$R = \frac{\dot{m}_c c_{pc}}{\dot{m}_h c_{ph}};\ A_o = N_t \pi (OD_t) L$$

Calculate $\dfrac{U_o A_o}{\dot{m}_c c_{pc}}$; use Figure 9.12; read $S = \dfrac{t_2 - t_1}{T_1 - t_1}$

$$t_2 = S(T_1 - t_1) + t_1$$

$$T_2 = T_1 - R(t_2 - t_1)$$

Heat balance

$$\mathrm{LMTD} = \frac{(T_1 - t_2) - (T_2 - t_1)}{\ln[(T_1 - t_2)/(T_2 - t_1)]} \qquad\qquad \text{(counterflow)}$$

Read F_1 from Figure 9.12.

$$q = \dot{m}_c c_{pc}(t_2 - t_1) = \dot{m}_h c_{ph}(T_1 - T_2) = U_o A_o F_t\,\mathrm{LMTD}$$

Area required to transfer heat

R_{di} and R_{do} from Table 9.1

$$\frac{1}{U} = \frac{1}{U_o} + R_{di} + R_{do}$$

$$A_o = \frac{q}{U F_t \text{ LMTD}}$$

$$L = A_o/[N_t \pi (OD_t)]$$

VII. Friction Factors

$$\text{Re}_t = \frac{V_t ID_t}{v_t} = \frac{\dot{m}_t}{\rho_t A_t} \frac{ID_t}{v_t}$$

$$\frac{\varepsilon}{ID_t} \text{ or smooth-wall tubing}$$

f_t from Figure 6.14 or curve fit equation

$$\text{Re}_s = \frac{V_s D_e}{v_s} = \frac{\dot{m}_s}{\rho_s A_s} \frac{D_e}{v_s}$$

$$f_s = \exp(0.576 - 0.19 \ln \text{Re}_s)$$

VIII. Velocities

$$V_t = \frac{\dot{m}_t}{\rho_t A_t} \qquad V_t = \frac{\dot{m}_t}{\rho_s A_s}$$

IX. Pressure Drops

$$\Delta p_t = \frac{\rho_t V_t^2}{2 g_c} \left(\frac{f_t L N_p}{ID_t} + 4 N_p \right)$$

$$\Delta p_s = \frac{\rho_s V_s^2}{2 g_c} \frac{D_s}{D_e} f_s (N_b + 1)$$

X. Summary of Requested Information

See chapter for effectiveness–NTU method of analysis.

9.7.3 Crossflow Heat Exchangers Suggested Order of Calculations

1. Discuss the problem.
2. List assumptions.
3. Obtain fluid properties.

4. Obtain necessary physical dimensions of the exchanger.
5. Perform the calculations for each fluid (subscripts: c = cooler fluid or fluid being heated; h = warmer fluid or fluid being cooled):

I. Heat Balance

$$q = \dot{m}_h c_{ph}(T_1 - T_2) = \dot{m}_c c_{pc}(t_2 - t_1) = UAF_t \text{ LMTD}$$

II. Log-Mean Temperature Difference for Counterflow

$$\text{LMTD} = \frac{(T_1 - t_2) - (T_2 - t_1)}{\ln[(T_1 - t_2)/(T_2 - t_1)]}$$

III. Frontal Areas for Each Fluid Stream

$$A_h, A_c$$

IV. Correction Factor

$$S = \frac{t_2 - t_1}{T_1 - t_1} \left.\begin{array}{c} \\ \\ \\ \\ \end{array}\right\} \begin{array}{l} \text{Figure 9.21a} \\ \text{Figure 9.22a} \to F_t \\ \text{Figure 9.23} \end{array}$$

$$R = \frac{T_1 - T_2}{t_2 - t_1}$$

V. Overall Coefficient-Area Product

$$UA = \frac{q}{F_t \text{ LMTD}}$$

VI. Capacitance Values: $(\dot{m}c_p)_{\min}, (\dot{m}c_p)_{\max}$

$$\text{Capacitance ratio} = \frac{(\dot{m}c_p)_{\min}}{(\dot{m}c_p)_{\max}}$$

VII. Number of Transfer Units

$$\text{NTU} = \frac{UA}{(\dot{m}c_p)_{\min}}$$

VIII. Effectiveness

$$\frac{(\dot{m}c_p)_{\min}}{(\dot{m}c_p)_{\max}} \left.\begin{array}{c} \\ \\ \\ \end{array}\right\} \begin{array}{l} \text{Figure 9.21b} \\ \text{Figure 9.22b} \to \varepsilon \\ \text{Figure 9.22c} \end{array}$$

$$\text{NTU}$$

IX. Outlet Temperatures

For $\dot{m}_c c_{pc} = (\dot{m} c_p)_{min}$

$$t_2 = \varepsilon(T_1 - t_1) + t_1$$

$$T_2 = T_1 - \frac{\dot{m}_c c_{pc}}{\dot{m}_h c_{ph}}(t_2 - t_1)$$

For $\dot{m}_h c_{ph} = (\dot{m} c_{pc})_{min}$

$$T_2 = T_1 - \varepsilon(T_1 - t_1)$$

$$t_2 = \frac{\dot{m}_h c_{ph}}{\dot{m}_c c_{pc}}(T_1 - T_2) + t_1$$

X. Summary of Requested Information

9.8 PROBLEMS

9.8.1 DOUBLE-PIPE HEAT EXCHANGERS

1. In calculating the overall heat-transfer coefficient, the resistance of the tube-wall material is considered negligible (see Equations 9.2 and 9.3). Determine whether this is actually the case for the following data:

$$h_i = 300 \text{ BTU}(/\text{hr}\cdot\text{ft}^2\cdot{}^\circ\text{R})$$

$$h_o = 320 \text{ BTU}(/\text{hr}\cdot\text{ft}^2\cdot{}^\circ\text{R})$$

2 nominal schedule 40 pipe of 1C steel

2. Repeat Problem 1 for the following data:

$$h_i = 1\,000 \text{ W}/(\text{m}^2\cdot\text{K})$$

$$h_o = 1\,200 \text{ W}/(\text{m}^2\cdot\text{K})$$

2-1/2 standard type K copper tubing

3. Repeat Problem 1 for the following data:

$$h_i = 1500 \text{ BTU}/(\text{hr}\cdot\text{ft}^2\cdot{}^\circ\text{R})$$

$$h_o = 100 \text{ BTU}/(\text{hr}\cdot\text{ft}^2\cdot{}^\circ\text{R})$$

3 standard type K copper tubing

4. Repeat Problem 1 for the following data:

$$h_i = 5\ 000\ \text{W/(m}^2\cdot\text{K)}$$

$$h_o = 5\ 500\ \text{W/(m}^2\cdot\text{K)}$$

   1 nominal schedule 40 pipe made of type 304 stainless steel

5. A hot fluid at 160°F enters a double-pipe heat exchanger and is cooled to 80°F. A cool fluid at 50°F is warmed in the exchanger to a temperature of 70°F. Calculate the log-mean temperature difference for
   (a) Counterflow
   (b) Parallel flow
6. A cool fluid is heated in a double-pipe heat exchanger from 15° to 75°C while a warmer fluid is cooled from 100° to 90°C. Calculate the LMTD for
   (a) Counterflow
   (b) Parallel flow
7. Although not discussed in this chapter, we examine in this problem what effect a fluid that changes phase has on the log-mean temperature difference. Consider a fluid that enters a double-pipe heat exchanger at 110°C condenses, then leaves still at 110°C. A cooler fluid enters the exchanger at 50°C and is heated to 75°C. Find the LMTD for
   (a) Counterflow
   (b) Parallel flow
8. Repeat Problem 7 for phase-change temperature of 40°F and a warmer fluid-temperature variation of 100°F at inlet to 50°F at outlet.
9. Calculate the LMTD for (a) counterflow and for (b) parallel flow for the following temperatures:
   Hot fluid inlet temperature = 200°F
   Hot fluid outlet temperature = 100°F
   Cold fluid inlet temperature = 50°F
   Cold fluid outlet temperature = 150°F
10. In Example 9.3, ethylene glycol is placed in the annulus and water in the pipe. Repeat the calculations, but with water in the annulus and ethylene glycol in the pipe.
11. Cutting oil is used in a machining operation. The oil acts as a lubricant and as a coolant, so the oil temperature gets quite high. It is proposed to cool the oil in a double-pipe heat exchanger from an inlet temperature of 60°C with a mass flow rate of 2 300 kg/hr. Water is used as the cooling medium and is available at a temperature of 20°C with a flow rate of 2 500 kg/hr. An available exchanger is made of 2 × 1-1/4 type M copper tubing, 2 m long.
    (a) Determine the overall coefficient of the exchanger for this application.
    (b) What is the outlet temperature of the oil?
    Assume properties of the oil to be the same as those of unused engine oil, and use counterflow.
12. Ammonia is used in an industrial process. The ammonia in liquid form is available at 14°F but is needed at a temperature of 32°F. The low inlet temperature may lead to trouble if water is used as a heating medium, so liquid carbon dioxide is used instead. The liquid carbon dioxide is available at 50°F and a mass flow rate of 10,000 lbm/hr. A number of double-pipe heat exchangers made of 2-1/2 × 1-1/4 type K copper tubing 20 ft long are available for this application. For an ammonia flow rate of 10,000 lbm/hr, determine the outlet temperature of the carbon dioxide fluid stream and the pressure

drops. Evaluate fluid properties at the average of the inlet temperatures, and use counterflow with ammonia in the inner tube. Take the fouling factors to be 0.0005 ft²·hr·°R/BTU for each stream.

13. Reverse the direction of either fluid stream in Example 9.3 so that parallel flow exists and repeat the calculations. Calculate the LMTD for each case.

14. Calculate LMTD for the fluids of Example 9.3 and check the heat-transfer-rate calculation by using

$$q = U_o A_o \text{LMTD}$$

Calculate also U for the exchanger when it is fouled and the corresponding outlet temperatures. Take the fouling factor for ethylene glycol to be the same as that for water.

15. In an oxygen-producing operation, air is cooled to a very low temperature until each component liquefies and is then separated from the mixture. Once oxygen is obtained in this fashion, it is then allowed to vaporize. While in the vapor state, the oxygen must be heated to a standard temperature so that it can be accurately metered for bottling purposes. The bottling machine automatically fills a number of vessels simultaneously and requires 80 lbm/hr of oxygen at 90°F. Air is used as the heating medium. The air is available at 95°F with a mass flow rate of 120 lbm/hr, while the oxygen has an inlet temperature of 70°F. Several 3 × 2 schedule 40 double-pipe exchangers 8 ft long, made of galvanized steel, are available for this service.
    (a) Determine how many are required.
    (b) Determine the overall coefficient of all exchangers taken together.
    (c) Determine the pressure drop for each stream. Assume a counterflow arrangement.

16. Tests on a new air-cooling unit are to be conducted. A 4 × 3 heat exchanger made of type M copper tubing 3 m long is used to cool air that is available at 320 K. The mass flow of air required is 110 kg/hr. Methyl chloride (a liquid refrigerant) is the cooling medium, and it is available at –10°C with a flow rate of 100 kg/hr.
    (a) Determine the outlet temperature of each fluid.
    (b) Determine the overall heat-transfer coefficient.
    (c) Determine the pressure drop for each stream.
    (d) Determine the overall coefficient after 1 year has elapsed.
    (e) Determine the outlet temperatures after 1 year has elapsed.
    Evaluate air properties at 300 K and the refrigerant properties at 10°C.

17. Figure P9.1 depicts what is called a *hairpin* exchanger, in which two double-pipe exchangers are connected at one end with U-shaped connectors. The inner tube is very long and is bent into a U-shape. The shell or outer tube is bolted to a connector at the end and completely encloses the inner tube. It is desired to use one or more hairpin exchangers for cooling liquid butane (mass flow rate is 9800 lbm/hr, specific heat is 0.58 BTU/(lbm·°R), absolute viscosity is 0.14 centipoise, thermal conductivity is 0.075 BTU/(hr·ft·°R), and sp. gr. = 0.56) from an inlet temperature of 155° to 100°F. Water is the cooling medium, and it is available at 85°F with a mass flow rate of 24,000 lbm/hr.
    (a) Determine the surface area required to effect this heat transfer.
    (b) A number of 4 × 3/4 schedule 40 hairpin exchangers are available in 20-ft lengths. How many are required for this application?
    (c) Calculate the overall heat-transfer coefficient for all exchangers.
    (d) Determine the pressure drop for each stream.
    According to the manufacturer, the U-bends supply an additional 2.73 ft² to the heat-transfer area. Use counterflow.

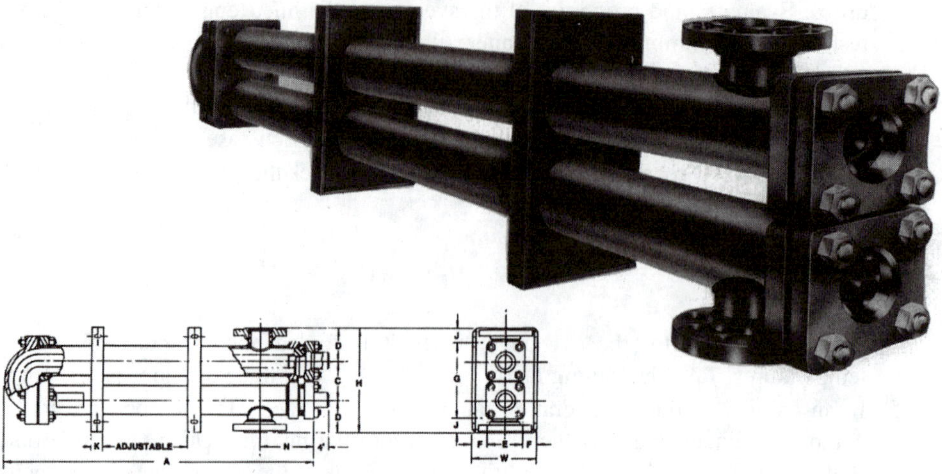

DOUBLE PIPE (any metal)

PHYSICAL DIMENSIONS

SHELL SIZE	OVERALL LENGTH NOM+A	℄ NOZZ N	WIDTH W	HEIGHT H	CENTER C	NOZZLE D	MOVABLE BRACKET DIMENSION E	F	G	J	K	SHELL NOZZLE SIZE	TUBE NOZZLE CONN.
1½″+2″+2½″	NOM + 8″	7½″	8″	13½″	4½″	4½″	5″	1½″	9½″	2″	2¾″	1½″-300#	1½″ W.E.
3″	NOM + 8¾″	7¾″	8″	13½″	4½″	4½″	5″	1½″	9½″	2″	2¾″	2″-300#	2″ W.E.
3½″ + 4″	NOM + 10″	9″	10″	16″	6″	5″	7″	1½″	10″	3″	2¾″	3″-300#	3″ W.E.
5″ + 6″	NOM + 12¾″	9¾″	12″	24″	8″	8″	8″	2″	18″	3″	2⅞″	4″-150#	4″ W.E.
8″	NOM + 15¾″	10¾″	15″	30″	12″	9″	9″	3″	24″	3″	2⅞″	4″-150#	4″-W.E.
10″	NOM + 17″	12″	18″	35″	15″	10″	12″	3″	27″	4″	2⅞″	AS REQ.	AS REQ.
12″	NOM + 19″	14″	20″	40″	18″	11″	14″	3″	32″	4″	2⅞″	AS REQ.	AS REQ.
16″	NOM + 21″	16″	24″	48″	24″	12″	19″	3″	38″	5″	2⅞″	AS REQ.	AS REQ.

BARE TUBE

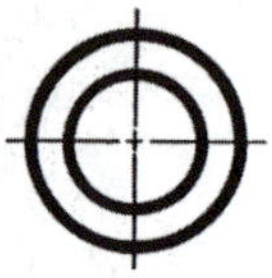

FIGURE P9.1 A double-pin hairpin heat exchanger. *Source:* courtesy of R.W. Holland, Inc.

18. Use the effectiveness–NTU method to calculate the outlet temperatures in Example 9.3.
19. Use the effectiveness–NTU method to calculate outlet temperatures for the fluids in Example 9.4 when the exchanger is fouled.
20. Water at a flow rate of 9,000 lbm/hr is to be heated from 40° to 95°F in a double-pipe counterflow heat exchanger by (engine) oil having an inlet temperature of 150°F with a flow rate of 12,000 lbm/hr. What area is required to do this? Take the overall heat-transfer coefficient to be 600 BTU/(hr·ft²·°R).

9.8.2 Shell-and-Tube Heat Exchangers

21. In Example 9.5, raw water is routed through the tubes, and steam through the shell. Exchange the fluid paths and rework the problem. Compose a tabular summary of the calculations.
22. In a distillation column, crude oil is heated and separated into various components. Kerosene [c_p = 2 530 J/(kg·K), sp. gr. = 0.74, μ = 0.4 cp, and k_f = 0.132 W/(m·K)] at a temperature of 200°C and a flow rate of 30 000 kg/hr is cooled by crude oil [c_p = 2 050 J/(kg·K), sp. gr. = 0.8, μ = 3.6 cp, k_f = 0.133 W/(m·K)] at a temperature of 38°C and a mass flow rate of 60 000 kg/hr. The available exchanger is a shell-and-tube type. It has a 23-1/4-in-ID shell and contains 1-in-OD tubes, 13 BWG, laid out on a 1-1/4-in triangular pitch. The tubes are 16 ft long, and the shell contains baffles spaced 2 ft apart. Determine the outlet temperature of each fluid and the corresponding pressure drops.
23. Is the correction factor F_t for Problem 22 high enough to warrant the use of the exchanger?
24. Example 9.5 deals with a steam–raw-water problem in which the two fluids exchange heat in a 1–2 shell-and-tube exchanger. The steam is routed through the shell and makes

one pass through the exchanger. The raw water is routed through the tubes and makes two passes through the exchanger. Repeat the calculations for outlet temperatures and pressure drops if the tube fluid makes
(a) four passes
(b) six passes
(c) eight passes
Compare the results.

25. Sulfur dioxide is used in the manufacture of sulfuric acid. During one phase of the process 35,000 lbm/hr of sulfur dioxide is to be heated from a temperature of 50° to 90°F (90°F is desired, but the actual outlet temperature must be calculated). Water is the heating medium and is available at a temperature of 95°F with a flow rate of 32,000 lbm/hr. A 12-in-ID 1–2 shell-and-tube exchanger containing baffles spaced every 2 ft is available for this service. The exchanger contains 1-in-OD, 13-BWG tubes 14 ft long and laid out on a 1-1/4-in square pitch.
(a) Determine the outlet temperatures and whether the exchanger will deliver sulfur dioxide at 90°F.
(b) Find the overall exchanger coefficient when clean.
(c) Find the overall exchanger coefficient when fouled.
(d) Calculate F_t for each stream.
(e) Calculate the pressure drop for each stream.
(f) Take the fouling factor for sulfur dioxide to be the same as that for a refrigerant liquid and determine the outlet temperatures when the exchanger becomes fouled.

26. Liquid carbon dioxide is to be heated from 0° to 20°C in a 1–2 shell-and-tube heat exchanger. Saturated liquid ammonia is available at a flow rate of 112 500 kg/hr at a temperature of 50°C. A 25-in-ID exchanger having 3/4-in, 10-BWG tubes laid out on a 1-1/4-in triangular pitch is available (N_t = 282, S_T = 3.175 cm). The tubes are 12 ft long, and the exchanger contains baffles spaced 2 ft apart.
(a) Determine the required flow rate of the carbon dioxide when clean.
(b) Determine the ammonia outlet temperature when clean.
(c) Find the overall heat-transfer coefficient for the exchanger when clean.
(d) Find the overall heat-transfer coefficient for the exchanger when fouled. (Take both fluids to have fouling coefficients equal to those for refrigerant liquids.)
(e) Calculate also the pressure drops for each stream.
(f) Find the fluid outlet temperatures for the fluids when the exchanger is fouled. Place the ammonia in the tubes.
Hint: Work through the suggested calculation procedure up to the point where the overall heat-transfer coefficient is expressed in terms of the mass flow rate of the carbon dioxide. Then work backwards through the remaining steps as per the following:
    (1) Assume a mass flow rate.
    (2) Calculate outlet temperatures: S, R, LMTD.
    (3) Determine F_t and U_o.
    (4) Use the heat balance to find U_o and compare results.

27. A 31-in-ID 1–2 shell-and-tube exchanger contains 3/4-in-OD, 12-BWG tubes laid out on a 1-in triangular pitch. The exchanger has baffles spaced 2 ft apart and is used to heat ethylene glycol (inlet temperature of 70°F and a mass flow rate of 300,000 lbm/hr). The heating medium is water, which is available at 170°F with a mass flow rate of 180,000 lbm/hr. Calculate the expected outlet temperatures for tube lengths of
(a) 4 ft
(b) 6 ft
(c) 8 ft
(d) 10 ft

(e) 12 ft

Evaluate properties of both fluids at 140°F for all cases.

28. A 1–2 shell-and-tube heat exchanger is used to heat 62 000 kg/hr of water from 20°C to about 60°C (60°C is desired, but actual outlet temperature most be calculated). Hot water at 100°C is available at a flow rate that can vary from 10 000 to 50 000 kg/hr. A 21-1/4-in-ID exchanger containing 3/4-in-OD, 12-BWG tubes 3 m long and laid out on a 1-in triangular pitch is available for this service. That exchanger contains baffles that are spaced 50 cm apart. Determine how the heat-transfer rate and the water outlet temperatures vary with the hot water mass flow rate.

29. A 1–2 shell-and-tube exchanger has a shell ID of 13-1/4 in and contains 3/4-in-OD, 13-BWG tubes 10 ft long and laid out on a 1-in square pitch. It contains baffles spaced 1 ft apart. At most, the exchanger can have 90 tubes, but tubes can be removed as necessary to change the pressure drop or the outlet temperature. This effect is investigated in this problem. A sugar solution [c_p = 0.86 BTU/(lbm·°R), sp. gr. = 1.08, μ = 1.3 cp, k_f = 0.34 BTU/(hr·ft·°R)] with a flow rate of 40,000 lbm/hr is to be heated from an inlet temperature of 75°F by water available at 120°F with a mass flow rate of 30,000 lbm/hr. Place the fluid with the higher mass flow rate into the passage with the greater area. Determine the outlet temperature of each fluid if the exchanger contains

    (a) 90 tubes
    (b) 80 tubes
    (c) 70 tubes
    (d) 60 tubes
    (e) 50 tubes

30. Determine the pressure drop for each stream of Problem 29 for the five cases.

9.8.3 CROSSFLOW HEAT EXCHANGERS

31. The radiator (crossflow heat exchanger with both fluids unmixed) of an engine is used to cool water. Water enters the exchanger at a temperature of 230°F with a flow rate that depends on engine revolutions per minute *when the flow passageway is unobstructed.* However, the engine thermostat is inserted into the flow passage, and the thermostat keeps the outlet temperature at 180°F. Air enters the exchanger at a temperature of 78°F, leaves at 130°F, and has a flow rate that depends on the speed of the car. The frontal dimensions of the radiator on the air side are 18 × 18 in. Determine the *UA* product for the radiator for a velocity of 30 mph. Evaluate air properties at 540°R.

32. A crossflow heat exchanger consisting of plates and fins is shown in Figure P9.2. It is designed for and tested with air as both the heating and cooling medium. When the exchanger is used as a device whose function is to provide *cooled* air, test data are as follows:

Air being cooled	Item	Air providing the cooling effect
155	Inlet temp, °F	95
110	Outlet temp, °F	—
13.0	Flow rate, lbm/min	23.9
11.5 × 8.5 in	Frontal areas	4.38 × 8.5 in

Evaluate all properties at 540°R.

(a) Determine the *UA* product for the exchanger.

(b) Calculate the exit temperatures of the fluids. Both fluids are unmixed.

FIGURE P9.2   A crossflow heat exchanger for cooling or heating a compartment.

33. Problem 32 deals with the exchanger of Figure P9.2 when the exchanger is used as a cooling device. That particular exchanger can also be used to provide heated air in an air-to-air configuration. Test data are as follows:

Air being heated	Item	Air providing the heating effect
313 K	Inlet temperature	366 K
357 K	Outlet temperature	—
5.90	Flow rate, kg/min	20.9
11.5 × 8.5 in	Frontal areas	4.38 × 8.5 in

(a) Determine the *UA* product for the exchanger.
(b) Calculate the exit temperatures of the fluids.
Both fluids are unmixed. Evaluate properties at 350 K.

34. Figure P9.3 shows a crossflow heat exchanger marketed as a mobile oil cooler. Data provided by the manufacturer for one particular model is as follows:

  Oil (assume unused engine oil) flow rate is 10 gpm
  Air inlet velocity is 1000 ft/min
  Air inlet area is 15 x 16 in

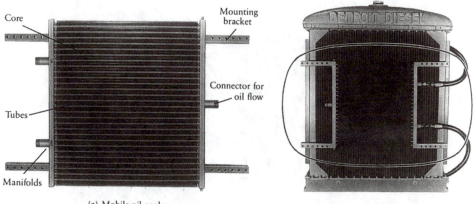

(a) Mobile oil cooler

(b) Suggested mounting

FIGURE P9.3   *Source:* courtesy of Young Radiator Co.

    Air inlet temperature is 70°F

    Oil inlet temperature is 170°F

    Measured heat-transfer rate is 380 BTU/min

(a) Determine the *UA* product for this exchanger.

(b) Calculate the fluid outlet temperatures.

Evaluate oil properties at 140°F and air properties at 540°R.

35. Figure P9.4 shows a 6.5 × 6.5-in crossflow heat exchanger used as an oil cooler. Data obtained by manufacturer's tests on the device are as follows:

Oil (assume unused engine oil)	Item	Air
—	Inlet temperature	109°F
230°F	Outlet temperature	200°F
36 lbm/min	Mass flow rate	32 lbm/min

(a) Determine the unknown temperature in the above data summary.

(b) Calculate the *UA* product for the exchanger. Evaluate oil properties at 212°F and air properties at 630°R.

36. A crossflow heat exchanger is used to cool water whose inlet temperature is 80°C and whose flow rate is 2 kg/s. The cooling medium is air. The air inlet temperature is 20°C, the air outlet temperature is 60°C, and the air has a flow rate of 10 kg/s. Which type of exchanger (mixed–mixed, mixed–unmixed, or unmixed–unmixed) has the greatest effectiveness? Evaluate air properties at 540°R and water properties at 100°C.

9.8.4 MISCELLANEOUS PROBLEMS AND EQUATIONS

37. Beginning with Equation 9.5, derive Equation 9.12a in detail.

38. Beginning with Equation 9.5, derive Equation 9.12c in detail.

39. Start with Equation 9.11a and verify that Equation 9.16 is correct.

40. Begin with the appropriate equation, and derive Equation 9.17.

41. Show that NTU is dimensionless.

42. Beginning with Equation 9.44, derive Equation 9.48 in detail.

43. Show that, for parallel flow in a double-pipe heat exchanger, the effectiveness is given by

$$\varepsilon = \frac{1 - \exp\{-\mathrm{NTU}[1 + (\dot{m}c_p)_{min}/(\dot{m}c_p)_{max}]\}}{1 + (\dot{m}c_p)_{min}/(\dot{m}c_p)_{max}}$$

FIGURE P9.4 An oil cooler.

44. It is often necessary to decide what type of exchanger to use for a given service. Consider an application where glycerin with a flow rate of 3 500 kg/hr is to be heated from 20° to 40°C by water whose flow rate and inlet temperature are 2 000 kg/hr and 50°C. Calculate the *UA* product required of a heat exchanger if it is
    (a) A double-pipe counterflow
    (b) A 1–2 shell-and-tube
    (c) A crossflow with both fluids unmixed

45. Repeat Problem 44 for the following data:

Fluid	Inlet temperature	Outlet temperature	Flow rate
Oxygen	720°R	—	7000 ft³/min at 720°R
Nitrogen	450°R	630°F	5000 ft³/min at 450°R

10 Condensation and Vaporization Heat Transfer

CONTENTS

10.1 INTRODUCTION

Condensation and boiling are convection processes associated with a phase change of a fluid. These processes occur at solid-fluid interfaces, and latent heat-transfer effects are involved. In either condensation or boiling, the magnitude of the heat transferred at the solid surface is sufficient to sustain the phase change.

It was remarked that condensation and boiling are convection processes. They are classified as such because there is a fluid motion near a solid surface. The change in phase, however, is unique and occurs without a change in fluid temperature. Heat-transfer rates (and coefficients) are much higher than those encountered in other convection problems.

Examples of systems in which condensation and vaporization both occur are conventional home air conditioners and refrigerators. Superheated vapor leaves the compressor and is condensed partially or entirely in a crossflow heat exchanger, usually the finned-tube type. The condensed refrigerant is then routed through a throttle to another finned-tube crossflow heat exchanger, where the fluid vaporizes. From there, the refrigerant vapor goes back to the compressor. The advantage of using condensation and vaporization in such a system is that the fluid encounters small temperature changes as it is circulated, but it still transfers a large quantity of heat. A knowledge of condensation and boiling heat-transfer processes is essential for the proper design of such systems.

10.2 CONDENSATION HEAT TRANSFER

Consider a solid surface at some temperature T_w in contact with a vapor at an elevated temperature T_∞. If the surface temperature T_w is less than the saturation temperature of the vapor, then the vapor will begin to condense on the surface. That is, the vapor will undergo a phase change, forming a liquid that will flow downward along the surface under the action of gravity. If the liquid does not wet the surface, the condensate takes the form of droplets that fall downward in a random fashion (just like water droplets along the side of a waxed automobile surface). This is called *dropwise condensation*; the droplets do not exhibit an affinity for the surface and do not coat it. Steam is the only pure vapor known to condense in a dropwise manner, and special conditions must be met for this to occur.

Another way in which a vapor condenses on a surface is by forming a smooth film. The liquid wets the surface, which becomes completely covered with liquid. During the process, known as *filmwise condensation*, the liquid film falls under the action of gravity and grows in thickness with increasing distance. A temperature gradient exists across the film from the surface temperature to the saturation temperature at the free surface. The film thus represents a thermal resistance to heat transfer.

In dropwise condensation, droplets form and fall away, leaving a large portion of the surface area bare and able to transfer heat. There is very little thermal resistance in dropwise condensation compared to filmwise condensation. In fact, dropwise condensation gives rise to heat-transfer rates that are 10 times those encountered in filmwise condensation.

Dropwise condensation is preferred because of the higher heat-transfer rate. However, it is difficult to maintain because surfaces become wetted after prolonged exposure to a condensing vapor. Various surface coatings and vapor additives (which in effect contaminate the fluid system) have been tried to maintain dropwise condensation, but such methods have not always been successful. Figure 10.1. shows tubes covered with Teflon® (a fluorocarbon resin that is a product of the DuPont Co.) in a condensing type of heat exchanger. The Teflon coating promotes dropwise rather than filmwise condensation, with the result that there is improved heat transfer during the process.

10.2.1 Laminar Film Condensation on a Vertical Flat Surface

Figure 10.2 is a sketch of a vertical flat surface of width W in contact with a vapor that is condensing in a filmwise fashion. Our objective is to formulate a description for this process. The film is two dimensional and has a thickness δ at any z location. Shown also are the x, y, and z axes and an element within the film in contact with the vapor. The element has a volume of $(\delta - y)W \, dz$. The forces acting on the element are due to gravity (acting downward in the positive z direction), buoyancy (in the negative z direction), and friction due to viscosity (negative z direction).

A force balance on a nonaccelerating element gives

$$\text{Gravity force} = \text{Buoyancy} + \text{Friction}$$

or

FIGURE 10.1 Dropwise condensation on Teflon-covered tubes. *Source:* courtesy of the Condensing Heat Exchanger Corp.

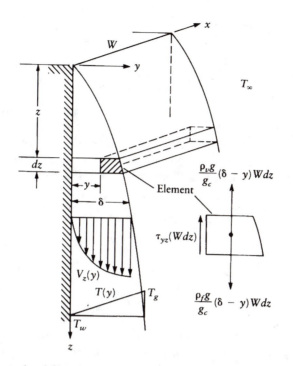

FIGURE 10.2 Two-dimensional filmwise condensation on a vertical flat surface.

$$\frac{\rho_f g}{g_c}(\delta - y)W \; dz = \frac{\rho_v g}{g_c}(\delta - y)W \; dz + \tau_{yz}(Wdz) \tag{10.1}$$

where ρ_f is the liquid density and ρ_v is the vapor density evaluated at the ambient vapor temperature T_∞. The saturation temperature is T_g, and for condensation to occur, it is less than the ambient temperature T_∞. We have not used ρ_g as the vapor density because in thermodynamics the g subscript refers to a saturated vapor condition. For this problem, the vapor is superheated at T_∞, while at the liquid-vapor interface the fluid is saturated. Assuming laminar flow of the film, we write

$$\tau_{yz} = \mu\frac{dV_z}{dy}$$

Substituting into Equation 10.1, simplifying, and rearranging, we get

$$\frac{dV_z}{dy} = \frac{g}{\mu g_c}(\delta - y)(\rho_f - \rho_v)$$

Integrating gives for the z-directed velocity

$$V_z = \frac{g}{\mu g_c}(\rho_f - \rho_v)(\delta y - y^2/2) + C_1$$

Applying the boundary condition of zero velocity at the wall ($y = 0$, $V_z = 0$), we obtain for the constant of integration

$$C_1 = 0$$

The velocity profile now becomes

$$V_z = \frac{\delta^2 g}{v_f}(1 - \rho_v/\rho_f)(y/\delta - y^2/2\delta^2) \tag{10.2}$$

This profile is graphed as $V_z/[\delta^2 g(1 - \rho_v/\rho_f)]$ vs. y/δ in Figure 10.3. In deriving this equation, we began with a force balance on an element of condensate. It is possible to obtain the velocity profile by starting with the Navier-Stokes Equations. This is reserved as an exercise.

Equation 10.2 is now integrated over the cross section (normal to the direction of the z-directed velocity V_z) to obtain the mass flow rate of condensate $\dot{m}_f$. Thus,

$$\dot{m}_f = \iint \rho_f V_z dA = \int_0^W \int_0^\delta \frac{\rho_f g}{v_f}\left(1 - \frac{\rho_v}{\rho_f}\right)\left(\delta y - \frac{y^2}{2}\right)dy \; dx$$

Integrating, the mass flow becomes

$$\dot{m}_f = \frac{W\delta^3 g \rho_f}{3 v_f}\left(1 - \frac{\rho_v}{\rho_f}\right) \tag{10.3}$$

To obtain an equation for the heat transferred, a form for the temperature profile must be assumed. The heat flow at the wall within the area $W \, dz$ is given by

$$q_y = -k_f(W \, dz)\frac{\partial T}{\partial y}\bigg|_{y=0}$$

We assume a linear temperature profile within the film that varies from T_w at the wall surface to T_g, the saturation temperature, at the liquid-vapor interface. The heat flow becomes

$$q_y = -k_f W \, dz \; \frac{T_w - T_g}{\delta} \tag{10.4}$$

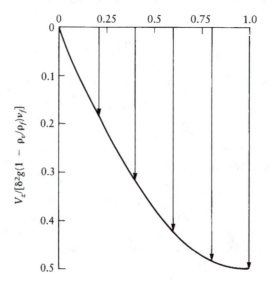

FIGURE 10.3 Dimensionless velocity profile for laminar flow of a condensate film.

The amount of condensate added between z and $z + dz$ is given by

$$\text{Added condensate in } dz = \dot{m}_f\big|_{z+dz} - \dot{m}_f\big|_z = \frac{d\dot{m}_f}{dz}dz = \frac{d\dot{m}_f}{d\delta}\frac{d\delta}{dz}dz = \frac{d\dot{m}_f}{d\delta}d\delta$$

Substituting from Equation 10.3, we get

$$\text{Added condensate in } dz = \frac{d\dot{m}_f}{d\delta}d\delta = \frac{W\rho_f\delta^2 g}{v_f}(1 - \rho_v/\rho_f)d\delta \tag{10.5}$$

With regard to our element, the energy added to the film by condensation within dz equals the incremental mass flow from condensation (given by Equation 10.5) times the specific enthalpy of condensation of the vapor, h_{fg}. It is this same added energy that must be removed at the wall. Therefore,

$$\text{Added condensate} \times h_{fg} = q_y$$

$$\frac{W\rho_f\delta^2 g}{v_f}(1 - \rho_v/\rho_f)h_{fg}d\delta = k_f W dz\frac{T_g - T_w}{\delta}$$

Simplifying and rearranging gives

$$\delta^3 d\delta = \frac{k_f v_f(T_g - T_w)}{\rho_f g h_{fg}(1 - \rho_v/\rho_f)}$$

Integrating, we obtain

$$\delta^4 = \frac{4k_f v_f z(T_g - T_w)}{\rho_f g h_{fg}(1 - \rho_v/\rho_f)} + C_1$$

The condensate film begins its growth at the top of the wall. The boundary condition is that at $z = 0$, $\delta = 0$. The integration constant therefore becomes $C_1 = 0$, and the film thickness then is

$$\delta = \left[\frac{4k_f v_f z(T_g - T_w)}{\rho_f g h_{fg}(1 - \rho_v/\rho_f)}\right]^{1/4} \tag{10.6}$$

At the wall, we essentially have a convection problem, so we can express the heat transferred in terms of a local convection coefficient for the element as

$$h_z W\, dz(T_w - T_g) = -k_f W\, dz\frac{T_g - T_w}{\delta}$$

or

$$h_z = k_f/\delta \tag{10.7a}$$

Substituting from Equation 10.6 gives

$$h_z = -k_f\left[\frac{\rho_f g h_{fg}(1 - \rho_v/\rho_f)}{4k_f v_f z(T_g - T_w)}\right]^{1/4} \tag{10.7b}$$

The local Nusselt number then is

$$\mathrm{Nu}_z = \frac{h_z z}{k_f} = \left[\frac{\rho_f g z^3 h_{fg}(1 - \rho_v/\rho_f)}{4 k_f v_f (T_g - T_w)}\right]^{1/4} \tag{10.8}$$

The average convection coefficient is obtained by integrating the local expression (Equation 10.7b) over the entire surface:

$$\bar{h}_L = \frac{1}{LW} \int_0^W \int_0^L h_z \, dz \, dx = \frac{1}{L} \int_0^L k_f \left[\frac{\rho_f g h_{fg}(1 - \rho_v/\rho_f)}{4 k_f v_f z (T_g - T_w)}\right]^{1/4} dz$$

Solving, we get

$$\bar{h}_L = \frac{4 k_f}{3} \left[\frac{\rho_f g h_{fg}(1 - \rho_v/\rho_f)}{4 k_f v_f L (T_g - T_w)}\right]^{1/4} \tag{10.9}$$

The Nusselt number based on the average convection coefficient is

$$\mathrm{Nu}_L = \frac{\bar{h}_L L}{k_f} = \frac{4}{3}\left[\frac{\rho_f g h_{fg}(1 - \rho_v/\rho_f) L^3}{4 k_f v_f (T_g - T_w)}\right]^{1/4} \tag{10.10}$$

The above equations apply only to laminar flow of the condensate film. If the plate is quite large, it is possible that turbulence effects will set in. Recall that turbulence exists when disturbances within the flow are no longer damped out by viscous forces but instead are propagated within the fluid. The mixing action that is characteristic of turbulent flow causes higher (than laminar) heat-transfer rates to exist. The criterion for determining whether the film is laminar or turbulent is the Reynolds number,

$$\mathrm{Re} = \frac{V D_h}{v_f} \tag{10.11a}$$

where V is the average film velocity and D_h is the hydraulic diameter (or characteristic length), defined as $4A/P$. For the vertical surface of Figure 10.2, the flow area is $A = \delta W$, and the wetted perimeter is $P = W$. Therefore, $D_h = 4\delta$, and the Reynolds number becomes

$$\mathrm{Re}_\delta = \frac{V(4\delta)}{v_f} = \frac{\dot{m}_f}{\rho_f A} \frac{4\delta}{v_f}$$

or

$$\mathrm{Re}_\delta = \frac{4 \dot{m}_f}{W \rho_f v_f} \tag{10.11b}$$

For laminar flow of the condensate film, the Reynolds number must be less than 1 800.

The Reynolds number can be written in terms of the convection coefficient. The heat-transfer rate is

$$q = \bar{h}_L A_s (T_g - T_w) \tag{10.12}$$

where A_s is the surface area of the plate in contact with the film. The heat transferred is also given by

$$q = \dot{m}_f h_{fg}$$

Equating these expressions and solving for mass flow gives

$$\dot{m}_f = \frac{\bar{h}_L A_s(T_g - T_w)}{h_{fg}} \tag{10.13}$$

Substituting into Equation 10.11b, we get

$$Re_L = \frac{4\bar{h}_L A_s(T_g - T_w)}{W h_{fg}\rho_f v_f} \tag{10.11c}$$

It was mentioned that, when the Reynolds number is less than 1 800, laminar conditions exist. Other phenomena occur at much lower Reynolds numbers, however. At values of Re_L as low as 30–40, ripples begin to appear in the film surface; these again are an indication of a mixing action. Experimental values of the convection coefficient can be as much as 20% higher than those predicted by Equation 10.9 for laminar flow.

Example 10.1

Steam at a pressure of 100 psi $(T_g = 327.81°F$, $h_{fg} = 888.8$ BTU/lbm from saturated steam tables) and a temperature of 400°F $(v_v = 4.937$ ft³/lbm from superheated steam tables) is in contact with a vertical plate that is 2 ft tall, 3 ft wide, and maintained at 325°F. Assuming that film-wise condensation occurs, graph film thickness and local convection coefficient as a function of height. Calculate the heat-transfer rate and the amount of steam condensed.

Solution

The plate-surface temperature is lower than the saturation temperature, so condensation will occur. The film will grow in thickness with increasing distance. We cannot calculate the Reynolds number at plate end to determine if the flow is laminar or turbulent because at this point the mass flow rate of condensate is unknown.

Assumptions

1. Filmwise condensation is occurring.
2. The flow is steady.
3. Properties are constant and are to be evaluated at $(328 + 325)/2 = 326.5°F \cong 320°F$.

From Appendix Table C.11, for water at 320°F

$\rho_f = 0.909(62.4)$ lbm/ft³ $k_f = 0.393$ BTU/(lbm·ft·°R)
$c_p = 1.037$ BTU/(lbm·°R) $\alpha = 6.70 \times 10^{-3}$ ft²/hr
$v = 0.204 \times 10^{-5}$ ft²/s $Pr = 1.099$

The vapor density is $\rho_v = 1/4.93 = 0.202$ lbm/ft³. The film thickness is given by Equation 10.6:

$$\delta = \left[\frac{4k_f v_f z(T_g - T_w)}{\rho_f g h_{fg}(1 - \rho_v/\rho_f)}\right]^{1/4} = \left\{\frac{4(0.393/3600)(0204 \times 10^{-5})(327.81 - 325)z}{0.909(62.4)(32.2)(888.8)1 - 0.202/[0.909(62.4)]}\right\}^{1/4}$$

Note that the ratio p_v/p_f is negligibly small. Solving,

$$\delta = 1.98 \times 10^{-4} z^{1/4} \qquad \text{(i)}$$

Calculations made with this equation are displayed in Table 10.1 and graphed in Figure 10.4. As shown in the calculations, the film thickness grows to less than 0.003 inches in 2 ft.

The local convection coefficient is given in Equation 10.7a as

$$h_z = k_f/\delta - 0.393/\delta \qquad \text{(ii)}$$

Calculations of the local convection coefficient are given in Table 10.1 and plotted in Figure 10.4. Equations 10.7b and 10.9 for the convection coefficients are

TABLE 10.1
Growth of and Heat-Transfer Coefficient for the
Condensate Film of Example 10.1

z, ft	δ (Eq. i), ft	(in)	h_z (Eq. ii), BTU/(hr·ft²·°R)
0	0	(0)	∞
0.2	1.32×10^{-4}	(0.0016)	2980
0.4	1.57	(0.0019)	2503
0.6	1.74	(0.0021)	2260
0.8	1.87	(0.0022)	2100
1.0	1.98	(0.0024)	1985
1.2	2.07	(0.0025)	1900
1.4	2.15	(0.0026)	1830
1.6	2.22	(0.0027)	1770
1.8	2.29	(0.0028)	1720
2.0	2.36	(0.0028)	1665

$$h_z = k_f \left[\frac{\rho_f g h_{fg}(1 - \rho_v/\rho_f)}{4 k_f v_f z (T_g - T_w)} \right]^{1/4}$$

$$\bar{h}_L = \frac{4 k_f}{3} \left[\frac{\rho_f g h_{fg}(1 - \rho_v/\rho_f)}{4 k_f v_f L (T_g - T_w)} \right]^{1/4}$$

Comparing these two equations, we conclude that

$$\bar{h}_L = \frac{4}{3} h_z \bigg|_{z=L}$$

Therefore, at plate end, where from Table 10.1 h_z = 1665 BTU/(hr·ft²·°R), we find

$$\bar{h}_L = \frac{4}{3}(1665) = 2220 \text{ BTU/(hr·ft}^2\cdot°R)$$

The amount of steam condensed is given by Equation 10.13:

$$\dot{m}_f = \frac{\bar{h}_L A_s (T_g - T_w)}{h_{fg}} = \frac{2220(2)(3)(327.81 - 325)}{888.8}$$

or

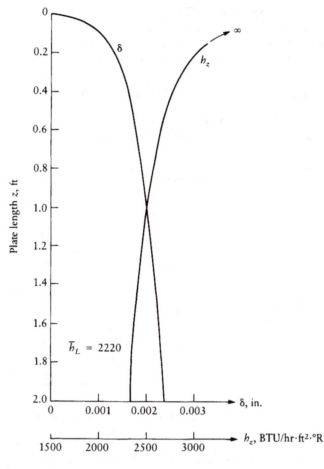

FIGURE 10.4 Condensate thickness and convection coefficient variation with distance for the data of Example 10.1.

$$\dot{m}_f = 42.1 \text{ lbm/hr}$$

The heat-transfer rate (= the heat transferred to the wall from condensation = the heat that must be removed from the wall to maintain it at 325°F) is

$$q = \dot{m}_f h_{fg} = 42.1(888.8)$$

$$q = 3.74 \times 10^4 \text{ BTU/hr}$$

As a check on the validity of the equations used, the Reynolds number at plate end must be calculated:

$$\text{Re}_\delta = \frac{4\dot{m}_f}{W\rho_f v_f} = \frac{4(42.1)}{3(0.909)(62.4)(0.204 \times 10^{-5})(3600)}$$

$$\text{Re}_\delta = 135 < 1800$$

Therefore, the film is laminar and the equations used do apply.   ❑

It is tempting to try to use the above correlations to analyze a problem such as steam condensing on a mirror when a shower is in use. This type of problem is known as condensation of one fluid (steam) in the presence of a noncondensable gas (air). It is a different problem entirely from what we have studied and involves derivations outside the scope of this text.

10.2.2 TURBULENT FILM CONDENSATION ON VERTICAL FLAT SURFACE

For turbulent flow of a condensing film on a vertical flat surface, we rely on experimental results to determine a solution. The convection coefficient is given by

$$\bar{h}_L = 0.007\,7 k_f \left[\frac{g(1-\rho_v/\rho_f)}{v_f^2}\right]^{1/3} \mathrm{Re}_\delta^{0.4} \tag{10.14}$$

where

$$\mathrm{Re}_\delta = \frac{4\dot{m}_f}{W\rho_f v_f} > 1\,800 \tag{10.11b}$$

$$q = \bar{h}_L A_s (T_g - T_w) \tag{10.12}$$

and

$$\dot{m}_f = q/\mathrm{h}_{fg} \tag{10.13}$$

10.2.3 LAMINAR FILM CONDENSATION ON AN INCLINED FLAT SURFACE

Figure 10.5 shows an inclined flat surface in contact with a vapor that is condensing in a filmwise fashion. Shown also are the y and z axes and expected velocity and temperature profiles. The plate is inclined at an angle θ with the horizontal. Performing a force balance on an element within the

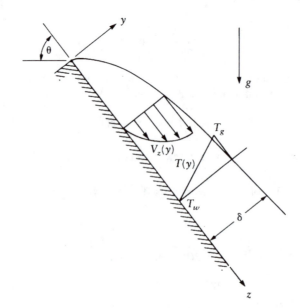

FIGURE 10.5   Laminar flow of a condensing liquid film on an inclined flat plate of width W (into the page).

film would yield a velocity profile similar to that obtained for a vertical plate, except the gravity term becomes $g \sin \theta$. Thus, for the inclined plate problem, we replace g in the vertical-plate solution with $g \sin \theta$, obtaining

$$V_z = \frac{\delta^2 g \sin\theta}{v_f}\left(1 - \frac{\rho_v}{\rho_f}\right)\left(\frac{y}{\delta} - \frac{y^2}{\delta^2}\right) \tag{10.15}$$

$$\dot{m}_f = \frac{W\delta^3 g \sin\theta\rho_f}{3v_f}\left(1 - \frac{\rho_v}{\rho_f}\right) = \frac{\bar{h}_L A_s(T_g - T_w)}{h_{fg}} \tag{10.16}$$

$$\delta = \frac{[4k_f v_f z(T_g - T_w)]}{\rho_f g \; \sin\theta h_{fg}(1 - \rho_v/\rho_f)} \tag{10.17}$$

$$h_z = k_f/\delta \tag{10.18}$$

$$\bar{h}_L = \frac{4}{3}h_z\Big|_{z=L} \tag{10.19}$$

$$q = \bar{h}_L A_s(T_g - T_w) \tag{10.4}$$

$$\text{Re}_\delta = \frac{4\dot{m}_f}{W\rho_f v_f} \tag{10.11b}$$

Properties at $(T_g + T_w)/2$

The correlations for an inclined plate should be used with caution at small angles and do not apply if $\theta = 0°$.

10.2.4 FILM CONDENSATION ON A VERTICAL TUBE

Vertical tubes are used in some condenser designs instead of horizontal tubes, especially when space is at a premium. The vertical-plate correlations can be used for a vertical tube if the film thickness is much less than the tube outside radius (i.e., curvature of the tube can be neglected). Except for Reynolds number, the equations are the same as those for a vertical flat plate:

$$\delta = \left[\frac{4k_f v_f z(T_g - T_w)}{\rho_f g h_{fg}(1 - \rho_v/\rho_f)}\right]^{1/4} \tag{10.6}$$

$$h_z = k_f/\delta \tag{10.7a}$$

$$\bar{h}_L = \frac{4}{3}h_z\Big|_{z=L}$$

$$q = \bar{h}_L A_s(T_g - T_w) \tag{10.4}$$

$$\dot{m}_f = q/h_{fg} \tag{10.16}$$

$$\text{Re}_\delta = \frac{4\dot{m}_f}{\pi D\rho_f v_f} \quad \text{(where D = outside diameter of tube)} \tag{10.20}$$

Liquid properties at $(T_g + T_w)/2$

10.2.5 FILM CONDENSATION ON A HORIZONTAL TUBE AND ON A HORIZONTAL TUBE BANK

Figure 10.6 is a sketch of a condensate film on a horizontal tube. The tube outside-surface temperature is T_w, and the vapor temperature is T_∞ ($> T_w$). If the surface temperature is less than the saturation temperature of the vapor, then condensation occurs. The film begins growing from the top of the tube and completely surrounds the tube. The film finally falls from the tube in the manner shown in Figure 10.6b. We can formulate equations for a description of the film just as was done for the vertical plate. The result for the average heat-transfer coefficient is

$$\bar{h}_D = 0.728 \left[\frac{g\rho_f(1 - \rho_v/\rho_f)k_f^3 h_{fg}}{v_f(T_g - T_w)OD_t} \right]^{1/4} \qquad \text{(one tube) (10.21)}$$

where OD_t is the outside diameter of the tube. If an array of tubes exists, such as within a shell-and-tube exchanger used as a condenser, then the liquid condensed from one tube will combine with that of a lower tube, as shown in Figure 10.7. Equation 10.21, if modified slightly, will apply to the system. The average convection coefficient for a bank of N_t tubes is

$$\bar{h}_D = 0.728 \left[\frac{g\rho_f(1 - \rho_v/\rho_f)k_f^3 h_{fg}}{v_f(T_g - T_w)N_t OD_t} \right]^{1/4} \qquad (10.22)$$

The equations for heat transfer and mass flow are

$$q = \bar{h}_D A_s (T_g - T_w) \qquad (10.4)$$

$$A_s = \pi(OD_t)LN_t$$

$$\dot{m}_f = q/h_{fg} \qquad (10.16)$$

Example 10.2

Steam at atmospheric pressure ($T_g = 100°C$, $h_{fg} = 2\ 257.06$ kJ/kg) and a temperature of 150°C ($v_v = 1.936\ 4$ m³/kg) is in contact with a horizontal tube through which a cooling fluid is circulated. The

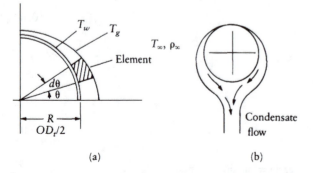

FIGURE 10.6 Film condensation on a horizontal tube: (a) film begins growing from the top of the tube, and (b) film flow pattern.

FIGURE 10.7 Flow pattern of film condensation on two in-line tubes.

tube is 1 nominal schedule 40, 1 m long, and has an outside-surface temperature that is maintained at 60°C. Determine both the heat that the cooling fluid must remove and the condensation rate.

Solution

The steam will condense in a filmwise fashion and fall from the tube.

Assumptions

1. The system is at steady state.
2. Properties are constant and evaluated at $(100 + 60)/2 = 80°C$.

From Appendix Table C.11, for water at 80°C,

$\rho_f = 974$ kg/m³ $k_f = 0.668$ W/(m·K)
$c_p = 4\ 196$ J/kg·K $\alpha = 1.636 \times 10^{-7}$ m²/s
$v = 0.374 \times 10^{-6}$ m/s   Pr = 2.22

Also, for the vapor, $\rho_v = 1/v_v = 0.516$ kg/m³. From Appendix Table F.1, for 1 nominal schedule 40 pipe, $OD_t = 3.340$ cm.

The average heat-transfer coefficient is given by Equation 10.21:

$$\bar{h}_D = 0.782\left[\frac{g\rho_f(1 - \rho_v/\rho_f)k_f^3 h_{fg}}{v_f(T_g - T_w)OD_t}\right]^{1/4}$$

Substituting and solving, we get

$$\bar{h}_D = 0.728\left[\frac{9.81(974)(1 - 0.516/947)(0.668)^3(2\ 257\ 060)}{0.364 \times 10^{-6}(100 - 60)(0.03340)}\right]^{1/4}$$

$$\bar{h}_D = 1.072 \times 10^4 \text{ W/(m}^2\text{·K)}$$

The heat-flow rate is found with Equation 10.4

$$q = \bar{h}_D(\pi)(OD_t)(L)(T_g - T_w)$$

$$= 1.072 \times 10^4(\pi)(0.033\ 40)(1)(100 - 60)$$

$$q = 4.5 \times 10^4 \text{ W}$$

The rate at which the steam condenses is calculated by using Equation 10.16:

$$\dot{m}_f = q/h_{fg} = \frac{4.5 \times 10^4}{2\ 257\ 060}$$

$$\dot{m}_f = 0.02 \text{ kg/s} = 72 \text{ kg/hr}$$

❑

10.2.6 FILM CONDENSATION WITHIN HORIZONTAL TUBES

Condensation inside tubes is an important problem because of its applications. One example already mentioned is the condenser of an air conditioning unit. Superheated refrigerant vapor leaves a compressor and enters the condenser, which (in most designs) is a finned-tube crossflow heat exchanger. The refrigerant condenses within the tubes and makes its way to the throttle of the system. Such systems are quite complex where mathematical modeling is concerned. The flow rate of the vapor strongly influences the heat-transfer rate.

During the condensation process, a mixture of vapor and liquid travels through the tube. At high flow rates, where turbulent flow exists, the liquid (in the form of droplets) and the vapor travel together as a uniform mixture in the core of the tube, with a liquid film condensing along the tube surface. For low flow rates, however, condensation occurs in the way sketched in Figure 10.8. The vapor is at some elevated temperature T_∞, and the tube wall is at T_w, lower than the saturation temperature T_g of the vapor. Vapor condenses on the tube surface and collects in the lower portion of the tube under the action of gravity. The condensate flow resembles open-channel flow. For low vapor velocities characterized by

$$\mathrm{Re}_D = \frac{V_v ID_t}{v_v} < 35\,000 \tag{10.23}$$

the Chato Equation has been derived:

$$\bar{h}_D = 0.555 \left[\frac{g\rho_f(1 - \rho_v/\rho_f)k_f^3 h'_{fg}}{v_f(T_g - T_w)OD_t} \right]^{1/4} \tag{10.24}$$

where V_v is the average velocity of the vapor, ID_t is the inside diameter of the tube, and

$$h'_{fg} = h_{fg} + \frac{3}{8}c_{pf}(T_g - T_w) \tag{10.25}$$

Reynolds number evaluated at the inlet temperature
Liquid properties at $(T_g + T_w)/2$

For higher flow rates, the flow becomes quite complicated. Correlations are available in the literature, however.[*]

10.3 BOILING HEAT TRANSFER

Boiling is the process of vaporization when it occurs at a solid-liquid interface. Boiling occurs when the surface temperature T_w exceeds the saturation temperature T_g of the liquid. Boiling data can be obtained with a device such as that illustrated in Figure 10.9. The device consists of a well

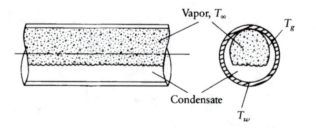

FIGURE 10.8 Condensation in a tube conveying a two-phase mixture at a low flow rate.

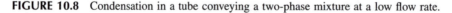

* See "Film Condensation," by W.M. Rohsenow in *Handbook of Heat Transfer Fundamentals,* edited by W.M. Rohsenow, J.P. Hartnett, and E.N. Ganic, Chapter 11, part 1, McGraw-Hill, New York, 1973.

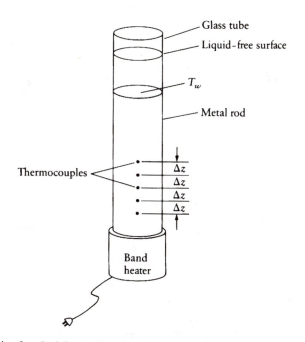

FIGURE 10.9 A device for obtaining boiling data. The metal must be well insulated.

insulated metal rod that has a band heater attached. The metal rod has a glass cylinder bonded to it, forming a liquid container with glass sides and a metal bottom. Thermocouples are located within the metal rod at equal intervals. One thermocouple is located at the metal surface in contact with the liquid. With the band heater on, one-dimensional, steady heat flow from the heater to the water will occur because of the insulation. The heat flux through the metal is given by Fourier's Law:

$$\frac{q_z}{A} = -k\frac{\partial T}{\partial z} = -k\frac{\Delta T}{\Delta z}$$

With temperature difference between adjacent thermocouples (ΔT) measured, the heat flux is easily calculated. This is the heat flux delivered to the liquid, which must be known in boiling experiments. The convection coefficient is then calculated with

$$\bar{h}_c = \frac{q_z/A}{T_w - T_g}$$

There are different boiling regimes, depending on the difference between the surface temperature T_w and the saturation temperature T_g. Data on various boiling regimes are summarized in what is known as a *boiling curve*. Figure 10.10 is a boiling curve for water. The sketch is a log-log graph of the heat transferred or the convection coefficient as a function of the *temperature excess, ΔT_e* (= surface temperature − saturation temperature = $T_w - T_g$). Although the curve is for water, other liquids will exhibit similar trends. The horizontal axis of the graph is divided into various regions labeled A through F. When we have a temperature excess that lies in region A, natural-convection currents are set up within the liquid. Such currents give rise to fluid motion near the heated surface. The liquid becomes slightly superheated, rises to the surface, and subsequently evaporates. In region A, the surface temperature exceeds the saturation temperature of the liquid by an amount that ranges from 0.1° to about 5°F (0.1° to about 2°C) when the fluid is water.

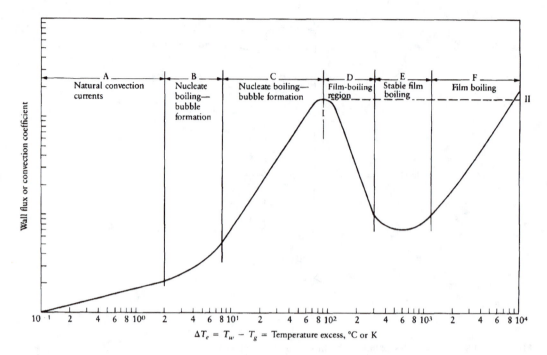

FIGURE 10.10 Boiling curve for water in which the convection coefficient (or heat-transfer rate) is graphed as a function of the temperature excess.

When the temperature excess lies within regions B and C, we have what is called *nucleate boiling*. Vapor bubbles are created at small cavities in the solid surface. The bubbles grow to a certain size that is a function of the surface tension of the liquid at the vapor-liquid interface and the temperature and the pressure. The bubbles may either collapse on the solid surface or expand and detach from the surface, depending on the temperature excess. After rising, the bubbles can become dissipated within the liquid (region B) or rise to the free surface of the liquid (region C).

Experiments have shown that the vapor inside a bubble is not necessarily at the same temperature as that of the surrounding liquid. Consider the spherical bubble of Figure 10.11. Inside, the vapor is at a pressure p_v, while outside, the liquid pressure is p. The force due to pressure is given by the product of pressure difference and area. Further, the pressure force must be balanced by the force due to surface tension, that is,

$$(p_v - p)\pi R^2 = 2\pi R\sigma$$

$$(p_v - p) = 2\sigma/R = 4\sigma/D \tag{10.26}$$

The surface tension acts over the interface length given by $2\pi R$. Now consider this bubble as being within a liquid that is being heated, and assume that the temperature of the vapor inside is the saturation temperature corresponding to the pressure p_v. If the surrounding liquid is at the saturation

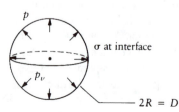

FIGURE 10.11 Pressure and surface-tension forces acting on a bubble.

temperature corresponding to p, then the liquid temperature is less than the vapor temperature. Therefore heat is transferred from the vapor to the liquid, the vapor condenses, and the bubble collapses (region B of Figure 10.10). If the bubble is to grow in size and be buoyed up to the free surface, the vapor must receive heat from the liquid. The liquid must then be superheated. If the liquid temperature exceeds the vapor temperature, the bubble rises (region C). Region B exists over a temperature excess that ranges from about 5° to about 11°F (~2° to ~6°C), while region C exists over the range ~11° to ~90°F (~9° to ~50°C) when the fluid is water.

At the end of region C, the heat-transfer coefficient (and the heat flux) reaches a maximum. Any further increase in the temperature excess ΔT_e ($= T_w - T_g$) gives a *decrease* in the convection coefficient and correspondingly a decrease in the heat flux. Thus, at point I in Figure 10.10, the heat flux is called the *critical heat flux;* in general, it is in excess of 3.2×10^5 BTU/(hr·ft²) (= 1 MW/m²).

Region D corresponds to what is called the *film-boiling region.* The system may oscillate between film boiling and nucleate boiling. In film boiling, bubbles begin forming so rapidly that they form a covering vapor over the heating surface and prevent more liquid from flowing in to replace them. The vapor blanket is an additional resistance to heat transfer because heat must be conducted through it to reach the liquid and provide the desired vaporizing effect. The heat per unit area transferred to the liquid is reduced even though the temperature excess has increased. The heat source (the metal bar of Figure 10.9, for example) must absorb the increased energy that does not find its way to the liquid. The heat-source temperature therefore increases while the system operates in region D, and the heat-transfer coefficient decreases with increases in temperature excess. For water, region D exists over the temperature excess that ranges from ~90° to ~700°F (~9° to ~120°C).

In region E, we still find the film-boiling phenomenon that we have in region D. In region F, however, the system has become stable. The convection coefficient has reached a local minimum. Further increases in the temperature excess yield an almost nonchanging value of the convection coefficient. Region E exists over a temperature excess that ranges from ~700° to ~1000°F (~120° to ~700°C) for water.

In region F, the film-boiling phenomenon still exists. Now radiation through the vapor blanket comes into play, and the heat transferred to the liquid again increases with increasing temperature excess. Eventually, when the temperature excess reaches the point labeled as II, the system reaches a condition in which the convection coefficient equals that at point I. By this time, however, the heating surface has probably burned out (or melted). For water, region F exists for temperature excesses greater than ~1000°F (~700°C).

Point I is the point of maximum heat flux. Operating beyond it gives rise to instabilities in the system. Therefore it is usually considered good practice to operate near or below the temperature excess that corresponds to point I. Following now are equations that have been derived or developed for various boiling regimes.

10.3.1 NUCLEATE POOL BOILING

The recommended correlation for nucleate pool boiling is

$$q'' = \frac{q}{A} = \rho_f v_f h_{fg} \left[\frac{g \rho_f (1 - \rho_g/\rho_f)}{\sigma g_c} \right]^{1/2} \left[\frac{c_{pf}(T_w - T_g)}{C_w h_{fg} \mathrm{Pr}_f^{1.7}} \right]^3 \tag{10.27}$$

where the f subscript refers to saturated liquid, the g subscript refers to saturated vapor, T_w is the surface temperature, σ is the surface tension, h_{fg} is the enthalpy of vaporization, and C_w is a constant that depends on the surface-liquid combination. Values of C_w for water and some selected surfaces are provided in Table 10.2. Values of C_w for other liquid–surface combinations are available in the literature.[*] The surface tension of water and its enthalpy of vaporization are provided in Table 10.3. Values for other substances are available in the literature.[†] Equation 10.27 relates the heat flux to the cube of the temperature excess.

TABLE 10.2
Values of the Dimensionless Constant C_w for Use in Equation 10.27 for Various Surface–Type–Water Combinations

Surface	C_w
Brass	0.006 0
Copper	
Scored	0.006 8
Polished	0.012 8
Platinum	0.013 0
Stainless steel	
Chemically etched	0.013 3
Ground and polished	0.008 0
Mechanically polished	0.013 2
Teflon (pitted)	0.005 8

Source: data from various sources.

10.3.2 NUCLEATE POOL BOILING CRITICAL HEAT FLUX

Point I of Figure 10.10 is a critical point. The heat flux corresponding to point I is called the *critical heat flux.* It was remarked that operating at or below this point is considered good practice, because there is an inherent danger in the instabilities present when operating beyond it. The Zuber Equation has been derived from theoretical considerations to predict the critical heat flux:

$$q_{crit}'' = \frac{\pi}{24}h_{fg}\rho_g\left[\frac{\sigma g g_c \rho_f(1 - \rho_g/\rho_f)}{\rho_g^2}\right]^{1/4}[1 + \rho_g/\rho_f]^{1/2} \qquad (10.28)$$

The constant is $\pi/24 = 0.131$, but if it is changed to 0.18, there is better agreement with experimental data. Also, because for most liquids $\rho_g/\rho_f \ll 1$, Equation 10.28 can be simplified to

$$q_{crit}'' = 0.18h_{fg}[\sigma g g_c \rho_f \rho_g^2]^{1/4} \qquad (10.29)$$

The simplified Zuber Equation is independent of temperature excess and heat-source geometry. Thus, the critical heat flux is the same whether the liquid is heated in a pan or by a submerged wire, for example. The theoretical value of the critical heat flux depends *only* on fluid properties.

Example 10.3

Waterless cookware is a name applied to stainless steel-clad aluminum pots and pans. Consider one such pan, as shown in Figure 10.12. The pan is used to boil water, and the water depth at the time of interest is 2.54 cm. The pan is placed on an electric stove, and the heating element raises

---

[*] "Heat Transfer to Liquids Boiling under Pressure," by M.T. Cichelli and C.F. Bonilla, *Trans. AlChE*, v. 41, p. 755, 1945. "Evaluation of Constants for the Rohsenow Pool-Boiling Correlation," by R.I. Vachon, G.H. Nix, and G.e. Tanger, J. *Heat Trans.*, v. 90, p. 239–1968.
[†] *Handbook of Tables for Applied Engineering Science,* edited by R.E. Bolz and G.L. Tuve, CRC Press, pp. 90–94, 1973.

TABLE 10.3
Selected Properties of Water as a Function of Temperature

Temperature T		Saturation pressure p		Enthalpy of vaporization h_{fg}		Surface tension σ		Specific gravity		Specific heat of liquid c_{pf}		Kinematic viscosity of liquid ν_f		Prandtl number of liquid Pr_f
K	°R	kPa	psia	kJ/kg	BTU/lbm	N/m	lbf/ft	Liquid	Vapor	J/(kg·K)	BTU/lbm·°R	m²/s	ft²/s	
273.15	492	0.611	0.0886	2 502	1076	0.075 5	0.00517	1.000	0.000 004 85	4 217	1.007	1.75×10^{-6}	1.88×10^{-5}	12.99
300	540	3.531	0.5122	2 438	1049	0.071 7	0.00487	0.997	0.000 0256	4 179	0.998	0.858×10^{-6}	0.923×10^{-5}	5.83
350	630	41.63	6.039	2 317	996.8	0.063 2	0.00433	0.974	0.000 260	4 195	1.002	0.374×10^{-6}	0.402×10^{-5}	2.29
373.15	672	101.33	14.7	2 257	971.0	0.058 9	0.00404	0.958	0.000 597	4 217	1.007	0.291×10^{-6}	0.313×10^{-5}	1.76
400	720	245.5	35.61	2 183	939.1	0.053 6	0.00367	0.937	0.001 368	4 256	1.017	0.231×10^{-6}	0.249×10^{-5}	1.34
450	810	931.9	135.2	2 024	870.7	0.042 9	0.00294	0.890	0.004 808	4 400	1.051	0.171×10^{-6}	0.184×10^{-5}	0.99
500	900	2 640	383.0	1 825	785.1	0.031 6	0.00217	0.831	0.013 05	4 660	1.113	0.142×10^{-6}	0.153×10^{-5}	0.86
550	990	6 119	887.7	1 564	672.8	0.019 7	0.00135	0.756	0.031 55	5 240	1.252	0.128×10^{-6}	0.138×10^{-5}	0.87
600	1080	12 350	1791.6	1 176	505.9	0.008 4	0.000576	0.649	0.072 99	7 000	1.672	0.125×10^{-6}	0.135×10^{-5}	1.14

ρ_f = sp. gr. liquid × 1 000 kg/m³ = sp. gr. liquid × 62.4 lbm/ft³

Source: Handbook of Tables for Applied Engineering Science, edited by R.E. Bolz and G.L. Tuve, CRC Press, 1973.

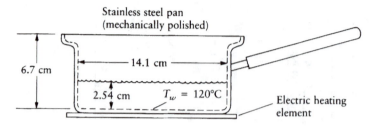

FIGURE 10.12   Stainless steel-clad aluminum pan.

the surface temperature of the pan (in contact with the water) to 120°C. Calculate (a) the power input to the water for boiling to occur, (b) the evaporation rate of water, and (c) the critical heat flux.

Solution

The heat input from the heater goes into boiling the water and into heating the pan and its handle. The answer we find for part (a) is the heat flux delivered to the water.

Assumptions

1. The system is at steady state.
2. The free surface of the water is exposed to atmospheric pressure, and the increased pressure at the bottom of the pan due to the weight of 2.54 cm of water is negligible.
3. The temperature of the water is its saturation temperature corresponding to atmospheric pressure (= 100°C).
4. Properties are constant.

From Table 10.3, for water at 100°C = 373 K,

$\rho_f = 958$ kg/m^3 $\rho_g = 0.596$ kg/m^3
$c_{pf} = 4\ 217$ J/(kg·K) $h_{fg} = 2\ 257\ 000$ J/kg
$v_f = 2.91 \times 10^{-7}$ m^2/s $\text{Pr}_f = 1.76$
$\sigma = 0.058\ 9$ N/m

The temperature excess is $T_w - T_g = 120 - 100 = 20°C$, which is in the nucleate boiling regime, so Equation 10.27 applies:

$$\frac{q}{A} = \rho_f v_f h_{fg} \left[\frac{g\rho_f(1 - \rho_g/\rho_f)}{\sigma g_c} \right]^{1/2} \left[\frac{c_{pf}(T_w - T_g)}{C_w h_{fg} \text{Pr}_f^{1.7}} \right]^3$$

From Table 10.2, for mechanically polished stainless steel, $C_w = 0.013\ 2$. Recognizing that $\rho_g/\rho_f \ll 1$, we substitute into the above equation to obtain

$$\frac{q}{A} = 958(2.91 \times 10^{-7})(2\ 257\ 000) \times \left[\frac{9.81(958)}{0.058\ 9} \right]^{1/2} \left[\frac{4\ 217(20)}{0.013\ 2(2\ 257\ 000)(1.76)^{1.7}} \right]^3$$

or

$$\frac{q}{A} = 3.19 \times 10^5 \text{ W/m}^2$$

The area A is the pan inside-bottom surface in contact with liquid:

$$A = \pi(0.141)^2/4 = 1.56 \times 10^{-2} \text{ m}^2$$

(a) The power delivered to the water then is
$$q = 3.19 \times 10^5 (1.56 \times 10^{-2})$$
$$q = 4.98 \times 10^3 \text{ W} = 4.98 \text{ kW}$$

(b) The water evaporation rate is given by

$$\dot{m}_f = \frac{q}{h_{fg}} = \frac{4.98 \times 10^3}{2\ 257\ 000}$$

$$\dot{m}_f = 2.21 \times 10^{-3} \text{kg/s} = 7.95 \text{ kg/hr}$$

(c) The critical heat flux is given by Equation 10.29:

$$q''_{cr} = 0.18 h_{fg} [\sigma g g_c \rho_f \rho_g^2]^{1/4}$$

$$= 0.18(2\ 257\ 000)[0.058\ 9(9.81)(958)(0.596)^2]^{1/4}$$

$$q''_{cr} = 1.52 \times 10^6 \text{ W/m}^2$$

We are operating below the critical heat flux.   ❑

10.4 SUMMARY

This chapter has presented a discussion of condensation and boiling. Film condensation was discussed in some detail. Correlations were given for condensation on flat and on cylindrical surfaces. Boiling was discussed, and a boiling curve for water was provided. The different boiling regimes were discussed in some detail. Correlations for pool boiling and for critical heat flux were provided. The correlations of this chapter are summarized in Table 10.4.

TABLE 10.4
Summary of Correlations of Chapter 10

		Equation numbers						
Problem	Geometry	V_z	$\dot{m}_f$	δ	h_z	$\bar{h}_L$	Re	q
Filmwise condensation	Vertical plate (laminar film)	10.2	10.3	10.6	10.7b	10.9	10.11b	10.12
	Vertical plate (turbulent film		10.13			10.4	10.11	10.12
	Inclined plate (laminar film)	10.15	10.16	10.17	10.18	10.19	10.11	10.12
	Vertical tube (laminar film)		10.16	10.6	10.7a	10.9	10.20	10.12
	Outside surface of horizontal							
	tubes and tube banks		10.16			10.21		
						10.22		10.4
	Inside surface of horizontal tubes					10.24	10.23	
Boiling heat transfer	Nucleate pool boiling							10.27
	Critical heat flux							10.28
								12.29

10.5 PROBLEMS

10.5.1 FILMWISE CONDENSATION–FLAT PLATES

1. The velocity profile for the film of Figure 10.2 was derived from a force balance performed on an element within the film. Derive the velocity profile by completing the following steps (with reference to the element within the film of Figure 10.2):

    (a) Give reasons to show that $V_x = V_y = 0$, and that V_z is nonzero.

    (b) Show that V_z varies only with the y coordinate.

    (c) Show that the Navier–Stokes Equations (Chapter 5) reduce to

$$0 = -\frac{dp}{dz} + \mu_f\left(\frac{d^2 V_z}{dy^2}\right) + \frac{\rho_f g_z}{g_c}$$

    (d) Figure P10.1 shows an element with pressures acting on it. Perform a force balance and show that

$$\frac{dp}{dz} = \frac{\rho_v g_z}{g_c}$$

    What simplifying assumptions had to be made?

    (e) Substitute into the differential equation and rearrange to obtain

$$\frac{d^2 V_z}{dy^2} = -\frac{\rho_f g_z}{\mu_f g_c}(1 - \rho_v/\rho_f)$$

    (f) Integrate twice to obtain

$$V_z = -\frac{g_z y^2}{2 v_f}(1 - \rho_v/\rho_f) + C_1 y + C_2$$

    (g) Explain the meaning of the following boundary conditions:

    B.C.1. $y = 0$, $V_z = 0$
    B.C.2. $y = \delta$, $dV_z/dy = 0$

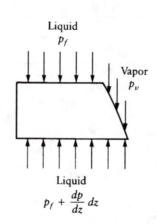

Liquid
p_f

Vapor
p_v

Liquid
$p_f + \dfrac{dp}{dz}\, dz$

FIGURE P10.1

(h) Apply the boundary conditions and derive Equation 10.2 for the velocity profile.

2. Verify the derivation of Equation 10.3 and that the equation is dimensionally correct.
3. Show that $\bar{h}_L = 4/3h_z|_{z=L}$ for laminar film condensation on a vertical flat surface.
4. The condensation number, Co, is a dimensionless group defined as

$$Co = \bar{h}_L \left[\frac{v_f^2}{k_f^3(1 - \rho_v/\rho_f)g} \right]^{1/3}$$

(a) Verify that it is dimensionless.
(b) Show that for filmwise condensation on a vertical flat plate, we can write

$$Co = 1.47(Re_\delta)^{-1/3}$$

where $Re = 4\dot{m}_f/W\rho_f v_f < 1\ 800$.

5. Graph velocity vs. y at $z = 1$ ft for the film of Example 10.1. Does the profile resemble that of Figure 10.3?
6. Steam at a pressure of 200 kPa $(T_g = 120°C, h_{fg} = 2\ 201\ 900$ J/kg) and a temperature of 150°C $(v_v = 0.959\ 6$ m³/kg) is in contact with a plate that is 1 m tall and maintained at 110°C. Assuming that filmwise condensation occurs, graph film thickness and local convection coefficient as a function of height. Calculate also the Reynolds number at plate end. The width of the wall is 0.5 m.
7. Steam at a pressure of 2.5 MPa $(T_g = 224°C, h_{fg} = 1\ 841\ 000$ J/kg) and a temperature of 300°C $(v_v = 0.098\ 90$ m³/kg) is in contact with a 0.5-m-tall vertical wall maintained at a temperature that can be made to vary from 120° to 220°C. Determine and graph the condensation rate at plate end as a function of wall temperature. Assume filmwise condensation and a plate width of 0.5 m.
8. Steam at a pressure of 2000 psia $(T_g = 636°F, h_{fg} = 465.7$ BTU/lbm) and a temperature of 700°C $(v_v = 0.2488$ ft³/lbm) is in contact with a vertical wall that is 12 in. wide. The wall is maintained at a temperature that can be made to vary from 550° to 630°F. Assuming filmwise condensation occurs, determine and graph the plate length required for the film to become turbulent as a function of temperature. Evaluate properties at 572°F.
9. The *condensation number*, Co, is a dimensionless group defined as

$$Co = \bar{h}_L \left[\frac{v_f^2}{k_f^3(1 - \rho_v/\rho_f)g} \right]^{1/3}$$

(a) Verify that it is dimensionless.
(b) Show that, for turbulent film condensation on a vertical surface, we have

$$Co = 0.007\ 7Re_\delta^{0.4}$$

where $Re = 4\dot{m}_f/W\rho_f v_f < 1\ 800$.

10. Freon-12 at a pressure of 21.4 psia $(T_g = -5°F, h_{fg} = 69.291$ BTU/lbm) and a temperature of 20°F $(v_v = 2$ ft³/lbm) is in contact with a vertical wall that is 12 ft tall and 2 ft wide. Determine the heat that is transferred to the wall if it is maintained at −10°F. Assume that filmwise condensation occurs.

11. Freon-12 at a pressure of 40 psia $(T_g = 25°F, h_{fg} = 65.946$ BTU/lbm) and a temperature of 40°F $(v_v = 1.026$ ft^3/lbm) is in contact with a 14-ft-tall vertical wall that is maintained at a temperature that can be varied from 0° to 23°F. Assuming filmwise condensation, determine and graph the variation of heat transferred to the wall as a function of wall temperature. The width of the wall is 1 ft.

12. Steam at a pressure of 200 kPa $(T_g = 120°C, h_{fg} = 2\ 201\ 900$ J/kg) and a temperature of 150°C $(v_v = 0.959\ 6$ m^5/kg) is in contact with a plate maintained at 115°C. The plate is inclined at an angle with the horizontal that can be made to vary from 90° to 45°. Assuming filmwise condensation, determine and graph
    (a) Film thickness at plate end
    (b) Overall convection coefficient
    (c) Heat transferred
    (d) Condensation rate at plate end as a function of plate angle
    Take the plate length × width dimensions to be 1 m × 0.5 m.

13. Ammonia vapor at 48 psia $(T_g = 20°F, h_{fg} = 553$ BTU/lbm) and 30°F is in contact with a wall inclined at 70° with the horizontal and maintained at 10°F. Assuming filmwise condensation, how long a wall is required for the film to undergo transition? Calculate the condensation rate at transition. The plate width is 1 ft.

10.5.2 CONDENSATION ON TUBES

14. A vertical tube 2 m tall is in contact with methyl chloride vapor at a pressure of 427.25 kPa $(T_g = 288.8$ K, $h_{fg} = 3.872 \times 10^5$ J/kg) and a temperature of 350 K. The tube is 4 standard type K copper tubing. Assuming filmwise condensation occurs, determine the film thickness at the end of the tube, the condensation rate, and the heat transferred to the tube. The tube is at a temperature of 280 K. How long a tube is required for the film to undergo transition?

15. Determine whether Equation 10.21 is dimensionally correct.

16. The *condensation number,* Co, is a dimensionless group defined as

$$\text{Co} = \bar{h}_L \left[\frac{v_f^2}{k_f^3(1 - \rho_v/\rho_f)g} \right]^{1/3}$$

    (a) Verify that it is dimensionless.
    (b) Show that, for condensation on a horizontal cylinder, we can write

$$\text{Co} = 1.514(\text{Re}_\delta)^{-1/3}$$

    where $\text{Re}_\delta = 4\dot{m}_f/\pi D \rho_f v_f$.

17. Ethylene glycol vapor at 14.7 psia $(T_g = 387°F, h_{fg} = 344$ BTU/lbm) and 450°F is being condensed on the outside surface of a horizontal tube. The tube is 1 standard type K copper that is 4 ft long. The outside-surface temperature of the tube is 380°F. Determine the heat-transfer rate and the condensation rate.

18. Steam at 200 kPa $(T_g = 120°C, h_{fg} = 2\ 201.9$ kJ/kg) and 150°C $(v_v = 0.959\ 6$ m^3/kg) is to be condensed on the outside surface of a tube. The tube itself is made of 1/2 standard type K copper tubing 2 m long, and the outside surface temperature can be made to vary from 80 to 115°C. Determine and graph the condensation rate as a function of tube-surface temperature.

19. Methyl chloride vapor at 427.25 kPa $(T_g = 288.8$ K, $h_{fg} = 3.872 \times 10^5$ J/kg) is to be condensed on the outside surface of a horizontal tube. The tube-surface temperature is

280 K. The tube diameter must be selected on the basis of condensation rate. Determine and graph condensation rate as a function of the tube diameter for 1/2 standard, 3/4 standard, 1 standard, 1-1/2 standard, and 2 standard, all type K copper tubing 1 m long.

20. After passing through a turbine, steam is to be condensed. Steam enters the condenser at 5 psia ($T_g = 163°F$, $h_{fg} = 1000.4$ BTU/lbm) and 200°F ($v_v = 78.14$ ft³/lbm). The condenser can contain any number of tubes ranging from 50 to 100. All tubes have an outside diameter of 3/4 in exactly and are 12 ft long. Determine and graph the condensation rate as a function of the number of tubes. Take the tube outside-surface temperature to be 155°F.

21. Suppose the steam of Example 10.2 is condensed on the same pipe, but in this case the pipe axis is vertical. Rework the example for a vertical pipe and compare the results.

10.5.3 BOILING

22. A 1/16-in diameter bubble is submerged in 4 in of water. The water temperature in the vicinity of the bubble is 220°F.
    (a) Use the hydrostatic equation [$p = p_{atm} + pg(\text{depth})/g_c$] to calculate the pressure at the bubble.
    (b) Use Equation 10.26 to calculate the pressure of the vapor inside the bubble.
    (c) In the range of interest, saturation temperature and pressure are related by

$$T = 0.282p_v + 345$$

    where T is in K and the corresponding vapor pressure p_v is in kPa. Determine the corresponding vapor temperature for the pressure in part (b).
    (d) Will the bubble collapse or rise to the surface? Evaluate liquid properties at 212°F.

23. The water of Example 10.3 is to be boiled in another container. Rework the problem if the pan bottom is made of
    (a) Polished copper
    (b) Stainless steel with a Teflon pitted coating. Keep all other parameters the same

24. Using the data of Table 10.3, graph the variation of the critical heat flux with temperature. Use the unit system of your choice.

25. Verify that Equation 10.29 is dimensionally correct.

11 Introduction to Radiation Heat Transfer

CONTENTS

11.1 INTRODUCTION

Radiation heat transfer is entirely different from heat transfer by conduction or convection. A material medium is not required for two objects to exchange heat by radiation. Radiation is an electromagnetic mechanism by which energy is transported at the speed of light through regions that are devoid of substance. For a net radiant interchange of energy between two bodies to occur, there must be a temperature difference. Furthermore, the rate at which the radiant interchange occurs is proportional to the difference of the fourth powers of their *absolute* temperatures.

Our objective in considering the radiation mode is in being able to calculate an energy-transfer or a heat-transfer rate for various problems of interest. For a better understanding of our model for radiation, it will be necessary to borrow from several different disciplines:

1. *Electromagnetic theory.* Radiation can be treated mathematically as if it were a traveling wave that contains energy and exerts a pressure. This wave-like nature is described by using electromagnetic wave theory.
2. *Statistical mechanics.* Radiant energy is transported over a range of wavelengths. Statistical mechanics helps to describe how this energy is distributed over the spectrum of wavelengths (i.e., whether most of the energy from the sun, for example, is radiated over a few wavelengths or over many wavelengths).
3. *Quantum mechanics.* All substances radiate and receive radiant energy continuously. Associated with these processes are atomic and molecular activities that are modeled with the principles of quantum mechanics.
4. *Thermodynamics.* Quantum mechanics gives a description of molecular processes associated with radiation. Thermodynamics, on the other hand, gives a description of how bulk properties behave.

The physics of radiation will be discussed in this chapter. Definitions of basic concepts will be presented, and at times it will become easy to forget that our objective is to calculate a heat-transfer rate. The amount of information indigenous to the field of radiation can be overwhelming. While this chapter presents definitions and laws that are important in radiation, the next chapter deals with the calculation of radiation heat-transfer rates.

11.2 ELECTROMAGNETIC RADIATION SPECTRUM

The principles of quantum mechanics provide a quantitative description of how radiation is produced by a solid body. Our interest does not require such detail, so a qualitative description will suffice. Suppose energy is supplied to a solid body as, for example, in the case of a tungsten filament in a light bulb. Some of the constituent atoms and molecules of the body are transformed into an excited state or a higher energy level. The atoms and molecules have a tendency for spontaneously returning to their lower energy level, and in so doing, energy in the form of electromagnetic radiation is emitted from the body. The atoms and molecules have electronic, rotational, and vibrational energies that become affected during the process. The emitted radiation may result from random changes in these energies, and therefore radiation in the form of waves is emitted over a range of wavelengths.

The radiation spectrum encompasses many forms: radio waves, microwaves, infrared waves, ultraviolet waves, X-rays, and gamma rays. We are concerned with thermal radiation. Figure 11.1

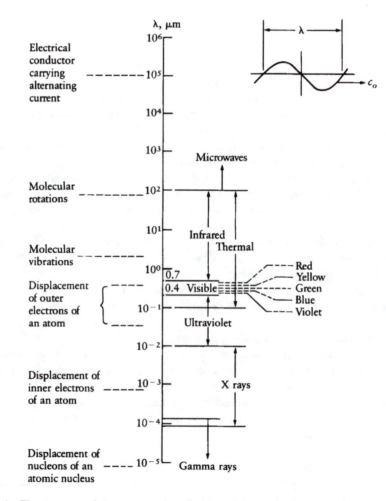

FIGURE 11.1   The spectrum of electromagnetic radiation and the mechanisms associated with various types.

illustrates the spectrum of electromagnetic radiation. Wavelengths of various components are shown. Also listed are the mechanisms (roughly) associated with the various types of radiation. The only difference between the various kinds of radiation is the range of wavelengths.

Figure 11.1 also shows a wave with the wavelength λ labeled on it. In a vacuum, all forms of radiation or radiant energy travel at the speed of light c_o, where

$$c_o = 2.998 \times 10^8 \text{ m/s} = 9.836 \times 10^8 \text{ ft/s} = 3.541 \times 10^{12} \text{ ft/hr}$$

The wavelength λ of a particular electromagnetic wave is related to the wave frequency v by

$$\lambda = c_o/v \tag{11.1}$$

In the visible range of electromagnetic radiation, the radiation our eyes interprets as being red in color is longer in wavelength than any of the other colors listed.

It is also possible to describe electromagnetic radiation as if it were propagated as particles. For an electromagnetic wave of frequency v, we associate with it a photon. A photon is a particle having zero charge, zero mass, and an energy e given by

$$e = hv^* \tag{11.2}$$

where h is Planck's constant, given as

$$h = 6.625\,6 \times 10^{-34} \text{ J·s} = 6.284 \times 10^{-37} \text{ BTU·s}\ ^\dagger$$

Combining Equations 11.1 and 11.2, we get

$$e = \frac{hc_o}{\lambda} \tag{11.3}$$

Because b and c_o are constants, it is seen that longer wavelengths are associated with smaller photon energies. Referring to Figure 11.1, we conclude that radiation emitted as a result of molecular rotations are associated with smaller photon energies than are those emitted due to molecular vibrations.

11.3 EMISSION AND ABSORPTION AT THE SURFACE OF AN OPAQUE SOLID

Radiant energy that is incident on an opaque solid is either reflected or absorbed. The total incident energy per unit area and per unit time is denoted as q_i''. The portion of this total energy per area per time that is absorbed is q_a''. The fraction of incident energy that is absorbed is called the *absorptivity* α:

$$\alpha = \frac{q_a''}{q_i''} \tag{11.4a}$$

---

* h and v are unfortunate choices but are widely recognized in the literature. They are not to be confused with the convection coefficient or the kinematic viscosity.

† This value and the value of other universal constants that we encounter in radiation are listed in the Summary section of this chapter.

If we are interested only in the radiant energy that exists within a certain wavelength range, say, λ to $\lambda + d\lambda$, then we define the fraction of incident radiation of wavelength λ that is absorbed as the *monochromatic* or *specular absorptivity*:

$$\alpha_\lambda = \frac{q_{a\lambda}''}{q_{i\lambda}''} \tag{11.4b}$$

where $q_{a\lambda}''$ is the absorbed energy per area per time having a wavelength ranging from λ to $\lambda + d\lambda$, and $q_{i\lambda}''$ is the incident energy per area per time in the same wavelength range.

For any real body, the spectral absorptivity α_λ is less than unity and will vary with wavelength. How the absorptivity varies with wavelength is difficult to evaluate and may change considerably over a period of time for a real surface. Consequently, it is useful to model real bodies as though they were what are called *gray bodies*. A gray body has a value of spectral absorptivity that is a constant for all wavelengths. A gray body will always absorb the same fraction of incident radiation at all wavelengths. The spectral absorptivity α_λ, although constant for a gray body, is less than unity. The limiting case of a gray body is that for which the spectral absorptivity is unity for all wavelengths. Such a body is called a *black body*.

A black body has an ideal surface (from a radiation viewpoint), which possesses the following properties:

1. A black body absorbs all incident radiation.
2. No surface can emit more energy than a black body at any given temperature and wavelength.
3. A black body emits radiant energy that depends on wavelength and temperature but is independent of direction.

All solid surfaces emit radiant energy, and the total radiant energy emitted per unit area and per unit time is denoted as q_e''. Having already defined a black body as being a perfect emitter, we denote the total energy emitted per area per time by a black body as q_b''. The ratio of total energy emitted by a real body to that emitted by a black body (both at the same temperature) is called the *emissivity*:

$$\varepsilon = \frac{q_e''}{q_b''} \tag{11.5a}$$

As discussed in Section 11.2, energy is radiated over a range of wavelengths. If we wish to consider emitted radiation over a narrow range rather than the total, we can define what is called the *monochromatic* or *spectral emissivity* as

$$\varepsilon_\lambda = \frac{q_{e\lambda}''}{q_{b\lambda}''} \tag{11.5b}$$

Emissivity is less than unity for real, nonfluorescing surfaces. Emissivity is unity for a black surface.

The spectral nature of real bodies is a feature that complicates the description of radiant emissions. Another aspect of real bodies is that they may emit radiant energy preferentially in one direction over another. Figure 11.2 is a polar graph of emitted energy vs. angle for a typical real body. As shown, the emitted energy normal to the surface is a maximum. To denote that emitted energy can vary with wavelength and direction, we use the notation $q_{e\lambda\theta}''$ or $q_{e\lambda\phi}''$. If the emissivity is to include a dependence on wavelength and direction, then we write

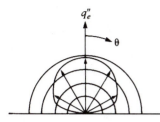

FIGURE 11.2 Variation of emitted energy per area per time unit with direction for a real body.

$$\varepsilon_{\lambda\theta\phi} = \frac{q_{e\lambda\theta\phi}''}{q_{i\lambda\theta\phi}''} \tag{11.4c}$$

The absorptivity of a surface can also vary with direction. To denote a variation with wavelength and direction, we write

$$\alpha_{\lambda\theta\phi} = \frac{q_{a\lambda\theta\phi}''}{q_{i\lambda\theta\phi}''}$$

We introduced the terms *monochromatic* and *spectral* to refer to wavelength dependence. The term *diffuse* is now introduced in reference to a direction dependence. A diffuse surface emits radiant energy in an equal amount in all directions. Appendix Table E.1 lists the emissivity of various metals. We will examine these topics in greater detail in subsequent sections.

11.4 RADIATION INTENSITY

We defined heat transfer by radiation as a bulk property, calling emitted radiation, for example, q_e''. A more fundamental emitted energy is called the intensity of radiation, from which q_e'' can be obtained. Radiation intensity is defined in this section.

Emission of radiant energy from a surface element travels in all directions. The most convenient description of such phenomena makes use of spherical coordinates. Figure 11.3 shows how angles and areas in spherical coordinates relate to those in polar cylindrical coordinates. Figure 11.3a is of a plane angle $d\theta$ on a polar graph. The arc length dL is

$$dL = r\,d\theta \tag{11.6}$$

The unit of the plane angle $d\theta$ is radians (rad).

Figure 11.3b is of a spherical coordinate system. The x, y, and z axes appear, but r, θ, and ϕ are the appropriate coordinates. They too are shown in the figure. The differential area on a hemispherical surface is dA_s. The angle θ is measured from the z axis to rays that connect the origin to dA_s. The angle ϕ is measured from the x axis as shown. The differential area dA_s has dimensions of $r\,d\theta$ by $r\sin\theta\,d\phi$. The solid angle formed by the rays from the origin to dA_s is $d\omega$. In a manner analogous to that for polar coordinates (see Equation 11.6), we relate the surface area to the solid angle by

$$dA_s = r^2\,d\omega \tag{11.7}$$

The unit of $d\omega$ is the steradian (sr), and it is treated in the same way as the radian. In terms of the zenith angle θ and the azimuthal angle ϕ, we can rewrite Equation 11.7 as

$$(r\sin\theta\,d\phi)r\,d\theta = r^2\,d\omega$$
$$d\omega = \sin\theta\,d\theta\,d\phi \tag{11.8}$$

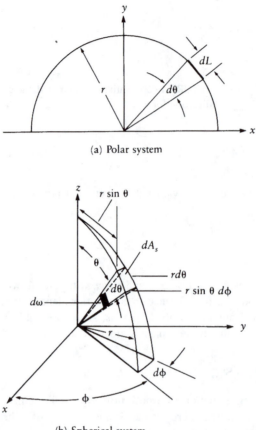

(a) Polar system

(b) Spherical system

FIGURE 11.3 Polar and spherical coordinate systems.

Although we have represented a surface element as being four sided, in radiation heat transfer, we use as our element what is called a *pencil ray* representation. This is illustrated in Figure 11.4. We say that for a given θ and ϕ, radiation is emitted from differential area dA_1 in a manner that resembles the sharpened end of a wooden pencil. The relations for area $[dA_n = (r \sin \theta \, d\phi)(r \, d\theta)]$ and solid angle (Equation 11.8) are the same as for Figure 11.3b, however, because we are dealing with differential areas.

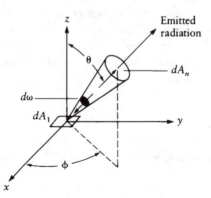

FIGURE 11.4 A pencil ray.

Figure 11.4 illustrates radiation emitted in the θ, ϕ direction from differential area dA_1 to dA_n. The rate at which radiant energy is transferred must include the quantity of energy transferred, the geometry, the solid angle $d\omega$, and the wavelength. These factors are combined into what is called the *intensity of radiation*. Intensity I is defined as the rate at which radiant energy is emitted per unit area of the emitting surface normal to the θ, ϕ direction, per unit solid angle $d\omega$ about the direction, per unit wavelength $d\lambda$. In equation form, we have

$$I_{e\lambda\theta\phi} = \frac{dq_e}{(dA_1 \cos\theta)d\omega\, d\lambda} \tag{11.9}$$

The subscripts on the intensity (i.e., the spectral intensity) indicate that the radiant energy is emitted (e), that it is spectral (depends on wavelength λ), and that it may radiate preferentially in one direction over another (a function of the direction defined as θ, ϕ). The term $dA_1 \cos\theta$ is related to the geometry of the system, and an interpretation is sketched in Figure 11.5. The area dA_1, as viewed from dA_n, appears to have an area given by $dA_1 \cos\theta$. The energy per unit wavelength $dq_e/d\lambda$ can be rewritten as $dq_{e\lambda}$, which still retains the functional dependence that the emitted energy has on wavelength. The dimensions of $dq_{e\lambda}$ are (F·L)/(T·L) [W/m or BTU/hr·ft]. The dimensions of dq_e are F·L/T. Substituting from Equation 11.8 and simplifying, Equation 11.9 becomes

$$I_{e\lambda\theta\phi} = \frac{dq_{e\lambda}}{dA_1 \cos\theta \sin\theta\, d\theta\, d\phi} \tag{11.10}$$

If the functional form of the spectral and directional intensity is known, then the rate at which energy is emitted by radiation can be found by integration of Equation 11.10:

$$q_e'' = \int_0^\infty \int_0^{2\pi} \int_0^{\pi/2} I_{e\lambda\theta\phi} \cos\theta\, \sin\theta\, d\theta\, d\phi\, d\lambda \tag{11.11}$$

Equation 11.11 gives the total heat flux emitted over all directions and over all wavelengths. Note that the solid angle $d\omega$, when integrated over all directions (0 to $\pi/2$ for θ and 0 to 2π for ϕ), yields a hemisphere. Thus, the heat flux of Equation 11.11 is called the *total* or *hemispherical emissive power*, with dimensions of F·L/(T·L^2) (W/m^2 or BTU/hr·ft^2). In many texts, the notation E is used rather than q_e''.

We have already defined a diffuse emitter as a surface for which the emitted radiation is independent of direction. For a diffuse surface, then, $I_{e\lambda\theta\phi}$ reduces to $I_{e\lambda}$ and can be removed from the integral on θ and the one on ϕ. Performing the integration, we get

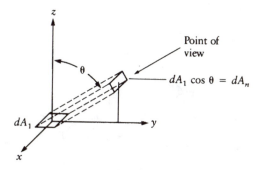

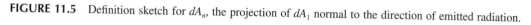

FIGURE 11.5 Definition sketch for dA_n, the projection of dA_1 normal to the direction of emitted radiation.

$$q_e'' = \int_0^\infty I_{e\lambda}\left[\int_0^{2\pi}\int_0^{\pi/2}\cos\theta\,\sin\theta\,d\theta\,d\phi\right]d\lambda$$

or

$$q_e'' = \pi\int_0^\infty I_{e\lambda}d\lambda \qquad (11.12)$$

In terms of wavelength-dependent emissive power, we could write

$$q_{e\lambda}'' = \pi I_{e\lambda} \qquad (11.13)$$

The constant in Equations 11.12 and 11.13 is π and not 2π. Furthermore, it has the unit sr.

Example 11.1

Consider the system sketched in Figure 11.6. Shown is a differential area dA_1 having dimensions of 1×1 in. It is known that this surface emits radiation and has an associated intensity given by

$$I_{e\theta} = 2000(1 - 0.4\sin^2\theta)\ \text{BTU/(hr·ft}^2\text{·sr)}$$

Two other differential areas, dA_2 and dA_3, both of dimensions 1×1 in, are located as shown. (a) Calculate the value of the solid angle subtended by surfaces dA_2 and dA_3 with respect to dA_1. (b) What is the intensity of emission from dA_1 in the direction of the other areas? (c) What is the rate at which radiation emitted by dA_1 is intercepted by the other areas?

Solution

The area dA_1 is emitting radiation in all directions according to the above equation. Also, the intensity is independent of wavelength λ and angle ϕ, so these integrations need not be performed. The surface is not a diffuse emitter.

Assumptions

1. Surfaces dA_1, dA_2, and dA_3 can be approximated as differential areas.
2. The system is at steady state.

(a) The solid angle is calculated with Equation 11.7:

$$d\omega = \frac{dA_n}{r^2} = \frac{dA_i\cos\beta_i}{r^2}$$

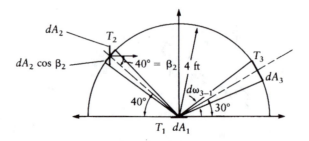

FIGURE 11.6   Radiation from dA_1 in all directions, some of which is intercepted by dA_2 and dA_3.

where β is the angle between the surface normal of a receiver surface and the line connecting the two surfaces.

For area dA_2,

$$d\omega_{2-1} = \frac{dA_2 \cos 40}{r^2} = \frac{\left[\frac{(1)(1)}{144}\right](\cos 40)}{4^2}$$

$$d\omega_{2-1} = 3.32 \times 10^{-4} \text{ sr}$$

Similarly,

$$d\omega_{3-1} = \frac{dA_3 \cos 0}{r^2} = \frac{\left[\frac{(1)(1)}{144}\right]}{16}$$

$$d\omega_{3-1} = 4.34 \times 10^{-4} \text{ sr}$$

(b) Now θ is measured from the z axis of Figure 11.6. Accordingly, for dA_2, $\theta_2 = 50°$, while for dA_3, $\theta_3 = 60°$. The intensity of radiation emitted from dA_1 in the direction of dA_2 is

$$I_{e\theta_2} = 2000(1 - 0.4 \sin^2 50) = 1530 \text{ BTU/(hr·ft}^2\text{·sr)}$$

Similarly, for dA_3

$$I_{e\theta_3} = 2000(1 - 0.4 \sin^2 60) = 1400 \text{ BTU/(hr·ft}^2\text{·sr)}$$

(c) The above intensities are what has left dA_1 in the respective directions. The amount intercepted is given by Equation 11.9:

$$dq_{1-i} = I_i dA_1 \cos\theta \, d\omega_{i-1}$$

We need not integrate, because we have differential areas (i.e., $dA_i \ll r^2$). For dA_2,

$$dq_{1-2} = 1530 \frac{(1)(1)}{144} \cos 50 (3.32 \times 10^{-4})$$

or

$$dq_{1-2} = 1.74 \times 10^{-3} \text{ BTU/hr}$$

Similarly,

$$dq_{1-3} = 1400 \frac{(1)(1)}{144} \cos 60 (4.34 \times 10^{-4})$$

$$dq_{1-3} = 2.11 \times 10^{-3} \text{ BTU/hr} \qquad \square$$

11.5 IRRADIATION AND RADIOSITY

The previous section dealt with intensity of radiation, where intensity was defined as the rate at which radiant energy is emitted over a given wavelength interval $d\lambda$ in the θ, ϕ direction per unit area per unit solid angle $d\omega$. The area referred to in that definition is the area of the emitting surface normal to the θ, ϕ direction (i.e., it is the projected area). We then related the intensity to the total emissive power q_e''.

Also associated with intensity is the radiation that is incident on a surface. Figure 11.7 shows radiation that is incident on a differential area dA_1. The intensity of the radiation is a function of wavelength and direction, so it is denoted as $I_{i\lambda\theta\phi}$.

We now define a radiative heat flux called the *irradiation* q_i'', which is the radiation incident on the surface coming from all directions. In terms of the irradiation, the intensity of incident radiation is defined as the rate at which radiant energy is received per unit area of the receiving surface per unit solid angle per unit wavelength interval (analogous to the intensity of emitted radiation). In equation form, we write

$$I_{i\lambda\theta\phi} = \frac{dq_i}{dA_1\cos\theta \; d\omega \; d\lambda} \tag{11.14a}$$

Noting that $dq_i/d\lambda = dq_{i\lambda}$ and that $d\omega = \sin\theta \; d\theta \; d\phi$ (Equation 11.8), we can rewrite Equation 11.4a as

$$I_{i\lambda\theta\phi} = \frac{dq_{i\lambda}}{dA_1\cos\theta \; \sin\theta \; d\theta \; d\phi} \tag{11.14b}$$

The dimension of the spectral incident irradiation $dq_{i\lambda}$ is (F·L)/(T·L) [W/m or BTU/hr·ft]. If the functional form of the spectral directional incident intensity $I_{i\lambda\theta\phi}$ is known, then the irradiation flux $q_i/dA_1 = q_i''$ can be found by integration:

$$q_i'' = \int_0^\infty \int_0^{2\pi} \int_0^{\pi/2} I_{i\lambda\theta\phi}\cos\theta \; \sin\theta \; d\theta \; d\phi \; d\lambda \tag{11.15}$$

Equation 11.15 gives the total irradiation received over all wavelengths and all directions. Again, when the solid angle $d\omega$ is integrated over all directions, a hemisphere results. Thus, the irradiative heat flux of Equation 11.15 is called the *total irradiation,* with dimensions of F·L/(L²·T) [W/m or BTU/hr·ft²]. In many texts, G is used rather than q_i''.

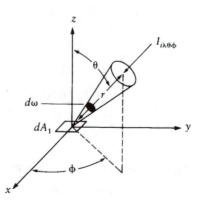

FIGURE 11.7 Radiation incident on a differential area.

Equation 11.15 reduces to a simpler form for a diffuse surface, where the intensity does not vary with direction. For a diffuse surface, $I_{i\lambda}$ can be removed from the integral on θ and the one on ϕ. The result is

$$q_i'' = \int_0^\infty I_{i\lambda}\left[\int_0^{2\pi}\int_0^{\pi/2}\cos\theta\,\sin\theta\,d\theta\,d\phi\right]d\lambda$$

or

$$q_i'' = \pi\int_0^\infty I_{i\lambda}d\lambda \tag{11.16}$$

In terms of a spectral irradiative flux, we write

$$q_{i\lambda}'' = \pi I_{i\lambda} \tag{11.17}$$

where the constant π has units of sr.

Example 11.2

Figure 11.8 shows a differential area dA_1 having dimensions of 2×2 cm. Measurements indicate that this surface is a diffuse emitter with an intensity of emitted radiation equal to 1 000 W/(m²·sr). A second differential area dA_2 is located as shown. (a) Calculate the value of the solid angle subtended by dA_2 with respect to dA_1. (b) What is the rate at which radiation emitted by dA_1 is intercepted by dA_2? (c) What is the irradiation associated with dA_2? Both areas have the same dimensions.

Solution

The area dA_1 is emitting radiation in all directions, and it is a diffuse emitter. The area dA_2 intercepts a small portion of the radiant energy.

Assumptions

1. Surfaces dA_1 and dA_2 can be treated as differential areas.
2. The system is at steady state.

(a) The solid angle is calculated with Equation 11.7:

$$d\omega = \frac{dA_n}{r^2} = \frac{dA_i\cos\beta_i}{r^2}$$

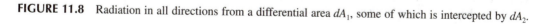

FIGURE 11.8 Radiation in all directions from a differential area dA_1, some of which is intercepted by dA_2.

The angle β is $0°$ because the surface normal of dA_2 is directed at dA_1. Substituting,

$$d\omega_{2-1} = \frac{dA_2 \cos 0}{r^2} = \frac{(0.02)(0.02)(1)}{(1)^2}$$

$$d\omega_{2-1} = 4.0 \times 10^{-4} \text{ sr}$$

(b) The intensity of radiation leaving dA_1 in any direction is 1 000 W/(m²·sr). The amount intercepted by dA_2 is given by

$$dq_{1-2} = I_{e1} \, dA_1 \, \cos\theta \, d\omega_{2-1}$$

Substituting,

$$dq_{1-2} = 1\,000(0.02)(0.02)\cos 30(4 \times 10^{-4})$$

$$dq_{1-2} = 1.39 \times 10^{-4} \text{ W}$$

(c) The irradiation associated with dA_2 can be found by dividing the incident radiation by the receiver area:

$$dq''_{i2} = \frac{dq_{1-2}}{dA_2} = \frac{1.39 \times 10^{-4}}{(0.02)(0.02)}$$

$$dq''_{i2} = 3.46 \times 10^{-1} \text{ W/m}^2 \qquad\qquad □$$

When an opaque solid is irradiated, some of the incident radiative flux is absorbed, and some is reflected. If the solid is transparent or translucent, then some of the incident radiative flux is absorbed, some is reflected, and some is transmitted through the material. This latter and more general case is illustrated in Figure 11.9a.

Figure 11.9b shows the definition of another radiative flux, called the *radiosity*, dq_j or q_j. A translucent material is receiving radiation on both sides. All the radiation leaving one of its surfaces (which includes emitted, reflected, and transmitted components) is defined as the radiosity. Although

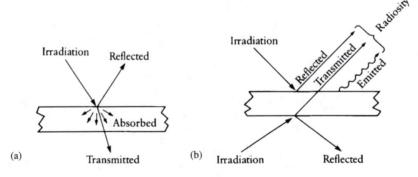

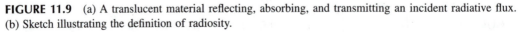

FIGURE 11.9 (a) A translucent material reflecting, absorbing, and transmitting an incident radiative flux. (b) Sketch illustrating the definition of radiosity.

Figure 11.9b shows the components of radiosity as single rays, it should be remembered that radiosity leaves the surface in all directions and over all wavelengths. Of course, if the material is opaque, then the transmitted component is zero. Radiosity is generally different from emissive power.

We can now define an intensity associated with the radiosity, $I_j = I_{\text{reflected}} + I_{\text{emitted}} + I_{\text{transmitted}}$. It is defined as the rate at which radiation leaves a surface per unit area of that surface per unit solid angle per unit wavelength interval. In equation form,

$$I_{j\lambda\theta\phi} = \frac{dq_j}{dA_1 \cos\theta\; d\omega\; dA}$$

With $d\omega = \sin\theta\; d\theta\; d\phi$ from Equation 11.8 and $dq/d\lambda = dq_{j\lambda}$ (the spectral radiosity), the above equation becomes

$$I_{j\lambda\theta\phi} = \frac{dq_{j\lambda}}{dA_1 \cos\theta\; \sin\theta\; d\theta\; d\phi} \tag{11.18}$$

The dimension of the spectral radiosity is (F·L)/(T·L) (W/m or BTU/hr·ft). If the functional form of the spectral, directional intensity is known, then the total radiosity is found with

$$q_j'' = \int_0^\infty \int_0^{2\pi} \int_0^{\pi/2} I_{j\lambda\theta\phi} \cos\theta\; \sin\theta\; d\theta\; d\phi \tag{11.19}$$

Equation 11.19 is the total or hemispherical radiosity, which is all the radiation leaving a surface in all directions and over all wavelengths. Again, integration of the solid angle $d\omega$ over all directions yields a hemisphere. The dimension of radiosity is (F·L)/(L²·T) (W/m² or BTU/hr·ft²). In many texts, J rather than q_j'' is used for radiosity.

For a diffuse opaque surface, that is, a diffuse reflector and a diffuse emitter, the intensity is no longer a function of direction, and the radiosity becomes

$$q_j'' = \int_0^\infty I_{j\lambda} \left[\int_0^{2\pi} \int_0^{\pi/2} \cos\theta\; \sin\theta\; d\theta\; d\phi \right] d\lambda$$

or

$$q_j'' = \pi \int_0^\infty I_{j\lambda} d\lambda \tag{11.20}$$

The spectral radiosity then is

$$q_{j\lambda}'' = \pi I_{j\lambda} \tag{11.21}$$

where the coefficient π has the unit of sr.

Having defined the radiation fluxes of significance, we can now express emissivity as

$$\varepsilon = \frac{\int_0^\infty \varepsilon_\lambda q_{b\lambda}'' \, d\lambda}{\int_0^\infty q_{b\lambda}'' \, d\lambda} \tag{11.22}$$

where ε_λ is the monochromatic or spectral emissivity, used in the integral to evaluate the total or hemispherical emissivity ε. The black-body radiation flux is q_b'' . Similarly, for absorptivity

$$\alpha = \frac{\int_0^\infty \alpha_\lambda q_{i\lambda}'' \, d\lambda}{\int_0^\infty q_{i\lambda}'' \, d\lambda} \tag{11.23}$$

where $q_{i\lambda}''$ is the spectral irradiative heat flux.

In the discussion that centered about Figure 11.8, reflectivity and transmissivity were introduced. *Transmissivity* of a body is the fraction of incident energy that is transmitted through a body. The total transmissivity is found with

$$\tau = \frac{\int_0^\infty \tau_\lambda q_{i\lambda}'' \, d\lambda}{\int_0^\infty q_{i\lambda}'' \, d\lambda} \tag{11.24}$$

where, like absorptivity and emissivity, τ_λ is the monochromatic or spectral transmissivity, and $q_{i\lambda}''$ is the spectral irradiative heat flux. For an opaque body, transmissivity is zero.

The reflectivity ρ of a surface is defined as the fraction of incident energy reflected by the surface:

$$\rho = \frac{\int_0^\infty \rho_\lambda q_{i\lambda}'' \, d\lambda}{\int_0^\infty q_{i\lambda}'' \, d\lambda} \tag{11.25}$$

While emissivity is dependent on the *emitted* flux (q_b'' of Equation 11.22), absorptivity, transmissivity, and reflectivity all involve the *incident* flux (q_i'' of Equations 11.23, 11.24, and 11.25). So when incident energy impinges on a surface, it is absorbed, transmitted, and/or reflected (see Figure 11.8). Therefore,

$$\alpha q_i + \tau q_i + \rho q_i = q_i$$

or

$$\alpha + \tau + \rho = 1 \tag{11.26a}$$

On a monochromatic or spectral basis, we can also write

$$\alpha_\lambda + \tau_\lambda + \rho_\lambda = 1 \tag{11.26b}$$

For an opaque surface, $\tau = 0$ and $\alpha + \rho = 1$. For a perfect reflector, $\alpha = 0$, $\tau = 0$, and $\rho = 1$. A black body absorbs all incident energy, so $\tau = \rho = 0$ and $\alpha = 1$. Also, for a black body, $\varepsilon = \alpha$ when both properties are evaluated at the same temperature.

11.6 RADIATION LAWS

In this section, we present Kirchhoff's Law, the Stefan-Boltzmann Law, Planck's Distribution Law, and Wien's Displacement Law.

11.6.1 KIRCHHOFF'S LAW

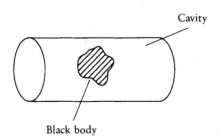

Cavity

Black body

FIGURE 11.10 A cavity with isothermal walls containing a black body at the
same temperature.

Earlier in this chapter, we discussed emissivity and absorptivity—both total and spectral. We can deduce further information on these properties by performing a thought experiment. We define a cavity as an enclosure with isothermal walls. A cavity could be something as simple as an empty but sealed soft-drink can whose walls are at a constant temperature. Consider a cavity in which there is located a small back body whose temperature equals that of the enclosure walls (see Figure 11.10). Recall that a black body is a perfect emitter and a perfect absorber and has a diffuse surface. Inside the enclosure, we have no temperature difference. This means that there is no *net* energy exchange between the black body and the walls *even though each surface is radiating energy.* Thus, with no temperature difference, we conclude that the walls of the cavity are receiving heat by radiation from the black body and are radiating an equal amount back to the black body. In other words, the energy impinging on the black-body surface will just equal the energy emitted by the black body:

$$q''_{cav} = q_b''$$ (11.27)

Thus, radiation emitted by a black body is the same as the equilibrium radiation intensity within a cavity when both are at the same temperature. Next, we replace the black body with a nonblack body whose temperature equals that of the cavity walls. Again, because there is no temperature difference, a net exchange of energy does not occur. The energy absorbed by the nonblack body is equal to the absorptivity of the body times the energy coming from the cavity (= $\alpha q''_{cav}$). This must equal the energy leaving the surface of the nonblack body q_e''. Thus,

$$\alpha q''_{cav} = q_e''$$

Comparing to Equation 11.27 yields the result that

$$\alpha = \frac{q_e''}{q_b''}$$ (11.28)

However, Equation 11.5 gives the definition of emissivity as

$$\varepsilon = \frac{q_e''}{q_b''}$$ (11.5)

and we conclude that

$$\varepsilon = \alpha$$ (11.29a)

The above result is known as Kirchhoff's Law, which states that, at a given temperature, the absorptivity and emissivity of any solid surface are equal when the surface emits just what it absorbs. Equation 11.29a is valid also on a spectral basis:

$$\varepsilon_\lambda = \alpha_\lambda$$ (11.29b)

11.6.2 Stefan-Boltzmann Law

We will now deduce a relationship between radiation emitted from a black body and its temperature. We will use thermodynamics and borrow without proof some results of electromagnetic field theory and statistical mechanics. A black body enclosed by a cavity has associated with it an (internal) energy density u_ρ. From electromagnetic theory the energy density (the internal energy per unit volume) of the cavity is given by

$$u_\rho = \frac{4q_b''}{c_o} \tag{11.30}$$

where c_o is the speed of light in a vacuum and q_b'' is the total energy emitted per unit area per unit time by the black body. The dimension of energy density is $(F \cdot L)/L^3$ (J/m^3 or BTU/ft^3). From statistical mechanics, we can consider (mathematically) the radiation impinging on the cavity walls from the black body as being a gas. The gas is made up of photons, each having an energy $h\nu$ and momentum $h\nu/c_o$. The photon gas will exert a pressure on the walls of the cavity given by

$$p = u_\rho/3 \tag{11.31}$$

Equation 11.31 is dimensionally correct, because the energy density has dimensions of $F \cdot L/L^3$, which reduces to F/L^2. From thermodynamics we write without derivation the following relationship for a gas:

$$\left(\frac{\partial U}{\partial \mathcal{V}}\right)_T = T\left(\frac{\partial p}{\partial T}\right)_v - p \tag{11.32}$$

where U is the internal energy, $\mathcal{V}$ is volume, and p is pressure. This equation is applied to the photon gas. The energy density u_ρ is related to the internal energy by

$$U = \mathcal{V}u_\rho$$

Therefore,

$$\frac{\partial U}{\partial \mathcal{V}} = u_\rho$$

Substitution of this equation and Equation 11.31 into Equation 11.32 yields a differential equation for u_ρ:

$$u_\rho = \frac{T}{3}\frac{du_\rho}{dT} - \frac{u_\rho}{3} \tag{11.33}$$

where we use ordinary differential notation because the energy density is presumed to depend only on temperature. Integration of Equation 11.33 gives

$$u_\rho = BT^4 \tag{11.34}$$

where B is a constant of integration. Combining with Equation 11.30, we obtain a relationship between the heat emitted by a black body and the black-body temperature:

$$q_b'' = \frac{Bc_o}{4}T^4 \qquad (11.35a)$$

Thus, q_b'' varies with the fourth power of the (absolute) temperature of the black body. This is called the Stefan-Boltzmann Law. With thermodynamics, we have been able to predict this relationship between heat-transfer rate and temperature, but not the value of the constant. It has been determined experimentally that the total energy emitted from a black surface is

$$q_b'' = \sigma T^4 \qquad (11.35b)$$

where σ [$=5.675 \times 10^{-8}$ W/(m$^2 \cdot$K^4) $= 0.1714 \times 10^{-8}$ BTU/(hr$\cdot$ft$^2 \cdot {}^\circ$R^4)] is called the Stefan-Boltzmann constant. For nonblack bodies, the rate of energy emission is given by

$$q_e'' = \varepsilon \sigma T^4 \qquad (11.36)$$

where emissivity ε was discussed earlier.

It was just mentioned that the Stefan-Boltzmann constant was determined experimentally, which implies that a black body is at our disposal. Solids that behave as black bodies (ideal emitters) do not exist, but we can make an excellent imitation by using a cavity. Consider an empty cavity in which we have made a very small hole. The hole itself, *even though there is no material surface present*, behaves very nearly like a black surface. Figure 11.11 illustrates what happens to an incident beam that impinges on the hole. The beam enters the cavity and is reflected many times before it leaves through the same hole. Therefore, the energy of the incident beam becomes almost completely absorbed by the cavity walls. The following equation gives the emissivity of a hole ε_h in a rough-walled enclosure as

$$\varepsilon_h \cong \frac{\varepsilon}{\varepsilon + f(1-\varepsilon)} \qquad (11.37)$$

where the emissivity of the cavity walls is ε and the fraction of the internal cavity-surface area removed by the hole is f.

Example 11.3

A beer can has internal measurements of 4-1/2 in tall and a diameter of 2-1/2 in. It is made of aluminum (rolled and polished) and is punctured on one of its ends with a round hole of diameter

Incident beam

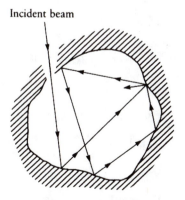

FIGURE 11.11 Approximating a black surface with a hole in a cavity.

1/8 in. (a) Calculate the emissivity of the hole. The can is placed in an oven and heated to a temperature of 150°F. (b) What is the rate of radiant emission from the hole?

Solution

We will use Equation 11.37 to calculate the emissivity.

Assumptions

1. The can is heated uniformly.
2. The can radiates in a steady fashion.
3. The can loses heat only by radiation from the small hole.

From Appendix Table E.1, we read, for rolled and polished aluminum, that $\varepsilon = 0.039$. This value is far from being equal to that for a black body. The inside surface area is

$$A_s = 2(\pi D^2/4) + \pi DL = \frac{\pi(2.5/12)^2}{2} + \pi(2.5/12)(4.5/12) = 0.0682 + 0.245$$

or

$$A_s = 0.314 \text{ ft}^2$$

The area of a 1/8-in hole is $\pi\{1/[(8)(12)]^2\}/4 = 8.522 \times 10^{-5}$ ft². The fraction of area removed is then

$$f = \frac{8.522\times10^{-5}}{0.314} = 2.714\times10^{-4}$$

(a) The emissivity of the hole is found with Equation 11.37:

$$\varepsilon_h = \frac{0.039}{0.039 + 2.714 \times 10^{-4}(1 - 0.039)}$$

or

$$\varepsilon_h = 0.9934$$

where 1.0 would result for a black body.

(b) The Stefan-Boltzmann Law for black-body radiation is

$$q_b'' = \sigma T^4$$

For the hole, we write

$$q_e'' = \varepsilon\sigma T^4 = 0.9934[0.1714 \times 10^{-8} \text{ BTU/(hr·ft}^2\cdot°\text{R})](150 + 460)^4$$

or

$$q_e'' = 235 \text{ BTU/(hr·ft}^2)$$

Multiplying by the area of the hole gives

$$q_e = 235(8.522 \times 10^{-5})$$

$$q_e = 2.01 \times 10^{-2} \text{ BTU/hr}$$

The heated can would lose heat to the surroundings also by natural convection and by radiation from its outside surface. ❏

11.6.3 Planck's Distribution Law

During the early 1900s, several efforts were made at developing equations that would predict the radiated energy flux from a black body. The Rayleigh-Jeans Formula predicted the energy flux for low-frequency (v) radiation, while the Wien Equation applied to high-frequency radiation. The Planck Equation was originally intended to connect or interpolate between the low- and high-frequency formulas. The Planck Equation is

$$q''_{b\lambda} = \frac{2\pi c_o^2 h}{\lambda^5} \frac{1}{\exp[c_o h/\lambda KT] - 1} = \pi I_{b\lambda} \tag{11.38}$$

where $q''_{b\lambda}$ is the spectral black-body emission per unit area per unit wavelength, c_o is the speed of light in a vacuum, h is Planck's constant (= $6.625\,6 \times 10^{-34}$ J·s = 6.286×10^{-37} BTU·s = 1.7433×10^{-40} BTU·hr), K is the Boltzmann constant (= $1.380\,5 \times 10^{-3}$ J/K = 7.274×10^{-7} BTU/°R), and T is the temperature of the surface in absolute units. The Planck Equation led to the development of quantum theory and quantum statistics. The Planck distribution is often written in a form that combines the constants:

$$q''_{b\lambda} = \pi I_{b\lambda} = \frac{C_1}{\lambda^5 [\exp(C_2/\lambda T) - 1]} \tag{11.39}$$

where

$$C_1 = 2\pi h c_o^2 = 3.741\,1 \times 10^{-16} \text{ W·m}^2 = 3.820 \times 10^{-18} \text{ BTU·ft}^2/\text{s}$$

and

$$C_2 = hc_o/K = 1.439 \times 10^{-2} \text{ m·K} = 8.500 \times 10^{-2} \text{ ft·°R}$$

Planck's Law contains three variables: the spectral heat rate $q''_{b\lambda}$, wavelength λ, and absolute temperature T. A graph of Planck's distribution over the thermal radiation spectrum is shown in Figure 11.12, where the spectral rate of emission for a black body is plotted as a function of wavelength for various temperatures. The spectral black-body emittance $q''_{b\lambda}$ varies from 10^7 to 10^{12} and has units of energy density (= energy per area per wavelength). Wavelength varies from 10^{-7} to 10^{12} m, and temperature values range from (but are not limited to) 273 to 2 600 K. Also shown on the graph is the region of visible radiation. (It is interesting to note that if the filament of a light bulb radiated like a black body at 500 K, then very little of the radiant energy would be within the visible region. So in a darkened room, we might not see it until its temperature were raised much higher.) The Planck Equation predicts energy vs. wavelength and that the peak of maximum energy shifts toward shorter wavelengths with increasing temperature. The Planck distribution as stated in Equation 11.39 is of spectral black-body emission. To obtain the total or hemispherical emission from a black body, we must integrate Equation 11.39 over all wavelengths. Thus,

$$q''_b = \int_0^\infty q''_{b\lambda} \, d\lambda = \int_0^\infty \frac{C_1 \, d\lambda}{\lambda^5 [\exp(C_2/\lambda T) - 1]}$$

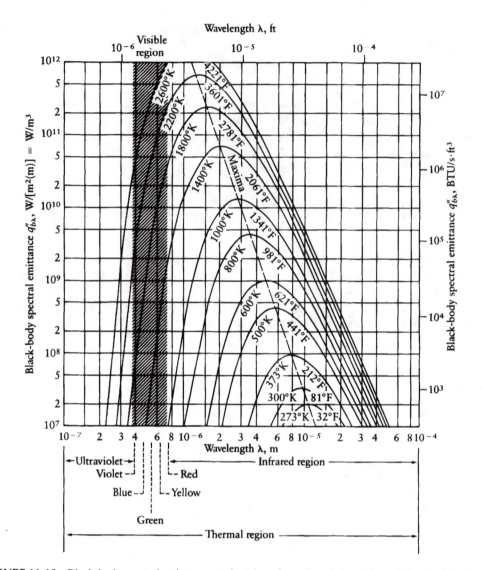

FIGURE 11.12 Black-body spectral emittance as a function of wavelength according to Planck's Distribution Law. *Source: from Handbook of Tables for Applied Engineering Science,* edited by R.E. Bolz and G.L. Tuve, 2nd ed., CRC Press, 1973, p. 208, with permission.

or, in terms of universal constants,

$$q_b'' = \int_0^\infty \frac{2\pi h c_o^2 d\lambda}{\lambda^5 [\exp(hc_o/K\lambda T) - 1]} \tag{11.40}$$

We change the variable of integration from λ to x (to make integration easier), where

$$x = hc_o/K\lambda T$$

$$\lambda = hc_o/KTx$$

$$d\lambda = -\frac{hc_o(KT)dx}{K^2T^2x^2}$$

In terms of x, Equation 11.40 becomes

$$q_b'' = \int_0^\infty \frac{2\pi hc_o^2(-hc_oKT\,dx)}{K^2T^2x^2(h^5c_o^5/K^5T^5x^5)[\exp(x)-1]}$$

or

$$q_b'' = \frac{2\pi K^4 T^4}{h^3 c_o^2}\int_0^\infty \frac{x^3\,dx}{\exp(x)-1}$$

This integral is evaluated by expanding the denominator in $\exp(x)$ and integrating term by term. The integral then becomes $\pi^4/15$. The result for total black-body emission is

$$q_b'' = \left(\frac{2}{15}\frac{\pi^5 K^4}{h^3 c_o^2}\right)T^4 \tag{11.41}$$

Equation 11.41 is recognized as the Stefan-Boltzmann Law, where the (Stefan-Boltzmann) constant σ is

$$\sigma = \frac{2}{15}\frac{\pi^5 K^4}{h^3 c_0^2} = \frac{C_1\pi^4}{15C_2^4} \tag{11.42}$$

In some instances, it is necessary to know the percentage of black-body radiation that lies within a certain wavelength interval at a given temperature. With regard to Figure 11.12, this means that we would select a given temperature curve and integrate between the limits (on wavelength) of interest. This is illustrated in Figure 11.13. To find the energy flux associated with the shaded region, we integrate the Planck distribution over the interval from λ_1 to λ_2:

$$q_b''(\lambda_1 \text{ to } \lambda_2) = \int_{\lambda_1}^{\lambda_2} q_{b\lambda}''\,d\lambda$$

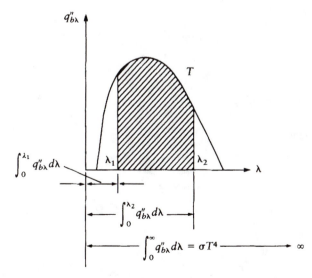

FIGURE 11.13 Energy existing over a given wavelength interval for a known black-body temperature.

However, because of the difficulty encountered in integrating the Planck Equation, the above equation is rewritten as

$$q_b''(\lambda_1 \text{ to } \lambda_2) = \int_0^{\lambda_2} q_{b\lambda}'' \, d\lambda - \int_0^{\lambda_1} q_{b\lambda}'' \, d\lambda$$

The values of the integrals are shown in Figure 11.13. It is convenient to express the black-body energy flux over the interval from λ_1 to λ_2 in terms of a percentage of total energy as

$$\frac{q_b''(\lambda_1 \text{ to } \lambda_2)}{q_b''(0 \text{ to } \infty)} = \frac{\int_0^{\lambda_2} q_{b\lambda}'' \, d\lambda - \int_0^{\lambda_1} q_{b\lambda}'' \, d\lambda}{\sigma T^4} = \begin{array}{l} \text{(Percentage of black-body} \\ \text{radiation that lies in the} \\ \text{wavelength interval } \lambda_1 \text{ to } \lambda_2) \end{array} \qquad (11.43)$$

Table 11.1 provides a tabular summary of the fraction of black-body radiation that lies in the range 0 to λ (actually λT) as a function of λT. The integrals of Equation 11.43 are more conveniently evaluated if the variable is changed from λ to λT. The usefulness of the chart is illustrated in the following example.

Example 11.4

A light bulb radiates like a black body at 2 800 K. Determine the percentage of total emitted energy that lies in the visible range.

Solution

A light bulb is not actually a black body, but it does radiate in all directions. We can solve this problem by using Table 11.1.

Assumptions

1. The light-bulb-emission spectra follow Planck's Law.
2. The visible region exists over 4×10^{-7} m $\leq \lambda \leq 7 \times 10^{-7}$ m.
3. The system is at steady state.

We calculate for the light bulb:

$$\lambda_1 T = 4 \times 10^{-7}(2\ 800) = 1.12 \times 10^{-3} \text{ m·K}$$

and

$$\lambda_2 T = 7 \times 10^{-7}(2\ 800) = 1.96 \times 10^{-3} \text{ m·K}$$

From Table 11.1 we read

$$\frac{\int_0^{\lambda_1 T} q_{b\lambda}'' \, d\lambda}{\sigma T^4} = 0.005\ 1$$

and

$$\frac{\int_0^{\lambda_2 T} q_{b\lambda}'' \, d\lambda}{\sigma T^4} = 0.065$$

TABLE 11.1
Fraction of Total Radiation Emitted over a Given Wavelength-Temperature Product

$\lambda T \times 10^3$ m·K	$\dfrac{\int_0^{\lambda T} q''_{b\lambda}\, d\lambda}{\sigma T^4}$	$\lambda T \times 10^3$ m·K	$\dfrac{\int_0^{\lambda T} q''_{b\lambda}\, d\lambda}{\sigma T^4}$
0.2	0.341796×10^{-26}	6.2	0.754187
0.4	0.186468×10^{-11}	6.4	0.769282
0.6	0.929299×10^{-7}	6.6	0.783248
0.8	0.164351×10^{-4}	6.8	0.796180
1.0	0.320780×10^{-3}	7.0	0.808160
1.2	0.213431×10^{-2}	7.2	0.819270
1.4	0.779084×10^{-2}	7.4	0.829580
1.6	0.197204×10^{-1}	7.6	0.839157
1.8	0.393449×10^{-1}	7.8	0.848060
2.0	0.667347×10^{-1}	8.0	0.856344
2.2	0.100897	8.5	0.874666
2.4	0.140268	9.0	0.890090
2.6	0.183135	9.5	0.903147
2.8	0.227908	10.0	0.914263
3.0	0.273252	10.5	0.923775
3.2	0.318124	11.0	0.931956
3.4	0.361760	11.5	0.939027
3.6	0.403633	12	0.945167
3.8	0.443411	13	0.955210
4.0	0.480907	14	0.962970
4.2	0.516046	15	0.969056
4.4	0.548830	16	0.973890
4.6	0.579316	18	0.980939
4.8	0.607597	20	0.985683
5.0	0.633786	25	0.992299
5.2	0.658011	30	0.995427
5.4	0.680402	40	0.998057
5.6	0.701090	50	0.999045
5.8	0.720203	75	0.999807
6.0	0.737864	100	1.000000

Note: Values from the λT column are read as follows: $\lambda T \times 10^3$ m·K = 0.2; therefore, $\lambda T = 0.2 \times 10^{-3}$ m·K.

The percentage of total emitted energy that lies in the visible range is

$$\frac{\int_{\lambda_1 T}^{\lambda_2 T} q''_{b\lambda}\, d\lambda}{\sigma T^4} = 0.065 - 0.005\ 1 = 0.060 = 6.0\% \qquad \square$$

11.6.4 WIEN'S DISPLACEMENT LAW

Another feature of the Planck Equation is that it shows a maximum for a given temperature, as indicated in each line of Figure 11.12. We can obtain an important relation associated with the maximum in the Planck distribution. To obtain the maximum, we differentiate Equation 11.35 with respect to λ and set the result equal to zero. The operation is simplified greatly if we first rewrite Planck's Equation in terms of $x\ (= hc_o/K\lambda T)$:

$$q''_{b\lambda} = \frac{C_1}{\lambda^5[\exp(C_2/\lambda T)-1]} = \frac{2\pi hc_o^2}{\lambda^5[\exp(hc_o/K\lambda T)-1]}$$

or

$$q''_{b\lambda} = \frac{2\pi K^5 T^5}{h^4 c_o^3} \frac{x^5}{\exp(x)-1} \tag{11.44}$$

Differentiating and setting the result equal to zero, we get

$$\frac{dq''_{b\lambda}}{dx} = \frac{2\pi K^5 T^5}{h^4 c_o^3} \frac{[\exp(x)-1]5x^4 - x^5[\exp(x)]}{[\exp(x)-1]^2} = 0$$

Simplifying,

$$5[\exp(x_{max})-1] = x_{max}\exp(x_{max}) \tag{11.45}$$

The solution is found by trial and error to be

$$x_{max} = \frac{hc_o}{K\lambda T} = 4.965\ 1$$

Rearranging and substituting for the constants, we obtain

$$\lambda_{max}T = \frac{hc_o}{4.965\ 1\ K}$$

or

$$\lambda_{max}T = 2.898 \times 10^{-3}\ \text{m·K} \tag{11.46}$$

Equation 11.46 is known as Wien's Displacement Law. It predicts that as temperature increases, the wavelength (denoted as λ_{max}) corresponding to the maximum spectral emissive black-body power $q''_{b\lambda}$ decreases. In other words, the peak of each curve in Figure 11.12 shifts toward shorter wavelengths with increasing temperature. The locus of the product $\lambda_{max}T$ is shown in the figure as the dashed line labeled "maxima." Wien's displacement law is useful for estimating the temperature of remotely located thermal radiation sources such as the sun.

Example 11.5

Measurements indicate that the maximum intensity of radiation from the sun is at a wavelength of $\lambda_{max} = 0.5$ μm. Estimate the surface temperature of the sun and the emitted heat flux.

Solution

The sun emits energy in all directions and is not necessarily a perfect emitter. Nevertheless, Wien's displacement law can be used to solve this problem.

Assumptions

1. The sun's surface emits like a black body (for lack of knowledge of the emissivity).
2. The system is at steady state.

From Wien's Displacement Law we write

$$\lambda_{max} T = 2.898 \times 10^{-3} \text{ m·K}$$

With λ_{max} given,

$$T = \frac{2.898 \times 10^{-3}}{0.5 \times 10^{-6}}$$

or

$$T = 5\ 800 \text{ K}$$

The heat flux is given by the Stefan-Boltzmann Equation:

$$q_b'' = \sigma T^4 = [5.675 \times 10^{-8} \text{ W/(m}^2 \cdot \text{K}^4)](5\ 800 \text{ K})^4$$

or

$$q_b'' = 6.422 \times 10^7 \text{ W/m}^2 \qquad\qquad \square$$

11.7 CHARACTERISTICS OF REAL SURFACES

Having examined the behavior of ideal surfaces, we now turn our attention to real surfaces. Our interest is in modeling the characteristics of real surfaces in terms of the ideal surfaces discussed.

Figure 11.14 shows the usefulness of the Planck distribution. Consider a black body at some temperature *T*. It will emit thermal radiation according to the black-body curve of Figure 11.14. A real surface at the same temperature *T* would emit energy according to the real-surface curve of the figure. We approximate real-body behavior by multiplying the black-body curve by the emissivity to obtain the gray-body emission curve. When we use the concept of emissivity, we are accounting for deviations from Planck's Law.

Also, for the gray body it is evident that ε_λ is a constant and equal to ε. Furthermore, for the gray body, α_λ = a constant = α = ε_λ = ε, τ_λ = a constant = τ, and ρ_λ = a constant = ρ. These relationships are written here without proof.

We have previously defined three different forms of the emissivity associated with a surface:

$\varepsilon_{\lambda\theta\phi}$ = a function of wavelength and direction (spectral, directional)
ε_λ = a function of wavelength (spectral)
ε = total emissivity

Data on the variation of the emissivity with direction is shown on a polar graph. Figure 11.15 shows a polar graph of emissivity vs. angle for two general classes of materials: conductors and

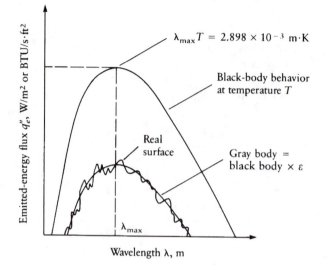

FIGURE 11.14   Gray-body approximation of a red surface.

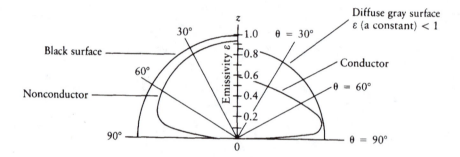

FIGURE 11.15   Variety of emissivity with viewing angle for two general classes of materials.

nonconductors. One quadrant is devoted to each material. The graphs are symmetric about the z axis and are of hemispherical representations. Thus, when a differential area of a nonconductor is viewed from an angle of 45°, the emissivity (= ratio of energy actually emitted to that emitted if the surface were that of a black body at the same temperature) is about 0.8 according to the graph of Figure 11.15. Both conductors and nonconductors have emissivities that vary greatly with angle. Both emissivity curves approach zero as the viewing angle approaches 90°. Another feature worth noting is how great the deviations are from black-body and diffuse-gray-body characteristics.

   Emissivity data are also presented in terms of normal emissivity vs. wavelength. The normal emissivity is that measured by a device located normal to the surface of interest (along the z axis of Figure 11.15, for example). Figure 11.16 gives normal spectral emissivity $\varepsilon_{n\lambda}$ data vs. wavelength for a number of surfaces. In general, data on numerous materials indicate that the emissivity of metallic surfaces is low; they approximate black surfaces poorly. Metals such as zinc and aluminum form oxides on their surfaces that act to protect the metal from further oxidation. It has been found that such layers may significantly increase the emissivity of the surface.

   The emissivity of conductors in general increases with increasing temperature. This feature is illustrated in Figure 11.17. The vertical axis is the total normal emissivity (over all wavelengths and measured normal to the surface).

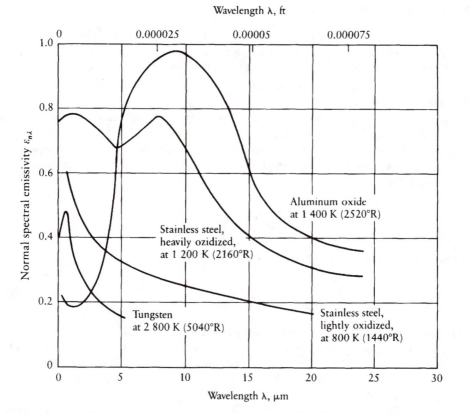

FIGURE 11.16 Variation of normal spectral emissivity with wavelength for several surfaces.

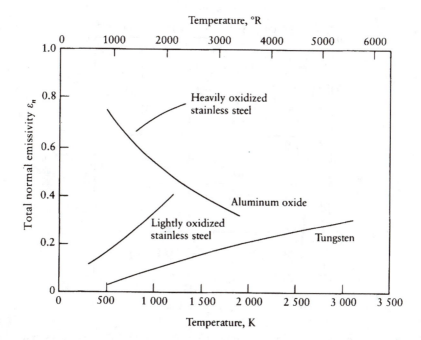

FIGURE 11.17 Variation of the total normal emissivity ε_n with temperature.

11.7.1 Absorptivity and Reflectivity

Figure 11.18 shows the variation of spectral normal reflectivity $\rho_{n\lambda}$ and absorptivity $\alpha_{n\lambda}$ for a variety of surfaces. Note that, over the wavelength range for which there is data, black paint is an almost perfect absorber. It has a high absorptivity and a low reflectivity. Also, white paint is a poor absorber over shorter wavelengths but improves at long wavelengths. Snow (not shown) exhibits properties similar to those of white paint.

As seen in Figure 11.18, reflectivity and absorptivity are shown as the vertical axes of the graph. As indicated by the way these data are presented, a material that has a high reflectivity has a low absorptivity, and vice versa. This is easily understood from the discussion that centered about Figure 11.9a. An incident beam is reflected, absorbed, and/or transmitted. We concluded that, for translucent surfaces,

$$\alpha + \rho + \tau = 1 \qquad (11.26a)$$

For opaque surfaces (such as those in Figure 11.18),

$$\alpha + \rho = 1 \qquad (11.47)$$

For many real surfaces, it is often assumed that emissivity and absorptivity are equal. A consequence of this assumption is that total radiation or total absorption are dependent solely on the temperature difference. Experiment has shown, however, that emissivity and absorptivity vary greatly with wavelength. Transmittivity also varies greatly with wavelength, which is the basis for radiation and optical filters.

Consider a surface being heated by a high-temperature source—for example, the sun heating the surface of an automobile. The sun is a high-temperature radiation source. The automobile is

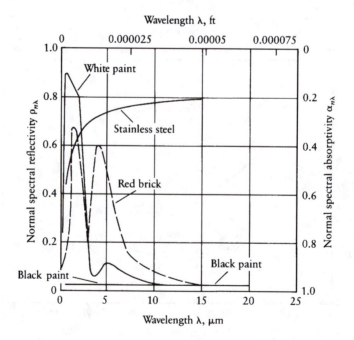

FIGURE 11.18 Variation of normal spectral reflectivity $\rho_{\lambda n}$ and normal spectral absorptivity $\alpha_{\lambda n}$ with wavelength for various opaque surfaces.

losing heat by convection to the surrounding air, but also by radiation. Compared to the sun, the automobile is a low-temperature radiation source. There is probably a net heat gain by the automobile, because the solar absorptance (meaning the radiation from a high-temperature source such as the sun) is many times higher than the low-temperature emissivity.

Radiation heat transfer is not something we can visualize. The tendency may be to believe that radiation phenomena behave in the same way as our eyes interpret visual appearance and color. Something that appears white to our eyes, however, may behave like a black body. Titanium white or snow are both white in color but behave as black bodies at low temperature (long wavelengths). Color and visual appearance are not always good guides to how a material behaves with respect to radiation heat transfer.

Other aspects that influence radiation are thin surface films that can develop or adhere to a body. For example, a thin film of oxide (as mentioned earlier) or of oil on the surface of a metal greatly increases its emissivity at low temperatures. Some dark pigment mixed into a white (in color) paint greatly increases its solar absorptance. These and many other factors influence decisions regarding the color of roofs, tank walls, car roofs, solar heaters, and batteries. Table 11.2 gives some data on low-temperature emissivity and solar (high-temperature-source) absorptivity.

TABLE 11.2
Low-Temperature Emissivity and High-Temperature Absorptivity of Various Surfaces

Surface material	Emissivity at atmospheric temperatures	Solar absorptivity
Highly polished "white" metals, gold, yellow brass	0.02–0.08	0.1–0.4
Clean, "dark" metals	0.1–0.35	0.3–0.6
Metallic-pigment paints	0.35–0.55	0.4–0.6
White, nonmetal surfaces	0.7–0.9	0.1–0.35
Dark-colored nonmetals	0.7–0.9	0.45–0.8
Black paint, asphalt, carbon, water	0.85–0.95	0.7–0.9

Source: from *Handbook of Tables for Applied Engineering Science,* edited by R.E. Bolz and G.L. Tuve, CRC Press, 2nd edition, 1973, p. 212, with permission.

11.7.2 TRANSMISSIVITY

A number of substances are selective in the energy they transmit. Glass is an example of such a material. Over certain wavelengths, glass will transmit little or no radiation. Over other wavelengths, clear, polished glass will transmit up to 90% of the incident energy. Table 11.3 gives the spectral transmittance of glass in the ultraviolet, visible, and infrared regions. Data are provided for clear polished, gray-tinted, bronze-tinted, and blue-green 1/4-in plate. As indicated by the numbers, glass is selective. At wavelengths below 320×10^{-9} m, very little radiative energy is transmitted. The numbers listed as "total solar" are obtained by integration of the numbers just above each total in the interval of interest.

Example 11.6

An automobile parked in the sun has a clear, polished plate-glass windshield approximately 1/4 in. thick. The car receives radiation in the amount of 349 BTU/(hr·ft^2) (= 1 100 W/m^2) from the sun (assumed to be a black body), whose temperature is 10,400°R. (a) Calculate the rate at which the sun's radiant energy is transmitted through the glass windshield. The interior of the car is considered to be a black body that radiates at 100°F. (b) Calculate the rate at which radiant energy from the car interior is transmitted through the glass windshield.

TABLE 11.3
Spectral Transmittance of Glass

Wavelength, nm*	Clear, polished 1/4-in plate	Gray tint, 1/4-in Solargray®	Bronze tint, 1/4-in Solarbronze®	Blue-green, 1/4-in Solex®
Ultraviolet region				
300	1.1	0.0	0.0	0.0
320	0.3	0.0	0.0	0.0
340	29.4	2.1	0.3	0.7
360	74.8	33.3	22.8	33.3
380	<u>79.3</u>	<u>44.2</u>	<u>34.1</u>	<u>44.3</u>
Total solar	68.0	34.0	26.0	35.0
Visible region				
400	87.2	53.8	48.5	70.6
440	87.3	43.1	44.3	71.6
480	89.3	40.6	43.3	79.0
520	89.5	38.1	45.2	79.8
560	88.7	46.1	54.2	76.5
600	87.0	38.3	51.7	69.4
640	84.4	38.6	51.5	60.3
680	81.3	49.4	54.4	50.1
720	<u>77.9</u>	<u>59.6</u>	<u>55.7</u>	<u>40.5</u>
Total solar	88.0	41.0	51.0	75.0
Infrared region				
800	71.4	51.3	46.7	27.1
900	65.8	43.6	39.0	18.8
1 000	63.4	40.5	36.3	15.8
1 100	62.9	38.9	35.2	15.3
1 200	63.8	38.2	35.8	16.3
1 400	69.9	44.4	43.7	21.1
1 600	73.8	54.3	55.3	32.6
1 800	75.9	56.3	57.8	40.0
2 000	<u>77.1</u>	<u>58.5</u>	<u>58.9</u>	<u>40.3</u>
Total solar	67.0	46.0	42.0	22.0

* One nanometer = 1×10^{-9} meters = 1×10^{-3} micrometers.

Source: from *Handbook of Tables for Applied Engineering Science,* edited by R.E. Bolz and G.L. Tuve, CRC Press, 2nd edition, 1973, p. 684, with permission.

Solution

We can use data from Table 11.3 for 1/4-in clear glass. We will use only average values given over the various wavelength intervals. Note that the sun is a high-temperature source, while the car interior is a low-temperature source.

Assumptions

1. We will consider only radiation from the sun and no other sources.
2. The sun is a black body radiating at 10,400°R = 5 800 K.
3. The data for glass given in Table 11.3 apply, and the average values over the wavelength intervals give sufficiently accurate results.
4. At wavelengths beyond those listed in Table 11.3, glass is opaque.
5. The car interior is a black body radiating at 100 + 460 = 560°R = 311 K.

(a) We will formulate the solution method for the first wavelength interval in detail. The total transmissivity is defined as

$$\tau = \frac{\int_0^\infty \tau_\lambda q''_{i\lambda}\, d\lambda}{\int_0^\infty q''_{i\lambda}\, d\lambda}$$

where τ_λ is the spectral transmissivity and $q''_{i\lambda}$ is the spectral radiosity per unit area. Over the interval of 300×10^{-9} m $\leq \lambda \leq 380 \times 10^{-9}$ m, spectral transmissivity is constant and equal to 68% or 0.68 (from Table 11.3). Moreover, radiation from only the sun, which is considered to be a black body, gives $q''_{i\lambda} = q''_{b\lambda}$. So the equation for transmissivity over the interval of interest becomes

$$\tau = \frac{0.68 \int_{300\text{ nm}}^{380\text{ nm}} q_{b\lambda}\, d\lambda}{\sigma T^4}$$

With the data of Table 11.1 in mind, we first calculate

$$\lambda_1 T = 300 \times 10^{-9}(5\ 800) = 1.74 \times 10^{-3}\ \text{m·K}$$

and

$$\lambda_2 T = 380 \times 10^{-9}(5\ 800) = 2.204 \times 10^{-3}\ \text{m·K}$$

Data from Table 11.1 give, for the transmissivity

$$\tau = 0.68(0.101 - 0.033\ 4) = 0.046\ 0$$

The energy transmitted from the sun through the glass is

$$q''_{in} = \tau q''_i = 0.046(1\ 100)$$

or

$$q''_{in} = 50.6\ \text{W/m}^2$$

(b) For a black-body source at $560°\text{R} = 311$ K, we find

$$\lambda_1 T = 300 \times 10^{-9}(311) = 9.44 \times 10^{-5}$$

$$\lambda_2 T = 380 \times 10^{-9}(311) = 1.18 \times 10^{-4}$$

Table 11.1 gives negligibly small values of the corresponding integrals. Therefore, for the car interior,

$$q''_{out} = 0.68(0) = 0$$

The calculations for the other wavelength intervals are summarized in Table 11.4. As indicated by the totals of Table 11.4, the heat from the sun transmitted through the windshield is many times

TABLE 11.4
Solution Chart for the Glass of Example 11.5

Wavelength interval, nm	τ_λ Table 11.3	$\lambda_1 T$ m·K	$\lambda_2 T$ m·K	Table 13.1 $\frac{\int_0^{\lambda_1 T} q''_{b\lambda}\,d\lambda}{\sigma T^4}$	$\frac{\int_0^{\lambda_2 T} q''_{b\lambda}\,d\lambda}{\sigma T^4}$	$\tau = \tau_\lambda \frac{\int_{\lambda_1 T}^{\lambda_2 T} q''_{b\lambda}\,d\lambda}{\sigma T^4}$	$q'' = q_i''$
Sun: $T = 5\ 800$ K							
300–380	0.68	1.74×10^{-3}	2.204×10^{-3}	0.101	0.334	0.158	50.6
400–720	0.88	2.32	4.18	0.121	0.514	0.346	380
800–2 000	0.67	4.64	1.16	0.585	0.940	0.238	262
						Total	693
Car interior: $T = 311$ K							
300–380	0.68	0.933×10^{-4}	1.18×10^{-4}	0	0	0	0
400–720	0.88	2.24	1.24	0	0	0	0
800–2 000	0.67	2.49	6.22	0	0	0	0
						Total	0

higher than that transmitted out from the car interior. The interior of a car left in the sun becomes quite warm, because the glass transmits about 80% of the incident thermal radiation into the passenger compartment while effectively very little is transmitted out. Anyone entering a car that has been left in the sun has experienced this result. (Remember, however, that the car can lose heat to the environment by convection, and that the calculations made in this example include the assumption of black-body behavior.)

Thus, glass transmits a large percentage of incident short-wavelength (high-temperature) radiation and transmits practically no long-wavelength (low-temperature) radiation. This is called the *greenhouse effect,* and it explains why an automobile passenger compartment or a greenhouse or a solar-collector interior becomes and remains warm on a sunny day, even during times when the ambient air temperature is low. ❑

11.8 SUMMARY

This chapter has served as an introduction to radiation heat transfer. We defined emissivity, absorptivity, reflectivity, and transmissivity. We stated and worked with radiation laws: Kirchhoff's Law, Planck's Distribution Law, and Wien's Displacement Law. We finally examined characteristics of a few real surfaces and saw that real surfaces do depart significantly from ideal-surface behavior. We defined also the concepts of a black body, a gray body, and a diffuse surface. Some of the definitions encountered are summarized in Table 11.5. During the course of developing some of the radiation laws, we encountered a number of universal constants: the Stefan-Boltzmann constant, Planck's constant, the Boltzmann constant, and the speed of light in a vacuum. These constants are listed in Table 11.6.

11.9 PROBLEMS

11.9.1 Radiation Spectrum

1. Electricity for home use has a frequency of 60 cycles per second (the unit for cycles per second is the hertz, Hz). What is the wavelength if it is delivered at the speed of light?
2. The typical wavelength of broadcast radio waves is 300 m. What is the corresponding frequency?

TABLE 11.5
Definitions Associated with Surface Properties

Subscripts	
λ	wavelength-dependent property (monochromatic or spectral)
θ, ϕ	direction-dependent property
n	property measured with a receiver located at a right angle to the surface of interest
i	incident radiation
e	emitted radiation
j	radiosity
b	black-body properties

Absorptivity = α	= ratio of absorbed radiation to that which is incident on the surface
Emissivity = ε	= ratio of radiation emitted by a surface to the radiation emitted by a black body or surface at the same temperature
Reflectivity = ρ	= ratio of reflected radiation to that which is incident on the surface
Transmissivity = τ	= ratio of transmitted radiation to that which is incident on the surface

Property relations	
Opaque surface	$\tau = 0;\ \alpha + \rho = 1,\ \alpha_\lambda + \rho_\lambda = 1$
Translucent or transparent surface	$\alpha + \rho + \tau = 1;\ \alpha_\lambda + \rho_\lambda + \tau_\lambda = 1$
Diffuse surface	$\varepsilon \neq \varepsilon(\theta, \phi);\ \alpha \neq \alpha(\theta, \phi)$

Radiation laws

Kirchoff's Law relates emissivity and absorptivity for surfaces irradiated by a black body: $\varepsilon_\lambda = \alpha_\lambda;\ \varepsilon = \alpha$.

Planck's Law relates spectral emission of radiant energy for a black body to wavelength: Equation 11.38 and Figure 11.12.

Stefan-Boltzmann's Law relates emissive power of a black body to its temperature: $q_b'' = \sigma T^4$.

Wein's Displacement Law gives the locus of peaks of black-body emission curves: Equation 11.46 and Figure 11.12.

Surface types

Black body is (a) an ideal emitter and emits according to Stefan-Boltzmann's Law. No surface emits more radiant energy than a black body when both are the same temperature; and (b) an ideal absorber that absorbs all incident energy.

Diffuse surface is one whose emissivity and absorptivity are independent of direction.

Gray surface is one whose emissivity and absorptivity are independent of wavelength.

TABLE 11.6
Universal Radiation Constants

Speed of light in a vacuum

$c_o = 2.997\ 76 \times 10^8$ m/s $= 9.83517 \times 10^8$ ft/s $= 3.5407 \times 10^{12}$ ft/hr

Planck's constant

$h = 6.625\ 6 \times 10^{-34}$ J·s $= 6.2759 \times 10^{-37}$ BTU·s $= 1.7433 \times 10^{-40}$ BTU·hr

Boltzmann's constant

$K = 1.380\ 5 \times 10^{-23}$ J/K $= 7.274 \times 10^{-27}$ BTU/(hr·°R)

Stefan-Boltzmann constant

$\sigma = 5.675 \times 10^{-8}$ W/(m²·K⁴) $= 0.1714 \times 10^{-8}$ BTU/(hr·ft²·°R⁴)

Constants in Planck's distribution equation

$C_1 = 2\pi h c_o^2 = 3.741\ 1 \times 10^{-16}$ W·m² $= 3.814 \times 10^{-18}$ BTU·ft²/s

$C_2 = hc_o/K = 1.439 \times 10^{-2}$ m·K $= 8.500 \times 10^{-2}$ ft·°R

3. Television channels (14–83) are delivered as waves having a typical wavelength of 14.7 in. What is the corresponding frequency?

4. The following table is of the visible spectrum. Each color is listed with a corresponding wavelength. Calculate the frequency for each color.

Color	Wavelength
Violet	0.41×10^{-6} m
Blue	0.47×10^{-6} m
Green	0.52×10^{-6} m
Yellow	0.58×10^{-6} m
Orange	0.60×10^{-6} m
Red	0.65×10^{-6} m

5. Use the data provided in the table of Problem 4 to calculate the photon energy associated with color as predicted by Equation 11.3. Are longer wavelengths associated with smaller photon energies?

6. Referring to Equations 11.1 and 11.2, what are the SI and English Engineering units for frequency v and for energy e?

11.9.2 RADIATION INTENSITY

7. Consider a sphere that is 1 ft in diameter. An area on the surface of the sphere is located with the following specifications:

$$\theta = 30°, \, d\theta = 1°$$

$$\phi = 60°, \, d\phi = 1°$$

Referring to Figure 11.3, calculate the length of each side of the surface element, its area, and the solid angle subtended by the area.

8. A sphere has a radius of 1 m, and an area on its surface is located at $\theta = 20°$, $\phi = 45°$. Also known is that $d\theta = 0.5°$ and $d\phi = 0.25°$. Calculate the area of the surface element and the solid angle subtended.

9. A sphere has a surface area element located at $\theta = 70°$, $\phi = 10°$, with $d\theta = d\phi = 1°$. The area of the element is 0.002 ft². Calculate the radius of the sphere.

10. Derive Equation 11.12.

11. Graph the intensity equation of Example 11.1 vs. θ on a polar graph. Is the surface diffuse?

12. Figure P11.1 shows a differential area dA_1 of dimensions 2×2 cm emitting energy at an intensity given by

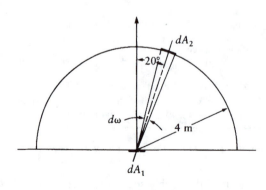

FIGURE P11.1

$$I_e = 1\ 000\ \text{W/(m}^2\cdot\text{sr)}$$

A second differential area dA_2 of dimensions 3×3 cm is located 4 m from dA_1, as shown. (a) Calculate the value of the solid angle subtended by dA_2 with respect to dA_1. (b) Calculate the rate at which radiation emitted by dA_1 is intercepted by dA_2.

13. Figure P11.2 shows a differential area dA_1 of dimensions 1×2 in emitting energy. A second differential area dA_2 of dimensions $1\text{-}1/2 \times 1\text{-}1/2$ in is located 3 ft away, as shown. A sensor at dA_2 intercepts radiation from dA_1. When the reading at dA_2 is 1×10^{-2} BTU/hr, what is the intensity of radiation emitted from dA_1? How would the intensity change if the area of dA_1 were doubled?

14. Figure P11.3 shows two areas, dA_1 and dA_2. The area dA_1 is a disk of radius 1 cm and emits radiant energy with an associated intensity given by

$$I_{e\theta} = 1\ 800(1 - 0.9 \sin\theta)\ \text{W/(m}^2\cdot\text{sr)}$$

The area dA_2 is 2 m from dA_1 and oriented as shown. What is the radius of the area dA_2 if the intercepted energy at dA_2 is 2×10^{-2} W?

15. Figure P11.4 shows two areas, dA_1 and dA_2. Area dA_1 is a disk having a radius of 1 in, and it emits radiation with an intensity of 1500 BTU/hr·ft^2·sr. Area dA_2 is also a disk of radius 1 in and is also emitting radiation. The centers of the disks are separated by a distance of 3 ft. If the net heat exchanged between dA_1 and dA_2 is zero, what is the intensity associated with area dA_2? Both areas are diffuse.

16. Consider the system described in Example 11.1 and shown in Figure 11.6. If the net heat exchanged between dA_1 and dA_2 is zero, what is the intensity associated with dA_2?

17. Repeat Problem 16 for areas dA_1 and dA_3.

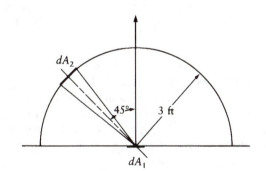

FIGURE P11.2

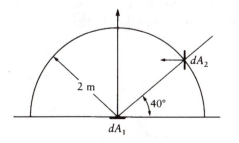

FIGURE P11.3

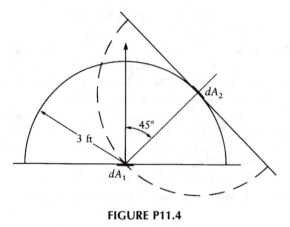

FIGURE P11.4

11.9.3 RADIATION AND RADIOSITY

18. Verify the integration leading to Equation 11.16.
19. What is the irradiation associated with area dA_2 of Problem 12?
20. What is the irradiation associated with area dA_2 of Problem 13?
21. What is the irradiation associated with area dA_2 of Problem 14?
22. A piece of transparent material is irradiated with energy. The material reflects 30% of the incident energy and has an emissivity of 0.35. If the surface is gray and emits energy diffusely, what is its transmissivity?
23. A material has an emissivity of 0.6. It is gray, opaque, and a diffuse emitter. What is its reflectivity?
24. A material has an emissivity equal to double its reflectivity. Its reflectivity is double its transmissivity. The material is gray and diffuse. Determine its reflectivity.

11.9.4 RADIATION LAWS

25. Verify that Equation 11.34 satisfies Equation 11.33.
26. A small canister used for holding camera film is sketched in Figure P11.5. The internal dimensions of the canister are 5 cm tall and 3.2 cm in diameter. The canister is made of plastic that has an emissivity of 0.85. A small hole 1 mm in diameter is drilled through the cap. What is the emissivity of the hole?
27. A shoe box has internal dimensions of 11-1/2 × 5-1/4 × 3-1/2 in. A 3/16-in-diameter hole is made in the box with a paper punch. Will the hole have a greater emissivity if the interior of the box is painted white or black?
28. Verify that the constant C_1 in Planck's Equation 11.39 equals $2\pi c_o^2$.
29. Verify that the constant C_2 in Planck's Equation 11.39 equals hc_o/K.
30. Verify that the numerical value of the Stefan-Boltzmann constant is given by Equation 11.42.

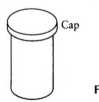

Cap

FIGURE P11.5

31. What percentage of total emitted energy radiated by the sun (a black body at 5 800 K) lies in the visible region 4×10^{-7} m $\leq \lambda \leq 7 \times 10^{-7}$ m?

32. What wavelength is associated with the maximum intensity of radiation emitted by the light bulb of Example 11.4?

33. Graph Planck's Equation over the thermal range for the sun, assuming it to be a black body at 5 800 K. Show the visible region in the graph and note where λ_{max} occurs with respect to the visible region.

34. The Rayleigh-Jeans distribution equation applies for low-frequency radiation and can be derived from Planck's Equation 11.39. Derive the Rayleigh-Jeans distribution by performing the following steps:

    (a) An exponential term can be expanded in a series according to

    $$e^x = 1 + x + \frac{x^2}{2!} + \frac{x^3}{3!} + \dots$$

    Expand the denominator of the Planck distribution, and for low frequency (= long wavelength because $c_o = v\lambda$) note that $hc_o/K\lambda T \ll 1$. Show that

    $$\exp(hc_o/K\lambda T) - 1 \cong hc_o/K\lambda T$$

    (b) Substitute into the Planck Equation to obtain the Rayleigh-Jeans Formula:

    $$q''_{b\lambda} = \frac{2\pi c_o KT}{\lambda^4}$$

    (c) Select any temperature line in Figure 11.12 and determine the range of applicability for the Rayleigh-Jeans Equation.

35. The Wien Formula is an equation that applies for high-frequency radiation, and it can be derived from the Planck Equation 11.39. When frequency v is high, λ is short ($c_o = \lambda v$); therefore

    $$hc_o/K\lambda T \ll 1$$

    and

    $$\exp(hc_o/K\lambda T) - 1 \cong \exp(hc_o/K\lambda T)$$

    The denominator of the Planck Equation can be easily simplified. For high frequency the Planck Equation becomes

    $$q''_{b\lambda} = \frac{2\pi hc_o^2}{\lambda^5 \exp(hc_o/K\lambda T)}$$

    which is known as the Wien Distribution Equation. Select any temperature line in Figure 11.12 and determine the range of applicability for the equation.

11.9.5 CHARACTERISTICS OF REAL SURFACES

36. For a number of materials emissivity varies with direction but not strongly with wavelength, a characteristic that is investigated in this problem. Emissivity is defined in Equation 11.22 as

$$\varepsilon = \frac{\int_0^\infty \varepsilon_\lambda q''_{b\lambda} \, d\lambda}{\int_0^\infty q''_{b\lambda} \, d\lambda}$$

where ε_λ is the spectral emissivity and $q''_{b\lambda}$ is the black-body radiation flux. If we wish to account for the directional variation of emissivity, we can do so by redefining emissivity in terms of intensity. From Equation 11.5a,

$$\varepsilon = \frac{q''_e}{q''_b}$$

Substituting from Equation 11.11, we get

$$\varepsilon = \frac{\int_0^\infty \int_0^{2\pi} \int_0^{\pi/2} I_{e\lambda\theta\phi} \cos\theta \, \sin\theta \, d\theta \, d\phi \, d\lambda}{\pi I_b}$$

where I_b the intensity of radiation from a black body. Now, if the directional spectral intensity $I_{e\lambda\theta\phi}$ independent of wavelength, we can write

$$\varepsilon = \frac{\int_0^{2\pi} \int_0^{\pi/2} I_{e\theta\phi} \cos\theta \, \sin\theta \, d\theta \, d\phi}{\pi I_b}$$

or

$$\varepsilon = \frac{1}{\pi} \int_0^{2\pi} \int_0^{\pi/2} (I_{e\theta\phi}/I_b) \cos\theta \, \sin\theta \, d\theta \, d\phi$$

In terms of the directional emissivity, the above integrand becomes

$$\varepsilon = \frac{1}{\pi} \int_0^{2\pi} \int_0^{\pi/2} \varepsilon_{\theta\phi} \cos\theta \, \sin\theta \, d\theta \, d\phi \tag{i}$$

If the directional emissivity $\varepsilon_{\theta\phi}$ is independent of ϕ, we get

$$\varepsilon = 2 \int_0^{\pi/2} \varepsilon_\theta \cos\theta \, \sin\theta \, d\theta \tag{ii}$$

Figure P11.6 shows the directional emissivity of wood on a polar graph. As indicated, the emissivity varies only with θ, so Equation ii above applies. Determine the hemispherical emissivity ε of wood by completing the following steps:
(a) Obtain values of ε_θ vs. θ from the graph.
(b) Calculate $\varepsilon_\theta \cos\theta \sin\theta$ from the obtained values.
(c) Graph $\varepsilon_\theta \cos\theta \sin\theta$ vs. θ, where θ varies from 0 to $\pi/2$.
(d) Find the area underneath the curve by any convenient method, from counting squares to a numerical scheme.
(e) Use the area under the curve to find the total emissivity according to Equation ii.
37. Repeat Problem 36 for copper oxide, whose directional emissivity is shown in Figure P11.7.
38. Repeat Problem 36 for aluminum, whose directional emissivity is shown in Figure P11.8.
39. Repeat Problem 36 for chromium, whose directional emissivity is shown in Figure P11.8.

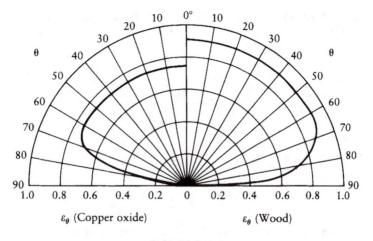

ε_θ (Copper oxide) ε_θ (Wood)

FIGURE P11.6

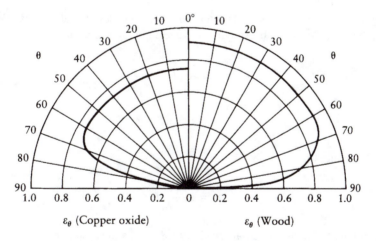

ε_θ (Copper oxide) ε_θ (Wood)

FIGURE P11.7

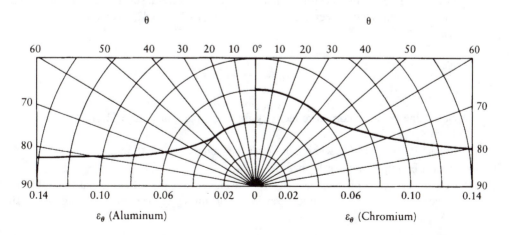

ε_θ (Aluminum) ε_θ (Chromium)

FIGURE P11.8

40. Figure 11.16 gives the spectral normal emissivity of several materials. We can integrate out the dependence of emissivity on wavelength by using Equation 11.22 evaluated over a finite wavelength interval:

$$\varepsilon_n = \frac{\int_{\lambda_1}^{\lambda_2} \varepsilon_{n\lambda} q''_{b\lambda} \, d\lambda}{\int_{\lambda_1}^{\lambda_2} q''_{b\lambda} \, d\lambda}$$

The above equation reduces to

$$e_n = \frac{1}{\lambda_2 - \lambda_1}\int_{\lambda_1}^{\lambda_2} \varepsilon_{n\lambda} d\lambda$$

Determine the total normal emissivity for the aluminum oxide of Figure 11.16 over the interval 1 μm ≤ λ ≤ 20 μm by using the above equation.

41. Repeat Problem 40 for the stainless steel (800 K) of Figure 11.16 over the same wavelength interval.

42. Figure 11.18 gives the spectral normal absorptivity of several substances. We can integrate out the dependence of absorptivity on wavelength by integration over a finite wavelength interval:

$$\alpha_n = \frac{\int_{\lambda_1}^{\lambda_2} \alpha_{n\lambda} q''_{i\lambda} d\lambda}{\int_{\lambda_1}^{\lambda_2} q''_{i\lambda} d\lambda}$$

which reduces to

$$\alpha_n = \frac{1}{\lambda_2 - \lambda_1}\int_{\lambda_1}^{\lambda_2} \varepsilon_{n\lambda} d\lambda$$

Determine the total normal absorptivity for the stainless steel of Figure 11.18 over the interval 1 μm ≤ λ ≤ 15 μm.

43. Repeat Problem 42 for red brick of Figure 11.18 over 1 μm ≤ λ ≤ 20 μm.

44. Figure 11.18 gives the spectral normal reflectivity of several substances. The dependence on wavelength can be integrated out (as in Problems 40 and 42) to obtain the total normal reflectivity:

$$\rho_n = \frac{1}{\lambda_2 - \lambda_1}\int_{\lambda_1}^{\lambda_2} \rho_{n\lambda} d\lambda$$

Determine the total normal reflectivity for black paint of Figure 11.18 over the interval 1 μm ≤ λ ≤ 20 μm.

45. Repeat Problem 44 for white paint over the same wavelength interval.

46. Repeat Example 11.6 for bronze-tint glass.

47. Construct a graph of transmissivity vs. wavelength for the gray-tint glass of Table 11.3 in the visible region. Determine the average transmissivity over the visible region by using

$$\tau = \frac{1}{\lambda_2 - \lambda_1}\int_{\lambda_1}^{\lambda_2} \tau_\lambda d\lambda$$

12 Radiation Heat Transfer between Surfaces

CONTENTS

12.1   INTRODUCTION

Chapter 11 was devoted to a discussion of surface properties: emissivity, absorptivity, transmissivity, and reflectivity. Details of spherical coordinates were also discussed, as well as the importance of geometry and surface behavior. The concepts of a black body, a gray body, and a diffuse surface were presented.

There was little mention of actual heat-transfer rates in the preceding chapter, however. The Stefan-Boltzmann Law was stated but not used in deriving a general equation for heat transfer between surfaces by radiation. In this chapter we shall examine a number of problems and calculate energy-transfer rates. We will see how the assumption of black-body or gray-body behavior affects the calculations. We will also learn what part geometry plays in the overall formulation.

Radiation heat transfer can be significant in a number of the problems discussed thus far in this text. In most of those problems we assumed that radiation heat transfer is negligible. For example, a steam pipe routed horizontally through a room loses heat by convection to the air and by radiation to the surroundings. Under certain conditions, the radiative loss equals the convective loss. To assume that radiation losses are negligible in this case can lead to an error of 50%. The engineer must therefore develop an intuition regarding those special problems in which radiation cannot be neglected. Hopefully, this chapter will provide the reader with enough experience to begin to recognize such problems.

12.2   THE VIEW FACTOR

The view factor (also called configuration factor, shape factor, or geometry factor) is a geometry term for a system in which two surfaces exchange energy by radiation. Radiation waves travel in straight lines, and if one surface cannot "see" another, then there is no *direct* radiation from the first surface to the second. Consider a fire in a fireplace. Anyone sitting where a view of the fire is totally obstructed receives no energy from the fire directly. Conversely, anyone sitting in view of the fire receives heat directly from the fire by radiation. Furthermore, someone sitting close to the fire receives more heat than someone sitting far away. Someone sitting just in front of the fire receives more heat than someone sitting off to either side. As indicated by this example, position (or the geometry) between the source (fire) and receiver is important in determining the radiative energy-transfer rate. The view factor expresses the geometry of position that exists between source and receiver.

12.2.1 VIEW FACTOR BETWEEN TWO DIFFERENTIAL ELEMENTS

Figure 12.1 shows two surfaces, labeled A_1 and A_2, in view of each other. A differential element on A_1 is defined as dA_1, with a similar definition for dA_2. Our objective will be to derive an equation for the view factor for the differential areas. Note the x, y, z axes and the unit normals n_1 and n_2 drawn for each respective area. The angle between the differential area and the unit normal is β. The distance between dA_1 and dA_2 is L. The values of β_1, β_2, and L will vary, depending on where we select dA_1 and dA_2 to be on each respective surface.

From Chapter 11, the radiation that leaves dA_1 that is intercepted by dA_2 is given by

$$dq_{d1-d2} = I_1 \cos\beta_1 dA_1 d\omega_{2-1} \tag{12.1}$$

where I_1 is the intensity of radiation associated with area dA_1, and $d\omega_{2-1}$ is the solid angle subtended by dA_2 when viewed from dA_1. By definition,

$$d\omega_{2-1} = \frac{\cos\beta_2 dA_2}{L^2}$$

For a black surface, we can write for the intensity

$$I_{b1} = \frac{\sigma T_1^4}{\pi}$$

Substituting these equations into Equation 12.1 gives

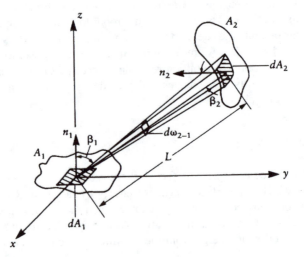

FIGURE 12.1 Two areas in full view of each other.

$$dq_{d1-d2} = \frac{\sigma T_1^4 \cos \beta_1 \cos \beta_2 dA_1 dA_2}{\pi \, L^2} \tag{12.2}$$

Following the same lines of reasoning, the radiation that leaves surface dA_2 in the direction of dA_1 is

$$dq_{d2-d1} = \frac{\sigma T_2^4 \cos \beta_1 \cos \beta_2 dA_1 dA_2}{\pi \, L^2} \tag{12.3}$$

Note that Equations 12.2 and 12.3 are for black surfaces. With respect to dA_1, the net heat transfer by radiation between dA_1 and dA_2 then is

$$dq_{d1 \to d2} = dq_{d1-d2} - dq_{d2-d1}$$

where the notation "$\to$" has been introduced in the subscript to denote a net exchange of heat. Substituting, we get

$$dq_{d1 \to d2} = \sigma(T_1^4 - T_2^4) \frac{\cos \beta_1 \cos \beta_2 dA_1 dA_2}{\pi L^2} \tag{12.4a}$$

On a per-unit area basis

$$\frac{dq_{d1 \to d2}}{dA_1} = \sigma(T_1^4 - T_2^4) \frac{\cos \beta_1 \cos \beta_2 dA_2}{\pi L^2} \tag{12.4b}$$

The above formulation relates the *net* heat exchanged per unit area between two elemental black surfaces ($dq_{d1 \to d2}/dA_1$) to the difference of the fourth power of the surface temperature and to a geometry term that contains the angles β_1 and β_2, as well as areas and distance between the areas. The geometry term of Equation 12.4b is defined as the view factor:

$$dF_{d1-d2} = \frac{\cos \beta_1 \cos \beta_2 dA_2}{\pi L^2} \tag{12.5}$$

Equation 12.5 indicates that the area dA_1 is the surface of interest and that the net heat is exchanged with respect to dA_1. It is possible to reformulate the above equations and make dA_2 the surface of interest. The result becomes

$$\frac{dq_{d2 \to d1}}{dA_2} = \sigma(T_2^4 - T_1^4) \frac{\cos \beta_1 \cos \beta_2 dA_1}{\pi L^2} \tag{12.6}$$

and we can write

$$dF_{d1-d2} = \frac{\cos \beta_1 \cos \beta_2 dA_1}{\pi L^2} \tag{12.7}$$

Comparison of Equations 12.5 and 12.7 shows that a relationship exists between the view factors. Equation 12.5 multiplied by dA_1 equals Equation 12.7 multiplied by dA_2. Thus,

$$dA_1 dF_{d1-d2} = dA_2 dF_{d2-d1} = \frac{\cos \beta_1 \cos \beta_2 dA_1 dA_2}{\pi L^2} \tag{12.8}$$

The above equation is called the *reciprocity relation*. Now the heat exchanged between the black elemental areas along path L is written as

$$dq_{d1 \to d2} = \sigma(T_1^4 - T_2^4)dF_{d1-d2}dA_1 = \sigma(T_1^4 - T_2^4)dF_{d2-d1}dA_2$$

12.2.2 View Factor Between a Differential Element and a Finite Area

Returning now to Figure 12.1, consider that we are seeking the view factor between area dA_1 and A_2. An equation can be derived by starting with an energy balance or by integration of Equation 12.5 over A_2. The result is

$$F_{d1-2} = \iint\limits_{A_2} \frac{\cos \beta_1 \cos \beta_2 dA_2}{\pi L^2} \tag{12.9}$$

Note that, if Equation 12.5 is integrated over A_2, then the view factor is no longer a differential quantity.

Equation 12.9 is the view factor written with dA_1 as the surface of interest. When A_2 is the surface of interest, on the other hand, a different view factor results. We can perform an energy balance or more simply use reciprocity to obtain a relationship. The result is

$$A_2 dF_{2-d1} = dA_1 F_{d1-2} \tag{12.10}$$

At first glance, the notation can seem confusing, but it is consistent. When the receiver is a differential area, only then is the view factor written as dF. The heat exchanged between dA_1 and A_2 therefore becomes

$$dq_{d1 \to 2} = \sigma(T_1^4 - T_1^4)dA_1 F_{d1-2} = \sigma(T_1^4 - T_2^4)A_2 dF_{2-d1}$$

An assumption inherent but not obvious in the above equation is that area A_2 is isothermal and at temperature T_2.

Example 12.1

Figure 12.2a shows two parallel plane areas that are located a distance H apart. One is a circular disk, and the other is a differential area. Determine the view factor F_{d1-2}.

Solution

Consistent with the analysis centered about Figure 12.1, we first select a coordinate system. As shown in Figure 12.2b, the origin is at the differential area. We now select a representative differential element on the finite area and draw unit normals for dA_1 and dA_2. The unit normals are connected with a line that is L long. The β angles are identified.

Assumptions

1. The surfaces are diffuse emitters and reflectors.
2. Uniform radiosity is associated with each surface.
3. Both areas are isothermal.

We can begin with Equation 12.9 derived for two differential areas and integrate over A_2 to obtain the desired result:

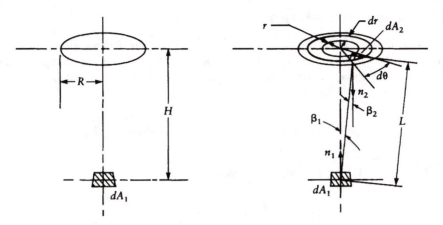

FIGURE 12.2 The two areas of Example 12.1.

$$F_{d1-2} = \int\int_{A_2} \frac{\cos\beta_1 \cos\beta_2 dA_2}{\pi L^2}$$
(i)

Each term of the integrand is evaluated as follows:

$$dA_2 = r \, dr \, d\theta$$

$$L = \sqrt{H^2 + r^2}$$

$$\cos\beta_1 = \cos\beta_2 = \frac{H}{L} = \frac{H}{\sqrt{H^2 + r^2}}$$

Substituting into Equation (i) gives

$$F_{d1-2} = \int_0^{2\pi}\int_0^R \frac{H^2 r dr d\theta}{\pi(H^2 + r^2)^2}$$

Integrating over r while holding θ constant, we get

$$F_{d1-2} = \frac{H^2}{\pi}\int_0^{2\pi}\left[-\frac{1}{H^2+r^2}\right]_0^R d\theta = \frac{1}{2\pi}\int_0^{2\pi}\frac{R^2 d\theta}{H^2 + R^2}$$

With H and R both independent of θ we can integrate again to obtain

$$F_{d1-2} = \frac{R_2}{2\pi(H^2 + R^2)}[\theta]_0^{2\pi}$$

which simplifies to

$$F_{d1-2} = \frac{R^2}{H^2 + R^2} = \frac{1}{1 + (H/R)^2}$$

Notice that H and R are always positive, so the view factor is positive and always less than 1. Also, as H approaches 0 or R approaches ∞, the view factor approaches 1 (see Figure 12.2). ❏

12.2.3 VIEW FACTORS FOR TWO FINITE AREAS

Referring to Figure 12.1, we wish to derive an equation for the view factor between the finite areas A_1 and A_2. Equations can be derived by performing an energy balance and evaluating the net energy exchanged between the two areas. Alternatively, we can start with Equation 12.5 and integrate. The result is

$$F_{1-2} = \frac{1}{A_1} \iint_{A_1} \iint_{A_2} \frac{\cos \beta_1 \cos \beta_2 dA_2 dA_1}{\pi L^2} \qquad (12.11)$$

The view factor from A_2 to A_1 is found in a similar way to be

$$F_{2-1} = \frac{1}{A_2} \iint_{A_1} \iint_{A_2} \frac{\cos \beta_1 \cos \beta_2 dA_2 dA_1}{\pi L^2} \qquad (12.12)$$

The reciprocity relation for these view factors is

$$A_1 F_{1-2} = A_2 F_{2-1} \qquad (12.13)$$

The energy radiated from area A_1 that reaches A_2 then is

$$q_{1-2} = \sigma T_1^4 A_1 F_{1-2}$$

Likewise, for A_2 to A_1,

$$q_{2-1} = \sigma T_2^4 A_2 F_{2-1}$$

The net energy exchanged then is

$$q_{1 \to 2} = \sigma(T_1^4 - T_2^4) A_1 F_{1-2} = \sigma(T_2^4 - T_1^4) A_2 F_{2-1}$$

The equations written above for view factors and heat-transfer rates are tabulated in the Summary section of this chapter. Figures 12.3, 12.4, 12.5, 12.6, and 12.7 give the view factor for several geometries. For other geometries the reader is referred to the literature.[*]

12.3 METHODS FOR EVALUATING VIEW FACTORS

The evaluation of view factors by integration can be quite cumbersome and, in many cases where an integral cannot be evaluated, impossible. Consequently, a number of alternative methods have been devised. These include view-factor algebra, the crossed-string method, contour integration, experimental methods, and graphical methods. We will examine only the first two of these alternative methods (to direct integration as per Equation 12.9, for example). For a review of the other methods, the interested reader is referred to the literature.[†]

12.3.1 VIEW-FACTOR ALGEBRA FOR PAIRS OF SURFACES

Figure 12.8 shows two surfaces A_1 and A_2. The area A_2 is divided into A_3 and A_4. Suppose that the configuration factors F_{1-2} and F_{1-4} are known and that F_{3-1} is desired. By definition, the fraction

---

[*] An excellent source is NACA TN 2836, *Radiant-Interchange Configuration Factors*, by D.C. Hamilton and W.R. Morgan.
[†] NACA TN 2836, *Radiant-Interchange Configuration Factors*, by D.C. Hamilton and W. K. Morgan; *Thermal Radiation Heat Transfer*, by R. Siegel and J. R. Howell, McGraw-Hill, 1972, Ch. 7.

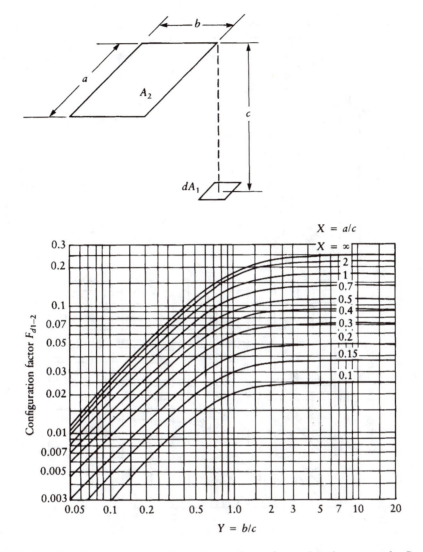

FIGURE 12.3 Configuration-factor curves for a plane point surface and a place rectangle. *Source:* from NACA TN 2836.

of total energy emitted by A_1 that is intercepted by A_2 is F_{1-2}. From an energy balance, we conclude that the energy emitted by A_1 that is intercepted by A_3 and added to that intercepted by A_4 must also equal F_{1-2}. We can therefore write

$$F_{1-2} = F_{1-3} + F_{1-4}$$

With F_{1-2} and F_{1-4} known, we get

$$F_{1-3} = F_{1-2} - F_{1-4}$$

We have from reciprocity

$$A_{3-1}F_{3-1} = A_1 F_{1-3}$$

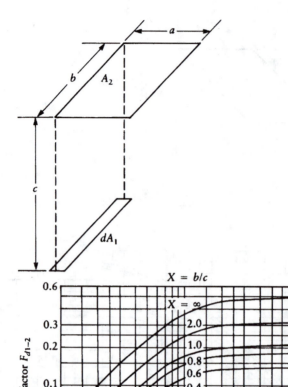

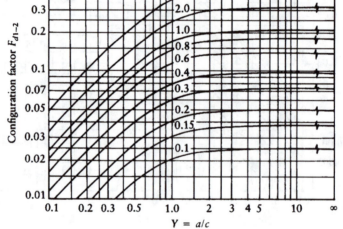

FIGURE 12.4 Configuration-factor curves for a line source and a plane rectangle. *Source:* from NACA TN 2836.

and so

$$F_{3-1} = \frac{A_1}{A_3}(F_{1-2} - F_{1-4})$$

In the opposite case, where A_2 is the source and A_1 is the receiver, it is useful to think in terms of energy quantities. Referring to Figure 12.8, the energy leaving A_2 that arrives at A_1 is proportional to A_2F_{2-1}. Also, this same energy equals the sum of the energies that leave both A_3 and A_4 and arrive at A_1. We therefore write

$$(A_3 + A_4)F_{(3+4)-1} = A_3F_{3-1} + A_4F_{4-1} \qquad (12.14)$$

which is an important relationship.

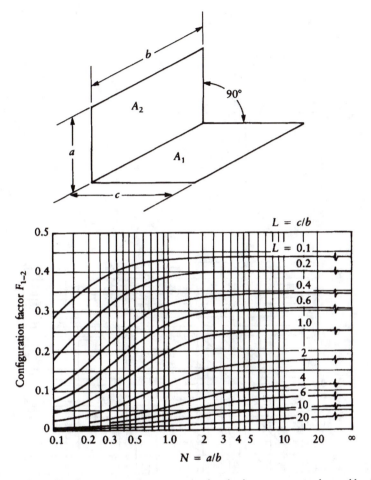

FIGURE 12.5   Configuration-factor curves for two rectangles sharing a common edge and having an included angle of 90°.

Example 12.2

Figure 12.9 is of two planes A_1 and A_2 that share a common edge. The area A_1 is divided into areas A_3 and A_4, while A_2 is divided into A_5 and A_6. Determine the configuration factor $F_{6\text{-}3}$.

Solution

We can use view-factor algebra to solve this problem. The desired view factor will ultimately be expressed in terms of known view factors, which are

$$F_{1\text{-}5} \text{ [or } F_{(3+4)\text{-}5}], \quad F_{4\text{-}5}, \quad F_{4\text{-}2} \text{ [or } F_{4\text{-}(5+6)}], \quad F_{1\text{-}2} \text{ [or } F_{(3+4)\text{-}(5+6)}]$$

Figure 12.5 gives the values of these known view factors. We begin by writing

$$F_{(3+4)\text{-}(5+6)} = F_{(3+4)\text{-}5} + F_{(3+4)\text{-}6}$$

$$= F_{(3+4)\text{-}5} + \frac{A_6}{A_3 + A_4} F_{6\text{-}(3+4)}$$

or

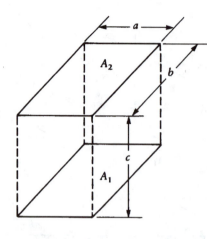

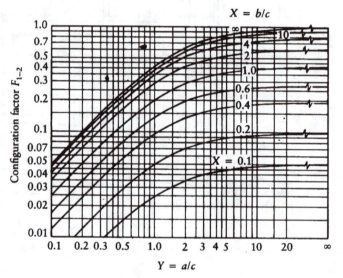

FIGURE 12.6 Configuration-factor curves for two identical directly apposed rectangles. *Source:* from NACA TN 2836.

$$F_{(3+4)-(5+6)} = F_{(3+4)-5} + \frac{A_6}{A_3 + A_4}(F_{6-3} + F_{6-4})$$

Rearranging and solving for F_{6-3}, we obtain

$$F_{6-3} = [F_{(3+4)-(5+6)} - F_{(3+4)-5}]\frac{A_3 + A_4}{A_6} - F_{6-4} \qquad \text{(i)}$$

All terms on the right-hand side are known except F_{6-4}. Working with F_{6-4}, we express it as

$$(A_5 + A_6)F_{(5+6)-4} = A_5 F_{5-4} + A_6 F_{6-4}$$

$$F_{6-4} = \frac{A_5 + A_6}{A_6}F_{(5+6)-4} - \frac{A_5}{A_6}F_{5-4}$$

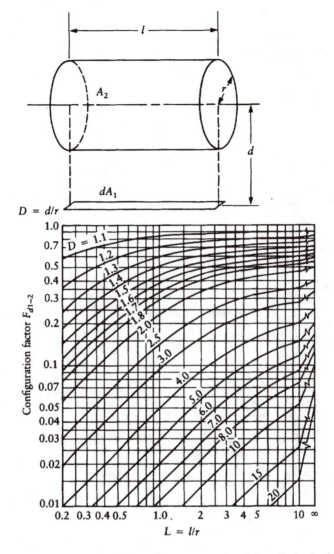

FIGURE 12.7 Configuration-factor curves for a line source and a right cylinder having the same length. *Source:* from NACA TN 2836.

Substituting into Equation (i) above gives

$$F_{6-3} = [F_{(3+4)-(5+6)} - F_{(3+4)-5}]\frac{A_3 + A_4}{A_6} - \frac{A_5 + A_6}{A_6}F_{(5+6)-4} + \frac{A_5}{A_6}F_{5-4}$$

We now use Figure 12.5 to evaluate each term in the above equation. The calculations are summarized in Table 12.1. Note that in the third and fourth sketches, a and c have reversed their roles from the first and second sketches because area 4 in the third and fourth sketches is the receiver area.

Substituting into the equation gives

$$F_{6-3} = (0.2 - 0.14)\frac{0.5 + 0.5}{0.5} - \frac{0.5 + 0.5}{0.5}(0.14) + \frac{0.5}{0.5}(0.23)$$

$$F_{6-3} = 0.07$$

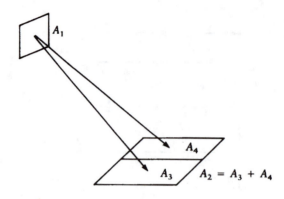

FIGURE 12.8 Energy exchange between pairs of surfaces.

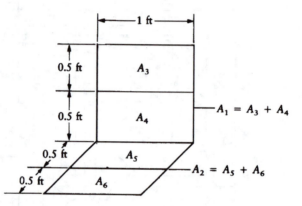

FIGURE 12.9 Two plane areas that share a common edge.

By using reciprocity, we could now calculate F_{3-6}:

$$A_3 F_{3-6} = A_6 F_{6-3}$$

Because $A_3 = A_6$,

$$F_{3-6} = F_{6-3} = 0.07$$

Care must be exercised in using the charts, especially in identifying the source and receiver areas for each component view factor. Alternatively, factors such as $F_{(5+6)-4}$ can be rewritten as

$$F_{(5+6)-4} = \frac{A_4}{A_5 + A_6} F_{4-(5+6)}$$

which will render A_2 as the receiver and require no interchange of a and c.   ❑

Example 12.3

Figure 12.10a is of a differential area located directly below an L-shaped area. The differential area is maintained at 200°F while the finite area above it is at 100°F. Determine the heat transferred by radiation from dA_1 to A_2.

TABLE 12.1
Summary of the Calculations of Example 12.2

Sketch	F_{i-j}	$N = a/b$	$L = c/b$	Configuration Factor (Figure (12.5))	
	$F_{(3+4)-(5-6)}$	1	1	0.2	$A_3 = 0.5$ ft^2 $A_4 = 0.5$ ft^2 $A_5 = 0.5$ ft^2 $A_6 = 0.5$ ft^2
	$F_{(3+4)-5}$	0.5	1	0.14	
	$F_{(5+6)-4}$	0.5	1	0.14	
	F_{5-4}	0.5	0.5	0.23	

Solution

We have an equation for heat transferred between two black surfaces in terms of the view factor.

Assumptions

1. Both areas are diffuse emitters and absorbers.
2. Both surfaces are black.

The heat-transfer rate is written as

$$dq_{1 \to 2} = \sigma(T_1^4 - T_2^4)dA_1 F_{d1-2}$$

With the size of dA_1 unknown, we write

$$\frac{dq_{d1 \to 2}}{dA_1} = \sigma(T_1^4 - T_2^4)F_{d1-2}$$

All terms on the right-hand side are known except the view factor. We can evaluate the view factor by applying Figure 12.3.

Figure 12.10b shows the same areas, but with A_2 divided into overlapping areas A_3 and A_4, A_5 being common to both A_3 and A_4. The view factor is

$$F_{d1-2} = F_{d1-3} + F_{d1-4} - F_{d1-5}$$

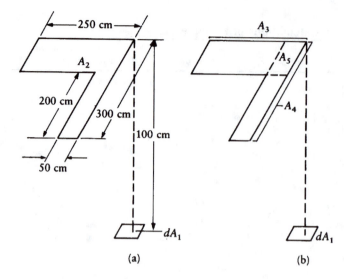

FIGURE 12.10   Energy transfer between a differential area and a finite area.

From Figure 12.3 we can evaluate each term as

F_{d1-3}: $a = 100$ cm $b = 250$ cm $c = 100$ cm
            $X = a/c = 1.0$ $Y = b/c = 2.5$ $F_{d1-3} = 0.17$
F_{d1-4}: $a = 300$ cm $b = 50$ cm $c = 100$ cm
            $X = 3.0$ $Y = 0.5$ $F_{d1-4} = 0.11$
F_{d1-5}: $a = 100$ cm $b = 50$ cm $c = 100$ cm
            $X = 1.0$ $Y = 0.5$ $F_{d1-5} = 0.09$

Thus,

$$F_{d1-2} = 0.17 + 0.11 - 0.09 = 0.19$$

So only 0.19 or 19% of the energy leaving dA_1 reaches A_2. The net heat transferred then is

$$\frac{dq_{d1 \to 2}}{dA_1} = (0.1714 \times 10^{-8} \text{ BTU/hr·ft}^2 \cdot {}^\circ\text{R}^4)(660^4 - 560^4){}^\circ\text{R}^4(0.19)$$

where temperature is expressed in absolute units. Solving,

$$\frac{dq_{d1 \to 2}}{dA_1} = 29.7 \text{ BTU/(hr·ft}^2)$$

12.3.2 VIEW-FACTOR ALGEBRA FOR ENCLOSURES

We now extend the discussion of view-factor algebra to include enclosures. The insides of an oven and of a room and of a refrigerator are examples of enclosures. For an enclosure consisting of N surfaces, the energy leaving any surface inside must be incident on all the surfaces making up the enclosure. This includes the surface itself if it is concave and can in effect "see" itself. All fractions of energy leaving the kth surface and reaching the other surfaces must total unity. Therefore,

$$F_{k-1} + F_{k-2} + F_{k-3} + \ldots + F_{k-k} + \ldots + F_{k-N} = 1$$

or

$$\sum_{m=1}^{N} F_{k-m} = 1$$

The view factor F_{k-k} is zero for convex and for flat surfaces.

Example 12.4

Figure 12.11 shows the cross section of a triangular oven. Painted parts move through on a conveyor over the bottom surface. The two inclined surfaces contain heaters that provide radiant energy to aid in the drying of the paint. Determine the heat transferred to the conveyed parts for the conditions shown.

Solution

The heaters provide energy to the bottom surface, and as parts there become warm, they radiate energy and lose energy by convection to the surrounding air as well as by conduction to the conveyor. The temperatures shown will exist only at one section within the system. The surfaces are real surfaces and do not exhibit black-surface behavior.

Assumptions

1. Conduction and convection effects are negligible.
2. The system is at steady state.
3. The oven length is very long compared to the dimensions of the cross section (so that no energy is radiated out the ends).
4. The surfaces are all black with respect to radiation heat transfer.

The energy exchanges of interest are from A_2 to A_1 and A_3 to A_1. The net energy exchanged between A_2 and A_1 is

$$q_{1 \to 2} = \sigma(T_2^4 - T_1^4)A_2 F_{2-1} = \sigma(T_2^4 - T_1^4)A_1 F_{1-2}$$

Between A_3 and A_1 we have

$$q_{3 \to 1} = \sigma(T_3^4 - T_1^4)A_3 F_{3-1} = \sigma(T_3^4 - T_1^4)A_1 F_{1-3}$$

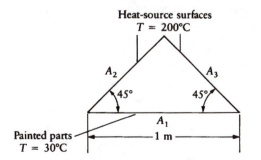

FIGURE 12.11 A paint-drying oven that is triangular in cross section.

Due to symmetry, we know that these heat-transfer rates are equal. We now evaluate each of the view factors. We write Equation 12.15 for surface A_1 as

$$F_{1-2} + F_{1-3} = 1$$

$$A_1 F_{1-2} + A_1 F_{1-3} = A_1 \tag{i}$$

For surfaces A_2 and A_3,

$$A_2 F_{2-1} + A_2 F_{2-3} = A_2 \tag{ii}$$

$$A_3 F_{3-1} + A_3 F_{3-2} = A_3 \tag{iii}$$

By using reciprocity on Equation (ii), we get

$$A_1 F_{1-2} + A_2 F_{2-3} = A_2 \tag{iv}$$

Likewise, for Equation (iii)

$$A_1 F_{1-3} + A_2 F_{2-3} = A_3 \tag{v}$$

Equations (i), (iv), and (v) are three relations with three unknowns, which we now solve simultaneously. Subtracting (v) from (iv) gives

$$A_1 F_{1-2} - A_1 F_{1-3} = A_2 - A_3$$

Adding this result to Equation (i), we obtain

$$A_1 F_{1-2} + A_1 F_{1-2} = A_1 + A_2 - A_3$$

Solving for the view factor, we get

$$F_{1-2} = \frac{A_1 + A_2 - A_3}{2A_1} = \frac{L_1 + L_2 - L_3}{2L_1}$$

where L is the width of area A_i. Each side of the triangular cross section has the same length (into the page).

Substituting into Equation (i) and solving for F_{1-3} gives

$$A_1 \left(\frac{A_1 + A_2 - A_3}{2A_1} \right) + A_1 F_{1-3} = A_1$$

or

$$F_{1-3} = \frac{A_1 - A_2 + A_3}{2A_1} = \frac{L_1 - L_2 + L_3}{2L_1}$$

Using reciprocity, we finally get

$$A_2 F_{2-1} = A_1 F_{1-2} = \frac{A_1 + A_2 - A_3}{2}$$

$$A_3 F_{3-1} = A_1 F_{1-3} = \frac{A_1 - A_2 + A_3}{2}$$

The heat-transfer equation $(A_2 \text{ to } A_1)$ becomes

$$q_{2 \to 1} = \sigma(T_2^4 - T_1^4)\frac{A_1 + A_2 - A_3}{2}$$

Because the length (into the page) is unknown, we will calculate the heat flux:

$$\frac{q_{2 \to 1}}{A_2} = \sigma(T_2^4 - T_1^4)\frac{A_1/A_2 + 1 - A_3/A_2}{2}$$

or in terms of width of each side,

$$\frac{q_{2 \to 1}}{A_2} = \sigma(T_2^4 - T_1^4)\frac{L_1/L_2 + 1 - L_3/L_2}{2}$$

The widths are

$$L_1 = 1 \text{ m}$$

$$L_2 = L_3 = (\sin 45°)(1 \text{ m}) = 0.707 \text{ m}$$

Substituting,

$$\frac{q_{2 \to 1}}{A_2} = (5.67 \times 10^{-8})(473^4 - 303^4)\left(\frac{1/0.707 + 1 - 0.707/0.707}{2}\right)$$

$$\frac{q_{2 \to 1}}{A_3} = 1.67 \times 10^3 \text{ W/m}^2$$

and

$$\frac{q_{3 \to 1}}{A_3} = 1.67 \times 10^3 \text{ W/m}^2 \qquad \square$$

12.3.3 CROSSED-STRING METHOD

The crossed-string method is a useful alternative method for determining the view factor for two-dimensional systems. Long grooves, for example, that extend infinitely far along one coordinate lend themselves readily to solution by the crossed-string method. Figure 12.12 shows a two-dimensional system consisting of many surfaces (not necessarily an enclosure) in which there may be blockage, and one or more surfaces may be convex or concave or flat. Shown also are a number of straight dashed lines that connect points labeled with lower-case letters. Our objective is to determine F_{1-2}.

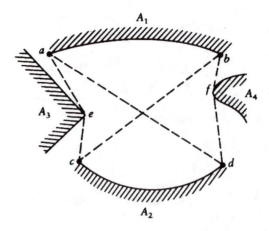

FIGURE 12.12 Illustration of the crossed-string method for finding the view factor between two-dimensional surfaces.

The dashed lines connect points located at the extremities of the areas of interest around the blocking surfaces (if any). In effect, we have formed some enclosures. For example, *aecba* is an enclosure. In Example 12.4, we obtained a configuration factor for a three-sided enclosure. Following similar lines of reasoning for *aecba,* we can write

$$A_{ab}F_{ab-aec} = \frac{A_{ab} + A_{aec} - A_{cb}}{2} \tag{12.16}$$

Furthermore, for *adfba* we get

$$A_{ab}F_{ab-dfb} = \frac{A_{ab} + A_{dfb} - A_{ad}}{2} \tag{12.17}$$

Also, because we have formed an enclosure,

$$F_{ab-aec} + F_{ab-cd} + F_{ab-dfb} = 1 \tag{12.18}$$

Substituting Equations 12.16 and 12.17 into 12.18 gives

$$\frac{A_{ab} + A_{aec} - A_{cb}}{2A_{ab}} + F_{ab-cd} + \frac{A_{ab} + A_{dfb} - A_{ad}}{2A_{ab}} = 1$$

Rearranging and simplifying, we obtain

$$A_{ab}F_{ab-cd} = \frac{A_{cb} + A_{ad} - A_{aec} - A_{dfb}}{2}$$

In terms of the area A_1,

$$A_1F_{1-2} = \frac{A_{cb} + A_{ad} - A_{aec} - A_{dfb}}{2}$$

Because all surfaces extend into the page the same distance, we can write

$$L_1 F_{1-2} = \frac{L_{cb} + L_{ad} - L_{aec} - L_{dfb}}{2} \tag{12.19}$$

It does not matter what cross-sectional shape A_2 has. It could be sinusoidal or saw-toothed or flat; as long as it subtends the same solid angle when viewed from A_1, the result would be the same.

The dashed lines of Figure 12.12 can be imagined to be pieces of string stretched from the endpoints of interest—hence the name *crossed-string method*. The right-hand side of Equation 12.19 is

$$L_i F_{i-j} = \frac{1}{2} (\text{Total formed by the sum of the lengths of the crossed string}$$
$$\text{minus the sum of the lengths of the uncrossed strings}$$
$$\text{attached to endpoints of areas } A_i \text{ and} A_j) \tag{12.20}$$

The method works even if the two surfaces of interest share a common edge.

Example 12.5

Two very long plates are directly opposed, as shown in Figure 12.13. The upper plate is maintained at 100°F while the lower plate is at 0°F, and both have black surfaces. Determine the heat exchanged between the two plates.

Solution
Figure 12.13b shows the two plates with dimensions and labeled endpoints.

Assumptions

1. The plate temperatures are uniform.
2. Convection is neglected.

The heat-transfer rate is given by

$$q_{1 \to 2} = \sigma(T_1^4 - T_2^4) A_1 F_{1-2} = \sigma(T_1^4 - T_2^4) A_2 F_{2-1}$$

The view factor can be found with the crossed-string method:

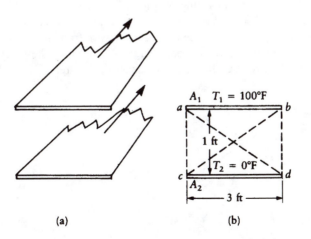

(a)   (b)

FIGURE 12.13 Two opposed infinite plates.

$$L_1 F_{1-2} = \frac{1}{2}(\text{Crossed strings} - \text{Uncrossed strings})$$

$$= \frac{1}{2}(ad + bc - ac - bd)$$

The length of the crossed strings is $\sqrt{9+1} = 3.16$ ft. Substituting,

$$L_1 F_{1-2} = \frac{1}{2}(3.16 + 3.16 - 1 - 1) = 2.16 \text{ ft}$$

With

$$L_1 = 3 \text{ ft}$$

$$F_{1-2} = 2.16/3 = 0.72$$

The view factor is less than 1 for this geometry, but the product of length and view factor is not. The heat-transfer rate then is

$$\frac{q_{1 \to 2}}{A_1} = \sigma(T_1^4 - T_2^4)F_{1-2} = 5.67 \times 10^{-8}(560^4 - 460^4)(0.72)$$

$$\frac{q_{1 \to 2}}{A_1} = 2.19 \times 10^3 \text{ W/m}^2$$

❏

12.4 RADIATION HEAT TRANSFER WITHIN AN ENCLOSURE OF BLACK SURFACES

In this section, we consider an enclosure consisting of surfaces that behave ideally (black surfaces). Each surface is maintained at a constant temperature. If the isothermal condition cannot be met by a surface, then that surface can be subdivided into a number of smaller surfaces, each at a different but constant temperature.

Figure 12.14 is an enclosure consisting of a number of surfaces. We will perform an energy balance on the kth surface. The enclosure contains N areas, and each is identified along with a corresponding temperature. The kth surface receives energy q_k from an external source to maintain it at T_k. Also, the kth surface receives radiant energy from each of the other surfaces and emits energy back (indicated by arrows). The energy received by the area A_k from the other surfaces is

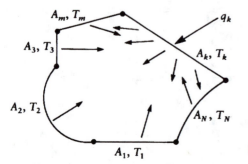

FIGURE 12.14 An enclosure consisting of N surfaces.

$$\sigma T_1^4 A_1 F_{1-k} + \sigma T_2^4 A_2 F_{2-k} + \ldots + \sigma T_k^4 A_k F_{k-k} + \ldots + \sigma T_N^4 A_N T_{N-k}$$

or in summation form:

Energy received from other surfaces $= \displaystyle\sum_{m=1}^{N} \sigma T_m^4 A_m F_{m-k}$

The term $\sigma T_k^4 A_k F_{k-k}$ is included to account for the case where A_k is concave. The rate of emission from area A_k is $\sigma T_k^4 A_k$. Under steady-state conditions, we write the energy balance for surface k as

$$q_k + \sum_{m=1}^{N} \sigma T_m^4 A_m F_{m-k} = \sigma T_k^4 A_k \tag{12.21}$$

In addition, for an enclosure the sum of the view factors must total unity; that is,

$$F_{k-1} + F_{k-2} + \ldots + F_{k-k} + \ldots + F_{k-N} = \sum_{m=1}^{N} F_{k-m} = 1 \tag{12.22}$$

Combining with Equation 12.21 gives

$$q_k = \sigma T_k^4 A_k \sum_{m=1}^{N} F_{k-m} - \sum_{m-1}^{N} \sigma T_m^4 A_k F_{k-m}$$

or

$$q_k = \sigma A_k \sum_{m=1}^{N} (T_k^4 - T_m^4) F_{k-m} \tag{12.23a}$$

We have thus expressed the external heat addition required to maintain area A_k at temperature T_k in terms of a sum of exchanges between A_k and each surface of an enclosure. When Equation 12.23 is applied to an enclosure of N surfaces, then N simultaneous algebraic nonlinear (in temperature) equations result. Consider, for example, the enclosure sketched in Figure 12.15. It is cubical in shape and could be the interior of an oven or of a room. Associated with each surface of the enclosure is a heat-transfer rate q_m and a temperature T_m. According to the assumptions made in

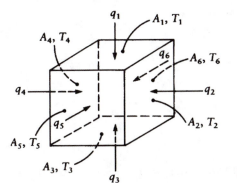

FIGURE 12.15   A cubical enclosure.

the derivation of Equation 12.23, each surface must be isothermal and uniform in temperature and behave like a black body. Applying Equation 12.23 to surface A_1 we get

$$k = 1: \quad q_1 = \sigma A_1 [(T_1^4 - T_2^4) F_{1-2} + (T_1^4 - T_3^4) F_{1-3} + (T_1^4 - T_4^4) F_{1-4}$$

$$+ (T_1^4 - T_5^4) F_{1-5} + (T_1^4 - T_6^4) F_{1-6}]$$

Example 12.6

Figure 12.16 is of a very long broiler whose proposed design is to be evaluated. The oven cross section is an equilateral triangle with one side insulated. The heater surface is maintained at 1000°R while the bottom is at 500°R. Determine the heat that must be supplied to each of the isothermal surfaces, and determine also the temperature of the insulated surface.

Solution

The heat source is radiating energy to each of the other surfaces. Because A_2 is insulated, the energy it receives is not taken out but is reradiated in all directions.

Assumptions

1. Each surface behaves like a black surface.
2. End effects are neglected; that is, the problem is two dimensional.
3. The system is at steady state.

We will need the view factor between each pair of surfaces. Due to symmetry, we see that the view factors between adjacent surfaces are all equal. By inspection, we conclude that the energy leaving surface A_1 is split equally between A_2 and A_3. Therefore,

$$F_{1-2} = F_{1-3} = F_{2-1} = \ldots = \frac{1}{2}$$

This can be proven by the crossed-string method and is reserved as an exercise.

We now apply Equation 12.23a to each surface:

$$k = 1: \quad q_1 = \sigma A_1 [(T_1^4 - T_2^4) F_{1-2} + (T_1^4 - T_3^4) F_{1-3}] \qquad \text{(i)}$$

$$k = 2: \quad q_2 = \sigma A_2 [(T_2^4 - T_1^4) F_{2-1} + (T_2^4 - T_3^4) F_{2-3}] \qquad \text{(ii)}$$

$$k = 3: \quad q_3 = \sigma A_3 [(T_3^4 - T_1^4) F_{3-1} + (T_3^4 - T_2^4) F_{3-2}] \qquad \text{(iii)}$$

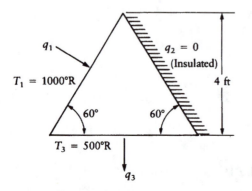

FIGURE 12.16   A very long triangular oven.

Substituting the following:

$$T_1 = 1000°R$$

$$T_3 = 500°R$$

$$q_2 = 0$$

the simultaneous equations rewritten in terms of fluxes become

$$\frac{q_1}{A_1} = 0.1714 \times 10^{-8}\left[(1000^4 - T_2^4)\frac{1}{2} + (1000^4 - 500^4)\frac{1}{2}\right]$$

$$0 = \left[(T_2^4 - 1000^4)\frac{1}{2} + (T_2^4 - 500^4)\frac{1}{2}\right]$$

$$\frac{q_3}{A_3} = 0.1714 \times 10^{-8}\left[(500^4 - 1000^4)\frac{1}{2} + (500^4 - T_2^4)\frac{1}{2}\right]$$

Solving the second equation for T_2 gives

$$-(T_2^4 - 1000^4) = T_2^4 - 500^4$$

$$T_2 = 853.7°R$$

Substituting this value into the first equation gives

$$q_1'' = q_1/A_1 = 0.1714 \times 10^{-8}(2.34 \times 10^{11} + 4.69 \times 10^{11})$$

or

$$q_1'' = 1200 \text{ BTU}/(\text{hr·ft}^2)$$

Similarly,

$$q_3'' = 0.1714 \times 10^{-8}(-4.69 \times 10^{11} - 2.34 \times 10^{11})$$

or

$$q_3'' = -1200 \text{ BTU}/(\text{hr·ft}^2)$$

The results are logical in that the heat entering the system (the oven itself) must equal that which leaves under steady-state conditions. ❑

12.5 RADIATION HEAT TRANSFER WITHIN AN ENCLOSURE OF DIFFUSE-GRAY SURFACES

Section 12.4 dealt with black-surface enclosures. Here we make a better approximation to the modeling of real surfaces. Several complications will arise, but the gray behavior allows for a number of simplifications.

For a gray surface that is diffuse, emissivity and absorptivity are independent of viewing angle and of wavelength. (The properties can depend on temperature, but in this analysis we assume they do not.) Furthermore, emissivity and absorptivity are equal for a gray surface. With absorptivity less than unity, some of the energy incident on a surface is reflected. The reflected energy is assumed to be uniform in all directions—a diffuse emitter and a diffuse reflector. Recall that the formulation presented earlier for view factors involved the condition of a diffuse uniform radiosity leaving a surface. That assumption is met here so that the view-factor charts can be used for the problems of this section.

It should be noted that the radiative heat balances and derived equations apply to unsteady-state as well as to steady-state conditions. The heat fluxes calculated are said to be instantaneous heat rates for unsteady problems.

Remember that we are approximating real-surface behavior by using diffuse-gray-surface models. Real surfaces are extremely difficult and tedious to work with, so it is important to understand the types of conditions that will give rise to substantial error in the gray-surface model. If temperatures of the surfaces differ greatly, then one surface may be emitting energy over a certain range of wavelengths (a function of its temperature) while receiving energy in a different wavelength region. Thus, the assumption of emissivity being equal to absorptivity may break down. If a highly polished surface is present, then the emitted energy varies greatly with angle, and the assumption of diffuse behavior breaks down. The diffuse-gray model does not apply to all problems.

Figure 12.17 shows an enclosure consisting of N surfaces. The surface we select for analysis is the kth surface. Associated with each surface is a temperature and an externally supplied (or removed) heat rate required to maintain the surface at the temperature specified. The formulation will be one where, if temperature is known, heat rate is to be calculated, and vice versa.

The radiative exchange occurring within the enclosure is quite complex. Radiation leaves a surface and travels to another. Part of the incident flux is absorbed and part is reflected. The reflected part travels to another surface and becomes absorbed and reflected. To work with the history of many radiant beams as they follow this process is not an attractive thought. Several methods of analysis exist, but we will formulate what is known as the *net radiation method.*

Figure 12.17 shows surface k with the associated heat rates sketched in. Surface k has a temperature of T_k and an area A_k and is supplied with a heat flux of $q_k''(=q_k/A_k)$ from an external source. Surface k receives radiation from all other surfaces. The incident flux on surface k is denoted as q_{ik}''. The amount reflected by surface k combined with the amount emitted by surface k is called the radiosity q_{jk}''. Adding these heat fluxes according to

Energy in = Energy out

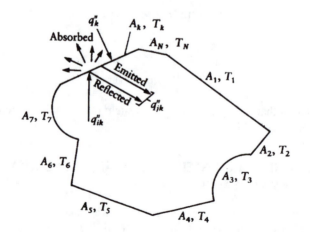

FIGURE 12.17   An enclosure of N diffuse-gray surfaces.

we obtain

$$q_k'' + q_{ik}'' = q_{jk}''$$

or

$$q_k'' A_k = (q_{jk}'' - q_{ik}'')A_k \qquad (12.24)$$

By definition, the energy leaving the surface (the radiosity) consists of reflected plus directly emitted components; that is,

$$q_{jk}'' = \rho_k q_{ik}'' + \varepsilon_k \sigma T_k^4 \qquad (12.25)$$

where ρ_k is the reflectivity of surface k, and ε_k is its emissivity. For an opaque surface, we write

$$\rho_k = 1 - \alpha_k$$

Because the surface is gray,

$$\rho_k = 1 - \varepsilon_k$$

Substituting into Equation 12.25 gives

$$q_{jk}'' = q_{ik}''(1 - \varepsilon_k) + \varepsilon_k \sigma T_k^4 \qquad (12.26)$$

We have an enclosure, so we can write an equation for the incident flux as being composed of the radiosity associated with all surfaces:

$$A_k q_{ik}'' = A_1 q_{j1}'' F_{1-k} + A_2 q_{j2}'' F_{2-k} + \ldots + A_m q_{jm}'' F_{m-k} + \ldots + A_k q_{jk}'' F_{k-k} + \ldots$$
$$+ \ldots + A_N q_{jN}'' F_{N-k} \qquad (12.27)$$

where the term containing F_{k-k} is included to account for concavity, if any, of surface k. From reciprocity, we have

$$A_1 F_{1-k} = A_k F_{k-1}$$
$$A_2 F_{2-k} = A_k F_{k-2}$$

This pattern continues. Equation 12.27 now becomes

$$A_k q_{ik}'' = A_k F_{k-1} q_{j1}'' + A_k F_{k-2} q_{j2}'' + \ldots + A_k F_{k-N} q_{jN}''$$

where the only area appearing is A_k. In summation notation, we write

$$q_{ik}'' = \sum_{m=1}^{N} F_{k-m} q_{jm}'' \qquad (12.28)$$

We now substitute Equation 12.28 into 12.24 to obtain

$$q_k = A_k \left(q_{jk}'' - \sum_{m-1}^{N} F_{k-m} q_{jm}'' \right) \tag{12.29a}$$

Next, we rearrange Equation 12.26 to Solve for q_{ik}'' and again substitute into Equation 12.24 to obtain a second equation for q_k:

$$q_{ik}'' = \frac{1}{1 - \varepsilon_k} (q_{jk}'' - \varepsilon_k \sigma T_k^4) \tag{12.26}$$

$$q_k = A_k \left(q_{jk}'' - \frac{1}{1 - \varepsilon_k} q_{jk}'' + \frac{\varepsilon_k \sigma T_k^4}{1 - \varepsilon_k} \right) \tag{12.24}$$

which simplifies to

$$q_k = \frac{A_k \varepsilon_k}{1 - \varepsilon_k} [\sigma T_k^4 - q_{jk}''] \tag{12.29b}$$

Equations 12.29a and b can be written for all N surfaces of an enclosure to obtain $2N$ simultaneous equations with $2N$ unknowns. The radiosity associated with the surfaces will make up N unknowns, while q_k and T_k will make up the remainder.

It is more convenient to combine Equations 12.29a and b to yield a single equation that applies to each surface of the enclosure. The variables with which we have to work are q_k, the external energy supplied to surface k; q_{jk}, the radiosity associated with surface k; and T_k, the temperature of surface k. In general, we want to keep T_k in the analysis so we can eliminate q_k or q_{jk} from the two equations. The advantage of keeping q_k is that it tells us how much heat is being supplied externally, which is related to the heat transferred through the enclosure. On the other hand, a measuring device aimed at a surface provides a reading that is a function of the radiosity associated with that surface. So in keeping q_{jk}, we can easily relate the calculated results to measured values.

12.5.1 SURFACE HEATING AND SURFACE TEMPERATURE

We now combine the descriptive equations to eliminate the radiosity term. Rearranging Equation 12.29b to solve for radiosity gives

$$q_{jk}'' = \sigma T_k^4 - q_k'' \frac{1 - \varepsilon_k}{\varepsilon_k} \tag{12.29b}$$

Substituting into Equation 12.29a, we get

$$q_k = A_k \left(q_{jk}'' - \sum_{m=1}^{N} F_{k-m} q_{jm}'' \right)$$

$$q_k'' = \left\{ \sigma T_k^4 - q_k'' \frac{(1 - \varepsilon_k)}{\varepsilon_k} - \sum_{m=1}^{N} F_{k-m} \left[\sigma T_m^4 - q_m'' \frac{(1 - \varepsilon_m)}{\varepsilon_m} \right] \right\} \tag{12.29a}$$

Rearranging and simplifying, we have

$$q_k'' \left(1 + \frac{1 - \varepsilon_k}{\varepsilon_k}\right) = \sigma T_k^4 - \sum_{m=1}^{N} F_{k-m} \sigma T_m^4 + \sum_{m=1}^{N} F_{k-m} q_m'' \frac{(1 - \varepsilon_m)}{\varepsilon_m}$$

or

$$\frac{q_k}{A_k \varepsilon_k} - \sum_{m=1}^{N} F_{k-m} \frac{q_m (1 - \varepsilon_m)}{A_m} \frac{(1 - \varepsilon_m)}{\varepsilon_m} = \sigma T_k^4 - \sum_{m=1}^{N} F_{k-m} \sigma T_m^4 \qquad (12.30)$$

Equation 12.30 is an energy balance for an enclosure consisting of N isothermal diffuse-gray surfaces. As an example of how the above equation is applied, consider a three-surface ($N = 3$) enclosure for which we want to perform an energy balance on the $k =$ first surface; we write

$$\frac{q_1''}{\varepsilon_1} - \left[F_{1-1} q_1'' \frac{(1 - \varepsilon_1)}{\varepsilon_1} + F_{1-2} q_2'' \frac{(1 - \varepsilon_2)}{\varepsilon_2} + F_{1-3} q_3'' \frac{(1 - \varepsilon_3)}{\varepsilon_3} \right]$$

$$= \sigma T_1^4 - (F_{1-1} \sigma T_1^4 + F_{1-2} \sigma T_2^4 + F_{1-3} \sigma T_3^4) \qquad (12.31)$$

Although this equation appears formidable, a solution can be obtained. Equation 12.30 must be applied to all the surfaces, which gives N equations. In addition, either temperature T_k or supplied heat flux q_k'' is known for each surface. The result then is N equations and N unknowns. (Methods for solving simultaneous algebraic equations were presented in Chapter 3.) Note that Equation 12.23, for an enclosure consisting of black surfaces, is a special case of Equation 12.30. If the emissivity terms of Equation 12.30 are set equal to 1, Equation 12.23 results.

Example 12.7

An oven manufacturer is investigating the possibility of eliminating insulation in the top of its industrial ovens and instead is proposing the use of a double panel with a vacuum between each piece. Figure 12.18 shows the top consisting of two sheet-steel surfaces (oxidized) separated by a 3-cm gap. In the worst case the temperature of the inside surface is 260°C, and correspondingly, an exterior-surface temperature of 50°C is acceptable (for safety reasons). Determine the heat lost by the oven through its top surface.

Solution

With an evacuated gap the heat lost by conduction and convection through the double panel is eliminated or at least considered negligible compared to the heat transferred by radiation. The energy transferred from the warmer panel is radiated in all directions, including the edges.

Assumptions

1. The sheet metal is thin enough that the temperatures given are the inside-surface temperatures.
2. Both sheet metal pieces are isothermal.
3. The plates are very large and spaced close together, so they can be considered infinite (i.e., edge effects are neglected).
4. The surfaces emit energy diffusely and can be considered gray.

As indicated in Figure 12.18, the warmer surface is A_1. A consequence of the third assumption is that all energy leaving A_1 is intercepted by A_2, and vice versa. Therefore, the view factors become

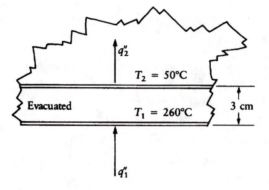

FIGURE 12.18   The double-panel top of an oven.

$$F_{1-2} = F_{2-1} = 1$$

Because the surfaces are flat,

$$F_{1-1} = F_{2-2} = 0$$

Equation 12.30 applies and, using it, we get

$$k = 1: \frac{q_1''}{\varepsilon_1} - F_{1-1}q_1'' \frac{(1-\varepsilon_1)}{\varepsilon_1} - F_{1-2}q_2'' \frac{(1-\varepsilon_2)}{\varepsilon_2} = \sigma T_1^4 - F_{1-1}\sigma T_1^4 - F_{1-2}\sigma T_2^4$$

$$k = 2: \frac{q_2''}{\varepsilon_2} - F_{2-1}q_1'' \frac{(1-\varepsilon_1)}{\varepsilon_1} - F_{2-2}q_2'' \frac{(1-\varepsilon_2)}{\varepsilon_2} = \sigma T_2^4 - F_{2-1}\sigma T_1^4 - F_{2-2}\sigma T_2^4$$

Substituting for the view factors, we reduce these equations to

$$\frac{q_1''}{\varepsilon_1} - q_2'' \frac{(1-\varepsilon_2)}{\varepsilon_2} = \sigma T_1^4 - \sigma T_2^4$$

$$\frac{q_2''}{\varepsilon_2} - q_1'' \frac{(1-\varepsilon_1)}{\varepsilon_1} = \sigma T_2^4 - \sigma T_1^4$$

We now solve by eliminating q_2'' from the equations. Solving the first equation for q_2'' gives

$$q_2'' = \frac{\varepsilon_2}{1-\varepsilon_2}\left[\frac{q_1''}{\varepsilon_1} + \sigma T_2^4 - \sigma T_1^4\right] \qquad (i)$$

Substituting into the second equation, we get

$$\frac{1}{1-\varepsilon_2}\left(\frac{q_1''}{\varepsilon_1} + \sigma T_2^4 - \sigma T_1^4\right) - q_1'' \frac{(1-\varepsilon_1)}{\varepsilon_1} = \sigma T_2^4 - \sigma T_1^4$$

After considerable manipulation, we finally obtain

$$q_1'' = \frac{\sigma(T_1^4 - T_2^4)}{1/\varepsilon_1 + 1/\varepsilon_2 - 1} \qquad (ii)$$

From Appendix Table E.1 for oxidized steel, $\varepsilon = 0.94$. Both surfaces are the same, so $\varepsilon_1 = \varepsilon_2 = 0.94$. With $T_1 = 260°C = 533$ K and $T_2 = 50°C = 323$ K, we substitute into Equation (ii) above to get

$$q_1'' = \frac{5.67 \times 10^{-8}(533^4 - 323^4)}{1/0.94 + 1/0.94 - 1}$$

or

$$q_1'' = 3\ 510\ \text{W/m}^2$$

If we substitute Equation (ii) into Equation (i), we will find that

$$q_2'' = -q_1'' = -3\ 510\ \text{W/m}^2$$

The result that $q_1'' = -q_2''$ is evident also from Figure 12.18.   ❏

Example 12.8

A serving dish containing cooked vegetables is removed from an oven and set out on a rack. The dish is 6 in tall and has a 12-in diameter, and its temperature upon removal from the oven is 350°F. The surroundings are at 70°F. Determine the net heat exchanged between the dish and the surroundings by radiation at the instant the dish is removed from the oven. Perform the calculations (a) if the dish and surroundings behave like black bodies, and again (b) if the dish has an emissivity of 0.82 and the surroundings have an emissivity of 0.93.

Solution

The serving dish will lose heat by convection to the ambient air as well as by radiation to the surroundings. Further air properties and the emissivity of the dish will change as the dish cools.

Assumptions

1. Convection is neglected and radiation is the only mode by which the dish loses heat.
2. The emissivity of the dish is constant.
3. The result applies to the instant in time when the dish temperature is 350°F (= 810°R) and the surroundings are at 70°F (= 530°R).

We identify the dish as surface A_1 with a surface area given by

$$A_1 = \pi D^2 (2)/4 + \pi DL = \frac{\pi(12)^2(2)}{4} + \pi(12)(6)$$

$$A_1 = 452\ \text{in}^2 = 3.14\ \text{ft}^2$$

(a) The area of the surroundings is infinite compared to the dish-surface area. Also, the view factor between the dish and the surroundings is unity (i.e., $F_{1-2} = 1$). We now apply Equation 12.23 to the dish, assuming it to have a black surface:

$$q_1 = \sigma A_1 [T_1^4 - T_2^4] F_{1-2}$$

With $T_1 = 810°R$ and $T_2 = 530°R$, we get

$$q_1 = 0.1714 \times 10^{-8} (3.14)(810^4 - 530^4)(1)$$

$$q_1 = 1890 \text{ BTU/hr} \qquad\qquad\qquad \text{(black-surface behavior)}$$

(b) For gray-surface behavior, we apply Equation 12.30 to get

$$k = 1: \qquad \frac{q_1}{A_1 \varepsilon_1} - \left[F_{1-1} \frac{q_1}{A_1} \frac{(1-\varepsilon_1)}{\varepsilon_1} + F_{1-2} \frac{q_2}{A_2} \frac{(1-\varepsilon_2)}{\varepsilon_2} \right] = \sigma T_1^4 - (F_{1-1} \sigma T_1^4 + F_{1-2} \sigma T_2^4)$$

$$k = 2: \qquad \frac{q_2}{A_2 \varepsilon_2} - \left[F_{2-1} \frac{q_1}{A_1} \frac{(1-\varepsilon_1)}{\varepsilon_1} + F_{2-2} \frac{q_2}{A_2} \frac{(1-\varepsilon_2)}{\varepsilon_2} \right] = \sigma T_2^4 - (F_{2-1} \sigma T_1^4 + F_{2-2} \sigma T_2^4)$$

Note that $F_{1-1} = 0$. Also, the second equation contains view-factor terms that we cannot determine without more information. Substituting into the first equation above gives

$$\frac{q_1}{3.14(0.82)} - \left[0 + (1) \frac{q_2}{A_2} \frac{(1-0.93)}{0.93} \right] = 0.1714 \times 10^{-8}(810^4) - [0 + (1)(0.1714 \times 10^{-8})(530^4)]$$

Ordinarily, we would use the second equation to evaluate q_2/A_2, but because $A_2 \to \infty$, $q_2/A_2 \to 0$. Simplifying the above equation, we get

$$0.388 q_1 = 737.8 - 135.2$$

$$q_1 = 1550 \text{ BTU/hr} \qquad\qquad \text{(gray-diffuse-surface behavior)}$$

This value is exactly the product of ε_1 and q_1 for black-surface behavior found in part (a). When one of the areas is infinite, this is usually the case for a two-surface enclosure. ❑

12.5.2 RADIOSITY AND SURFACE TEMPERATURE

We now examine the alternative solution technique in which we eliminate the heat flux from the energy-balance equations. The result is an equation written in terms of radiosity and surface temperature. Beginning with Equations 12.29a and b, we set them equal to one another to get

$$q_{jk}'' - \sum_{m=1}^{N} F_{k-m} q_{jm}'' = \frac{\varepsilon_k}{1-\varepsilon_k} (\sigma T_k^4 - q_{jk}'')$$

Rearranging, we obtain

$$q_{jk}'' - (1-\varepsilon_k) \sum_{m=1}^{N} F_{k-m} q_{jm}'' = \varepsilon_k \sigma T_k^4 \qquad\qquad (12.32)$$

As an example, for a system of two surfaces, we write

$$k = 1: \qquad q''_{j1} - (1 - \varepsilon_1)(F_{1-1}q''_{j1} + F_{1-2}q''_{j2}) = \varepsilon_1\sigma T_1^4 \qquad (12.33a)$$

$$k = 2: \qquad q''_{j2} - (1 - \varepsilon_2)(F_{2-1}q''_{j1} + F_{2-2}q''_{j2}) = \varepsilon_2\sigma T_2^4 \qquad (12.33b)$$

Remember that the above formulation applies to gray, isothermal surfaces that reflect and emit radiant energy diffusely.

12.5.3 Electrical Analogy

In an effort to ease the task of generating equations for the problem of heat transfer in an enclosure, an electrical analogy has been formulated. Consider, for example, the three surface enclosure of Figure 12.16. The equation that describes the energy transfer within is Equation 12.23a, which when applied yielded Equations i, ii, and iii of Example 12.6.

Figure 12.19 shows the same three surface enclosure with three black surfaces. Also shown is an electrical network within. Equation 12.23a is

$$q_k = \sigma A_k \sum_{m=1}^{N} (T_k^4 - T_m^4)F_{k-m} \qquad (12.23a)$$

Each surface has a black body emissive power defined as

$$E_b = \sigma T^4$$

which acts as an electrical potential. The area–view factor product $A_k F_{k-m}$ is the conductance between the nodes labeled as E_{bk} (= σT_k^4, the source node) and E_{bm} (= σT_m^4, receiver node). In the three-surface network of Figure 12.19, the net flow of heat is analogous to the flow of electricity. Rewriting Equation 12.23a in terms of the black body emissive power gives

$$q_k = \sum_{m=1}^{N} \frac{E_{bk} - E_{bm}}{R_{km}} \qquad (12.23b)$$

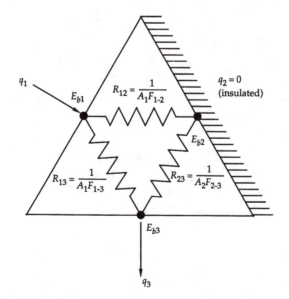

FIGURE 12.19 Electrical analogy for an enclosure consisting of three black surfaces.

where $R_{km} = 1/A_k F_{k-m}$. Applying this equation to the electrical network of Figure 12.19 yields

for $k = 1$
$$q_1 = \frac{E_{b1} - E_{b1}}{R_{11}} + \frac{E_{b1} - E_{b2}}{R_{12}} + \frac{E_{b1} - E_{b3}}{R_{13}}$$

Substituting, and noting that $F_{1-1} = 0$,

for $k = 1$
$$q_1 = 0 + \frac{\sigma T_1^4 - \sigma T_2^4}{1/(A_1 F_{1-2})} + \frac{\sigma T_1^4 - \sigma T_3^4}{1/(A_1 F_{1-3})}$$

which becomes, after simplification,

$$q_1 = \sigma A_1 [(T_1^4 - T_2^4) F_{1-2} + (T_1^4 - T_3^4) F_{1-3}]$$

which is the same result as Equation (i) of Example 12.6.

Figure 12.20 shows an equivalent circuit for a four-surface enclosure whose surfaces are all black. In drawing this circuit, first all surfaces are labeled 1 through 4, with black body emissive power E_b located as shown. Then resistors are drawn between one surface and all the others. The resistor values are identified and then Equation 12.23b is applied to generate the N (= 4 surfaces) simultaneous equations for the enclosure. For q_1, assuming the surfaces cannot "see" themselves (e.g., $F_{1-1} = 0$), we write

for $k = 1$
$$q_1 = \frac{E_{b1} - E_{b2}}{R_{12}} + \frac{E_{b1} - E_{b3}}{R_{13}} + \frac{E_{b1} - E_{b4}}{R_{14}}$$

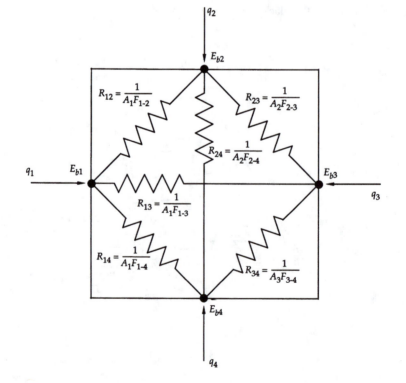

FIGURE 12.20 Electrical analogy for an enclosure consisting of four black surfaces.

We also write

for $k = 2$
$$q_2 = \frac{E_{b2} - E_{b1}}{R_{21}} + \frac{E_{b2} - E_{b3}}{R_{23}} + \frac{E_{b2} - E_{b4}}{R_{24}}$$

Note how the subscripts change when the source (E_b) changes. Substituting for the resistances in the surface 1 equation, and simplifying gives:

for $k = 1$
$$q_1 = \sigma A_1 [(T_1^4 - T_2^4)F_{1-2} + (T_1^4 - T_3^4)F_{1-3} + (T_1^4 - T_4^4)F_{1-4}]$$

Similar equations can be written for all surfaces.

For an enclosure consisting of gray surfaces, we refer back to Equation 12.29b:

$$q_k = \frac{A_k \varepsilon_k}{1 - \varepsilon_k} (\sigma T_k^4 - q_{jk}'') \qquad (12.29b)$$

To develop an electrical circuit model, we rewrite this equation as

$$q_k = \frac{E_{bk} - e_k}{r_k} \qquad (12.29c)$$

where

$$E_{bk} = \sigma T_k^4$$

$$e_k = q_{jk}''$$

and

$$r_k = \frac{1 - \varepsilon_k}{A_k \varepsilon_k}$$

In addition, we use the fact that the direct radiation exchange between any two opaque, diffuse surfaces is given by

$$q_{k-m} = \frac{e_k - e_m}{R_{km}} = A_k F_{k-m}(e_k - e_m) \qquad (12.30)$$

We consider the rate of radiation heat transfer q_k to be the current in a circuit between potentials E_{bk} and e_k, having a resistance r_k between them. We model the two gray surface enclosure of Example 12.7 (Figure 12.18) with the circuit shown in Figure 12.21. The combined effect of Equation 12.29c and 12.30 allows us to write directly:

$$q_{1-2} = \frac{E_{b1} - E_{b2}}{r_1 + R_{12} + r_2}$$

Substituting gives

$$q_{1-2} = \frac{\sigma T_1^4 - \sigma T_2^4}{\dfrac{1 - \varepsilon_1}{A_1 \varepsilon_1} + \dfrac{1}{A_1 F_{1-2}} + \dfrac{1 - \varepsilon_2}{A_2 \varepsilon_2}}$$

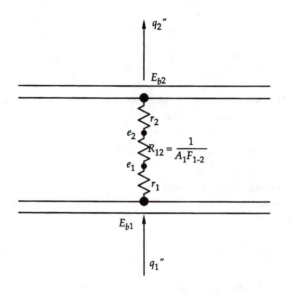

FIGURE 12.21   Electrical analogy for an enclosure consisting of two gray surfaces.

Note that if the surfaces were black, $\varepsilon_1 = \varepsilon_2 = 1$, and the first and third terms in the denominator of the right hand side of the preceding equation become 0. For the oven of Example 12.7, $A_1 = A_2, F_{1-2} = 1$, and so we get

$$\frac{q_1}{A_1} = q_1'' = \frac{\sigma T_1^4 - \sigma T_2^4}{\left(\dfrac{1}{\varepsilon_1} - 1\right) + \dfrac{1}{1} + \left(\dfrac{1}{\varepsilon_2} - 1\right)}$$

which finally becomes

$$\frac{q_1}{A_1} = q_1'' = \frac{\sigma T_1^4 - \sigma T_2^4}{\dfrac{1}{\varepsilon_1} + \dfrac{1}{\varepsilon_2} - 1}$$

This is the same result obtained in the example [Equation (ii)] without the lengthy algebraic manipulations.

Figure 12.22 shows the equivalent circuit for an enclosure of three gray diffuse surfaces. In this case, the difference between black surfaces and gray surfaces is more pronounced. Energy leaving surface 1 makes it way to surface 2 along two paths. One path is directly through R_{12}. the second is through R_{13} to e_3 and then through R_{23}. The second path means physically that energy leaves surface 1, is received by surface 2, and subsequently is re-radiated to surface 3. Hence, we refer to surface 2 as a *re-radiating surface*. Figure 12.22 has the resistor connecting e_3 and e_2 labeled as R_{23}, but if we consider energy being re-radiated from 3 to 2, we denote it as R_{32}.

In essence, we have two resistors in series: R_{13} and R_{23}, and these act in parallel with R_{12}. We would therefore write:

$$q_{1-2} = \frac{E_{b1} - E_{b2}}{r_1 + \left[\dfrac{1}{R_{12}} + \dfrac{1}{R_{13} + R_{32}}\right]^{-1} + r_2}$$

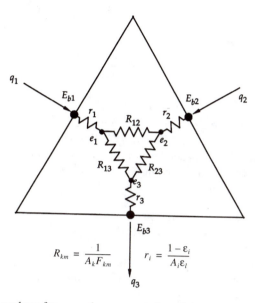

$$R_{km} = \frac{1}{A_k F_{km}} \qquad r_i = \frac{1 - \varepsilon_i}{A_i \varepsilon_i}$$

FIGURE 12.22 Electrical analogy for an enclosure consisting of three gray surfaces.

$$q_{1-3} = \frac{E_{b1} - E_{b3}}{r_1 + \left[\dfrac{1}{R_{13}} + \dfrac{1}{R_{12} + R_{23}}\right]^{-1} + r_3}$$

$$q_{2-3} = \frac{E_{b2} - E_{b3}}{r_2 + \left[\dfrac{1}{R_{23}} + \dfrac{1}{R_{21} + R_{13}}\right]^{-1} + r_3}$$

Note how the subscripts on the R_{km} terms are modified to take proper account of the heat flow direction. Substituting for the resistances gives

$$q_{1-2} = \frac{\sigma T_1^4 - \sigma T_2^4}{\dfrac{1 - \varepsilon_1}{A_1 \varepsilon_1} + \left[\dfrac{1}{1/A_1 F_{1-2}} + \dfrac{1}{1/A_1 F_{1-3} + 1/A_3 F_{3-2}}\right]^{-1}} + \frac{1 - \varepsilon_2}{A_2 \varepsilon_2}$$

$$q_{1-3} = \frac{\sigma T_1^4 - \sigma T_3^4}{\dfrac{1 - \varepsilon_1}{A_1 \varepsilon_1} + \left[\dfrac{1}{1/A_1 F_{1-3}} + \dfrac{1}{1/A_1 F_{1-2} + 1/A_2 F_{2-3}}\right]^{-1}} + \frac{1 - \varepsilon_3}{A_3 \varepsilon_3}$$

$$q_{2-3} = \frac{\sigma T_2^4 - \sigma T_3^4}{\dfrac{1 - \varepsilon_2}{A_2 \varepsilon_2} + \left[\dfrac{1}{1/A_2 F_{2-3}} + \dfrac{1}{1/A_2 F_{2-1} + 1/A_1 F_{1-3}}\right]^{-1}} + \frac{1 - \varepsilon_3}{A_3 \varepsilon_3}$$

The radiosities e_k ($= q_{km}''$) are found by applying a heat balance to all surfaces, or by analyzing the circuit of Figure 12.22. for surface 1, we get

$$q_1 = \frac{\text{potential}}{\text{resistance}} = \frac{E_{b1} - e_1}{r_1}$$

or

$$q_1 = \frac{\sigma T_1^4 - q_{j1}''}{\dfrac{1 - \varepsilon_1}{A_1 \varepsilon_1}} \tag{12.31}$$

So, in analyzing an enclosure like the one shown in Figure 12.22, we have three variables associated with each surface (e.g., for surface 1, q_1, T_1 and q_{j1}''). Knowing any two allows us to solve for the third. If we knew all the temperatures and heat transfer rates, we could write an equation like 12.31 for all surfaces and combine them to obtain equations for the radiosities. In general, the result would be an equation of the form:

$$q_{ji}'' = \varepsilon_i E_{bi} + (1 - \varepsilon_i) \sum_{i=1}^{n} q_{ji}'' F_{i-k}$$

For the three surface enclosure of Figure 12.31, we would write

$$q_{j1}'' = \varepsilon_1 \sigma T_1^4 + (1 - \varepsilon_1)(q_{j1} F_{1-1} + q_{j2} F_{1-2} + q_{j3} F_{1-3})$$

$$q_{j2}'' = \varepsilon_2 \sigma T_2^4 + (1 - \varepsilon_2)(q_{j1} F_{2-1} + q_{j2} F_{2-2} + q_{j3} F_{2-3})$$

$$q_{j3}'' = \varepsilon_3 \sigma T_3^4 + (1 - \varepsilon_3)(q_{j1} F_{3-1} + q_{j2} F_{3-2} + q_{j3} F_{3-3})$$

These equations would be solved simultaneously to find the radiosities q_{ji}'', and the results would be substituted into Equation 12.29b to determine the heat transfer rates q_i.

In some radiation enclosure problems, one of the surfaces might be insulated, for which $q_i = 0$. Such surfaces are also called adiabatic or refractory surfaces, because historically such topics were applied to furnaces. A surface lined with refractory brick is usually taken to be adiabatic.

12.6 SUMMARY

In this chapter, we have developed equations and procedures for evaluating the view factor. The view factor was defined in an equation that involved integration. Alternative methods were also introduced. Enclosure theory was presented for two cases. One case was for black surfaces, and the second was for gray-diffuse isothermal surfaces. Table 12.2 provides a summary of the important relations discussed in this chapter.

12.7 PROBLEMS

12.7.1 THE VIEW FACTOR AND HEAT TRANSFER RADIATION

1. Figure P12.1 shows an elemental area and a washer placed a distance H apart. Show that the view factor F_{d1-2} is given by

$$F_{d1-2} = \frac{R_1^2}{H^2 + R_1^2} - \frac{R_2^2}{H^2 + R_2^2}$$

Use the method of integration beginning with Equation 12.9.

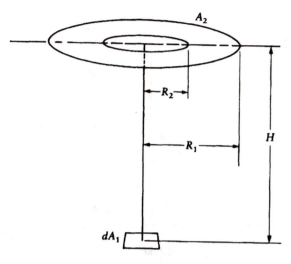

FIGURE P12.1

TABLE 12.2
Summary of Equations of Chapter 12

System	View factor	Net heat exchanged
Differential area $d1$ and Differential area $d2$	$dF_{d1-d2} = \dfrac{\cos\beta_1 \cos\beta_2 dA_2}{\pi L^2}$ $dF_{d2-d1} = \dfrac{\cos\beta_1 \cos\beta_2 dA_1}{\pi L^2}$	$dq_{d1 \to d2} = \sigma(T_1^4 - T_2^4)dF_{d1-d2}dA_1$ $= \sigma(T_1^4 - T_2^4)dF_{d2-d1}dA_2$
Differential area $d1$ and Finite area 2	$F_{d1-2} = \displaystyle\iint_{A_2} \dfrac{\cos\beta_1 \cos\beta_2 dA_1}{\pi L^2}$ $A_2 dF_{2-d1} = dA_1 F_{d1-2}$	$dq_{d1 \to 2} = \sigma(T_1^4 - T_2^4)dA_1 F_{d1-2}$ $= \sigma(T_1^4 - T_2^4)A_2 dF_{2-d1}$
Finite area 1 and Finite area 2	$F_{1-2} = \dfrac{1}{A_1}\displaystyle\iint_{A_1}\iint_{A_2} \dfrac{\cos\beta_1 \cos\beta_2 dA_2 dA_1}{\pi L^2}$ $F_{2-1} = \dfrac{1}{A_2}\displaystyle\iint_{A_1}\iint_{A_2} \dfrac{\cos\beta_1 \cos\beta_2 dA_2 dA_1}{\pi L^2}$	$q_{1 \to 2} = \sigma(T_1^4 - T_2^4)A_1 F_{1-2}$ $= \sigma(T_1^4 - T_2^4)A_2 F_{2-1}$
For an enclosure consisting of N black surfaces:		$q_k = \sigma A_k \displaystyle\sum_{m=1}^{N} (T_k^4 - T_m^4)F_{k-m}$ (12.23)
For an enclosure consisting of N gray-diffuse surfaces:		$\dfrac{q_k}{A_k \varepsilon_k} - \displaystyle\sum_{m=1}^{N} F_{k-m}\dfrac{q_m(1-\varepsilon_m)}{A_m \varepsilon_m}$ $= \sigma T_k^4 - \displaystyle\sum_{m=1}^{N} F_{k-m}\sigma T_m^4$ (12.30)

2. The areas of Problem 1 behave as black bodies, with the temperature of dA_1 being 350 K, while that of A_2 is 100 K. Calculate the net heat exchanged between the surfaces. The washer dimensions are $R_1 = 1$ m and $R_2 = 0.5$ m. The distance between dA_1 and A_2 is 1 m.

3. Figure P12.2 shows two perpendicular planes that share a common edge. Use configuration-factor algebra to show that the view factor for surfaces 1 and 4 is given by

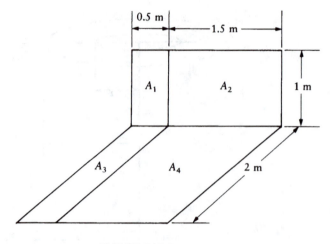

FIGURE P12.2

$$F_{1-4} = \frac{1}{2A_1}[(A_1 + A_2)F_{(1+2)-(3+4)} - A_1F_{1-3} - A_2F_{2-4}]$$

4. For the dimensions given in Figure P12.2 and the equation of Problem 3, evaluate F_{1-4} numerically.
5. Figure P12.3 shows two perpendicular plates that share a common edge. Use configuration-factor algebra to show that the view factor for surfaces 1 and 6 is given by

$$F_{1-6} = \frac{1}{2A_1}[(A_1 + A_2 + A_3 + A_4)F_{(1+2+3+4)-(5+6)}$$

$$- (A_2 + A_4)F_{(2+4)-6} - (A_1 + A_3)F_{(1+3)-5}$$

$$- (A_3 + A_4)F_{(3+4)-(5+6)} + A_4F_{4-6} + A_3F_{3-5}]$$

6. For the dimensions given in Figure P12.3 and the equation of Problem 5, evaluate F_{1-6} numerically.

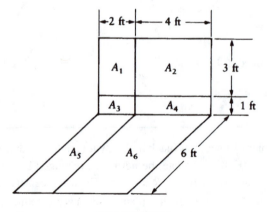

FIGURE P12.3

7. Example 12.2 deals with two perpendicular planes that share a common edge. In this problem, we investigate the effect that doubling an area has on the view factor.

   (a) Referring to Figure 12.9, suppose that A_3 is doubled in size so that is 1 ft tall and 1 ft wide. Determine the numerical value of F_{6-3}. Is it double that found in the example?

   (b) Referring again to Figure 12.9, suppose that A_6 is doubled (instead of A_3) so that its new dimensions are 1×1 ft. Calculate the numerical value of the view factor and determine if it is double that found in the example.

8. Radiant heaters are installed in the bathrooms of many southern homes. A heater consists of a wall-mount housing that contains two coiled wires through which an electric current is passed. The wires become red hot (520°C as measured from an actual heater) and radiate heat in all directions. Figure P12.4 illustrates a wall heater and the corresponding opposite wall.

   (a) Use Figure 12.3 to determine the view factor between the heater and the opposite wall, assuming that the heater area is very small compared to the wall itself.

   (b) Determine the net heat exchanged between the heater and the opposite wall for a temperature of 17°C. Assume both surfaces to be black and that each is diffuse.

9. Two pipes convey a liquid. One pipe conveys heated liquid to a process, and the other pipe is the cooled-liquid return. The pipes are laid parallel to each other and are very long. Figure P12.5 is a cross section of the two pipes. Using the crossed-string method, express the view factor F_{1-2} in terms of R (the outside radius of both pipes) and X by completing the following steps:

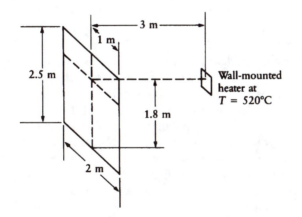

FIGURE P12.4

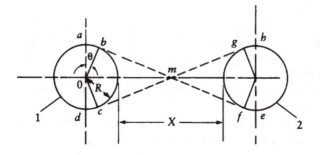

FIGURE P12.5

(a) Verify that the view factor can be written as

$$F_{1-2} = \frac{L_{abfe} - L_{de}}{\pi R}$$

(b) The length L_{de} is easily seen to be equal to $X + 2R$. Working with the right triangle *Obm*, evaluate L_{bm}. Add this to the arc length L_{ab} $(= R\theta)$ to show that

$$2L_{abm} = L_{abfe} = 2[(X/2 + R)^2 - R^2 + R\theta]$$

(c) The angle θ can be written as

$$\theta = \sin^{-1}[R/(X/2 + R)]$$

(d) Substitute the above relations into the equation written in part (a) and show that

$$F_{1-2} = \frac{2\{(X/2 + R)^2 - R^2 + R \sin^{-1}[R/(X/2 + R)]\} - (X + 2R)}{\pi R}$$

10. The pipes of Problem 9 are made of 4 nominal schedule 40 pipe (assume black surfaces) spaced at a center-to-center distance of 2 ft. At one location, one tube surface is at 200°F, while the other is at 75°F. Determine the net heat transferred between the two surfaces.

11. The two cylinders of Figure P12.5 are moved together so that they touch. Determine the view-factor relationship.

12. For the system of Problem 11, the pipes are both 6 nominal schedule 40 (the pipes are touching). Each pipe's surface can be approximated as being black and isothermal, where $T_1 = 100°C$ and $T_2 = 0°C$. Determine the heat-transfer rate between the two pipes.

13. A novel concept for room heating involves running heated water through pipes in the floor of a room. The floor reaches a temperature of 80°F. The ceiling of the room is the underside of the floor of an attic, and there is heat flow from the floor to the ceiling. The ceiling itself is approximately isothermal at 65°F. Neglect convective effects and calculate the net heat exchanged between the floor and the ceiling. Take both surfaces to be black and evaluate the view factor by using Figure 12.6. Room dimensions are $9 \times 12 \times 8$ ft tall.

14. *Burn victims experience difficulties in maintaining the proper body temperature because the physical senses that alert the body's thermostat are damaged. One way of providing comfort from cold to burn victims is to expose them to microwave radiation in an amount sufficient to provide necessary warmth. A microwave line source located in the ceiling of a room would radiate to a body lying on a patient cart. Determine the view factor between the line source and the body by assuming the body to be a cylinder 1.8 m long and 36 cm in diameter. The line source has the same length, and its distance from the body can be varied from 0.5 to 1.5 m above the centerline of the cylinder. Graph the view factor as a function of the distance between source and receiver. Use Figure 12.7 and assume both surfaces are black.

15. Example 12.5 deals with two opposed plates. Rework that example, but with the plates oriented as in Figure P12.6.

16. Prove that the configuration factors in Example 12.6 for adjacent surfaces equal 1/2 by the crossed-string method. Complete the following steps:

---

* Contributed by Dr. R.D. McAfee of the New Orleans V.A. Hospital.

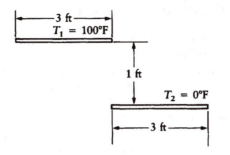

FIGURE P12.6

(a) Although any two adjacent sides share a common edge, two sides are drawn separated as shown in Figure P12.7b.
(b) Draw in the strings and evaluate their lengths, noting that the length from *a* to *c* is really zero.
(c) Evaluate the view factor, using Equation 12.20.
17. Figure P12.8 shows three surfaces that extend to infinity into the page.
    (a) Use the crossed-string method to evaluate F_{1-2}.
    (b) Following the same lines of reasoning as those outlined in Problem 17, evaluate F_{1-3}.
    (c) Determine Y required for F_{1-2} to equal F_{1-3}.

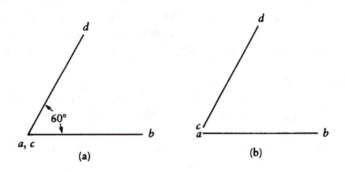

FIGURE P12.7

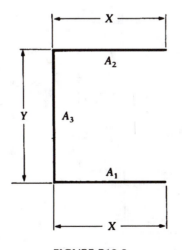

FIGURE P12.8

12.7.2 RADIATION IN AN ENCLOSURE OF BLACK SURFACES

18. Write all the radiation equations for the enclosure of Figure 12.15.
19. The sphere used in a pinball machine is made by a process that includes cooling the sphere from 500°F to room temperature. The sphere is in an isothermal enclosure maintained at 70°F. The sphere (at 500°F) is brought in and allowed to cool. Determine the net heat exchanged at the instant the sphere is exposed to the enclosure. Assume that the sphere and the enclosure behave like black bodies, and take the sphere diameter to be 1-1/4 in.
20. An ice cube of dimensions 3 × 3 × 3 cm at 0°C is removed from a freezer and placed in a room. The room itself is isothermal at 25°C. Determine the net heat exchanged between the room and the ice cube during the melting of the ice cube. Assume that the cube and the enclosure behave like black bodies.
21. A can of warm beer (4-1/2 inches tall × 2-1/2 inches diameter) is placed into the freezer compartment of a refrigerator. The can is assumed to radiate heat from all its surfaces to the enclosure (even though it may be set down on its end). The can is at 65°F while the freezer walls are at 30°F. Determine the net heat exchanged if the surfaces of the can and of the freezer behave like black bodies. Take the freezer-wall area to be very large.

12.7.3 RADIATION IN AN ENCLOSURE OF DIFFUSE-GRAY SURFACES

22. An enclosure that is a circular cylinder is sketched in Figure P12.9.
    (a) Write the equations relating the heat exchanged to temperature for each surface, using Equation 12.23, which reflects the assumption that the surfaces are black.
    (b) Write the equations relating the heat exchanged to the temperature for each surface, using Equation 12.30, which contains the assumptions that the surfaces are gray and emit and reflect diffusely.
    Compare the sets of equations.
23. Rework Problem 19, assuming that the surfaces are gray and emit diffusely. Take the emissivity of the sphere surface to be 0.06 (which is for chromium, from Appendix Table E.1).
24. Rework Problem 20, assuming that the surfaces are gray and emit diffusely.

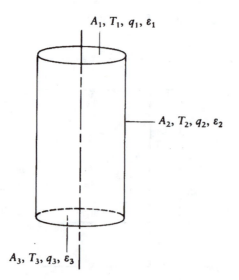

FIGURE P12.9

25. Rework Problem 21, assuming that the surfaces are gray and emit diffusely. Take the emissivity of the can to be 0.04 (which is for polished aluminum, from Appendix Table E.1).
26. Figure P12.10 shows a long oven that is semicircular in cross section with radiant heaters in the top. The top is at 400°C, and the heaters are drawing 500 W/m². Estimate the temperature of the bottom surface. The heaters are tungsten wire, and the receiver surface is red brick.
27. Example 12.7 deals with a two-panel enclosure. We applied Equation 12.30 to the panels and derived the following equation:

$$q_1'' = \frac{\sigma(T_1^4 - T_2^4)}{1/\varepsilon_1 + 1/\varepsilon_{2-1}} \qquad \text{(i)}$$

Recall that Equation 12.30 resulted from performing an energy balance on the surfaces of an enclosure, so Equation (i) above is essentially a statement of conservation of energy. In this problem, we will derive Equation (i) by following a different procedure. We will trace the path of a beam emanating from one surface and reflected by the other.

Figure P12.11 shows two parallel planes. The planes are gray and emit diffusely (therefore $\varepsilon = \alpha$ for each plane). The emissivity of plane 1 is ε_1, and its temperature is T_1. A beam emitted by plane 1 is denoted as $\varepsilon_1 \sigma T_1^4$. Upon arriving at plane 2, part of the beam is absorbed ($= \alpha_2 \varepsilon_1 \sigma T_1^4 = \varepsilon_2 \varepsilon_1 \sigma T_1^4$), and part of the beam is reflected $= (1 - \alpha_2)\varepsilon_1 \sigma T_1^4 = (1 - \varepsilon_2)\varepsilon_1 \sigma T_1^4$. The reflected portion then arrives back at plane 1. Part is absorbed and part is reflected. This process continues, as shown in the figure.

(a) Draw a similar beam-tracing representation for plane 2.
(b) Verify that the net heat exchanged between the planes is given by

$$q_1'' = \sigma(T_1^4 - T_2^4)\left[\, \varepsilon_1\varepsilon_2 + \varepsilon_1\varepsilon_2(1 - \varepsilon_1)(1 - \varepsilon_2) + \varepsilon_1\varepsilon_2(1 - \varepsilon_1)^2(1 - \varepsilon_2)^2 + \ldots \right]$$

or

$$q_1'' = \varepsilon_1\varepsilon_2\sigma(T_1^4 - T_2^4)\sum_{m=0}^{\infty}[(1 - \varepsilon_1)(1 - \varepsilon_2)]^m$$

It may be shown (although not necessary to do so here) that

$$\frac{1}{1 - (1 - \varepsilon_1)(1 - \varepsilon_2)} = \sum_{m=0}^{\infty}[(1 - \varepsilon_1)(1 - \varepsilon_2)]^m$$

Substitute into Equation (ii) above and simplify to obtain Equation (i).

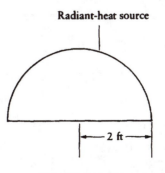

Radiant-heat source

— 2 ft —

FIGURE P12.10

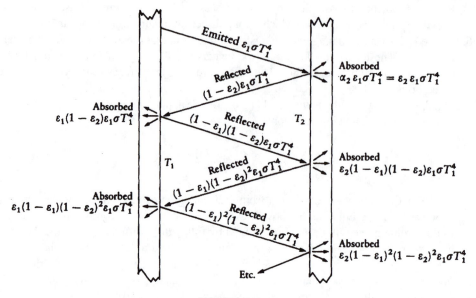

FIGURE P12.11

28. A radiation shield or screen is a third body placed between two bodies that are exchanging heat by radiation. A wire screen or a glass door placed between a fire in a fireplace and an observer is an example of a shield. With the shield in place the heat source exchanges heat directly with the shield. Then, as the shield gains energy, it exchanges heat with the original receiver. The performance of a shield is discussed in this problem.

    Consider two parallel gray planes (1 and 3) that emit diffusely, as shown in Figure P12.12. The first plane is at temperature T_1 and has an emissivity of ε_1. The properties of the receiver plane are labeled with a subscript 3. Another plane denoted with a subscript 2 is placed between the two and acts as a screen or radiation shield. The source plane exchanges heat with the shield according to Equation (ii) of Example 12.7:

$$q''_{1-2} = \frac{\sigma(T_1^4 - T_2^4)}{1/\varepsilon_1 + 1/\varepsilon_2 - 1}$$

Similarly,

$$q''_{2-3} = \frac{\sigma(T_2^4 - T_3^4)}{1/\varepsilon_2 + 1/\varepsilon_3 - 1}$$

(a) Eliminate T_2 from these equations and show that

$$q''_{1-3} = \frac{\sigma(T_1^4 - T_3^4)}{\left(\dfrac{1}{\varepsilon_1} + \dfrac{1}{\varepsilon_2} - 1\right) + \left(\dfrac{1}{\varepsilon_2} + \dfrac{1}{\varepsilon_3} - 1\right)} \qquad \text{(with shield)}$$

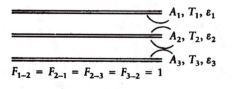

$F_{1-2} = F_{2-1} = F_{2-3} = F_{3-2} = 1$

FIGURE P12.12

Without the shield, we have

$$q''_{1-3} = \frac{\sigma(T_1^4 - T_3^4)}{1/\varepsilon_1 + 1/\varepsilon_3 - 1} \qquad \text{(without shield)}$$

(b) Show that the ratio of the radiant heat-transfer rate with a shield to that obtained without one is

$$\frac{q''_{1-3}(\text{with a shield})}{q''_{1-3}(\text{without a shield})} = \frac{1/\varepsilon_1 + 1/\varepsilon_3 - 1}{1/\varepsilon_1 + 2/\varepsilon_2 + 1/\varepsilon_3 - 2}$$

(c) Determine the value of the above ratio if all emissivities are equal. Does use of a shield reduce the heat received by plane 3 from that obtained without a shield?

29. A very long insulated steam pipe passes through a room. The outside-surface temperature of the insulation is 150°F, while the surroundings are at 65°F. Determine how much heat per unit pipe area is lost from the insulated pipe by radiation to the surroundings. The outside diameter of the insulation is 6 in., and its emissivity is 0.7.

12.8 PROJECT PROBLEMS

1. The introductory discussion of Problem 28 is about radiation shields. In this problem we examine cylindrical radiation shields. Consider a tube carrying a low-temperature fluid through a warm room.
   (a) Use Equation 12.30 for gray-diffuse surfaces to obtain the descriptive equation for heat exchange by radiation. Now consider a concentric rube arrangement where the space between the inner tube and the outer tube is evacuated. The inner tube carries a low-temperature fluid, and the tube gains energy as in part (a). However, the outer tube acts as a shield to the heat transferred by radiation.
   (b) Use Equation 12.30 again to write the descriptive heat-exchange equation. Determine the heat transferred from the surroundings to the tube for both cases and evaluate the screening effect of the shield.
2. (a) Determine the view factor F_{1-2} for the two infinitely long cylinders of Figure P12.13. Develop an equation similar to that obtained in Problem 9 and express F_{1-2} in terms

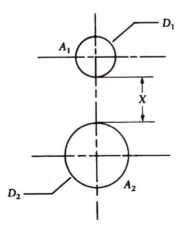

FIGURE P12.13

of D_1, D_2, and X. Evaluate F_{1-2} for the case where the two cylinders are touching. Follow the same or similar steps as those outlined in Problem 9.

(b) For this system, D_1 is that for 2 nominal schedule 40 pipe, D_2 is for 6 nominal schedule 40 pipe, and the pipes are touching. Each pipe's surface can be approximated as being black and isothermal, with $T_1 = 100°C$ and $T_2 = 0°C$. Determine the heat-transfer rate between the two pipes.

3. (a) Problem 19 deals with radiation from a sphere. Assume the sphere is in air whose temperature is 68°F. Calculate how much heat is lost by convection, neglecting radiation heat-transfer losses. Compare the results.

(b) Repeat for the pipe of Problem 29.

Appendix A: Tables

CONTENTS

TABLE A.1
Prefixes (the names of multiples and submultiples of SI units may be formed by application of the prefixes)

Factor by which unit is multiplied	Prefix	Symbol
10^{12}	tera	T
10^{9}	giga	G
10^{6}	mega	M
10^{3}	kilo	k
10^{2}	hecto	h
10	deka	da
10^{-1}	deci	d
10^{-2}	centi	c
10^{-3}	milli	m
10^{-6}	micro	μ
10^{-9}	nano	n
10^{-12}	pico	p
10^{-15}	femto	f
10^{-18}	atto	a

Source: From E. A. Mechtly, NASA SP-7012, 1973.

TABLE A.2
Conversion Factors Listed by Physical Quantity (see note at end of table)

To convert from	To	Multiply by
Acceleration		
foot/second^2	meter/second^2	-01 3.048*
inch/second^2	meter/second^2	-02 2.54*
Area		
foot^2	meter^2	-02 9.290 304*
inch^2	meter^2	-04 6.4516*
mile^2 (U.S. statute)	meter^2	$+06$ 2.589 988 110 336*
yard^2	meter^2	-01 8.361 273 6*
Convection coefficient or Film conductance		
BTU/hour·ft^2·°R	watt/(meter^2·K)	$+00$ 5.678
Density		
gram/centimeter^3	kilogram/meter^3	$+03$ 1.00*
lbm/inch^3	kilogram/meter^3	$+04$ 2.767 990 5
lbm/foot^3	kilogram/meter^3	$+01$ 1.601 846 3
Energy		
British thermal unit (thermochemical)	joule	$+03$ 1.054 350
calorie (thermochemical)	joule	$+00$ 4.184*
foot lbf	joule	$+00$ 1.355 817 9
kilocalorie (thermochemical)	joule	$+03$ 4.184*
kilowatt hour	joule	$+06$ 3.60*
watt hour	joule	$+03$ 3.60*

* Indicates an exact conversion. Numbers not followed by an asterisk are only approximate representations of definitions or are the results of physical measurements. The first two digits of each numerical entry represent a power of 10.

Source: From E. A. Mechtly, NASA SP-7012, 1973.

TABLE A.2 continued

To convert from	To	Multiply by
Specific Heat Capacity		
BTU/lbm·°F	joule/(kilogram·K)	+03 4.187
Speed		
foot/hour	meter/second	−05 8.466 666 6
foot/minute	meter/second	−03 5.08*
foot/second	meter/second	−01 3.048*
inch/second	meter/second	−02 2.54*
kilometer/hour	meter/second	−01 2.777 777 8
knot (international)	meter/second	−01 5.144 444 444
mile/hour (U.S. statute)	meter/second	−01 4.4704*
mile/minute (U.S. statute)	meter/second	+01 2.682 24*
mile/second (U.S. statute)	meter/second	+03 1.609 344*
Temperature		
Celsius	kelvin	$t_K = t_c + 273.15$
Fahrenheit	kelvin	$t_K = (5/9)(t_F + 459.67)$
Fahrenheit	Celsius	$t_c = (5/9)(t_F - 32)$
Rankine	kelvin	$t_K = (5/9)t_R$
Thermal conductivity		
BTU/hour·foot·°F	watt/(meter·K)	+00 1.731
Thermal diffusivity		
foot²/second	meter²/second	−02 9.29
foot²/hour	meter²/second	−05 2.581
Thermal resistance		
hour·°F/BTU	K/watt	+00 1.8958
Time		
day (mean solar)	second (mean solar)	+04 8.64*
hour (mean solar)	second (mean solar)	+03 3.60*
minute (mean solar)	second (mean solar)	+01 6.00*
month (mean calendar)	second (mean solar)	+06 2.628*
second (ephemeris)	second	+00 1.000 000 000
year (calendar)	second (mean solar)	+07 3.1536*
Viscosity		
centistoke	meter²/second	−06 1.00*
stoke	meter²/second	−04 1.00*
foot²/second	meter²/second	−02 9.290 304*
centipoise	newton·second/meter²	−03 1.00*
lbm/foot second	newton·second/meter²	+00 1.488 163 9
lbf second/foot²	newton·second/meter²	+01 4.788 025 8
poise	newton·second/meter²	−01 1.00*
Volume		
acre-foot	meter³	−03 1.233 481 9
barrel (petroleum, 42 gallons)	meter³	−01 1.589 873
cup	meter³	−04 2.365 882 365*

* Indicates an exact conversion. Numbers not followed by an asterisk are only approximate representations of definitions or are the results of physical measurements. The first two digits of each numerical entry represent a power of 10.

Source: From E. A. Mechtly, NASA SP-7012, 1973.

TABLE A.2 continued

To convert from	To	Multiply by
foot^3	meter^3	-02 2.831 684 659 2*
gallon (U.S. dry)	meter^3	-03 4.404 883 770 86*
gallon (U.S. liquid)	meter^3	-03 3.785 411 784*
inch^3	meter^3	-05 1.638 706 4*
liter	meter^3	-03 1.00*
ounce (U.S. fluid)	meter^3	-05 2.957 352 956 25*
pint (U.S. dry)	meter^3	-04 5.506 104 713 575*
pint (U.S. liquid)	meter^3	-04 4.731 764 73*
quart (U.S. dry)	meter^3	-03 1.101 220 942 715*
quart (U.S. liquid)	meter^3	-04 9.463 529 5
tablespoon	meter^3	-05 1.478 676 478 125*
teaspoon	meter^3	-06 4.928 921 593 75*
yard^3	meter^3	-01 7.645 548 579 84*

* Indicates an exact conversion. Numbers not followed by an asterisk are only approximate representations of definitions or are the results of physical measurements. The first two digits of each numerical entry represent a power of 10.

Source: From E. A. Mechtly, NASA SP-7012, 1973.

TABLE A.3
Temperature Conversions

K	°C	°R	°F
250	-23.15	450	-10
260	-13.15	468	8
270	-3.15	486	26
280	6.85	504	44
290	16.85	522	62
295	21.85	531	71
300	26.85	540	80
310	36.85	558	98
320	46.85	576	116
330	56.85	594	134
340	66.85	612	152
350	76.85	630	170
360	86.85	648	188
370	96.85	666	206
380	106.85	684	224
390	116.85	702	242
400	126.85	720	260
410	136.85	738	278
420	146.85	756	296

TABLE A.4
Hyperbolic Functions

Defnitions and identities of hyperbolic functions

The hyperbolic functions are usually defined in terms of exponential functions as

$$\sinh(u) = \frac{e^u - e^{-u}}{2} \tag{A.1}$$

$$\cosh(u) = \frac{e^u + e^{-u}}{2} \tag{A.2}$$

The hyperbolic functions are related to a hyperbola in a way that is analogous to how the trigonometric functions are related to a circle. Equations A.1 and A.2 can be combined to show that

$$\cosh^2(u) - \sinh^2(u) = 1 \tag{A.3}$$

The equation of a hyperbola is

$$x^2 - y^2 = 1 \tag{A.4}$$

and the correspondence between Equations A.3 and A.4 is

$$x = \cosh(u), \quad y = \sinh(u) \tag{A.5}$$

The above relationships are illustrated graphically in Figure A4.1. Shown are the x and y axes with a hyperbola described by Equation A.4. Point P is any point on the hyperbola, and M is the x intercept. Point P is located at $(x, y) = [\cosh(u), \sinh(u)]$. The value u is related to the area of the shaded sector OMP by

$$\text{Area } OMP = u/2$$

Thus, the x, y location of any point on the hyperbola is related through the hyperbolic functions of Equation A.5 to the area of the sector created.

Further definitions and identities relating hyperbolic functions and their derivatives follow.

Definitions:

$$\tanh(u) = \frac{\sinh(u)}{\cosh(u)} \qquad \text{sech}(u) = \frac{1}{\cosh(u)} \qquad \coth(u) = \frac{1}{\tanh(u)}$$

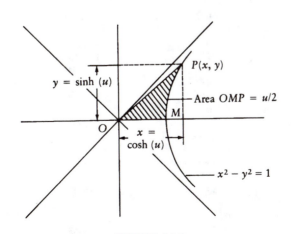

FIGURE A4.1

TABLE A.4
Hyperbolic Functions (continued)

Identities:

$$\sinh(-u) = -\sinh(u) \qquad\qquad \cosh(-u) = \cosh(u)$$

$$\tanh(-u) = -\tanh(u) \qquad\qquad \cosh^2(U) - \sinh^2(u) = 1$$

$$\tanh^2(u) + \mathrm{sech}^2(u) = 1$$

$$\sinh(u + v) = \sinh(u)\cosh(v) + \cosh(u)\sinh(v)$$

$$\cosh(u + v) = \cosh(u)\cosh(v) + \sinh(u)\sinh(v)$$

$$\sinh(u)\cosh(v) = 1/2\,\sinh(u + v) + 1/2\,\sinh(u - v)$$

$$\cosh(u)\cosh(v) = 1/2\,\cosh(u + v) + 1/2\,\cosh(u - v)$$

$$\sinh(u)\sinh(v) = 1/2\,\cosh(u + v) - 1/2\,\cosh(u - v)$$

$$\sinh(2u) = 2\sinh(u)\cosh(u)$$

$$\cosh(2u) = \cosh^2(u) + \sinh^2(u) = 2\cosh^2(u) - 1 = 1 + 2\sinh^2(u)$$

$$\tanh(2u) = \frac{2\tanh(u)}{1 + \tanh^2(u)}$$

$$\sinh^{-1}(u) = \ln(u + \sqrt{u^2 + 1})$$

$$\cosh^{-1}(u) = \ln(u + \sqrt{u^2 - 1}) \qquad u \geq 1$$

$$\tanh^{-1}(u) = \frac{1}{2}\ln\left(\frac{1 + u}{1 - u}\right) \qquad u^2 < 1$$

Derivatives:

$$\frac{d}{dx}\sinh(u) = \frac{du}{dx}\cosh(u) \qquad\qquad \frac{d}{dx}\cosh(u) = \frac{du}{dx}\sinh(u)$$

$$\frac{d}{dx}\tanh(u) = \frac{du}{dx}\mathrm{sech}^2(u) \qquad\qquad \frac{d}{dx}\sinh^{-1}(u) = \frac{1}{\sqrt{u^2 + 1}}\frac{du}{dx}$$

$$\frac{d}{dx}\cosh^1(u) = \frac{1}{\sqrt{u^2 - 1}}\frac{du}{dx}; \quad u > 1, \cosh^{-1}(u) > 0$$

$$\frac{d}{dx}\tanh^{-1}(u) = \frac{1}{1 - u^2}\frac{du}{dx}$$

Values of the hyperbolic functions

u	sinh(u)	cosh(u)	tanh(u)
0.00	0.00000	1.00000	0.00000
0.05	0.05002	1.00125	0.04996
0.10	0.10017	1.00500	0.09967
0.15	0.15056	1.01127	0.14889
0.20	0.20134	1.02007	0.19738
0.25	0.25261	1.03141	0.24492
0.30	0.30452	1.04534	0.29131
0.35	0.35719	1.06188	0.33638
0.40	0.41075	1.08107	0.37995
0.45	0.46534	1.10297	0.42190
0.50	0.52110	1.12763	0.46212
0.55	0.57815	1.15510	0.50052
0.60	0.63665	1.18547	0.53705
0.65	0.69675	1.21879	0.57167
0.70	0.75858	1.25517	0.60437
0.75	0.82232	1.29468	0.63515
0.80	0.88811	1.33744	0.66404
0.85	0.95612	1.38353	0.69107

TABLE A.4
Hyperbolic Functions (continued)

0.90	1.02652	1.43309	0.71630
0.95	1.09948	1.48623	0.73978
1.00	1.17520	1.54308	0.76159
1.05	1.25386	1.60379	0.78181
1.10	1.33565	1.66852	0.80050
1.15	1.42078	1.73742	0.81775
1.20	1.50946	1.81066	0.83365
1.25	1.60192	1.88842	0.84828
1.30	1.69838	1.97091	0.86172
1.35	1.79909	2.05833	0.87405
1.40	1.90430	2.15090	0.88535
1.45	2.01427	2.24884	0.89569
1.50	2.12928	2.35241	0.90515
1.55	2.24961	2.46186	0.91379
1.60	2.37557	2.57746	0.92167
1.65	2.50747	2.69952	0.92886
1.70	2.64563	2.82832	0.93541
1.75	2.79041	2.96419	0.94138
1.80	2.94217	3.10747	0.94681
1.85	3.10129	3.25853	0.95175
1.90	3.26816	3.41773	0.95624
1.95	3.44321	3.58548	0.96032
2.00	3.62686	3.76220	0.96403
2.05	3.81958	3.94832	0.96740
2.10	4.02186	4.14431	0.97045
2.15	4.23419	4.35067	0.97323
2.20	4.45711	4.56791	0.97574
2.25	4.69117	4.79657	0.97803
2.30	4.93696	5.03722	0.98010
2.35	5.19510	5.29047	0.98197
2.40	5.46623	5.55695	0.98368
2.45	5.75103	5.83732	0.98522
2.50	6.05020	6.13229	0.98661
2.55	6.36451	6.44259	0.98788
2.60	6.69473	6.76901	0.98903
2.65	7.04169	7.11234	0.99007
2.70	7.40626	7.47347	0.99101
2.75	7.78935	7.85328	0.99186
2.80	8.19192	8.25273	0.99263
2.85	8.61497	8.67281	0.99333
2.90	9.05956	9.11458	0.99396
2.95	9.52681	9.57915	0.99454
3.0	10.0179	10.0677	0.99505
3.5	16.5426	16.5728	0.99818
4.0	27.2899	27.3082	0.99933
4.5	45.0030	45.0141	0.99975
5.0	74.2032	74.2099	0.99991
5.5	122.34	122.35	0.99997
6.0	201.71	201.72	0.99999
6.5	332.57	332.57	1.0000
7.0	548.32	548.32	1.0000
7.5	904.02	904.02	1.0000
8.0	1490.5	1490.5	1.0000
8.5	2457.4	2457.4	1.0000
9.0	4051.5	4051.5	1.0000
9.5	6679.9	6679.9	1.0000
10.0	11013.2	11013.2	1.0000

TABLE A.5
Error Function or Probability Integral

z	erf(z)	z	erf(z)	z	erf(z)	z	erf(z)
0.00	0.00000	0.52	0.53789	1.04	0.85864	1.56	0.97262
0.01	0.01128	0.53	0.54646	1.05	0.86243	1.57	0.97360
0.02	0.02256	0.54	0.55493	1.06	0.86614	1.58	0.97454
0.03	0.03384	0.55	0.56332	1.07	0.86977	1.S9	0.97546
0.04	0.04511	0.56	0.57161	1.08	0.87332	1.60	0.97634
0.05	0.05637	0.57	0.57981	1.09	0.87680	1.61	0.97720
0.06	0.06762	0.58	0.58792	1.10	0.88020	1.62	0.97803
0.07	0.07885	0.59	0.59593	1.11	0.88353	1.63	0.97884
0.08	0.09007	0.60	0.60385	1.12	0.88678	1.64	0.97962
0.09	0.10128	0.61	0.61168	1.13	0.88997	1.65	0.98037
0.10	0.11246	0.62	0.61941	1.14	0.89308	1.66	0.98110
0.11	0.12362	0.63	0.62704	1.15	0.89612	1.67	0.98181
0.12	0.13475	0.64	0.63458	1.16	0.89909	1.68	0.98249
0.13	0.14586	0.65	0.64202	1.17	0.90200	1.69	0.98315
0.14	0.15694	0.66	0.64937	1.18	0.90483	1.70	0.98379
0.15	0.16799	0.67	0.65662	1.19	0.90760	1.71	0.98440
0.16	0.17901	0.68	0.66378	1.20	0.91031	1.72	0.98500
0.17	0.18999	0.69	0.67084	1.21	0.91295	1.73	0.98557
0.18	0.20093	0.70	0.67780	1.22	0.91553	1.74	0.98613
0.19	0.21183	0.71	0.68466	1.23	0.91805	1.75	0.98667
0.20	0.22270	0.72	0.69143	1.24	0.92050	1.76	0.98719
0.21	0.23352	0.73	0.69810	1.25	0.92290	1.77	0.98769
0.22	0.24429	0.74	0.70467	1.26	0.92523	1.78	0.98817
0.23	0.25502	0.75	0.71115	1.27	0.92751	1.79	0.98864
0.24	0.26570	0.76	0.71753	1.28	0.92973	1.80	0.98909
0.25	0.27632	0.77	0.72382	1.29	0.93189	1.81	0.98952
0.26	0.28689	0.78	0.73001	1.30	0.93400	1.82	0.98994
0.27	0.29741	0.79	0.73610	1.31	0.93606	1.83	0.99034
0.28	0.30788	0.80	0.74210	1.32	0.93806	1.84	0.99073
0.29	0.31828	0.81	0.74800	1.33	0.94001	1.85	0.99111
0.30	0.32862	0.82	0.75381	1.34	0.94191	1.86	0.99147
0.31	0.33890	0.83	0.75952	1.35	0.94376	1.87	0.99182
0.32	0.34912	0.84	0.76514	1.36	0.94556	1.88	0.99215
0.33	0.35927	0.85	0.77066	1.37	0.94731	1.89	0.99247
0.34	0.36936	0.86	0.77610	1.38	0.94901	1.90	0.99279
0.35	0.37938	0.87	0.78143	1.39	0.95067	1.91	0.99308
0.36	0.38932	0.88	0.78668	1.40	0.95228	1.92	0.99337
0.37	0.39920	0.89	0.79184	1.41	0.95385	1.93	0.99365
0.38	0.40900	0.90	0.79690	1.42	0.95537	1.94	0.99392
0.39	0.41873	0.91	0.80188	1.43	0.95685	1.95	0.99417
0.40	0.42839	0.92	0.80676	1.44	0.95829	1.96	0.99442
0.41	0.43796	0.93	0.81156	1.45	0.95969	1.97	0.99466
0.42	0.44746	0.94	0.81627	1.46	0.96105	1.98	0.99489
0.43	0.45688	0.95	0.82089	1.47	0.96237	1.99	0.99511
0.44	0.46622	0.96	0.82542	1.48	0.96365	2.00	0.99532
0.45	0.47548	0.97	0.82987	1.49	0.96489	2.20	0.99814
0.46	0.48465	0.98	0.83423	1.50	0.96610	2.40	0.99931
0.47	0.49374	0.99	0.83850	1.51	0.96727	2.60	0.99976
0.48	0.50274	1.00	0.84270	1.52	0.96841	2.80	0.99992
0.49	0.51166	1.01	0.84681	1.53	0.96951	3.00	0.99998
0.50	0.52049	1.02	0.85083	1.54	0.97058		
0.51	0.52924	1.03	0.85478	1.55	0.97162		

Notes: Error function = $\text{erf}(z) = \dfrac{2}{\sqrt{\pi}} \int_0^z \exp(-\lambda^2)\,d\lambda$. Complementary error function = $\text{erfc}\,(z) = 1 - \text{erf}(z)$.

TABLE A.6
Symbols and Units

Symbol	Definition	SI	Engineering
		Unit	
A	Area	m^2	ft^2
a	Acceleration	m/s^2	ft/s^2
Bi	Biot number	—	—
C_D	Drag coefficient	—	—
C_d	Local drag coefficient	—	—
c	Specific heat	$J/(kg{\cdot}K)$	$BTU/(lbm{\cdot}°R)$
c_p	Constant-pressure specific heat	$J/(kg{\cdot}K)$	$BTU/(lbm{\cdot}°R)$
c_v	Constant-volume specific heat	$J/(kg{\cdot}K)$	$BTU/(lbm{\cdot}°R)$
D	Diameter	m	ft
D_h	Hydraulic diameter	m	ft
Ec	Eckert number	—	—
F	Force	N	lbf
F_{i-j}	Shape factor	—	—
Fo	Fourier number	—	—
Gr	Grashof number	—	—
Gz	Graetz number	—	—
g	Gravitaitonal acceleration	m/s^2	ft/s^2
g_c	Conversion factor	—	$32.2\ ft{\cdot}lbm/(lbf{\cdot}s^2)$
h	Planck's constant	$J{\cdot}s$	$BTU{\cdot}hr$
h	Enthalpy	J/kg	BTU/lbm
$\bar{h}_c$, h_c, h_{avg}	Convection heat transfer coefficient	$W/(m^2{\cdot}K)$	$BTU/(hr{\cdot}ft^2{\cdot}°R)$
h_{fg}	Enthalpy of vaporization	J/kg	BTU/lbm
h_r	Radiation heat transfer coefficient	$W/(m^2{\cdot}K)$	$BTU/(hr{\cdot}ft^2{\cdot}°R)$
k, k_f	Thermal conductivity	$W/(m{\cdot}K)$	$BTU/(hr{\cdot}ft{\cdot}°R)$
L	Length	m	ft
m	Mass	kg	lbm
$\dot{m}$	Mass flow rate	kg/s	lbm/s
Nu	Nusselt number	—	—
NTU	Number of transfer units	—	—
Pe	Peclet number	—	—
Pr	Prandtl number	—	—
p	Pressure	Pa	lbf/in^2
Q	Volume flow rate	m^3/s	f^3/s
q	Heat transferred per time	W	BTU/hr
q''	Heat transferred per time per area	W/m^2	$BTU/(hr{\cdot}ft^2)$
R	Gas constant	$J/(kg{\cdot}K)$	$BTU/(lbm{\cdot}°R)$
R_u	Universal gas constant	$J/(kg{\cdot}K)$	$BTU/(lbm{\cdot}°R)$
R_k, R_c	Resistance to conduction	K/W	$°R{\cdot}hr/BTU$
R_c	Resistance to convection	K/W	$°R{\cdot}hr/BTU$
r	Radius or radial coordinate	m	ft

TABLE A.6
Symbols and Units (continued)

Symbol	Definition	SI	Engineering
		Unit	
Ra	Rayleigh number	—	—
Re	Reynolds number	—	—
S_L, S_T	Pitch-spacing parameters	m	ft
St	Stanton number	—	—
T	Temperature	K or °C	°R or °F
T_w	Wall or surface temperature	K or °C	°R or °F
T_∞ or T_a	Ambient temperature	K or °C	°R or °F
t	Time	s	s or hr
U	Overall heat transfer coefficient	W/(m²·K)	BTU/(hr·ft²·°R)
V	Velocity	m/s	ft/s
$\forall$	Volume	m³	ft³
v	Specific volume	m³/kg	ft³/lbm
W	Work	J	ft·lbf
dW/dt	Power	J/s	ft·lbf/s or HP
α	Thermal diffusivity	m²/s	ft²/s
α	Absorptivity	—	—
β	Coefficient of thermal expansion	1/K	1/°R
δ	Boundary layer thickness	m	ft
δ_t	Thermal boundary layer thickness	m	ft
ε	Emissivity	—	—
η_e	Efficiency of fin	—	—
η_f	Fin effectiveness	—	—
λ	Wavelength	m	ft
μ	Viscosity	N·s/m²	lbf·s/ft²
ν	Kinematic viscosity	m²/s	ft²/s
ρ	Density	kg/m	lbm/ft
ρ	Reflectivity	—	—
σ	Stefan-Boltzmann constant	W/(m²·K⁴)	BTU/(hr·ft²·°R⁴)
τ	Shear stress	Pa	lbf/in²
	Transmissivity	—	—

Thermal Property Tables

TABLE B.1
Thermal Properties of Selected Metallic Elements at 293 K (20°C) or 528°R (65°F)

Element	Specific gravity	Specific heat, c_p		Thermal conductivity, k		Diffusivity, α		Melting temperature	
		J/(kg·K)	BTU/(lbm·°R)	W/(m·K)	BTU/(hr·ft·°R)	$m^2/s \times 10^6$	$ft^2/s \times 10^3$	K	°R
Aluminum	2.702	896	0.214	236	136	97.5	1.05	933	1,680
Beryllium	1.850	1750	0.418	205	118	63.3	0.681	1550	2,790
Chromium	7.160	440	0.105	91.4	52.8	29.0	0.312	2118	3,812
Copper	8.933	383	0.0915	399	231	116.6	1.26	1356	2,441
Gold	19.300	129	0.0308	316	183	126.9	1.37	1336	2,405
Iron	7.870	452	0.108	31.1	18.0	22.8	0.245	1810	3,258
Lead	11.340	129	0.0308	35.3	20.4	24.1	0.259	601	1,082
Magnesium	1.740	1017	0.243	156	90.1	88.2	0.949	923	1,661
Manganese	7.290	486	0.116	7.78	4.50	2.2	0.0236	1517	2,731
Molybdenum	10.240	251	0.0600	138	79.7	53.7	0.578	2883	5,189
Nickel	8.900	446	0.107	91	52.6	22.9	0.246	1726	3,107
Platinum	21.450	133	0.0318	71.4	41.2	25.0	0.269	2042	3,676
Potassium	0.860	741	0.177	103	59.6	161.6	1.74	337	607
Silicon	2.330	703	0.168	153	88.4	93.4	1.01	1685	3,033
Silver	10.500	234	0.0559	427	247	173.8	1.87	1234	2,221
Tin	5.750	227	0.0542	67.0	38.7	51.3	0.552	505	909
Titanium	4.500	611	0.146	22.0	12.7	8.0	0.0861	1953	3,515
Tungsten	19.300	134	0.0320	179	103	69.2	0.745	3653	6,575
Uranium	19.070	113	0.0270	27.4	15.8	12.7	0.137	1407	2,533
Vanadium	6.100	502	0.120	31.4	18.1	10.3	0.111	2192	3,946
Zinc	7.140	385	0.0920	121	69.9	44.0	0.474	693	1,247

Source: Data from several sources.

Notes: Density $= \rho =$ specific gravity $\times 62.4$ lbm/ft$^3 =$ specific gravity $\times 1000$ kg/m^3
Diffusivity $= \alpha$; for aluminum, α m^2/s $\times 10^6 = 97.5$, so $\alpha = 97.5 \times 10^{-6}$ m^2/s
Also, $\alpha = k/\rho c_p$

TABLE B.2
Thermal Properties of Selected Alloys at 293 K (20°C) or 528°R (65°F)

Base metal	% Composition	Specific gravity	Specific heat, c_p		Thermal conductivity, k		Diffusivity, α	
			J/(kg·K)	BTU/(lbm·°R)	W/(m·K)	BTU/(hr·ft·°R)	m²/s × 10⁵	ft²/s × 10⁴
Aluminum								
Duralumin	94–96 Al, 3–5 Cu, tr Mg	2.787	833	0.199	164	94.7	6.676	7.187
Silumin	87 Al, 13 Si	2.659	871	0.208	164	94.7	7.099	7.642
Copper								
Al-bronze	95 Cu, 5 Al	8.666	410	0.0979	83	47.9	2.330	2.508
Bronze	75 Cu, 25 Sn	8.666	343	0.0819	26	15.0	0.859	0.925
Red brass	85 Cu, 9 Sn, 6 Zn	8.714	385	0.0920	61	35.2	1.804	1.942
Brass	70 Cu, 30 Zn	8.522	385	0.0920	111	64.1	3.412	3.673
German silver	62 Cu, 15 Ni, 22 Zn	8.618	394	0.0941	24.9	14.4	0.733	0.789
Constantan	60 Cu, 40 Ni	8.922	410	0.0979	22.7	13.1	0.612	0.659
Iron								
Cast iron	4 C	7.272	420	0.100	52	30.0	1.702	1.832
Wrought iron	0.5 CH	7.849	460	0.110	59	34.1	1.626	1.750
Steel								
Carbon steel	1 C	7.801	473	0.113	43	24.8	1.172	1.262
	1.5 C	7.753	486	0.116	36	20.8	0.970	1.04
Chrome steel	1 Cr	7.865	460	0.110	61	35.2	1.665	1.792
	5 Cr	7.833	460	0.110	40	23.1	1.110	1.195
	10 Cr	7.785	460	0.110	31	17.9	0.867	0.933
Chrome–nickel steel	15 Cr, 10 Ni	7.865	460	0.110	19	11.0	0.526	0.577
	20 Cr, 15 Ni	7.833	460	0.110	15.1	8.72	0.415	0.447
Nickel steel	10 Ni	7.945	460	0.110	26	15.0	0.720	0.775
	20 Ni	7.993	460	0.110	19	11.0	0.526	0.566
	40 Ni	8.169	460	0.110	10	5.78	0.279	0.300
	60 Ni	8.378	460	0.110	19	11.0	0.493	0.531
Nickel–chrome steel	80 Ni, 15 C	8.522	460	0.110	17	9.82	0.444	0.478
	40 Ni, 15 C	8.073	460	0.110	11.6	6.70	0.305	0.328
Manganese steel	1 Mn	7.865	460	0.110	50	28.9	1.388	1.494
	5 Mn	7.849	460	0.110	22	12.7	0.637	0.686
Silicon steel	1 Si	7.769	460	0.110	42	24.3	1.164	1.164
	5 Si	7.417	460	0.110	19	11.0	0.555	0.597
Stainless steel	Type 304	7.817	461	0.110	14.4	8.32	0.387	0.417
	Type 347	7.817	461	0.110	14.3	8.26	0.387	0.417
Tungsten steel	1 W	7.913	448	0.107	66	38.1	1.858	2.000
	5 W	8.073	435	0.104	54	31.2	1.525	1.642

Notes: Density = ρ = specific gravity × 62.4 lbm/ft³ = specific gravity × 1 000 kg/m³
Diffusivity = α; for Duralumin, $\alpha \times 10^5 = 6.676$ m²/s; so $\alpha = 6.676 \times 10^{-5}$ m²/s
Also, $\alpha = k/\rho c_p$

TABLE B.3
Thermal Properties of Selected Building Materials and Insulations at 293 K (20°C) or 528°R (65°F)

Material	Specific gravity	Specific heat, c_p		Thermal conductivity, k		Diffusivity, α	
		J/(kg·K)	BTU/(lbm·°R)	W/(m·K)	BTU/(hr·ft·°R)	m²/s × 10⁵	ft²/s × 10⁶
Asbestos	0.383	816	0.195	0.113	0.0653	0.036	3.88
Asphalt	2.120			0.698	0.403		
Bakelite	1.270			0.233	0.135		
Brick							
Carborundum (50% SiC)	2.200			5.82	3.36		
Common	1.800	840	0.201	0.38–0.52	0.22–0.30	0.028–0.034	3.0–3.66
Magnesite (50% MgO)	2.000			2.68	1.55		
Masonry	1.700	837	0.200	0.658	0.38	0.046	5.0
Silica (95% SiO₂)	1.900			1.07	0.618		
Cardboard				0.14–0.35	0.08–0.2		
Cement (hard)				1.047	0.605		
Clay (48.7% moist)	1.545	880	0.210	1.26	0.728	0.101	10.9
Coal (anthracite)	1.370	1260	0.301	0.238	0.137	0.013–0.015	1.4–1.6
Concrete (dry)	0.500	837	0.200	0.128	0.074	0.049	5.3
Cork board	0.150	1880	0.449	0.042	0.0243	0.015–0.044	1.6–4.7
Cork (expanded)	0.120			0.036	0.0208		
Earth (diatomaceous)	0.466	879	0.210	0.126	0.072	0.031	3.3
Earth (clay with 28% moist)	1.500			1.51	0.872		
Earth (sandy with 8% moist)	1.500			1.05	0.607		

TABLE B.3
(continued)

Material	Sp. gr.			k		α×10⁵	
Glass fiber	0.220			0.035	0.02		
Glass (window pane)	2.800	800	0.191	0.81	0.47	0.034	3.66
Glass (wool)	0.200	670	0.160	0.040	0.023	0.028	3.0
Granite	2.750			3.0	1.73		
Ice at 0°C	0.913	1830	0.437	2.22	1.28	0.124	13.3
Kapok	0.025			0.035	0.02		
Linoleum	0.535			0.081	0.047		
Mica	2.900			0.523	0.302		
Pine bark	0.342			0.080	0.046		
Plaster	1.800			0.814	0.47		
Plexiglas	1.180			0.195	0.113		
Plywood	0.590			0.109	0.063		
Polystyrene	1.050			0.157	0.0907		
Rubber							
Buna	1.250			0.465	0.269		
Ebonite	1.150	2009	0.480	0.163	0.0942	0.0062	0.67
Spongy	0.224			0.055	0.0318		
Sand							
Dry	1.640			0.582	0.336		
Moist				1.13	0.653		
Sawdust	0.215			0.071	0.041		
Wood							
Fir, pine, and spruce	0.444	2720	0.650	0.15	0.087	0.0124	1.33
Oak	0.705	2390	0.571	0.19	0.11	0.0113	1.22
Celotex	0.400			0.055	0.0318		
Fiber sheets	0.200			0.047	0.0172		
Wool	0.200			0.038	0.0220		

Notes: Density = ρ = specific gravity $\times$ 62.4 lbm/ft³ = specific gravity $\times$ 1000 kg/m³

Diffusivity = α; for asbestos, $\alpha \times 10^5 = 0.036$ m²/s; so $\alpha = 0.036 \times 10^{-5}$ m²/s

Also, $\alpha = k/\rho c_p$

Saturated Liquid Property Tables

TABLE C.1
Properties of Saturated Liquids: Ammonia NH_3.*

Temp, T		Specific gravity	Specific heat, c_p		Kinematic viscosity, ν		Thermal conductivity, k		Thermal diffusivity, α		Prandtl number, Pr	β	
°C	°F		$\frac{J}{kg \cdot K}$	$\frac{BTU}{lbm \cdot °R}$	m²/s	ft²/s	$\frac{W}{m \cdot K}$	$\frac{BTU}{hr \cdot ft \cdot °R}$	m²/s	ft²/hr		1/K	1/°R
−50	−58	0.703	4463	1.066	0.435×10^{-6}	0.468×10^{-5}	0.547	0.316	1.742×10^{-7}	6.75×10^{-3}	2.60		
−40	−40	0.691	4467	1.067	0.406	0.437	0.547	0.316	1.775	6.88	2.28		
−30	−22	0.679	4476	1.069	0.387	0.417	0.549	0.317	1.801	6.98	2.15		
−20	−4	0.666	4509	1.077	0.381	0.410	0.547	0.316	1.819	7.05	2.09		
−10	14	0.653	4564	1.090	0.378	0.407	0.543	0.314	1.825	7.07	2.07		
0	32	0.640	4635	1.107	0.373	0.402	0.540	0.312	1.819	7.05	2.05		
10	50	0.626	4714	1.126	0.368	0.396	0.531	0.307	1.801	6.98	2.04		
20	68	0.611	4798	1.146	0.359	0.386	0.521	0.301	1.775	6.88	2.02	2.45×10^{-3}	1.36×10^{-3}
30	86	0.596	4890	1.168	0.349	0.376	0.507	0.293	1.742	6.75	2.01		
40	104	0.580	4999	1.194	0.340	0.366	0.493	0.285	1.701	6.59	2.00		
50	122	0.564	5116	1.222	0.330	0.355	0.476	0.275	1.654	6.41	1.99		

* Source: Data taken from Analysis of Heat and Mass Transfer by E. R. G. Eckert and R. M. Drake, Jr., Taylor & Francis Inc. 1972. Used with permission.

TABLE C.2
Properties of Saturated Liquids: Carbon Dioxide CO_2.*

Temp, T		Specific gravity	Specific heat, c_p		Kinematic viscosity, ν		Thermal conductivity, k		Thermal diffusivity, α		Prandtl number, Pr	β	
°C	°F		J/kg·K	BTU/lbm·°R	m²/s	ft²/s	W/m·K	BTU/hr·ft·°R	m²/s	ft²/hr		1/K	1/°R
−50	−58	1.156	1840	0.44	0.119×10^{-6}	0.128×10^{-5}	0.0855	0.0494	0.4021×10^{-7}	1.558×10^{-3}	2.96		
−40	−40	1.117	1880	0.45	0.118	0.127	0.1011	0.0584	0.4810	1.864	2.46		
−30	−22	1.076	1970	0.47	0.117	0.126	0.1116	0.0654	0.5272	2.043	2.22		
−20	−4	1.032	2050	0.49	0.115	0.124	0.1151	0.0665	0.5445	2.110	2.12		
−10	14	0.983	2180	0.52	0.113	0.122	0.1099	0.0635	0.5133	1.989	2.20		
0	32	0.926	2470	0.59	0.108	0.117	0.1045	0.0604	0.4578	1.774	2.38		
10	50	0.860	3140	0.75	0.101	0.109	0.0971	0.0561	0.3608	1.398	2.80		
20	68	0.772	5000	1.2	0.091	0.098	0.0872	0.0504	0.2219	0.860	4.10	14.00×10^{-3}	3.67×10^{-3}
30	86	0.597	36400	8.7	0.080	0.086	0.0703	0.0406	0.0279	0.108	28.7		

TABLE C.3
Properties of Saturated Liquids: Dichlorodifluoromethane (Freon-12) CCl_2F_2.*

Temp, T		Specific gravity	Specific heat, c_p		Kinematic viscosity, ν		Thermal conductivity, k		Thermal diffusivity, α		Prandtl number, Pr	β	
°C	°F		J/kg·K	BTU/lbm·°R	m²/s	ft²/s	W/m·K	BTU/hr·ft·°R	m²/s	ft²/hr		1/K	1/°R
−50	−58	1.546	875.0	0.2090	0.310×10^{-6}	0.334×10^{-5}	0.067	0.039	0.501×10^{-7}	1.94×10^{-3}	6.2	2.63×10^{-3}	1.4×10^{-4}
−40	−40	1.518	884.7	0.2113	0.279	0.300	0.069	0.040	0.514	1.99	5.4		
−30	−22	1.489	895.6	0.2139	0.253	0.272	0.069	0.040	0.526	2.04	4.8		
−20	−4	1.460	907.3	0.2167	0.235	0.253	0.071	0.041	0.539	2.09	4.4		
−10	14	1.429	920.3	0.2198	0.221	0.238	0.073	0.042	0.550	2.13	4.0		
0	32	1.397	934.5	0.2232	0.214	0.230	0.073	0.042	0.557	2.16	3.8		
10	50	1.364	949.6	0.2268	0.203	0.219	0.073	0.042	0.560	2.17	3.6		
20	68	1.330	965.9	0.2307	0.198	0.213	0.073	0.042	0.560	2.17	3.5		
30	86	1.295	983.5	0.2349	0.194	0.209	0.071	0.041	0.560	2.17	3.5		
40	104	1.257	1001.9	0.2393	0.191	0.206	0.069	0.040	0.555	2.15	3.5		
50	122	1.215	1021.6	0.2440	0.190	0.204	0.067	0.039	0.545	2.11	3.5		

TABLE C.4
Properties of Saturated Liquids: Engine Oil (Unused).*

Temp, T		Specific gravity	Specific heat, c_p		Kinematic viscosity, v		Thermal conductivity, k		Thermal diffusivity, α		Prandtl number, Pr	β	
°C	°F		$\dfrac{J}{kg \cdot K}$	$\dfrac{BTU}{lbm \cdot °R}$	m²/s	ft²/s	$\dfrac{W}{m \cdot K}$	$\dfrac{BTU}{hr \cdot ft \cdot °R}$	m²/s	ft²/hr		1/K	1/°R
0	32	0.899	1796	0.429	0.004 28	0.0461	0.147	0.085	0.911×10^{-7}	3.53×10^{-3}	47 100		
20	68	0.888	1880	0.449	0.000 90	0.0097	0.145	0.084	0.872	3.38	10 400	0.70×10^{-3}	0.39×10^{-3}
40	104	0.876	1964	0.469	0.000 24	0.0026	0.144	0.083	0.834	3.23	2870		
60	140	0.864	2047	0.489	0.839×10^{-4}	0.903×10^{-3}	0.140	0.081	0.800	3.10	1050		
80	176	0.852	2131	0.509	0.375	0.404	0.138	0.080	0.769	2.98	490		
100	212	0.840	2219	0.530	0.203	0.219	0.137	0.079	0.738	2.86	276		
120	248	0.828	2307	0.551	0.124	0.133	0.135	0.078	0.710	2.75	175		
140	284	0.816	2395	0.572	0.080	0.086	0.133	0.077	0.686	2.66	116		
160	320	0.805	2483	0.593	0.056	0.060	0.132	0.076	0.663	2.57	84		

TABLE C.5
Properties of Saturated Liquids: Ethylene Glycol $C_2H_4(OH)_2$.*

Temp, T		Specific gravity	Specific heat, c_p		Kinematic viscosity, v		Thermal conductivity, k		Thermal diffusivity, α		Prandtl number, Pr	β	
°C	°F		$\dfrac{J}{kg \cdot K}$	$\dfrac{BTU}{lbm \cdot °R}$	m²/s	ft²/s	$\dfrac{W}{m \cdot K}$	$\dfrac{BTU}{hr \cdot ft \cdot °R}$	m²/s	ft²/hr		1/K	1/°R
0	32	1.130	2294	0.548	57.53×10^{-6}	61.92×10^{-5}	0.242	0.140	0.934×10^{-7}	3.62×10^{-3}	615		
20	68	1.116	2382	0.569	19.18	20.64	0.249	0.144	0.939	3.64	204	0.65×10^{-3}	0.36×10^{-3}
40	104	1.101	2474	0.591	8.69	9.35	0.256	0.148	0.939	3.64	93		
60	140	1.087	2562	0.612	4.75	5.11	0.260	0.150	0.932	3.61	51		
80	176	1.077	2650	0.633	2.98	3.21	0.261	0.151	0.921	3.57	32.4		
100	212	1.058	2742	0.655	2.03	2.18	0.263	0.152	0.908	3.52	22.4		

TABLE C.6
Properties of Saturated Liquids: Eutectic Calcium Chloride Solution (29.9% CaCl₂).*

Temp, T °C	Temp, T °F	Specific gravity	Specific heat, c_p J/kg·K	Specific heat, c_p BTU/lbm·°R	Kinematic viscosity, ν m²/s	Kinematic viscosity, ν ft²/s	Thermal conductivity, k W/m·K	Thermal conductivity, k BTU/hr·ft·°R	Thermal diffusivity, α m²/s	Thermal diffusivity, α ft²/hr	Prandtl number, Pr	β 1/K	β 1/°R
-50	-58	1.319	2608	0.623	36.35×10^{-6}	39.13×10^{-5}	0.402	0.232	1.166×10^{-7}	4.52×10^{-3}	312		
-40	-40	1.314	2635.6	0.6295	24.97	26.88	0.415	0.240	1.200	4.65	208		
-30	-22	1.310	2661.1	0.6356	17.18	18.49	0.429	0.248	1.234	4.78	139		
-20	-4	1.305	2688	0.642	11.04	11.88	0.445	0.257	1.267	4.91	87.1		
-10	14	1.300	2713	0.648	6.96	7.49	0.459	0.265	1.300	5.04	53.6		
0	32	1.296	2738	0.654	4.39	4.73	0.472	0.273	1.332	5.16	33.0		
10	50	1.291	2763	0.660	3.35	3.61	0.485	0.280	1.363	5.28	24.6		
20	68	1.286	2788	0.666	2.72	2.93	0.498	0.288	1.394	5.40	19.6		
30	86	1.281	2814	0.672	2.27	2.44	0.511	0.295	1.419	5.50	16.0		
40	104	1.277	2839	0.678	1.92	2.07	0.523	0.302	1.445	5.60	13.3		
50	122	1.272	2868	0.685	1.65	1.78	0.535	0.309	1.468	5.69	11.3		

TABLE C.7
Properties of Saturated Liquids: Glycerin C₃H₅(OH)₃.*

Temp, T °C	Temp, T °F	Specific gravity	Specific heat, c_p J/kg·K	Specific heat, c_p BTU/lbm·°R	Kinematic viscosity, ν m²/s	Kinematic viscosity, ν ft²/s	Thermal conductivity, k W/m·K	Thermal conductivity, k BTU/hr·ft·°R	Thermal diffusivity, α m²/s	Thermal diffusivity, α ft²/hr	Prandtl number, Pr	β 1/K	β 1/°R
0	32	1.276	2261	0.540	0.00831	0.0895	0.282	0.163	0.983×10^{-7}	3.81×10^{-3}	84.7×10^3		
10	50	1.270	2319	0.554	0.00300	0.0323	0.284	0.164	0.965	3.74	31.0		
20	68	1.264	2386	0.570	0.00118	0.0127	0.286	0.165	0.947	3.67	12.5	0.50×10^{-3}	0.28×10^{-3}
30	86	1.258	2445	0.584	0.00050	0.0054	0.286	0.165	0.929	3.60	5.38		
40	104	1.252	2512	0.600	0.00022	0.0024	0.286	0.165	0.914	3.54	2.45		
50	122	1.244	2583	0.617	0.00015	0.0016	0.287	0.166	0.893	3.46	1.63		

TABLE C.8
Properties of Saturated Liquids: Mercury Hg.*

Temp, T °C	Temp, T °F	Specific gravity	Specific heat, c_p J/kg·K	Specific heat, c_p BTU/lbm·°R	Kinematic viscosity, ν m²/s	Kinematic viscosity, ν ft²/s	Thermal conductivity, k W/m·K	Thermal conductivity, k BTU/hr·ft·°R	Thermal diffusivity, α m²/s	Thermal diffusivity, α ft²/hr	Prandtl number, Pr	β 1/K	β 1/°R
0	32	13.628	140.3	0.0335	0.124×10^{-6}	0.133×10^{-5}	8.20	4.74	42.99×10^{-7}	166.6×10^{-3}	0.0288		
20	68	13.579	139.4	0.0333	0.114	0.123	8.69	5.02	46.06	178.5	0.0249	1.82×10^{-4}	1.01×10^{-4}
50	122	13.505	138.6	0.0331	0.104	0.112	9.40	5.43	50.22	194.6	0.0207		
100	212	13.384	137.3	0.0328	0.0928	0.0999	10.51	6.07	57.16	221.5	0.0162		
150	302	13.264	136.5	0.0326	0.0853	0.0918	11.49	6.64	63.54	246.2	0.0134		
200	392	13.144	157.0	0.0375	0.0802	0.0863	12.34	7.13	69.08	267.7	0.0116		
250	482	13.025	135.7	0.0324	0.0765	0.0823	13.07	7.55	74.06	287.0	0.0103		
315.5	600	12.847	134.0	0.032	0.0673	0.0724	14.02	8.10	81.5	316	0.0083		

TABLE C.9
Properties of Saturated Liquids: Methyl Cloride Ch₃Cl.*

Temp, T °C	Temp, T °F	Specific gravity	Specific heat, c_p J/kg·K	Specific heat, c_p BTU/lbm·°R	Kinematic viscosity, ν m²/s	Kinematic viscosity, ν ft²/s	Thermal conductivity, k W/m·K	Thermal conductivity, k BTU/hr·ft·°R	Thermal diffusivity, α m²/s	Thermal diffusivity, α ft²/hr	Prandtl number, Pr	β 1/K	β 1/°R
-50	-58	1.052	1475.9	0.3525	0.320×10^{-6}	0.344×10^{-5}	0.215	0.124	1.388×10^{-7}	5.38×10^{-3}	2.31		
-40	-40	1.033	1482.6	0.3541	0.318	0.342	0.209	0.121	1.368	5.30	1.32		
-30	-22	1.016	1492.2	0.3564	0.314	0.338	0.202	0.117	1.337	5.18	2.35		
-20	-4	0.999	1504.3	0.3593	0.309	0.333	0.196	0.113	1.301	5.04	2.38		
-10	14	0.981	1519.4	0.3629	0.306	0.329	0.187	0.108	1.257	4.87	2.43		
0	32	0.962	1537.8	0.3673	0.302	0.325	0.178	0.103	1.213	4.70	2.49		
10	50	0.942	1560.0	0.3726	0.297	0.320	0.171	0.099	1.166	4.52	2.55		
20	68	0.923	1586.0	0.3788	0.293	0.315	0.163	0.094	1.112	4.31	2.63		
30	86	0.903	1616.1	0.3860	0.288	0.310	0.154	0.089	1.058	4.10	2.72		
40	104	0.883	1650.4	0.3942	0.281	0.303	0.144	0.083	0.996	3.86	2.83		
50	122	0.861	1689.0	0.4034	0.274	0.295	0.133	0.077	0.921	3.57	2.97		

TABLE C.10
Properties of Saturated Liquids: Surfur Dioxide SO$_2$.*

Temp, T		Specific gravity	Specific heat, c_p		Kinematic viscosity, ν		Thermal conductivity, k		Thermal diffusivity, α		Prandtl number, Pr	β	
°C	°F		$\frac{J}{kg \cdot K}$	$\frac{BTU}{lbm \cdot °R}$	m^2/s	ft^2/s	$\frac{W}{m \cdot K}$	$\frac{BTU}{hr \cdot ft \cdot °R}$	m^2/s	ft^2/hr		$1/K$	$1/°R$
−50	−58	1.560	1359.5	0.3247	0.484×10^{-6}	0.521×10^{-5}	0.242	0.140	1.141×10^{-7}	4.42×10^{-3}	4.24		
−40	−40	1.536	1360.7	0.3250	0.424	0.456	0.235	0.136	1.130	4.38	3.74		
−30	−22	1.520	1361.6	0.3252	0.371	0.399	0.230	0.133	1.117	4.33	3.31		
−20	−4	1.488	1362.4	0.3254	0.324	0.349	0.225	0.130	1.107	4.29	2.93		
−10	14	1.463	1362.8	0.3255	0.288	0.310	0.218	0.126	1.097	4.25	2.62		
0	32	1.438	1363.6	0.3257	0.257	0.277	0.211	0.122	1.081	4.19	2.38		
10	50	1.412	1364.5	0.3259	0.232	0.250	0.204	0.118	1.066	4.13	2.18		
20	68	1.386	1365.3	0.3261	0.210	0.226	0.199	0.115	1.050	4.07	2.00	1.94×10^{-3}	1.08×10^{-3}
30	86	1.359	1366.2	0.3263	0.190	0.204	0.192	0.111	1.035	4.01	1.83		
40	104	1.329	1367.4	0.3266	0.173	0.186	0.185	0.107	1.019	3.95	1.70		
50	122	1.299	1368.3	0.3268	0.162	0.174	0.177	0.102	0.999	3.87	1.61		

TABLE C.11
Properties of Saturated Liquids: Water H_2O.*

Temp, T		Specific gravity	Specific heat, c_p		Kinematic viscosity, ν		Thermal conductivity, k		Thermal diffusivity, α		Prandtl number, Pr	β	
°C	°F		$\dfrac{J}{kg \cdot K}$	$\dfrac{BTU}{lbm \cdot °R}$	m^2/s	ft^2/s	$\dfrac{W}{m \cdot K}$	$\dfrac{BTU}{hr \cdot ft \cdot °R}$	m^2/s	ft^2/hr		$1/K$	$1/°R$
0	32	1.002	4217	1.0074	1.788×10^{-6}	1.925×10^{-5}	0.552	0.319	1.308×10^{-7}	5.07×10^{-3}	13.6		
20	68	1.000	4181	0.9988	1.006	1.083	0.597	0.345	1.430	5.54	7.02	0.18×10^{-3}	0.10×10^{-3}
40	104	0.994	4178	0.9980	0.658	0.708	0.628	0.363	1.512	5.86	4.34		
60	140	0.985	4184	0.9994	0.478	0.514	0.651	0.376	1.554	6.02	3.02		
80	176	0.974	4196	1.0023	0.364	0.392	0.668	0.386	1.636	6.34	2.22		
100	212	0.960	4216	1.0070	0.294	0.316	0.680	0.393	1.680	6.51	1.74		
120	248	0.945	4250	1.015	0.247	0.266	0.685	0.396	1.708	6.62	1.446		
140	284	0.928	4283	1.023	0.214	0.230	0.684	0.395	1.724	6.68	1.241		
160	320	0.909	4342	1.037	0.190	0.204	0.680	0.393	1.729	6.70	1.099		
180	356	0.889	4417	1.055	0.173	0.186	0.675	0.390	1.724	6.68	1.004		
200	392	0.866	4505	1.076	0.160	0.172	0.665	0.384	1.706	6.61	0.937		
220	428	0.842	4610	1.101	0.150	0.161	0.652	0.377	1.680	6.51	0.891		
240	464	0.815	4756	1.136	0.143	0.154	0.635	0.367	1.639	6.35	0.871		
260	500	0.785	4949	1.182	0.137	0.148	0.611	0.353	1.577	6.11	0.874		
280	537	0.752	5208	1.244	0.135	0.145	0.580	0.335	1.481	5.74	0.910		
300	572	0.714	5728	1.368	0.135	0.145	0.540	0.312	1.324	5.13	1.019		

Gas Property Tables

TABLE D.1
Properties of Gases at Atmospheric Pressure (101.3 kPa = 14.7 psia): Air [Gas constant = 286.8 J/(kg·K) = 53.3 ft·lbf/lbm·°R; $\gamma = c_p/c_v = 1.4$].*

Temp, T (K)	Temp, T (°R)	Density, ρ (kg/m³)	Density, ρ (lbm/ft³)	Specific heat, c_p (J/kg·K)	Specific heat, c_p (BTU/lbm·°R)	Kinematic viscosity, ν (m²/s)	Kinematic viscosity, ν (ft²/s)	Thermal conductivity, k (W/m·K)	Thermal conductivity, k (BTU/hr·ft·°R)	Thermal diffusivity, α (m²/s)	Thermal diffusivity, α (ft²/hr)	Prandtl number, Pr
100	180	3.601	0.225	1026.6	0.245	1.923×10^{-6}	2.070×10^{-5}	0.009246	0.005342	0.02501×10^{-4}	0.0969	0.770
150	270	2.368	0.148	1009.9	0.241	4.343	4.674	0.013735	0.007936	0.05745	0.223	0.753
200	360	1.768	0.110	1006.1	0.240	7.490	8.062	0.01809	0.01045	0.10165	0.394	0.739
250	450	1.413	0.0882	1005.3	0.240	9.49	10.2	0.02227	0.01287	0.13161	0.510	0.722
300	540	1.177	0.0735	1005.7	0.240	15.68	16.88	0.02624	0.01516	0.22160	0.859	0.708
350	630	0.998	0.0623	1009.0	0.241	20.76	22.35	0.03003	0.01735	0.2983	1.156	0.697
400	720	0.883	0.0551	1014.0	0.242	25.90	27.88	0.03365	0.01944	0.3760	1.457	0.689
450	810	0.783	0.0489	1020.7	0.244	28.86	31.06	0.03707	0.02142	0.4222	1.636	0.683
500	900	0.705	0.0440	1029.5	0.245	37.90	40.80	0.04038	0.02333	0.5564	2.156	0.680
550	990	0.642	0.0401	1039.2	0.248	44.34	47.73	0.04360	0.02519	0.6532	2.531	0.680
600	1080	0.589	0.0367	1055.1	0.252	51.34	55.26	0.04659	0.02692	0.7512	2.911	0.680
650	1170	0.543	0.0339	1063.5	0.254	58.51	62.98	0.04953	0.02862	0.8578	3.324	0.682
700	1260	0.503	0.0314	1075.2	0.257	66.25	71.31	0.05230	0.03022	0.9672	3.748	0.684
750	1350	0.471	0.0294	1085.6	0.259	73.91	79.56	0.05509	0.03183	1.0774	4.175	0.686
800	1440	0.441	0.0275	1097.8	0.262	82.29	88.58	0.05779	0.03339	1.1951	4.631	0.689
850	1530	0.415	0.0259	1109.5	0.265	90.75	97.68	0.06028	0.03483	1.3097	5.075	0.692
900	1620	0.393	0.0245	1121.2	0.268	99.3	107	0.06279	0.03628	1.4271	5.530	0.696
950	1710	0.372	0.0232	1132.1	0.270	108.2	116.5	0.06525	0.03770	1.5510	6.010	0.699
1000	1800	0.352	0.0220	1141.7	0.273	117.8	126.8	0.06752	0.03901	1.6779	6.502	0.702
1100	1980	0.320	0.0120	1160	0.277	138.6	149.2	0.0732	0.0423	1.969	7.630	0.704
1200	2160	0.295	0.0184	1179	0.282	159.1	171.3	0.0782	0.0452	2.251	8.723	0.707
1300	2340	0.271	0.0169	1197	0.286	182.1	196.0	0.0837	0.0484	2.583	10.01	0.705
1400	2520	0.252	0.0157	1214	0.290	205.5	221.2	0.0891	0.0515	2.920	11.32	0.705
1500	2700	0.236	0.0147	1230	0.294	229.1	246.6	0.0946	0.0547	3.262	12.64	0.705
1600	2880	0.221	0.0138	1248	0.298	254.5	273.9	0.100	0.0578	3.609	13.98	0.705
1700	3060	0.208	0.0130	1267	0.303	280.5	301.9	0.105	0.0607	3.977	15.41	0.705
1800	3240	0.197	0.0123	1287	0.307	308.1	331.6	0.111	0.0641	4.379	16.97	0.704
1900	3420	0.186	0.0115	1309	0.313	338.5	364.4	0.117	0.0676	4.811	18.64	0.704
2000	3600	0.176	0.010	1338	0.320	369.0	397.2	0.124	0.0716	5.260	20.38	0.702
2100	3780	0.168	0.0105	1372	0.328	399.6	430.1	0.131	0.0757	5.715	22.15	0.700
2200	3960	0.160	0.0100	1419	0.339	432.6	465.6	0.139	0.0803	6.120	23.72	0.707
2300	4140	0.154	0.00955	1482	0.354	464.0	499.4	0.149	0.0861	6.540	25.34	0.710
2400	4320	0.146	0.00905	1574	0.376	504.0	542.5	0.161	0.0930	7.020	27.20	0.718
2500	4500	0.139	0.00868	1688	0.403	543.5	585.0	0.175	0.101	7.441	28.83	0.730

* *Source:* Data taken from a number of sources. See references at end of text.

TABLE D.2

Properties of Gases at Atmospheric Pressure (101.3 kPa = 14.7 psia): Carbon Dioxide [Gas constant = 188.9 J/(kg·K) = 35.11 ft·lbf/lbm·°R; $\gamma = c_p/c_v = 1.30$].*

Temp, T		Density, ρ		Specific heat, c_p		Kinematic viscosity, ν		Thermal conductivity, k		Thermal diffusivity, α		Prandtl number, Pr
K	°R	kg/m³	lbm/ft³	$\frac{J}{kg \cdot K}$	$\frac{BTU}{lbm \cdot °R}$	m²/s	ft²/s	$\frac{W}{m \cdot K}$	$\frac{BTU}{hr \cdot ft \cdot °R}$	m²/s	ft²/hr	
220	396	2.4733	0.1544	783	0.187	4.490×10^{-6}	4.833×10^{-5}	0.010805	0.006243	0.05920×10^{-4}	0.2294	0.818
250	450	2.1657	0.1352	804	0.192	5.813	6.257	0.012884	0.007444	0.07401	0.2868	0.793
300	540	1.7973	0.1122	871	0.208	8.321	8.957	0.016572	0.009575	0.10588	0.4103	0.770
350	630	1.5362	0.0959	900	0.215	11.19	12.05	0.02047	0.01183	0.14808	0.5738	0.755
400	720	1.3424	0.0838	942	0.225	14.39	15.49	0.02461	0.01422	0.19463	0.7542	0.738
450	810	1.1918	0.0744	980	0.234	17.90	19.27	0.02897	0.01674	0.24813	0.9615	0.721
500	900	1.0732	0.0670	1013	0.242	21.67	23.33	0.03352	0.01937	0.3084	1.195	0.702
550	990	0.9739	0.0608	1047	0.250	25.74	27.71	0.03821	0.02208	0.3750	1.453	0.685
600	1080	0.8938	0.0558	1076	0.257	30.02	32.31	0.04311	0.02491	0.4483	1.737	0.668

TABLE D.3

Properties of Gases at Atmospheric Pressure (101.3 kPa = 14.7 psia): Helium [Gas constant = 2 077 J/(kg·K) = 386 ft·lbf/lbm·°R; $\gamma = c_p/c_v = 1.66$].*

Temp, T		Density, ρ		Specific heat, c_p		Kinematic viscosity, ν		Thermal conductivity, k		Thermal diffusivity, α		Prandtl number, Pr
K	°R	kg/m³	lbm/ft³	$\frac{J}{kg \cdot K}$	$\frac{BTU}{lbm \cdot °R}$	m²/s	ft²/s	$\frac{W}{m \cdot K}$	$\frac{BTU}{hr \cdot ft \cdot °R}$	m²/s	ft²/hr	
33	60	1.4657	0.0915	5 200	1.242	3.42×10^{-6}	3.68×10^{-5}	0.0353	0.0204	0.04625×10^{-4}	0.1792	0.74
144	260	0.3380	0.0211	5 200	1.242	37.11	39.95	0.0928	0.0536	0.5275	2.044	0.70
200	360	0.2435	0.0152	5 200	1.242	64.38	69.30	0.1177	0.0680	0.9288	3.599	0.69
255	460	0.1906	0.0119	5 200	1.242	95.50	102.8	0.1357	0.0784	1.3675	5.299	0.70
366	660	0.13280	0.00829	5 200	1.242	173.6	186.9	0.1691	0.0977	2.449	9.490	0.71
477	860	0.10204	0.00637	5 200	1.242	269.3	289.9	0.197	0.114	3.716	14.40	0.72
589	1060	0.08282	0.00517	5 200	1.242	375.8	404.5	0.225	0.130	5.215	20.21	0.72
700	1260	0.07032	0.00439	5 200	1.242	494.2	531.9	0.251	0.145	6.661	25.81	0.72
800	1440	0.06023	0.00376	5 200	1.242	634.1	682.5	0.275	0.159	8.774	34.00	0.72
900	1620	0.05286	0.00330	5 200	1.242	781.3	841.0	0.298	0.172	10.834	41.98	0.72

TABLE D.4

Properties of Gases at Atmospheric Pressure (101.3 kPa = 14.7 psia): Hydrogen [Gas constant = 4 126 J/(kg·K) = 767 ft·lbf/lbm·°R; $\gamma = c_p/c_v = 1.405$].*

Temp, T		Density, ρ		Specific heat, c_p		Kinematic viscosity, ν		Thermal conductivity, k		Thermal diffusivity, α		Prandtl number, Pr
K	°R	kg/m³	lbm/ft³	$\frac{J}{kg \cdot K}$	$\frac{BTU}{lbm \cdot °R}$	m²/s	ft²/s	$\frac{W}{m \cdot K}$	$\frac{BTU}{hr \cdot ft \cdot °R}$	m²/s	ft²/hr	
30	54	0.847 22	0.05289	10 840	2.589	1.895×10^{-6}	2.040×10^{-5}	0.0228	0.0132	0.02493×10^{-4}	0.0966	0.759
50	90	0.509 55	0.03181	10 501	2.508	4.880	5.253	0.0362	0.0209	0.0676	0.262	0.721
100	180	0.245 72	0.01534	11 229	2.682	17.14	18.45	0.0665	0.0384	0.2408	0.933	0.712
150	270	0.163 71	0.01022	12 602	3.010	34.18	36.79	0.0981	0.0567	0.475	1.84	0.718
200	360	0.122 70	0.00766	13 540	3.234	55.53	59.77	0.1282	0.0741	0.772	2.99	0.719
250	450	0.098 19	0.00613	14 059	3.358	80.64	86.80	0.1561	0.0902	1.130	4.38	0.713
300	540	0.081 85	0.00511	14 314	3.419	109.5	117.9	0.182	0.105	1.554	6.02	0.706
350	630	0.070 16	0.00438	14 436	3.448	141.9	152.7	0.206	0.119	2.031	7.87	0.697
400	720	0.061 35	0.00383	14 491	3.461	177.1	190.6	0.228	0.132	2.568	9.95	0.690
450	810	0.054 62	0.00341	14 499	3.463	215.6	232.1	0.251	0.145	3.164	12.26	0.682
500	900	0.049 18	0.00307	14 507	3.465	257.0	276.6	0.272	0.157	3.817	14.79	0.675
550	990	0.044 69	0.00279	14 532	3.471	301.6	324.6	0.292	0.169	4.516	17.50	0.668
600	1080	0.040 85	0.00255	14 537	3.472	349.7	376.4	0.315	0.182	5.306	20.56	0.664
700	1260	0.034 92	0.00218	14 574	3.481	455.1	489.9	0.351	0.203	6.903	26.75	0.659
800	1440	0.030 60	0.00191	14 675	3.505	569	612	0.384	0.222	8.563	33.18	0.664
900	1620	0.027 23	0.00170	14 821	3.540	690	743	0.412	0.238	10.217	39.59	0.676
1000	1800	0.024 51	0.00153	14 968	3.575	822	885	0.440	0.254	11.997	46.49	0.686
1100	1980	0.022 27	0.00139	15 165	3.622	965	1039	0.464	0.268	13.726	53.19	0.703
1200	2160	0.020 50	0.00128	15 366	3.670	1107	1192	0.488	0.282	15.484	60.00	0.715
1300	2340	0.018 90	0.00118	15 575	3.720	1273	1370	0.512	0.296	17.394	67.40	0.733
1333	2400	0.018 42	0.00115	15 638	3.735	1328	1429	0.519	0.300	18.013	69.80	0.736

TABLE D.5
Properties of Gases at Atmospheric Pressure (101.3 kPa = 14.7 psia): Nitrogen [Gas constant = 296.8 J/(kg·K) = 55.16 ft·lbf/lbm·°R; $\gamma = c_p/c_v = 1.40$].*

Temp, T		Density, ρ		Specific heat, c_p		Kinematic viscosity, ν		Thermal conductivity, k		Thermal diffusivity, α		Prandtl number, Pr
K	°R	kg/m³	lbm/ft³	J/kg·K	BTU/lbm·°R	m²/s	ft²/s	W/m·K	BTU/hr·ft·°R	m²/s	ft²/hr	
100	180	3.4808	0.2173	1072.2	0.2561	1.971×10^{-6}	2.122×10^{-5}	0.009450	0.005460	0.025319×10^{-4}	0.09811	0.786
200	360	1.7108	0.1068	1042.9	0.2491	7.568	8.146	0.01824	0.01054	0.10224	0.3962	0.747
300	540	1.1421	0.0713	1040.8	0.2486	15.63	16.82	0.02620	0.01514	0.22044	0.8542	0.713
400	720	0.8538	0.0533	1045.9	0.2498	25.74	27.71	0.03335	0.01927	0.3734	1.447	0.691
500	900	0.6824	0.0426	1055.5	0.2521	37.66	40.54	0.03984	0.02302	0.5530	2.143	0.684
600	1080	0.5687	0.0355	1075.6	0.2569	51.19	55.10	0.04580	0.02646	0.7486	2.901	0.686
700	1260	0.4934	0.0308	1096.9	0.2620	65.13	70.10	0.05123	0.02960	0.9466	3.668	0.691
800	1440	0.4277	0.0267	1122.5	0.2681	81.46	87.68	0.05609	0.03241	1.1685	4.528	0.700
900	1620	0.3796	0.0237	1146.4	0.2738	91.06	98.02	0.06070	0.03507	1.3946	5.404	0.711
1000	1800	0.3412	0.0213	1167.7	0.2789	117.2	126.2	0.06475	0.03741	1.6250	6.297	0.724
1100	1980	0.3108	0.0194	1185.7	0.2382	136.0	146.4	0.06850	0.03958	1.8591	7.204	0.736
1200	2160	0.2851	0.0178	1203.7	0.2875	156.1	168.0	0.07184	0.04151	2.0932	8.111	0.748

TABLE D.6

Properties of Gases at Atmospheric Pressure (101.3 kPa = 14.7 psia): Oxygen [Gas constant = 260 J/(kg·K) = 48.3 ft·lbf/lbm·°R; $\gamma = c_p/c_v = 1.40$].*

Temp, T		Density, ρ		Specific heat, c_p		Kinematic viscosity, ν		Thermal conductivity, k		Thermal diffusivity, α		Prandtl number, Pr
K	°R	kg/m³	lbm/ft³	J/kg·K	BTU/lbm·°R	m²/s	ft²/s	W/m·K	BTU/hr·ft·°R	m²/s	ft²/hr	
100	180	3.9118	0.2492	947.9	0.2264	1.946×10^{-6}	2.095×10^{-5}	0.00903	0.00522	0.023876×10^{-4}	0.09252	0.815
150	270	2.6190	0.1635	917.8	0.2192	4.387	4.722	0.01367	0.00790	0.05688	0.2204	0.773
200	360	1.9559	0.1221	913.1	0.2181	7.593	8.173	0.01824	0.01054	0.10214	0.3958	0.745
250	450	1.5618	0.0975	915.7	0.2187	11.45	12.32	0.02259	0.01305	0.15794	0.6120	0.725
300	540	1.3007	0.0812	920.3	0.2198	15.86	17.07	0.02676	0.01546	0.22353	0.8662	0.709
350	630	1.1133	0.0695	929.1	0.2219	20.80	22.39	0.03070	0.01774	0.2968	1.150	0.702
400	720	0.9755	0.0609	942.0	0.2250	26.18	28.18	0.03461	0.02000	0.3768	1.460	0.695
450	810	0.8682	0.0542	956.7	0.2285	31.99	34.43	0.03828	0.02212	0.4609	1.786	0.694
500	900	0.7801	0.0487	972.2	0.2322	38.34	41.27	0.04173	0.02411	0.5502	2.132	0.697
550	990	0.7096	0.0443	988.1	0.2360	45.05	48.49	0.04517	0.02610	0.6441	2.496	0.700
600	1080	0.6504	0.0406	1004.4	0.2399	52.15	56.13	0.04882	0.02792	0.7399	2.867	0.704

TABLE D.7
Properties of Gases at Atmospheric Pressure (101.3 kPa = 14.7 psia): Water Vapor or Steam [Gas constant = 461.5 J/(kg·K) = 85.78 ft·lbf/lbm·°R; $\gamma = c_p/c_v = 1.33$].*

Temp, T		Density, ρ		Specific heat, c_p		Kinematic viscosity, ν		Thermal conductivity, k		Thermal diffusivity, α		Prandtl number, Pr
K	°R	kg/m³	lbm/ft³	$\frac{J}{kg \cdot K}$	$\frac{BTU}{lbm \cdot °R}$	m²/s	ft²/s	$\frac{W}{m \cdot K}$	$\frac{BTU}{hr \cdot ft \cdot °R}$	m²/s	ft²/hr	
380	684	0.5863	0.0366	2060	0.492	2.16×10^{-5}	2.33×10^{-4}	0.0246	0.0142	0.2036×10^{-4}	0.789	1.060
400	720	0.5542	0.0346	2014	0.481	2.42	2.61	0.0261	0.0151	0.2338	0.906	1.040
450	810	0.4902	0.0306	1980	0.473	3.11	3.35	0.0299	0.0173	0.307	1.19	1.010
500	900	0.4405	0.0275	1985	0.474	3.86	4.16	0.0339	0.0196	0.387	1.50	0.996
550	990	0.4005	0.0250	1997	0.477	4.70	5.06	0.0379	0.0219	0.475	1.84	0.991
600	1080	0.3652	0.0228	2026	0.484	5.66	6.09	0.0422	0.0244	0.573	2.22	0.986
650	1170	0.3380	0.0211	2056	0.491	6.64	7.15	0.0464	0.0268	0.666	2.58	0.995
700	1260	0.3140	0.0196	2085	0.498	7.72	8.31	0.0505	0.0292	0.772	2.99	1.000
750	1350	0.2931	0.0183	2119	0.506	8.88	9.56	0.0549	0.0317	0.883	3.42	1.005
800	1440	0.2739	0.0171	2152	0.514	10.20	10.98	0.0592	0.0342	1.001	3.88	1.010
850	1530	0.2579	0.0161	2186	0.522	11.52	12.40	0.0637	0.0368	1.130	4.38	1.019

Emissivity Tables

TABLE E.1
Normal Emissivity of Various Metals

Material	Surface condition	Temperature K	°R	Normal emissivity, ε
Aluminum	Polished plate	296	533	0.040
		498	896	0.039
	Rolled and polished	443	797	0.039
	Rough plate	298	536	0.070
Brass	Oxidized	611	1,100	0.22
	Polished	292	526	0.05
		573	1,030	0.032
	Tarnished	329	592	0.202
Chromium	Polished	423	761	0.058
Copper	Black oxidized	293	527	0.780
	Tarnished lightly	293	527	0.037
	Polished	293	527	0.030
Gold	Not polished	293	527	0.47
	Polished	293	527	0.025
Iron	Smooth oxidized	398	716	0.78
	Ground bright	293	527	0.24
	Polished	698	1,256	0.144
Lead	Gray oxidized	293	527	0.28
	Polished	403	725	0.056
Nickel	Oxidized	373	671	0.41
	Polished	373	671	0.045
Silver	Polished	293	527	0.025
Steel	Rough oxidized	313	563	0.94
	Ground	1 213	2,183	0.520
Tin	Bright	293	527	0.070
Tungsten	Filament	3 300	5,940	0.39
Zinc	Tarnished	293	527	0.25
	Polished	503	905	0.045

Source: Data taken from a number of sources. See references at end of text.

TABLE E.2
Normal or Total (Hemispherical) Emissivity of Various
Nonmetallic Solids

Surface	Temperature		Emissivity	
	K	°R	Normal	Total
Aluminum oxide	600	1080	0.69	
	1 000	1800	0.55	
Asphalt pavement	300	540		0.85–0.93
Building materials				
Asbestos sheet	300	540		0.93–0.96
Red brick	300	540		0.93–0.96
Gypsum or plasterboard	300	540		0.90–0.92
Wood	300	540		0.82–0.92
Concrete	300	540		0.88–0.93
Cloth	300	540		0.75–0.90
Glass, window	300	540		0.90–0.95
Ice	273	492		0.95–0.98
Paint				
Black	300	540		0.98
White	300	540		0.90–0.92
Paper, white	300	540		0.92–0.97
Skin	300	540		0.95
Snow	273	492		0.82–0.90
Soil	300	540		0.93–0.96
Teflon	300	540		0.85
	500	900		0.92
Water	300	540		0.96

Source: Data taken from a number of sources. See references at end of text.

Pipe and Tubing Tables

TABLE F.1
Dimensions of Wrought-Steel and Wrought-Iron Pipe*

| Pipe size | Outside diameter | | Schedule | Internal diameter | | Flow area | |
	in.	cm		ft	cm	ft²	cm²
$\frac{1}{8}$	0.405	1.029	40 (STD)	0.02242	0.683	0.0003947	0.3664
			80 (XS)	0.01792	0.547	0.0002522	0.2350
$\frac{1}{4}$	0.540	1.372	40 (STD)	0.03033	0.924	0.0007227	0.6706
			80 (XS)	0.02517	0.768	0.0004974	0.4632
$\frac{3}{8}$	0.675	1.714	40 (STD)	0.04108	1.252	0.001326	1.233
			80 (XS)	0.03525	1.074	0.0009759	0.9059
$\frac{1}{2}$	0.840	2.134	40 (STD)	0.05183	1.580	0.002110	1.961
			80 (XS)	0.04550	1.386	0.001626	1.508
			160	0.03867	1.178	0.001174	1.090
			(XXS)	0.02100	0.640	0.0003464	3.217
$\frac{3}{4}$	1.050	2.667	40 (STD)	0.06867	2.093	0.003703	3.441
			80 (XS)	0.06183	1.883	0.003003	2.785
			160	0.05100	1.555	0.002043	1.898
			(XXS)	0.03617	1.103	0.001027	9.555
1	1.315	3.340	40 (STD)	0.08742	2.664	0.006002	5.574
			80 (XS)	0.07975	2.430	0.004995	5.083
			160	0.06792	2.070	0.003623	3.365
			(XXS)	0.04992	1.522	0.001957	1.815
$1\frac{1}{4}$	1.660	4.216	40 (STD)	0.1150	3.504	0.01039	9.643
			80 (XS)	0.1065	3.246	0.008908	8.275
			160	0.09667	2.946	0.007339	6.816
			(XXS)	0.7467	2.276	0.004379	4.069
$1\frac{1}{2}$	1.900	4.826	40 (STD)	0.1342	4.090	0.01414	13.13
			80 (XS)	0.1250	3.810	0.01227	11.40
			160	0.1115	3.398	0.009764	9.068
			(XXS)	0.09167	2.794	0.006600	6.131
2	2.375	6.034	40 (STD)	0.1723	5.252	0.02330	21.66
			80 (XS)	0.1616	4.926	0.02051	19.06
			160	0.1406	4.286	0.01552	14.43
			(XXS)	0.1253	3.820	0.01232	11.46
$2\frac{1}{2}$	2.875	7.303	40 (STD)	0.2058	6.271	0.03325	30.89
			80 (XS)	0.1936	5.901	0.02943	27.35
			160	0.1771	5.397	0.02463	22.88
			(XXS)	0.1476	4.499	0.01711	15.90
3	3.500	8.890	40 (STD)	0.2557	7.792	0.05134	47.69
			80 (XS)	0.2417	7.366	0.04587	42.61
			160	0.2187	6.664	0.03755	34.88
			(XXS)	0.1917	5.842	0.02885	26.80

* Dimensions in English units obtained from ANSI B36.10-1979, *American National Standard Wrought Steel and Wrought Iron Pipe*, and reprinted with permission from the publisher: American Society of Mechanical Engineers. *Notes*: STD implies Standard; XS is extra strong; XXS is double extra strong.

TABLE F.1
(continued)

Pipe size	Outside diameter		Schedule	Internal diameter		Flow area	
	in.	cm		ft	cm	ft^2	cm^2
3½	4.000	10.16	40 (STD)	0.2957	9.012	0.06866	63.79
			80 (XS)	0.2803	8.544	0.06172	57.33
4	4.500	11.43	40 (STD)	0.3355	10.23	0.08841	82.19
			80 (XS)	0.3198	9.718	0.07984	74.17
			120	0.3020	9.204	0.07163	66.54
			160	0.2865	8.732	0.06447	59.88
			(XXS)	0.2626	8.006	0.05419	50.34
5	5.563	14.13	40 (STD)	0.4206	12.82	0.1389	129.10
			80 (XS)	0.4011	12.22	0.1263	117.30
			120	0.3803	11.59	0.1136	105.50
			160	0.3594	10.95	0.1015	94.17
			(XXS)	0.3386	10.32	0.09004	83.65
6	6.625	16.83	40 (STD)	0.5054	15.41	0.2006	186.50
			80 (XS)	0.4801	14.64	0.1810	168.30
			120	0.4584	13.98	0.1650	153.50
			160	0.4823	13.18	0.1467	136.40
			(XXS)	0.4081	12.44	0.1308	121.50
8	8.625	21.91	20	0.6771	20.64	0.3601	334.60
			30	0.6726	20.50	0.3553	330.10
			40 (STD)	0.6651	20.27	0.3474	322.70
			60	0.6511	19.85	0.3329	309.50
			80 (XS)	0.6354	19.37	0.3171	294.70
			100	0.6198	18.89	0.3017	280.30
			120	0.5989	18.26	0.2817	261.90
			140	0.5834	17.79	0.2673	248.60
			(XXS)	0.5729	17.46	0.2578	239.40
			160	0.5678	17.31	0.2532	235.30
10	10.750	27.31	20	0.8542	26.04	0.5730	332.60
			30	0.8447	25.75	0.5604	520.80
			40 (STD)	0.8350	25.46	0.5476	509.10
			60 (XS)	0.8125	24.77	0.5185	481.90
			80	0.7968	24.29	0.4987	463.40
			100	0.7760	23.66	0.4730	439.70
			120	0.7552	23.02	0.4470	416.20
			140 (XXS)	0.7292	22.23	0.4176	388.10
			160	0.7083	21.59	0.3941	366.10
12	12.750	32.39	20	1.021	31.12	0.8185	760.60
			30	1.008	30.71	0.7972	740.71
			(STD)	1.000	30.48	0.7854	729.70
			40	0.9948	30.33	0.7773	722.50
			(XS)	0.9792	29.85	0.7530	699.80
			60	0.9688	29.53	0.7372	684.90
			80	0.9478	28.89	0.7056	655.50
			100	0.9218	28.10	0.6674	620.20
			120 (XX)	0.8958	27.31	0.6303	585.80
			140	0.8750	26.67	0.6013	558.60
			160	0.8438	25.72	0.5592	519.60

TABLE F.1
(continued)

Pipe size	Outside diameter		Schedule	Internal diameter		Flow area	
	in.	cm		ft	cm	ft^2	cm^2
14	14.000	35.56	30 (STD)	1.104	33.65	0.9575	889.30
			160	0.9323	28.42	0.6827	634.40
16	16.000	40.64	30 (STD)	1.271	38.73	1.268	1178.00
			160	1.068	32.54	0.8953	831.60
18	18.000	45.72	(STD)	1.438	43.81	1.623	1507.00
			160	1.203	36.67	1.137	1056.00
20	20.000	50.80	20 (STD)	1.604	48.89	2.021	1877.00
			160	1.339	40.80	1.407	1307.00
22	22.000	55.88	20 (STD)	1.771	53.97	2.463	2288.00
			160	1.479	45.08	1.718	1596.00
24	24.000	60.96	20 (STD)	1.938	59.05	2.948	2739.00
			160	1.609	49.05	2.034	1890.00
26	26.000	66.04	(STD)	2.104	64.13	3.477	3230.00
28	28.000	71.12	(STD)	2.271	69.21	4.050	3762.00
30	30.000	76.20	(STD)	2.438	74.29	4.666	4335.00
32	32.000	81.28	(STD)	2.604	79.34	5.326	4944.00
34	34.000	86.36	(STD)	2.771	84.45	6.030	5601.00
36	36.000	91.44	(STD)	2.938	89.53	6.777	6295.00
38	38.000	96.52	—	3.104	94.61	7.568	7030.00
40	40.000	101.6	—	3.271	99.69	8.403	7805.00

* Dimensions in English units obtained from ANSI B36.10-1979, *American National Standard Wrought Steel and Wrought Iron Pipe*, and reprinted with permission from the publisher: American Society of Mechanical Engineers. *Notes*: STD implies Standard; XS is extra strong; XXS is double extra strong.

TABLE F.2
Dimensions of Seamless Copper Tubing*

Standard size	Outside diameter in.	Outside diameter cm	Type	Internal diameter ft	Internal diameter cm	Flow area ft^2	Flow area cm^2
$\frac{1}{4}$	0.375	0.953	K	0.02542	0.775	0.0005074	0.4717
			L	0.02625	0.801	0.0005412	0.5039
$\frac{3}{8}$	0.500	1.270	K	0.03350	1.022	0.0008814	0.8203
			L	0.03583	1.092	0.001008	0.9366
			M	0.03750	1.142	0.001104	1.024
$\frac{1}{2}$	0.625	1.588	K	0.04392	1.340	0.001515	1.410
			L	0.04542	1.384	0.001620	1.505
			M	0.04742	1.446	0.001766	1.642
$\frac{5}{8}$	0.750	1.905	K	0.05433	1.657	0.002319	2.156
			L	0.05550	1.691	0.002419	2.246
$\frac{3}{4}$	0.875	2.222	K	0.06208	1.892	0.003027	2.811
			L	0.06542	1.994	0.003361	3.123
			M	0.06758	2.060	0.003587	3.333
1	1.125	2.858	K	0.08292	2.528	0.005400	5.019
			L	0.08542	2.604	0.005730	5.326
			M	0.08792	2.680	0.006071	5.641
$1\frac{1}{4}$	1.375	3.493	K	0.1038	2.163	0.008454	7.858
			L	0.1054	3.213	0.008728	8.108
			M	0.1076	3.279	0.009090	8.444
$1\frac{1}{2}$	1.625	4.128	K	0.1234	3.762	0.01196	11.12
			L	0.1254	3.824	0.01235	11.48
			M	0.1273	3.880	0.01272	11.82
2	2.125	5.398	K	0.1633	4.976	0.02093	11.95
			L	0.1654	5.042	0.02149	19.97
			M	0.1674	5.102	0.02201	20.44
$2\frac{1}{2}$	2.625	6.668	K	0.2029	6.186	0.03234	30.05
			L	0.2054	6.262	0.03314	30.80
			M	0.2079	6.338	0.03395	40.17
3	3.125	7.938	K	0.2423	7.384	0.04609	42.82
			L	0.2454	7.480	0.04730	43.94
			M	0.2484	7.572	0.04847	45.03
$3\frac{1}{2}$	3.625	9.208	K	0.2821	8.598	0.06249	58.06
			L	0.2854	8.700	0.06398	59.45
			M	0.2883	8.786	0.06523	60.63
4	4.125	10.48	K	0.3214	9.800	0.08114	75.43
			L	0.3254	9.922	0.08317	77.32
			M	0.3279	9.998	0.08445	78.51
5	5.125	13.02	K	0.4004	12.21	0.1259	117.10
			L	0.4063	12.38	0.1296	120.50
			M	0.4089	12.47	0.1313	122.10

* Dimensions in English units obtained from ANSI/ASTM B88-78, "Standard Specifications for Seamless Copper Water Tube," and reprinted with permission from the publisher: American Society for Testing Materials.

Notes: Type K is for underground service and general plumbing; Type L is for interior plumbing; Type M is for use only with soldered fittings.

TABLE F.2
(continued)

Standard size	Outside diameter		Type	Internal diameter		Flow area	
	in.	cm		ft	cm	ft^2	cm^2
6	6.125	15.56	K	0.4784	14.58	0.1798	167.00
			L	0.4871	14.85	0.1863	173.20
			M	0.4901	14.39	0.1886	175.30
8	8.125	20.64	K	0.6319	19.26	0.3136	291.50
			L	0.6438	19.62	0.3255	302.50
			M	0.6488	19.78	0.3306	307.20
10	10.125	25.72	K	0.7874	24.00	0.4870	452.50
			L	0.8021	24.45	0.5053	469.50
			M	0.8084	24.64	0.5133	476.80
12	12.125	30.80	K	0.9429	28.74	0.6983	648.80
			L	0.9638	29.38	0.7295	677.90
			M	0.9681	29.51	0.7361	684.00

* Dimensions in English units obtained from ANSI/ASTM B88-78, "Standard Specifications for Seamless Copper Water Tube," and reprinted with permission from the publisher: American Society for Testing Materials.

Notes: Type K is for underground service and general plumbing; Type L is for interior plumbing; Type M is for use only with soldered fittings.

TABLE G.1
The Greek Alphabet

English spelling	Greek capital letters	Greek lower-case letters
Alpha	A	α
Beta	B	β
Gamma	Γ	γ
Delta	Δ	δ
Epsilon	E	ε
Zeta	Z	ζ
Eta	H	η
Theta	Θ	θ
Iota	I	ι
Kappa	K	κ
Lambda	Λ	λ
Mu	M	μ
Nu	N	ν
Xi	Ξ	ξ
Omicron	O	o
Pi	Π	π
Rho	P	ρ
Sigma	Σ	σ
Tau	T	τ
Upsilon	Υ	υ
Phi	Φ	ϕ
Chi	X	χ
Psi	Ψ	ψ
Omega	Ω	ω

Answers to Selected Odd-Numbered Problems

1.5	0.034 W/(m·°C)	3.21	q/L = 21 800 W/m
1.7	949 BTU/(hr·ft²)	3.23	q/L = 193.2 W/m
1.9	T3 = 25°C	3.25	q/L = 1 542 W/m
1.13	4.259 BTU/(hr·ft²)	3.31	S/L = 14.49 by equation
1.15	6.27 BTU/hr	3.33	d = 1.15
1.19	3.6 W/(m·K)		
1.23	576 W	4.1	Bi = 0.0023
1.25	8.1×10^5 BTU/(hr·ft²)	4.3	t = 0.207 hr = 12.45 min
1.27	432.9 BTU/hr	4.5	t = 0.168 hr = 10.1 min
1.29	14 130 W/m²	4.9	t = 36 min
1.31	938°R	4.13	q = 2 300 W
1.33	118°F	4.15	T = 67°C
1.35	T_w = 489°R = 29°F	4.21	$t = 1.914 \times 10^4$ s = 5.32 hr
1.37	T_w = 1800°R = 1340°F	4.27	t = 50 000 s = 13.9 hr
		4.29	t = 16.5 min
2.1	4.6×10^6 BTU/hr	4.37	t = 13.75
2.3	T_2 = 697°F	4.39	Left at 178°C; Right at 37°C
2.5	T_3 = 69.2°F		
2.7	$T_0 - T_3$ = 41.9°F	5.9	$\beta = (1.45 \times 10^{-3})/°R$
2.9	U = 0.889 W/(m²·K)		
2.11	4 133 W/m	6.3	z_t = 11 325 m
2.13	26.9 W/m	6.5	z_t = 20.4 ft
2.15	1.028	6.7	z_t = 3 109 m (Oil)
2.17	−50.2 BTU(hr·ft)	6.9	T_{wo} = 127°F
2.19	10.06	6.11	T_{wo} = 82.2°C
2.23	T_{max} = 1 322°C	6.13	T_{wo} = 38.3°F
2.25	T_{max} = 363.5°F	6.15	260 K
2.27	T_1 = 75°F	6.17	T_w = 97.8°C
2.31	5.076×10^{-4}°R·ft²/BTU	6.21	T_{bo} = 396 K
2.37	0.133	6.29	470.6 m
2.43	1.86 ft = 22.4 in	6.43	f = 0.31
2.49	$S_{opt} = 5.42 \times 10^{-4}$ ft = 0.0065 in	6.45	8
2.51	3.95		
2.55	$Q_{ac} = 7.912 \times 10^{-2}$ W	7.1	x = 1.6 m
2.57	Q_{ac} = 1.88 BTU/hr	7.3	x = 150 m
2.63	5.12	7.5	q_w = 60.2 W
2.65	38,200 BTU/hr	7.7	q_w = 13.9 W (one side)
		7.9	3.5°F
		7.17	12.4 lbf
3.11	q/L = 105.2 W/m	7.19	q_{13} = 16.7 W
3.13	S = 0.8	7.21	78.03 W/(m²·K)
3.17	q/L = 113 BTU/(hr·ft)	7.23	smooth = 16.2 lbf
3.19	q/L = 44.4 BTU/(hr·ft)		rough = 6.2 lbf

7.25   Drag = 0.27 lbf
7.31 q = 6330 BTU/hr

8.9 L = 0.040 5 m
8.11 L = 1.13 ft
8.13 q = 160 W
8.15 q = 13.0 W
8.19 q = 1.35 BTU/hr·ft^2
8.23 T_w = 121°F
8.25 q = 740 W
8.29 q = 786 BTU/hr
8.33 V = 2.54 ft/s
8.35 q = 607 W
8.37 q = 9480 BTU/hr

9.1   *% error* = (145–133)/145 = 8.3%
9.3   *% error* = (93.0–93.3)/93.3 = negligible
9.5   *LMTD* = 41.7°F
9.13 t_2 = 99.3°F
9.19   538.6 K for CO_2
9.31   *UA* = 0.334 BTU/(hr·°R)

9.33   *UA* = 229.8 W/K
9.35   *UA* = 480 BTU/hr·°R

10.13 L = 32.2 ft
10.17 21.5 lbm/hr

11.7 $d\omega$ = 1.524 × 10^{-4} sr
11.9 r = 2.64 ft
11.26 0.99998
11.37 0.91
11.39 0.071 Chromium
11.47 T = 0.45

12.7 F_{6-3} = 0.015
12.13 q_{1-2} = 419 BTU/hr
12.17 F_{1-2} = 1/2
12.19 q_{1-2} = 45 BTU/hr
12.21 q_{1-2} = 9.86 BTU/hr
12.23 q_1 = 2.7 BTU/hr
12.25 q_1 = 0.394 BTU/hr

Bibliography and Selected References

GENERAL REFERENCES

1. *Heat Transfer,* by Alan J. Chapman, Macmillan Publishing Co., 1987.
2. *Heat Transfer,* by J. P. Holman, 8th edition. McGraw-Hill, 1997.
3. *Fundamentals of Heat Transfer,* 3rd edition, by F. P. Incropera and D. P. DeWitt, John Wiley & Sons, 1996.
4. *Principles of Heat Transfer,* by Frank Kreith and Mark Bohn, 2nd edition. PWS Publishing, 1997.
5. *Heat Transfer,* 2nd edition, by A.F. Mills, Prentice-Hall, Inc., 1999.
6. *A Heat Transfer Textbook,* by John H. Lienhard, Prentice-Hall, Inc., 1981.
7. *Heat Transfer: A Basic Approach,* by M. Necati Özisik, McGraw-Hill, 1985.
8. *Heat Transfer,* by James Sucec, Wm. C. Brown Publishers, Dubuque, Iowa, 1985.
9. *Heat Transfer,* by Frank M. White, Addison-Wesley Publishing Co., 1984.
10. *Heat Transfer,* by Helmut Wolf, Harper & Row Publishers, 1983.

CHAPTER 1

Properties of Substances

1. *Handbook of Fundamentals,* published by American Society of Heating, Refrigerating and Air Conditioning Engineers (ASHRAE), Chapters 17 and 31, 1972.
2. *Thermophysical Properties of Matter, The TPRC Data Series,* edited by Y.S.Touloukian and C. Y. Ho. 13 volumes on thermophysical properties: specific heat, thermal conductivity, thermal diffusivity, and thermal expansion. Plenum Publishing Co., New York, 1970–1977.
3. *Handbook of Tables for Applied Engineering Science,* edited by Ray E. Bolz and George L. Tuve, 2nd edition. CRC Press, Boca Raton, FL, 1973.
4. *Mechanical Measurements*, by T. G. Beckwith, N. L. Buck, and R. D. Marangoni, 3rd edition. Addison-Wesley Publishing Co., Chapter 16 on measuring radiation properties. 1982.
5. *Experimental Methods for Engineers,* by J. P. Holman, 4th edition, McGraw-Hill, New York, Chapter 12 on measuring radiation properties. 1984.

SE Unit System

1. *Standard for Metric Practice,* published by ASTM, 100 Bar Harbor St., West Conshohocken, PA 19428; designated as ASTM E 380–76. A 35-page booklet that contains base units, derived units, prefixes, recommendations concerning units, rules for rounding off and conversion, terminology, and conversion factors.
2. *ASME Orientation and Guide for Use of SI (Metric) Units,* 8th edition, published by ASME, United Engineering Center, 345 East 47th St., New York, NY 10017; designated as ASME Guide SI-1.
3. NASA SP–7012, by E. A. Mechtly, 1973. This publication contains conversion factors and is the source for Appendix Tables A.1 and A.2.

CHAPTER 2

General Conduction Equations

1. *Analytical Methods in Conduction Heat Transfer*, by Glen E. Myers, McGraw-Hill, 1971.
2. *Conduction Heat Transfer*, by P. J. Schneider, Addison-Wesley Publishing Co., 1955.

3. *Boundary Value Problems of Heat Conduction*, by M. Necati & Özisik, International Textbook Co., 1968.
4. *Transport Phenomena*, by R. B. Bird, W. E. Stewart, and E. N. Lightfoot, John Wiley & Sons, Inc., 1960. Chapter 10 has derivations of a number of forms of the general energy equation.
5. See also the General References.

CHAPTER 3

Pipe and Tube Specifications

1. ANSI B36.10-1979, "American National Standard Wrought Steel and Wrought Iron Pipe," published by ASME.
2. ANSI/ASTM B88-78, "Standard Specifications for Seamless Copper Water Tube," published by ASTM.

Thermal Contact Resistance

1. "Thermal Conduction Contribution to Heat Transfer at Contacts," by E. Fried and published in *Thermal Conductivity*, R. P. Tye, editor. Volume 2, Academic Press, London, 1969.
2. "Conduction, Interface Resistance," by P. J. Schneider and published in *Handbook of Heat Transfer*, W. M. Rohsenow and J. P. Hartnett, editors. McGraw-Hill, New York, 1973.

Textbooks

1. *Analytical Methods in Conduction Heat Transfer*, by Glen F. Myers, McGraw-Hill, New York, 1971.
2. *Boundary Value Problems of Heat Conduction*, by M. Necati Özisik, International Textbook Co., Scranton, PA, 1968.
3. *Conduction Heat Transfer*, by P. J. Schneider, Addison-Wesley Publishing Co., Reading, MA, 1955.
4. See also the General References.

CHAPTER 4

Extended Surfaces

1. *Conduction Heat Transfer*, by P. J. Schneider, Addison-Wesley, Reading, MA, 1955.
2. *Extended Surface Heat Transfer*, by D. Q. Kern and A. D. Kraus, McGraw-Hill, New York, 1972.

Numerical Methods

1. *Analytical Methods in Conduction Heat Transfer*, by Glen F. Myers, McGraw-Hill, New York, 1971.
2. *Computer Aided Heat Transfer Analysis*, by J. A. Adams and D. F. Rogers, McGraw-Hill, New York, 1973. This text contains many computer programs written in BASIC.
3. *Heat Transfer Calculations, by Finite-Difference*, by G. M. Dusinberre, International Textbook Co., Scranton, PA, 1961.
4. "Methods of Obtaining Approximate Solutions," by P. Razelos and published in *Handbook of Heat Transfer*, W. M. Rohsenow and J. P. Hartnett, editors. McGraw-Hill, New York, 1973.

CHAPTER 5

Analytical Methods in Conduction

1. *Analytical Methods in Conduction Heat Transfer*, by Glen F. Myers, McGraw-Hill, New York, 1971.

2. *Boundary Value Problems of Heat Conduction*, by M. Necati Özisik, International Textbook Co., Scranton, PA, 1968.
3. *Conduction Heat Transfer*, by P. J. Schneider, Addison-Wesley Publishing Co., Reading, MA, 1955.

Alternative (to the Analytical) Methods in Conduction

1. See the conduction texts listed above.
2. *Principles of Heat Transfer*, by Frank Kreith, 2nd edition, International Textbook Co., Scranton, PA, 1965.

Conduction Shape Factors

1. *Conduction of Heat in Solids*, by H. S. Carslaw and J. C. Jaeger, 2nd edition, Oxford University Press, London, 1959.
2. "Shape Factors for Heat Conduction Through Bodies with Isothermal or Convective Boundary Conditions," *Trans ASHRAE*, v. 10, pp. 237–241, 1964.
3. *Fundamentals of Heat Transfer*, by S. S. Kutateladze, Academic Press, New York, 1963.
4. *Heat Transfer Data Book*, General Electric Co., Corporate Research and Development, Section 502, General Electric Co., Schenectady, New York, 1973.
5. F. Hahne and U. Grigull, *Int J Heat Transfer*, v. 18, pp. 751–767, 1975.
6. See also the General References.

Numerical Methods

1. *Analytical Methods in Conduction Heat Transfer*, by Glen E. Myers, McGraw-Hill, New York, 1971.
2. *Computer Aided Heat Transfer Analysis*, by J. A. Adams and D. F. Rogers, McGraw-Hill, New York, 1973. This text contains many computer programs written in BASIC.
3. *Heat Transfer Calculations by Finite-Difference*, by G. M. Dusinberre, International Textbook Co., Scranton, PA, 1961.
4. Methods of Obtaining Approximate Solutions," by P. Razelos and published in *Handbook of Heat Transfer*, by W. M. Rohsenow and J. P. Hartnett, editors, McGraw-Hill, New York, 1973.

CHAPTER 6

Analytical Methods in Unsteady Conduction Problems

1. *Analytical Methods in Conduction Heat Transfer*, by Glen F. Myers, McGraw-Hill, New York, 1971.
2. *Conduction Heat Transfer*, by P. J. Schneider, Addison-Wesley Publishing Co., Reading, MA, 1955.

Temperature Charts

1. "Temperature Charts for Induction and Constant Temperature Heating," by M. P. Heisler, *Trans ASME*, v. 69, pp. 227–236, 1947.
2. *Fundamentals of Heat Transfer*, by H. Gröber, S. Erk, and U. Grigull, McGraw-Hill, New York, 1961.

Numerical Methods

1. See the listing in the Chapter 5 references.
2. See also the General References.

Graphical Methods

1. *Principles of Heat Transfer*, by Frank Kreith, 2nd edition, International Textbook Co., Scranton, PA, 1965. This text has some in-depth information about the Schmidt plot.

2. "Graphical Solution of Unsteady Viscous Flow Problems," by T. G. Keith and W. S. Janna, IMechE & UMIST. *International Journal of Mechanical Engineering Education,* v. 5, no. 2, pp. 97–105, 1977. This reference has information about the Saul'ev method.

Properties of Foods

1. *Transport Processes and Unit Operations,* by Christie J. Geonkoplis, Allyn & Bacon, Inc., Boston, 1978. Appendix tables have some information as well as a listing of sources.
2. *New Cook Book,* published by Better Homes and Gardens, 1976. This reference has information (all empirical) on cooking times for many foods.

CHAPTER 7

General Convection References

1. *Convection Heat Transfer,* by V. S. Arpaci and P.S. Larsen, Prentice-Hall, Inc., Englewood Cliffs, NJ, 1984.
2. *Convection Heat Transfer,* by A. Bejan, John Wiley & Sons, New York, 1984.
3. *Convective Heat and Mass Transfer,* by W. M. Kays and M. F. Crawford, McGraw-Hill, 2nd ed., 1980.
4. *Convective Heat Transfer,* by Louis C. Burmeister, John Wiley & Sons, New York, 1983.
5. *Transport Phenomena,* by R. B. Bird, W. F. Stewart, and F. N. Lightfoot, John Wiley & Sons, New York, 1960. This text contains a derivation of the Navier-Stokes Equations (Chapter 3) and of the energy equation (Chapter 10). In addition, Chapter 13 contains an overview of convection heat transfer.

CHAPTER 8

1. "Heat Transfer to Laminar Flow in a Round Tube or Flat Conduit—The Graetz Problem Extended," by J. R. Sellars, M. Tribus, and J. S. Klein. *Trans ASME,* v. 78, 1956, p. 441.
2. "Steady Laminar Heat Transfer in a Circular Tube with Prescribed Wall Heat Flux," by R. Siegel, F. M. Sparrow, and T. M. Hallman. *App Sci Res,* Section A, v. 7, 1958, p. 386.
3. "Turbulent Heat Transfer in the Thermal Entrance Region of a Pipe with Uniform Heat Flux," by F. M. Sparrow, T. M. Hallman, and R. Siegel. *App Sci Res,* Section A, v. 7, 1957, p. 37.
4. "Laminar Flow Forced Convection in Rectangular Tubes," by S. H. Clark and W. M. Kays. *Trans ASME,* v. 75, 1953, p. 859.
5. "Analysis of Turbulent Heat Transfer and Flow in the Entrance Regions of Smooth Passages," by R. G. Deissler. *NACA TN* 3016, Oct.1953.
6. "Experimental Determination of the Thermal Entrance Length for the Flow of Water and Oil in Circular Pipes," by J. P. Hartnett. *Trans ASME,* v. 77, 1955, p. 1211.
7. *Publications Engineering,* by F. W. Dittus and L. M. K. Boelter, Univ of Cal Berkeley, v. 2, 1930, p. 443.
8. "Heat Transfer and Pressure Drops of Liquids in Tubes," by F. N. Sieder and G. F. Tate. *Ind & Engrg Chem,* v. 28, 1936, p. 1429.
9. "The Analogy Between Fluid Friction and Heat Transfer," by T. H. Von Karman. *Trans ASME,* v. 61, Nov. 1939.
10. "A Method of Correlating Forced Convection Heat Transfer Data and a Comparison with Fluid Friction," by A. P. Colburn. *Trans AIChE,* v. 29, 1933, p. 174.
11. "Heat Transfer to a Fluid Flowing Turbulently in a Smooth Pipe with Walls at Constant Temperature," by R.A. Seban and T. T. Shimazaki. *Trans ASME,* v. 73, 1951, p. 803.
12. "Friction and Heat Transfer Measurements to Turbulent Pipe Flow of Water (Pr = 7 and 8) at Uniform Heat Flux," by R. W. Allen and E. R. G. Eckert. *J Heat Trans,* set. C, v. 86, 1964, p. 301.
13. "Heat and Momentum Transfer in Smooth and Rough Tubes at Various Prandtl Numbers," by D. F. Dipprey and R. H. Sabersky. *Int J Heat Mass Transfer,* v. 6, 1963, p. 329.

14. *Laminar Flow: Forced Convection in Ducts*, by R. K. Shah and A. L. London, Academic Press, New York, 1978.

15. "Heat Transfer and Friction in Turbulent Pipe Flow with Variable Physical Properties," by B. S. Petukhov and published in *Advances in Heat Transfer*, J. P. Hartnett and T. F. Irvine, editors. Academic Press, New York, 1970. pp. 504–564.

16. "Friction Factors for Pipe Flow," by L. F. Moody. *Trans ASME,* v. 66, 1944, p. 671.

17. See also the General References.

18. See also the General Convection References (Chapter 7).

CHAPTER 9

1. "Investigation of Variation of Point Unit-Heat-Transfer Coefficient Around a Cylinder Normal to an Air Stream," by W. H. Geidt. *Trans ASME,* v. 71, 1949, pp. 375–381.

2. "Heat Transfer by Forced Convection from a Cylinder to Water in Cross-flow," by R. M. Fand. *lnt J Heat Mass Transfer,* v. 8, 1965, p. 995.

3. "Forced Convection Heat-Transfer Correlations for Flow in Pipes, Past Flat Plates, Single Cylinders, Single Spheres, and Flow in Packed Beds and Tube Bundles," by S. Whitaker. *AIChE J,* v. 18, 1972, p. 361.

4. "A Correlating Equation for Forced Convection from Gases and Liquids to a Circular Cylinder in Crossflow," by S. W. Churchill and M. Bernstein. *J Heat Transfer,* v. 99, 1977, pp. 300–306.

5. "Experiments on the Flow Over a Rough Surface," by H. W. Townes and R. H. Sabersky. *Int J Heat Mass Transfer,* v. 9, 1966, p. 729.

6. "Turbulent Heat Transfer from Smooth and Rough Surfaces," by R. A. Gowen and J. W. Smith. *lnt J Heat Mass Transfer,* v. 11, 1968, p. 1657.F

7. "Heat Transfer from Spheres to Flowing Media," by H. Kramers. *Physica,* v. 12, 1946, p. 61.

8. "Forced Convection Heat Transfer from an Isothermal Sphere to Water," by G. C. Vuet and G. Leppert. *J Heat Transfer,* series C, v. 83, 1961, p. 163.

9. "Hear Transfer from Spheres" up to Re = 6 x 10^6, by F. Achenbach. *Proc 6th Int Heat Transfer Conf,* v. 5, Hemisphere Publishing Co., Washington, D.C., 1978, pp. 341–346.

10. "Correlation and Utilization of New Data on Flow Resistance and Heat Transfer for Cross Flow of Gases Over Tube Banks," by F. D. Grimison. *Trans ASME,* v. 59, 1937, pp. 583–594.

11. "Heat Transfer and Flow Resistance in Cross Flow of Gases Over Tube Banks," by M. Jakob. *Trans ASME,* v. 60, 1983, p. 384.

12. Heat Transfer from Tubes in Cross Flow," by A. Zhukauskas. *Adv Heat Transfer,* v. 8, 1972, pp. 93–160.

13. See also the General References.

14. See also the General Convection References (Chapter 7).

CHAPTER 10

1. An Analysis of Laminar-Free-Convection Flow and Heat Transfer About a Flat Plate Parallel to the Direction of the Generating Body Force," by S. Ostrach. *NACA Tech Rep* 1111, 1953.

2. "An Experimental Investigation of Turbulent Natural Convection in Air at Low Pressure Along a Vertical Heated Flat Plate," by C. Y. Warner and V. S. Arpaci. *Int J Heat Mass Transfer,* v. 11, 1968, p. 397.

3. "Turbulent Natural Convection from a Vertical Plane Surface," by R. Cheesewright. *J Heat Transfer,* v. 90, Feb 1968, p. 1.

4. "Laminar Free Convection from a Vertical Flat Plate," by F. M. Sparrow and J. L. Gregg. *Trans ASME,* v. 78, 1956, p. 435.

5. "Correlating Equations for Laminar and Turbulent Free Convection from a Vertical Plate," by S. W. Churchill and H. H. S. Chu. *Int J Heat Mass Transfer,* v. 18, 1975, p. 1323.

6. "Natural Convection Flow, Instability, and Transition," by B. Gebhart. *ASME Paper* 69-HT-29, August 1969.

7. "An Analysis of Turbulent Free Convection Heat Transfer," by F. J. Bayley. *Proc Inst Mech Eng,* v. 169, no.20, 1955, p. 361.

8. "Analysis of Turbulent Free Convection Boundary Layer on a Flat Plate," by F. R. G. Eckert and T. W. Jackson. *NACA Rep* 1015, 1951.

9. "A General Method of Obtaining Approximate Solutions to Laminar and Turbulent Free Convection Problems," by G. D. Raithby and K. G. T. Hollands. *Advances in Heat Transfer,* Academic Press, New York, 1974.

10. "Natural Convection Local Heat Transfer on Constant Heat Flux Inclined Surfaces," by G. C. Vliet. *J Heat Transfer,* v. 91, 1969, p. 511.

11. "Turbulent Natural Convection on Upward and Downward Facing Inclined Constant Heat Flux Surfaces," by G. C. Vliet and D. C. Ross. *ASME Paper* 74-WA/HT-32.

12. "Natural Convection Boundary Layer Flow Over Horizontal and Slightly Inclined Surfaces," by L. Pera and B. Gebhart. *Int J Heat Mass Transfer,* v. 16, 1973, p. 1131.

13. "Patterns of Free Convection Flow Adjacent to Horizontal Heated Surfaces," by R. B. Husar and F. M. Sparrow. *Int J Heat Mass Transfer,* v. II, 1968, p. 1206.

14. "Natural Convection Above Unconfined Horizontal Surfaces," by Z. Rotern and L. Claassen. *J Fluid Mech,* v. 39, pt. 1, 1969, p. 173.

15. "Natural Convection on a Finite-Size Horizontal Plate," by J. V. Clifton and A. J. Chapman. *Int J Heat Mass Transfer,* v. 11, 1969, p. 1573.

16. "Free Convection Along the Downward-facing Surface of a Heated Horizontal Plate," by T. Aihara, Y. Yamada, and S. Endo. *Int J Heat Mass Transfer,* v. 15, 1972, p. 2535.

17. "Laminar Free Convection Heat Transfer from Downward-facing Horizontal Surfaces of Finite Dimension," by S. N. Singh, R. C. Birkebak, and R. M. Drake. *Prog Heat Mass Transfer,* v. 2, 1969, p. 87.

18. "Combined Natural and Forced Convection Heat Transfer from Horizontal Cylinders to Water," by R. M. Fand and K. K. Keswani. *Int J Heat Mass Transfer,* v. 16, 1973, p. 175.

19. "Correlating Equations for Laminar and Turbulent Free Convection from a Horizontal Cylinder," by S. W. Churchill and H. H. S. Chu. *Int J Heat Mass Transfer,* v. 18, 1975, p. 1049.

20. "The Overall Convective Heat Transfer from Smooth Circular Cylinders," by V. T. Morgan. *Adv Heat Transfer,* v. 11, 1975, p. 199.

21. "Experiments on Heat Transfer from Spheres Including Combined Natural and Forced Convection," by T. Yuge. *J Heat Transfer,* ser C, v 82, 1960, p. 214.

22. "Forced, Mixed and Free Convection Regimes," by B. Metais and F. R. C. Eckert. *J Heat Transfer,* series C, v. 86, 1964, p. 295.

23. "Combined Free and Forced Convection," by C. K. Brown and W. H. Gauvin. *Can J Chem Eng,* v. 43, no. 6, 1965, pp. 306 and 313.

24. "A Comprehensive Correlating Equation for Laminar, Assisting, Forced and Free Convection," by S. W. Churchill. *AIChE J,* v. 23, no. 1, 1977, p. 10.

25. See also the General References.

26. See also the General Convection References (Chapter 7).

CHAPTER 11

1. *Process Heat Transfer,* by Don Q. Kern, McGraw-Hill, 1950. This text contains information about heat exchangers, including double-pipe (Chapter 6), shell-and-tube type (Chapter 7), extended surfaces (Chapter 16), and cooling towers (Chapter 17). Details regarding flow arrangements for increased heat recovery are also presented (for example, increasing heat-transfer rates by connecting two or more exchangers in parallel or in series or in some combination). Condensation of single and mixed vapors is also discussed (Chapters 12 and 13).

2. *Compact Heat Exchangers,* by W. M. Kays and A. L. London, McGraw-Hill, 1964. This text contains information on crossflow heat exchangers.

3. See also the General References.

CHAPTER 12

Condensation

1. "Heat Transfer by Condensing Vapors on Vertical Tubes," by C. G. Kirk-bride. *Trans AIChE,* v. 30, 1934, p. 170.
2. "Dropwise Condensation: Some Factors Influencing the Validity of Heat Transfer Measurements," by E. Citakoglu and J. W. Rose. *Int J Heat Mass Transfer,* v. 11, 1968, p. 523.
3. "Drop Size Distributions and Heat Transfer in Dropwise Condensation," by C. Graham and P. Griffith. *Int J Heat Mass Transfer,* v. 16, 1973, p. 337.
4. "Film Condensation," by W. M. Rohsenow. Chapter 12 of *Handbook of Heat Transfer,* McGraw-Hill, New York, 1973.
5. "Condensing Heat Transfer within Horizontal Tubes," by W. W. Akers, H. A. Deans, and 0. K. Crosser. *Chem Eng Prog Symp Series,* v. 55, no.29, 1958, p. 171.
6. See also the General References.

Boiling

1. "A Method of Correlating Heat Transfer Data for Surface Boiling Liquids," by W. M. Rohsenow. *Trans ASME,* v. 74, 1952, p. 969.
2. "Correlation of Maximum Heat Flux Data for Boiling of Saturated Liquids," by W. M. Rohsenow and P. Griffith. *AIChE-ASME Heat Transfer Symposium,* Louisville, KY, 1955.
3. "On the Stability of Boiling Heat Transfer," by N. Zuber. *Trans ASME,* v. 80, 1958, p. 711.
4. "Heat Transfer in Stable Film Boiling," by L. A. Bromley. *Chem Eng Prog,* v. 46, 1950, p. 221.
5. "Heat Transfer to Water Boiling Under Pressure," by E. A. Farber and F. L. Scorah. *Trans ASME,* v. 70, 1948, p. 369.
6. "Generalized Correlation of Boiling Heat Transfer," by S. Levy. *J Heat Transfer,* v. 81C, 1959, pp. 37–42.
7. "Nucleation with Boiling Heat Transfer," by W. M. Rohsenow. *ASME Paper* 70-HT-18.
8. "The Peak Boiling Heat Flux on Horizontal Cylinders," by K. H. Sun and I.H. Lienhard. *Int J Heat Mass Transfer,* v. 13, 1970, p. 1425.
9. See also the General References.

CHAPTERS 13 AND 14

General Radiation References

1. *Thermal Radiation Heat Transfer* by R. Siegel and J. R. Howell. McGraw-Hill, 1972. An excellent work on radiation. The text is readable and contains much information.
2. *Radiation Heat Transfer*, by F. M. Sparrow and R. D. Cess. McGraw-Hill, New York, 1978. A very well-written textbook on radiation.

Shape Factors and Surface Properties

1. "Radiant Interchange Configuration Factors," by D. C. Hamilton and W. R. Morgan. *NACA Tech Note* 2836, 1952.
2. "Effects of Roughness of Metal Surfaces on Angular Distribution of Monochromatic Radiation," by R. D. Birkebak and F. R.G. Eckert. *J Heat Transfer,* v. 87, 1965, p. 85.
3. "Surface Roughness Effects on Radiant Transfer Between Surfaces," by R. G. Hering and T. F. Smith. *Int J Heat Mass Transfer,* v. 13, 1970, p. 725.
4. "Thermal Radiation Tables and Applications," by R. V. Dunkle. *Trans ASME,* v. 76, 1956, p. 549.
5. See also the General Radiation References above.
6. See also the General References.

Enclosure Theory

1. "An Enclosure Theory for Radiative Exchange Between Specular and Diffusely Reflecting Surfaces," by F. M. Sparrow, F. R. G. Eckert, and V. K. Johnson. *J Heat Transfer,* series C, v. 84, 1962, pp. 294–299.
2. "Radiation Analysis by the Network Method," by A. K. Oppenheim. *Trans ASME,* v. 78, 1956, pp. 725–735.
3. See also the General Radiation References above.
4. See also the General References.

APPENDIX TABLE DATA

1. *Handbook of Tables for Applied Engineering Science,* edited by R. F. Bolz and G. L. Tuve, 2nd edition, CRC Press, Boca Raton, FL, 1973. This is an excellent reference and source of property table data. Information is presented in conventional Engineering units as well as SI units.
2. *Heat and Mass Transfer,* by F. R. G. Eckert and R. M. Drake, Taylor & Francis, Inc., Philadelphia, PA, 1999.
3. *Introduction to Heat Transfer,* by A. I. Brown and S. M. Marco, 3rd edition, McGraw-Hill, New York, 1958.
4. *U.S. National Bureau of Standards Circular 564,* 1955.
5. *Handbook of Fluid Dynamics,* edited by T. Johnson, CRC Press, Boca Raton, FL, 1999.
6. See also the General References.

Index